T0299042

MULTIDIMENSIONAL REAL ANALYSIS I: DIFFERENTIATION

Already published; for full details see http://publishing.cambridge.org/stm/mathematics/csam/

11 J.L. Alperin *Local representation theory*
12 P. Koosis *The logarithmic integral I*
13 A. Pietsch *Eigenvalues and s-numbers*
14 S.J. Patterson *An introduction to the theory of the Riemann zeta-function*
15 H.J. Baues *Algebraic homotopy*
16 V.S. Varadarajan *Introduction to harmonic analysis on semisimple Lie groups*
17 W. Dicks & M. Dunwoody *Groups acting on graphs*
18 L.J. Corwin & F.P. Greenleaf *Representations of nilpotent Lie groups and their applications*
19 R. Fritsch & R. Piccinini *Cellular structures in topology*
20 H. Klingen *Introductory lectures on Siegel modular forms*
21 P. Koosis *The logarithmic integral II*
22 M.J. Collins *Representations and characters of finite groups*
24 H. Kunita *Stochastic flows and stochastic differential equations*
25 P. Wojtaszczyk *Banach spaces for analysts*
26 J.E. Gilbert & M.A.M. Murray *Clifford algebras and Dirac operators in harmonic analysis*
27 A. Frohlich & M.J. Taylor *Algebraic number theory*
28 K. Goebel & W.A. Kirk *Topics in metric fixed point theory*
29 J.E. Humphreys *Reflection groups and Coxeter groups*
30 D.J. Benson *Representations and cohomology I*
31 D.J. Benson *Representations and cohomology II*
32 C. Allday & V. Puppe *Cohomological methods in transformation groups*
33 C. Soulé et al *Lectures on Arakelov geometry*
34 A. Ambrosetti & G. Prodi *A primer of nonlinear analysis*
35 J. Palis & F. Takens *Hyperbolicity, stability and chaos at homoclinic bifurcations*
37 Y. Meyer *Wavelets and operators I*
38 C. Weibel *An introduction to homological algebra*
39 W. Bruns & J. Herzog *Cohen-Macaulay rings*
40 V. Snaith *Explicit Brauer induction*
41 G. Laumon *Cohomology of Drinfield modular varieties I*
42 E.B. Davies *Spectral theory and differential operators*
43 J. Diestel, H. Jarchow & A. Tonge *Absolutely summing operators*
44 P. Mattila *Geometry of sets and measures in euclidean spaces*
45 R. Pinsky *Positive harmonic functions and diffusion*
46 G. Tenenbaum *Introduction to analytic and probabilistic number theory*
47 C. Peskine *An algebraic introduction to complex projective geometry I*
48 Y. Meyer & R. Coifman *Wavelets and operators II*
49 R. Stanley *Enumerative combinatorics I*
50 I. Porteous *Clifford algebras and the classical groups*
51 M. Audin *Spinning tops*
52 V. Jurdjevic *Geometric control theory*
53 H. Voelklein *Groups as Galois groups*
54 J. Le Potier *Lectures on vector bundles*
55 D. Bump *Automorphic forms*
56 G. Laumon *Cohomology of Drinfeld modular varieties II*
57 D.M. Clarke & B.A. Davey *Natural dualities for the working algebraist*
59 P. Taylor *Practical foundations of mathematics*
60 M. Brodmann & R. Sharp *Local cohomology*
61 J.D. Dixon, M.P.F. Du Sautoy, A. Mann & D. Segal *Analytic pro-p groups, 2nd edition*
62 R. Stanley *Enumerative combinatorics II*
64 J. Jost & X. Li-Jost *Calculus of variations*
68 Ken-iti Sato *Lévy processes and infinitely divisible distributions*
71 R. Blei *Analysis in integer and fractional dimensions*
72 F. Borceux & G. Janelidze *Galois theories*
73 B. Bollobás *Random graphs*
74 R.M. Dudley *Real analysis and probability*
75 T. Sheil-Small *Complex polynomials*
76 C. Voisin *Hodge theory and complex algebraic geometry I*
77 C. Voisin *Hodge theory and complex algebraic geometry II*
78 V. Paulsen *Completely bounded maps and operator algebra*
79 F. Gesztesy & H. Holden *Soliton equations and their algebro-geometric solutions I*
80 F. Gesztesy & H. Holden *Soliton equations and their algebro-geometric solutions II*

MULTIDIMENSIONAL REAL ANALYSIS I: DIFFERENTIATION

J.J. DUISTERMAAT
J.A.C. KOLK
Utrecht University

Translated from Dutch by J. P. van Braam Houckgeest

CAMBRIDGE UNIVERSITY PRESS
Cambridge, New York, Melbourne, Madrid, Cape Town, Singapore, São Paulo

Cambridge University Press
The Edinburgh Building, Cambridge CB2 8RU, UK

Published in the United States of America by Cambridge University Press, New York

www.cambridge.org
Information on this title: www.cambridge.org/9780521551144

First published 2004

A catalogue record for this publication is available from the British Library

ISBN 978-0-521-55114-4 hardback

Transferred to digital printing 2007

To Saskia and Floortje

With Gratitude and Love

Contents

Preface

I prefer the open landscape under a clear sky with its depth of perspective, where the wealth of sharply defined nearby details gradually fades away towards the horizon.

This book, which is in two parts, provides an introduction to the theory of vector-valued functions on Euclidean space. We focus on four main objects of study and in addition consider the interactions between these. Volume I is devoted to differentiation. Differentiable functions on $\mathbf{R}^n$ come first, in Chapters 1 through 3. Next, differentiable manifolds embedded in $\mathbf{R}^n$ are discussed, in Chapters 4 and 5. In Volume II we take up integration. Chapter 6 deals with the theory of n-dimensional integration over $\mathbf{R}^n$. Finally, in Chapters 7 and 8 lower-dimensional integration over submanifolds of $\mathbf{R}^n$ is developed; particular attention is paid to vector analysis and the theory of differential forms, which are treated independently from each other. Generally speaking, the emphasis is on geometric aspects of analysis rather than on matters belonging to functional analysis.

In presenting the material we have been intentionally concrete, aiming at a thorough understanding of Euclidean space. Once this case is properly understood, it becomes easier to move on to abstract metric spaces or manifolds and to infinite-dimensional function spaces. If the general theory is introduced too soon, the reader might get confused about its relevance and lose motivation. Yet we have tried to organize the book as economically as we could, for instance by making use of linear algebra whenever possible and minimizing the number of ϵ–δ arguments, always without sacrificing rigor. In many cases, a fresh look at old problems, by ourselves and others, led to results or proofs in a form not found in current analysis textbooks. Quite often, similar techniques apply in different parts of mathematics; on the other hand, different techniques may be used to prove the same result. We offer ample illustration of these two principles, in the theory as well as the exercises.

A working knowledge of analysis in one real variable and linear algebra is a prerequisite. The main parts of the theory can be used as a text for an introductory course of one semester, as we have been doing for second-year students in Utrecht during the last decade. Sections at the end of many chapters usually contain applications that can be omitted in case of time constraints.

This volume contains 334 exercises, out of a total of 568, offering variations and applications of the main theory, as well as special cases and openings toward applications beyond the scope of this book. Next to routine exercises we tried also to include exercises that represent some mathematical idea. The exercises are independent from each other unless indicated otherwise, and therefore results are sometimes repeated. We have run student seminars based on a selection of the more challenging exercises.

In our experience, interest may be stimulated if from the beginning the student can perceive analysis as a subject intimately connected with many other parts of mathematics and physics: algebra, electromagnetism, geometry, including differential geometry, and topology, Lie groups, mechanics, number theory, partial differential equations, probability, special functions, to name the most important examples. In order to emphasize these relations, many exercises show the way in which results from the aforementioned fields fit in with the present theory; prior knowledge of these subjects is not assumed, however. We hope in this fashion to have created a landscape as preferred by Weyl,[1] thereby contributing to motivation, and facilitating the transition to more advanced treatments and topics.

[1] Weyl, H.: *The Classical Groups.* Princeton University Press, Princeton 1939, p. viii.

Acknowledgments

Since a text like this is deeply rooted in the literature, we have refrained from giving references. Yet we are deeply obliged to many mathematicians for publishing the results that we use freely. Many of our colleagues and friends have made important contributions: E. P. van den Ban, F. Beukers, R. H. Cushman, W. L. J. van der Kallen, H. Keers, M. van Leeuwen, E. J. N. Looijenga, D. Siersma, T. A. Springer, J. Stienstra, and in particular J. D. Stegeman and D. Zagier. We were also fortunate to have had the moral support of our special friend V. S. Varadarajan. Numerous small errors and stylistic points were picked up by students who attended our courses; we thank them all.

With regard to the manuscript's technical realization, the help of A. J. de Meijer and F. A. M. van de Wiel has been indispensable, with further contributions coming from K. Barendregt and J. Jaspers. We have to thank R. P. Buitelaar for assistance in preparing some of the illustrations. Without LaTeX, Y&Y TeX and Mathematica this work would never have taken on its present form.

J. P. van Braam Houckgeest translated the manuscript from Dutch into English. We are sincerely grateful to him for his painstaking attention to detail as well as his many suggestions for improvement.

We are indebted to S. J. van Strien to whose encouragement the English version is due; and furthermore to R. Astley and J. Walthoe, our editors, and to F. H. Nex, our copy-editor, for the pleasant collaboration; and to Cambridge University Press for making this work available to a larger audience.

Of course, errors still are bound to occur and we would be grateful to be told of them, at the e-mail address kolk@math.uu.nl. A listing of corrections will be made accessible through http://www.math.uu.nl/people/kolk.

Introduction

Motivation. Analysis came to life in the number space $\mathbf{R}^n$ of dimension n and its complex analog $\mathbf{C}^n$. Developments ever since have consistently shown that further progress and better understanding can be achieved by generalizing the notion of space, for instance to that of a manifold, of a topological vector space, or of a scheme, an algebraic or complex space having infinitesimal neighborhoods, each of these being defined over a field of characteristic which is 0 or positive. The search for unification by continuously reworking old results and blending these with new ones, which is so characteristic of mathematics, nowadays tends to be carried out more and more in these newer contexts, thus bypassing $\mathbf{R}^n$. As a result of this the uninitiated, for whom $\mathbf{R}^n$ is still a difficult object, runs the risk of learning analysis in several real variables in a suboptimal manner. Nevertheless, to quote F. and R. Nevanlinna: "The elimination of coordinates signifies a gain not only in a formal sense. It leads to a greater unity and simplicity in the theory of functions of arbitrarily many variables, the algebraic structure of analysis is clarified, and at the same time the geometric aspects of linear algebra become more prominent, which simplifies one's ability to comprehend the overall structures and promotes the formation of new ideas and methods".[2]

In this text we have tried to strike a balance between the concrete and the abstract: a treatment of differential calculus in the traditional $\mathbf{R}^n$ by efficient methods and using contemporary terminology, providing solid background and adequate preparation for reading more advanced works. The exercises are tightly coordinated with the theory, and most of them have been tried out during practice sessions or exams. Illustrative examples and exercises are offered in order to support and strengthen the reader's intuition.

Organization. In a subject like this with its many interrelations, the arrangement of the material is more or less determined by the proofs one prefers to or is able to give. Other ways of organizing are possible, but it is our experience that it is not such a simple matter to avoid confusing the reader. In particular, because the Change of Variables Theorem in Volume II is about diffeomorphisms, it is necessary to introduce these initially, in the present volume; a subsequent discussion of the Inverse Function Theorems then is a plausible inference. Next, applications in geometry, to the theory of differentiable manifolds, are natural. This geometry in its turn is indispensable for the description of the boundaries of the open sets that occur in Volume II, in the Theorem on Integration of a Total Derivative in $\mathbf{R}^n$, the generalization to $\mathbf{R}^n$ of the Fundamental Theorem of Integral Calculus on $\mathbf{R}$. This is why differentiation is treated in this first volume and integration in the second. Moreover, most known proofs of the Change of Variables Theorem require an Inverse Function, or the Implicit Function Theorem, as does our first proof. However, for the benefit of those readers who prefer a discussion of integration at

[2] Nevanlinna, F., Nevanlinna, R.: *Absolute Analysis.* Springer-Verlag, Berlin 1973, p. 1.

an early stage, we have included in Volume II a second proof of the Change of Variables Theorem by elementary means.

On some technical points. We have tried hard to reduce the number of ϵ–δ arguments, while maintaining a uniform and high level of rigor. In the theory of differentiability this has been achieved by using a reformulation of differentiability due to Hadamard.

The Implicit Function Theorem is derived as a consequence of the Local Inverse Function Theorem. By contrast, in the exercises it is treated as a result on the conservation of a zero for a family of functions depending on a finite-dimensional parameter upon variation of this parameter.

We introduce a submanifold as a set in $\mathbf{R}^n$ that can locally be written as the graph of a mapping, since this definition can easily be put to the test in concrete examples. When the "internal" structure of a submanifold is important, as is the case with integration over that submanifold, it is useful to have a description as an image under a mapping. If, however, one wants to study its "external" structure, for instance when it is asked how the submanifold lies in the ambient space $\mathbf{R}^n$, then a description in terms of an inverse image is the one to use. If both structures play a role simultaneously, for example in the description of a neighborhood of the boundary of an open set, one usually flattens the boundary locally by means of a variable $t \in \mathbf{R}$ which parametrizes a motion transversal to the boundary, that is, one considers the boundary locally as a hyperplane given by the condition $t = 0$.

A unifying theme is the similarity in behavior of global objects and their associated infinitesimal objects (that is, defined at the tangent level), where the latter can be investigated by way of linear algebra.

Exercises. Quite a few of the exercises are used to develop secondary but interesting themes omitted from the main course of lectures for reasons of time, but which often form the transition to more advanced theories. In many cases, exercises are strung together as projects which, step by easy step, lead the reader to important results. In order to set forth the interdependencies that inevitably arise, we begin an exercise by listing the other ones which (in total or in part only) are prerequisites as well as those exercises that use results from the one under discussion. The reader should not feel obliged to completely cover the preliminaries before setting out to work on subsequent exercises; quite often, only some terminology or minor results are required. In the review exercises we have primarily collected results from real analysis in one variable that are needed in later exercises and that might not be familiar to the reader.

Notational conventions. Our notation is fairly standard, yet we mention the following conventions. Although it will often be convenient to write column vectors as row vectors, the reader should remember that all vectors are in fact column vectors, unless specified otherwise. Mappings always have precisely defined domains and images, thus $f : \mathrm{dom}(f) \to \mathrm{im}(f)$, but if we are unable, or do not wish, to specify the domain we write $f : \mathbf{R}^n \supset\!\!\to \mathbf{R}^p$ for a mapping that is well-defined on some subset of $\mathbf{R}^n$ and takes values in $\mathbf{R}^p$. We write $\mathbf{N}_0$ for $\{0\} \cup \mathbf{N}$, $\mathbf{N}_\infty$ for $\mathbf{N} \cup \{\infty\}$, and $\mathbf{R}_+$ for $\{ x \in \mathbf{R} \mid x > 0 \}$. The open interval $\{ x \in \mathbf{R} \mid a < x < b \}$ in $\mathbf{R}$ is denoted by $]\, a, b\, [$ and not by (a, b), in order to avoid confusion with the element $(a, b) \in \mathbf{R}^2$.

Making the notation consistent and transparent is difficult; in particular, every way of designating partial derivatives has its flaws. Whenever possible, we write $D_j f$ for the j-th column in a matrix representation of the total derivative Df of a mapping $f : \mathbf{R}^n \to \mathbf{R}^p$. This leads to expressions like $D_j f_i$ instead of Jacobi's classical $\frac{\partial f_i}{\partial x_j}$, etc. As a bonus the notation becomes independent of the designation of the coordinates in $\mathbf{R}^n$, thus avoiding absurd formulae such as may arise on substitution of variables; a disadvantage is that the formulae for matrix multiplication look less natural. The latter could be avoided with the notation $D_j f^i$, but this we rejected as being too extreme. The convention just mentioned has not been applied dogmatically; in the case of special coordinate systems like spherical coordinates, Jacobi's notation is the one of preference. As a further complication, D_j is used by many authors, especially in Fourier theory, for the momentum operator $\frac{1}{\sqrt{-1}} \frac{\partial}{\partial x_j}$.

We use the following dictionary of symbols to indicate the ends of various items:

❑ Proof
❍ Definition
☆ Example

Chapter 1

Continuity

Continuity of mappings between Euclidean spaces is the central topic in this chapter. We begin by discussing those properties of the n-dimensional space $\mathbf{R}^n$ that are determined by the standard inner product. In particular, we introduce the notions of distance between the points of $\mathbf{R}^n$ and of an open set in $\mathbf{R}^n$; these, in turn, are used to characterize limits and continuity of mappings between Euclidean spaces. The more profound properties of continuous mappings rest on the completeness of $\mathbf{R}^n$, which is studied next. Compact sets are infinite sets that in a restricted sense behave like finite sets, and their interplay with continuous mappings leads to many fundamental results in analysis, such as the attainment of extrema as well as the uniform continuity of continuous mappings on compact sets. Finally, we consider connected sets, which are related to intermediate value properties of continuous functions.

In applications of analysis in mathematics or in other sciences it is necessary to consider mappings depending on more than one variable. For instance, in order to describe the distribution of temperature and humidity in physical space–time we need to specify (in first approximation) the values of both the temperature T and the humidity h at every $(x, t) \in \mathbf{R}^3 \times \mathbf{R} \simeq \mathbf{R}^4$, where $x \in \mathbf{R}^3$ stands for a position in space and $t \in \mathbf{R}$ for a moment in time. Thus arises, in a natural fashion, a mapping $f : \mathbf{R}^4 \to \mathbf{R}^2$ with $f(x, t) = (T, h)$. The first step in a closer investigation of the properties of such mappings requires a study of the space $\mathbf{R}^n$ itself.

1.1 Inner product and norm

Let $n \in \mathbf{N}$. The n-dimensional space $\mathbf{R}^n$ is the Cartesian product of n copies of the linear space $\mathbf{R}$. Therefore $\mathbf{R}^n$ is a linear space; and following the standard convention in linear algebra we shall denote an element $x \in \mathbf{R}^n$ as a column vector

$$x = \begin{pmatrix} x_1 \\ \vdots \\ x_n \end{pmatrix} \in \mathbf{R}^n.$$

For typographical reasons, however, we often write $x = (x_1, \ldots, x_n) \in \mathbf{R}^n$, or, if necessary, $x = (x_1, \ldots, x_n)^t$ where ^{t} denotes the transpose of the $1 \times n$ matrix. Then $x_j \in \mathbf{R}$ is the j-th *coordinate* or *component* of x.

We recall that the *addition* of vectors and the *multiplication* of a vector by a *scalar* are defined by components, thus for $x, y \in \mathbf{R}^n$ and $\lambda \in \mathbf{R}$

$$\begin{aligned} (x_1, \ldots, x_n) + (y_1, \ldots, y_n) &= (x_1 + y_1, \ldots, x_n + y_n), \\ \lambda(x_1, \ldots, x_n) &= (\lambda x_1, \ldots, \lambda x_n). \end{aligned}$$

We say that $\mathbf{R}^n$ is a *vector space* or a *linear space* over $\mathbf{R}$ if it is provided with this addition and scalar multiplication. This means the following. Vector addition satisfies the commutative group axioms: *associativity* $((x + y) + z = x + (y + z))$, *existence of zero* $(x + 0 = x)$, *existence of additive inverses* $(x + (-x) = 0)$, *commutativity* $(x + y = y + x)$; scalar multiplication is *associative* $((\lambda\mu)x = \lambda(\mu x))$ and *distributive* over addition in both ways, i.e. $(\lambda(x + y) = \lambda x + \lambda y$ and $(\lambda + \mu)x = \lambda x + \mu x)$. We assume the reader to be familiar with the basic theory of finite-dimensional linear spaces.

For mappings $f : \mathbf{R}^n \supset\!\!\to \mathbf{R}^p$, we have the *component functions* $f_i : \mathbf{R}^n \supset\!\!\to \mathbf{R}$, for $1 \le i \le p$, satisfying

$$f = \begin{pmatrix} f_1 \\ \vdots \\ f_p \end{pmatrix} : \mathbf{R}^n \supset\!\!\to \mathbf{R}^p.$$

Many geometric concepts require an extra structure on $\mathbf{R}^n$ that we now define.

Definition 1.1.1. The *Euclidean space* $\mathbf{R}^n$ is the aforementioned linear space $\mathbf{R}^n$ provided with the *standard inner product*

$$\langle x, y \rangle = \sum_{1 \le j \le n} x_j y_j \qquad (x,\ y \in \mathbf{R}^n).$$

In particular, we say that x and $y \in \mathbf{R}^n$ are mutually *orthogonal* or *perpendicular* vectors if $\langle x, y \rangle = 0$. ❍

The standard inner product on $\mathbf{R}^n$ will be used for introducing the notion of a distance on $\mathbf{R}^n$, which in turn is indispensable for the definition of limits of mappings defined on $\mathbf{R}^n$ that take values in $\mathbf{R}^p$. We list the basic properties of the standard inner product.

Lemma 1.1.2. *All x, y, $z \in \mathbf{R}^n$ and $\lambda \in \mathbf{R}$ satisfy the following relations.*

(i) Symmetry: $\langle x, y\rangle = \langle y, x\rangle$.

(ii) Linearity: $\langle \lambda x + y, z\rangle = \lambda\langle x, z\rangle + \langle y, z\rangle$.

(iii) Positivity: $\langle x, x\rangle \geq 0$, *with equality if and only if* $x = 0$.

Definition 1.1.3. The *standard basis* for $\mathbf{R}^n$ consists of the vectors

$$e_j = (\delta_{1j}, \dots, \delta_{nj}) \in \mathbf{R}^n \qquad (1 \leq j \leq n),$$

where δ_{ij} equals 1 if $i = j$ and equals 0 if $i \neq j$. ❍

Thus we can write

$$x = \sum_{1\leq j\leq n} x_j e_j \qquad (x \in \mathbf{R}^n). \tag{1.1}$$

With respect to the standard inner product on $\mathbf{R}^n$ the standard basis is *orthonormal*, that is, $\langle e_i, e_j\rangle = \delta_{ij}$, for all $1 \leq i, j \leq n$. Thus, $\|e_j\| = 1$, while e_i and e_j, for distinct i and j, are mutually orthogonal vectors.

Definition 1.1.4. The *Euclidean norm* or *length* $\|x\|$ of $x \in \mathbf{R}^n$ is defined as

$$\|x\| = \sqrt{\langle x, x\rangle}.$$ ❍

From this definition and Lemma 1.1.2 we directly obtain

Lemma 1.1.5. *All x, $y \in \mathbf{R}^n$ and $\lambda \in \mathbf{R}$ satisfy the following properties.*

(i) Positivity: $\|x\| \geq 0$, *with equality if and only if* $x = 0$.

(ii) Homogeneity: $\|\lambda x\| = |\lambda|\,\|x\|$.

(iii) Polarization identity: $\langle x, y\rangle = \frac{1}{4}(\|x+y\|^2 - \|x-y\|^2)$, *which expresses the standard inner product in terms of the norm.*

(iv) Pythagorean property: $\|x \pm y\|^2 = \|x\|^2 + \|y\|^2$ *if and only if* $\langle x, y\rangle = 0$.

Proof. For (iii) and (iv) note

$$\begin{aligned}\|x \pm y\|^2 &= \langle x \pm y, x \pm y\rangle = \langle x, x\rangle \pm \langle x, y\rangle \pm \langle y, x\rangle + \langle y, y\rangle \\ &= \|x\|^2 + \|y\|^2 \pm 2\langle x, y\rangle.\end{aligned}$$ ❑

Proposition 1.1.6 (Cauchy–Schwarz inequality). *We have*

$$|\langle x, y\rangle| \le \|x\|\,\|y\| \qquad (x,\ y \in \mathbf{R}^n),$$

with equality if and only if x and y are linearly dependent, that is, if $x \in \mathbf{R}y = \{\lambda y \mid \lambda \in \mathbf{R}\}$ or $y \in \mathbf{R}x$.

Proof. The assertions are trivially true if x or $y = 0$. Therefore we may suppose $y \neq 0$; otherwise, interchange the roles of x and y. Now

$$x = \frac{\langle x, y\rangle}{\|y\|^2}y + \Big(x - \frac{\langle x, y\rangle}{\|y\|^2}y\Big)$$

is a decomposition of x into two mutually orthogonal vectors, the former belonging to $\mathbf{R}y$ and the latter being perpendicular to y and thus to $\mathbf{R}y$. Accordingly it follows from Lemma 1.1.5.(iv) and (ii) that

$$\|x\|^2 = \frac{\langle x, y\rangle^2}{\|y\|^2} + \Big\|x - \frac{\langle x, y\rangle}{\|y\|^2}y\Big\|^2, \quad \text{so} \quad \Big\|x - \frac{\langle x, y\rangle}{\|y\|^2}y\Big\|^2 = \frac{\|x\|^2\|y\|^2 - \langle x, y\rangle^2}{\|y\|^2}.$$

The inequality as well as the assertion about equality are now immediate from Lemma 1.1.5.(i) and the observation that $x = \lambda y$ implies $|\langle x, y\rangle| = |\lambda|\,\|y\|^2 = \|x\|\,\|y\|$. ❑

The Cauchy–Schwarz inequality is one of the workhorses in analysis; it serves in proving many other inequalities.

Lemma 1.1.7. *For all $x,\ y \in \mathbf{R}^n$ and $\lambda \in \mathbf{R}$ we have the following inequalities.*

(i) Triangle inequality: $\|x \pm y\| \le \|x\| + \|y\|$.

(ii) $\|\sum_{1\le k\le l} x^{(k)}\| \le \sum_{1\le k\le l}\|x^{(k)}\|$, *where $l \in \mathbf{N}$ and $x^{(k)} \in \mathbf{R}^n$, for $1 \le k \le l$.*

(iii) Reverse triangle inequality: $\|x - y\| \ge |\,\|x\| - \|y\|\,|$.

(iv) $|x_j| \le \|x\| \le \sum_{1\le i\le n}|x_i| \le \sqrt{n}\|x\|$, *for $1 \le j \le n$.*

(v) $\max_{1\le j\le n}|x_j| \le \|x\| \le \sqrt{n}\max_{1\le j\le n}|x_j|$. *As a consequence*

$$\{x \in \mathbf{R}^n \mid \max_{1\le j\le n}|x_j| \le 1\} \subset \{x \in \mathbf{R}^n \mid \|x\| \le \sqrt{n}\}$$
$$\subset \{x \in \mathbf{R}^n \mid \max_{1\le j\le n}|x_j| \le \sqrt{n}\}.$$

In geometric terms, the cube in $\mathbf{R}^n$ about the origin of side length 2 is contained in the ball about the origin of diameter $2\sqrt{n}$ and, in turn, this ball is contained in the cube of side length $2\sqrt{n}$.

Proof. For (i), observe that the Cauchy–Schwarz inequality implies

$$\begin{aligned}\|x \pm y\|^2 &= \langle x \pm y, x \pm y\rangle = \langle x, x\rangle \pm 2\langle x, y\rangle + \langle y, y\rangle \\ &\le \|x\|^2 + 2\|x\|\,\|y\| + \|y\|^2 = (\|x\| + \|y\|)^2.\end{aligned}$$

Assertion (ii) follows by repeated application of (i). For (iii), note that $\|x\| = \|(x-y)+y\| \le \|x-y\| + \|y\|$, and thus $\|x-y\| \ge \|x\| - \|y\|$. Now interchange the roles of x and y. Furthermore, (iv) is a consequence of (see Formula (1.1))

$$\begin{aligned}|x_j| &\le \Big(\sum_{1\le j\le n} |x_j|^2\Big)^{1/2} = \|x\| = \|\sum_{1\le j\le n} x_j e_j\| \le \sum_{1\le j\le n} |x_j|\,\|e_j\| \\ &= \sum_{1\le j\le n} |x_j| = \langle (1, \dots, 1), (|x_1|, \dots, |x_n|)\rangle \le \sqrt{n}\|x\|.\end{aligned}$$

For the last inequality we used the Cauchy–Schwarz inequality. Finally, (v) follows from (iv) and

$$\|x\|^2 = \sum_{1\le j\le n} x_j^2 \le n\,(\max_{1\le j\le n} |x_j|)^2. \qquad \square$$

Definition 1.1.8. For x and $y \in \mathbf{R}^n$, we define the *Euclidean distance* $d(x, y)$ by

$$d(x, y) = \|x - y\|. \qquad \bigcirc$$

From Lemmas 1.1.5 and 1.1.7 we immediately obtain, for $x, y, z \in \mathbf{R}^n$,

$$d(x, y) \ge 0, \qquad d(x, y) = 0 \iff x = y, \qquad d(x, z) \le d(x, y) + d(y, z).$$

More generally, a *metric space* is defined as a set provided with a distance between its elements satisfying the properties above. Not every distance is associated with an inner product: for an example of such a distance, consider $\mathbf{R}^n$ with $d(x, y) = \max_{1\le j\le n} |x_j - y_j|$, or $d(x, y) = \sum_{1\le j\le n} |x_j - y_j|$. Some of the results to come, for instance, the Contraction Lemma 1.7.2, are applicable in a general metric space.

Occasionally we will consider the field $\mathbf{C}$ of complex numbers. We identify $\mathbf{C}$ with $\mathbf{R}^2$ via

$$z = \operatorname{Re} z + i \operatorname{Im} z \in \mathbf{C} \quad \longleftrightarrow \quad (\operatorname{Re} z, \operatorname{Im} z) \in \mathbf{R}^2. \tag{1.2}$$

Here $\operatorname{Re} z \in \mathbf{R}$ denotes the real part of z, and $\operatorname{Im} z \in \mathbf{R}$ the imaginary part, and $i \in \mathbf{C}$ satisfies $i^2 = -1$. Thus $z = z_1 + i z_2 \in \mathbf{C}$ corresponds with $(z_1, z_2) \in \mathbf{R}^2$. By means of this identification the complex-linear space $\mathbf{C}^n$ will always be regarded as a linear space over $\mathbf{R}$ whose dimension over $\mathbf{R}$ is twice the dimension of the linear space over $\mathbf{C}$, i.e. $\mathbf{C}^n \simeq \mathbf{R}^{2n}$.

The notion of limit, which will be defined in terms of the norm, is of fundamental importance in analysis. We first discuss the particular case of the limit of a sequence of vectors.

We note that the notation for sequences in $\mathbf{R}^n$ requires some care. In fact, we usually denote the j-th component of a vector $x \in \mathbf{R}^n$ by x_j, therefore the notation $(x_j)_{j\in\mathbf{N}}$ for a sequence of vectors $x_j \in \mathbf{R}^n$ might cause confusion, and even more so $(x_n)_{n\in\mathbf{N}}$, which conflicts with the role of n in $\mathbf{R}^n$. We customarily use the subscript $k \in \mathbf{N}$: thus $(x_k)_{k\in\mathbf{N}}$. If the need arises, we shall use the notation $(x_j^{(k)})_{k\in\mathbf{N}}$ for a sequence consisting of j-th coordinates of vectors $x^{(k)} \in \mathbf{R}^n$.

Definition 1.1.9. Let $(x_k)_{k\in\mathbf{N}}$ be a sequence of vectors $x_k \in \mathbf{R}^n$, and let $a \in \mathbf{R}^n$. The sequence is said to be *convergent*, with *limit* a, if $\lim_{k\to\infty} \|x_k - a\| = 0$, which is a limit of numbers in $\mathbf{R}$. Recall that this limit means: for every $\epsilon > 0$ there exists $N \in \mathbf{N}$ with

$$k \geq N \quad\Longrightarrow\quad \|x_k - a\| < \epsilon.$$

In this case we write $\lim_{k\to\infty} x_k = a$. ❍

From Lemma 1.1.7.(iv) we directly obtain

Proposition 1.1.10. *Let $(x^{(k)})_{k\in\mathbf{N}}$ be a sequence in $\mathbf{R}^n$ and $a \in \mathbf{R}^n$. The sequence is convergent in $\mathbf{R}^n$ with limit a if and only if for every $1 \leq j \leq n$ the sequence $(x_j^{(k)})_{k\in\mathbf{N}}$ of j-th components is convergent in $\mathbf{R}$ with limit a_j. Hence, in case of convergence,*

$$\lim_{k\to\infty} x_j^{(k)} = (\lim_{k\to\infty} x^{(k)})_j.$$

From the definition of limit, or from Proposition 1.1.10, we obtain the following:

Lemma 1.1.11. *Let $(x_k)_{k\in\mathbf{N}}$ and $(y_k)_{k\in\mathbf{N}}$ be convergent sequences in $\mathbf{R}^n$ and let $(\lambda_k)_{k\in\mathbf{N}}$ be a convergent sequence in $\mathbf{R}$. Then $(\lambda_k x_k + y_k)_{k\in\mathbf{N}}$ is a convergent sequence in $\mathbf{R}^n$, while*

$$\lim_{k\to\infty} (\lambda_k x_k + y_k) = \lim_{k\to\infty} \lambda_k \lim_{k\to\infty} x_k + \lim_{k\to\infty} y_k.$$

1.2 Open and closed sets

The analog in $\mathbf{R}^n$ of an open interval in $\mathbf{R}$ is introduced in the following:

Definition 1.2.1. For $a \in \mathbf{R}^n$ and $\delta > 0$, we denote the *open ball* of center a and radius δ by

$$B(a;\delta) = \{\, x \in \mathbf{R}^n \mid \|x - a\| < \delta \,\}.$$ ❍

Definition 1.2.2. A point a in a set $A \subset \mathbf{R}^n$ is said to be an *interior point* of A if there exists $\delta > 0$ such that $B(a; \delta) \subset A$. The set of interior points of A is called the *interior* of A and is denoted by $\text{int}(A)$. Note that $\text{int}(A) \subset A$. The set A is said to be *open* in $\mathbf{R}^n$ if $A = \text{int}(A)$, that is, if every point of A is an interior point of A. ❍

Note that $\emptyset$, the empty set, satisfies every definition involving conditions on its elements, therefore $\emptyset$ is open. Furthermore, the whole space $\mathbf{R}^n$ is open. Next we show that the terminology in Definition 1.2.1 is consistent with that in Definition 1.2.2.

Lemma 1.2.3. *The set $B(a; \delta)$ is open in $\mathbf{R}^n$, for every $a \in \mathbf{R}^n$ and $\delta \geq 0$.*

Proof. For arbitrary $b \in B(a; \delta)$ set $\beta = \|b - a\|$, then $\delta - \beta > 0$. Hence $B(b; \delta - \beta) \subset B(a; \delta)$, because for every $x \in B(b; \delta - \beta)$

$$\|x - a\| \leq \|x - b\| + \|b - a\| < (\delta - \beta) + \beta = \delta. \quad \square$$

Lemma 1.2.4. *For any $A \subset \mathbf{R}^n$, the interior $\text{int}(A)$ is the largest open set contained in A.*

Proof. First we show that $\text{int}(A)$ is open. If $a \in \text{int}(A)$, there is $\delta > 0$ such that $B(a; \delta) \subset A$. As in the proof of the preceding lemma, we find for any $b \in B(a; \delta)$ a $\beta > 0$ such that $B(b, \beta) \subset A$. But this implies $B(a; \delta) \subset \text{int}(A)$, and hence $\text{int}(A)$ is an open set. Furthermore, if $U \subset A$ is open, it is clear by definition that $U \subset \text{int}(A)$, thus $\text{int}(A)$ is the largest open set contained in A. ❑

Infinite sets in $\mathbf{R}^n$ may well have an empty interior; this happens, for example, for $\mathbf{Z}^n$ or even $\mathbf{Q}^n$.

Lemma 1.2.5. (i) *The union of any collection of open subsets of $\mathbf{R}^n$ is again open in $\mathbf{R}^n$.*

(ii) *The intersection of finitely many open subsets of $\mathbf{R}^n$ is open in $\mathbf{R}^n$.*

Proof. Assertion (i) follows from Definition 1.2.2. For (ii), let $\{ U_k \mid 1 \leq k \leq l \}$ be a finite collection of open sets, and put $U = \cap_{1 \leq k \leq l} U_k$. If $U = \emptyset$, it is open. Otherwise, select $a \in U$ arbitrarily. For every $1 \leq k \leq l$ we have $a \in U_k$ and therefore there exist $\delta_k > 0$ with $B(a; \delta_k) \subset U_k$. Then $\delta := \min\{ \delta_k \mid 1 \leq k \leq l \} > 0$ while $B(a; \delta) \subset U$. ❑

Definition 1.2.6. Let $\emptyset \neq A \subset \mathbf{R}^n$. An *open neighborhood* of A is an open set containing A, and a *neighborhood* of A is any set containing an open neighborhood of A. A neighborhood of a set $\{x\}$ is also called a neighborhood of the point x, for all $x \in \mathbf{R}^n$. ❍

Note that $x \in A \subset \mathbf{R}^n$ is an interior point of A if and only if A is a neighborhood of x.

Definition 1.2.7. A set $F \subset \mathbf{R}^n$ is said to be *closed* if its *complement* $F^c := \mathbf{R}^n \setminus F$ is open. ❍

The empty set is closed, and so is the entire space $\mathbf{R}^n$.

Contrary to colloquial usage, the words open and closed are not antonyms in mathematical context: to say a set is not open does not mean it is closed. The interval $[-1, 1[$, for example, is neither open nor closed in $\mathbf{R}$.

Lemma 1.2.8. *For every $a \in \mathbf{R}^n$ and $\delta \geq 0$, the set $V(a; \delta) = \{x \in \mathbf{R}^n \mid \|x-a\| \leq \delta\}$, the* closed ball *of center a and radius δ, is closed.*

Proof. For arbitrary $b \in V(a; \delta)^c$ set $\beta = \|a-b\|$, then $\beta-\delta > 0$. So $B(b; \beta-\delta) \subset V(a; \delta)^c$, because by the reverse triangle inequality (see Lemma 1.1.7.(iii)), for every $x \in B(b; \beta-\delta)$

$$\|a-x\| \geq \|a-b\| - \|x-b\| > \beta - (\beta - \delta) = \delta.$$

This proves that $V(a; \delta)^c$ is open. ❑

Definition 1.2.9. A point $a \in \mathbf{R}^n$ is said to be a *cluster point* of a subset A in $\mathbf{R}^n$ if for every $\delta > 0$ we have $B(a; \delta) \cap A \neq \emptyset$. The set of cluster points of A is called the *closure* of A and is denoted by $\overline{A}$. ❍

Lemma 1.2.10. *Let $A \subset \mathbf{R}^n$, then $(\overline{A})^c = \operatorname{int}(A^c)$; in particular, the closure of A is a closed set. Moreover, $\operatorname{int}(A)^c = \overline{A^c}$.*

Proof. Note that $A \subset \overline{A}$. To say that x is not a cluster point of A means that it is an interior point of A^c. Thus $(\overline{A})^c = \operatorname{int}(A^c)$, or $\overline{A} = (\operatorname{int}(A^c))^c$, which implies that $\overline{A}$ is closed in $\mathbf{R}^n$. Furthermore, by applying this identity to A^c we obtain the second assertion of the lemma. ❑

By taking complements of sets we immediately obtain from Lemma 1.2.4:

Lemma 1.2.11. *For any $A \subset \mathbf{R}^n$, the closure $\overline{A}$ is the smallest closed set containing A.*

Lemma 1.2.12. *For any $F \subset \mathbf{R}^n$, the following assertions are equivalent.*

(i) *F is closed.*

(ii) *$F = \overline{F}$.*

(iii) *For every sequence $(x_k)_{k\in\mathbf{N}}$ of points $x_k \in F$ that is convergent to a limit, say $a \in \mathbf{R}^n$, we have $a \in F$.*

Proof. (i) $\Rightarrow$ (ii) is a consequence of Lemma 1.2.11. Next, (ii) $\Rightarrow$ (iii) because a is a cluster point of F. For the implication (iii) $\Rightarrow$ (i), suppose that F^c is not open. Then there exists $a \in F^c$, thus $a \notin F$, such that for all $\delta > 0$ we have $B(a, \delta) \not\subset F^c$, that is, $B(a, \delta) \cap F \neq \emptyset$. Successively taking $\delta = \frac{1}{k}$, we can choose a sequence $(x_k)_{k\in\mathbf{N}}$ of points $x_k \in F$ with $\|x_k - a\| < \frac{1}{k}$. But then (iii) implies $a \in F$, and we have arrived at a contradiction. ❑

From set theory we recall *DeMorgan's laws*, which state, for arbitrary collections $\{A_\alpha\}_{\alpha\in\mathcal{A}}$ of sets $A_\alpha \subset \mathbf{R}^n$

$$\Big(\bigcup_{\alpha\in\mathcal{A}} A_\alpha\Big)^c = \bigcap_{\alpha\in\mathcal{A}} A_\alpha^c, \qquad \Big(\bigcap_{\alpha\in\mathcal{A}} A_\alpha\Big)^c = \bigcup_{\alpha\in\mathcal{A}} A_\alpha^c. \tag{1.3}$$

In view of these laws and Lemma 1.2.5 we find, by taking complements of sets,

Lemma 1.2.13. (i) *The intersection of any collection of closed subsets of $\mathbf{R}^n$ is again closed in $\mathbf{R}^n$.*

(ii) *The union of finitely many closed subsets of $\mathbf{R}^n$ is closed in $\mathbf{R}^n$.*

Definition 1.2.14. We define ∂A, the *boundary* of a set A in $\mathbf{R}^n$, by

$$\partial A = \overline{A} \cap \overline{A^c}.$$ ❍

It is immediate from Lemma 1.2.10 that

$$\partial A = \overline{A} \setminus \operatorname{int}(A). \tag{1.4}$$

Example 1.2.15. We claim $\overline{B(a;\delta)} = V(a;\delta)$, for every $a \in \mathbf{R}^n$ and $\delta > 0$. Indeed, any $x \in \overline{B(a;\delta)}$ is cluster point of $B(a;\delta)$, and thus certainly of the larger set $V(a;\delta)$, which is closed by Lemma 1.2.8. It then follows from Lemma 1.2.12 that $\overline{B(a;\delta)} \subset V(a;\delta)$. On the other hand, let $x \in V(a;\delta)$ and consider the line segment $\{\, a_t = a + t(x-a) \mid 0 \le t \le 1 \,\}$ from a to x. Then $a_t \in B(a;\delta)$ for $0 \le t < 1$, in view of $\|a_t - a\| = t\|x-a\| < \|x-a\| \le \delta$. Given arbitrary $\epsilon > 0$, we can find, as $\lim_{t\uparrow 1} a_t = x$, a number $0 \le t < 1$ with $a_t \in B(x;\epsilon)$. This implies $B(a;\delta) \cap B(x;\epsilon) \ne \emptyset$, which gives $x \in \overline{B(a;\delta)}$.

Finally, the boundary of $B(a;\delta)$ is $V(a;\delta) \setminus B(a;\delta) = \{\, x \in \mathbf{R}^n \mid \|x-a\| = \delta \,\}$, that is, it equals the *sphere* in $\mathbf{R}^n$ of center a and radius δ. ☆

Observe that all definitions above are given in terms of the collection of open subsets of $\mathbf{R}^n$. This is the point of view adopted in *topology* (ὁ τόπος = place, and ὁ λόγος = word, reason). There one studies topological spaces: these are sets X provided with a *topology*, which by definition is a collection $\mathcal{O}$ of subsets that are said to be open and that satisfy: $\emptyset$ and X belong to $\mathcal{O}$, and both the union of arbitrarily many and the intersection of finitely many, respectively, elements of $\mathcal{O}$ again belong to $\mathcal{O}$. Obviously, this definition is inspired by Lemma 1.2.5. Some of the results below are actually valid in a topological space.

The property of being open or of being closed strongly depends on the ambient space $\mathbf{R}^n$ that is being considered. For example, in $\mathbf{R}$ the segment $]-1, 1[$ and $\mathbf{R}$ are both open. However, in $\mathbf{R}^n$, with $n \ge 2$, neither the segment nor the whole line, viewed as subsets of $\mathbf{R}^n$, is open in $\mathbf{R}^n$. Moreover, the segment is not closed in $\mathbf{R}^n$ either, whereas the line is closed in $\mathbf{R}^n$.

In this book we often will be concerned with proper subsets V of $\mathbf{R}^n$ that form the ambient space, for example, the sphere $V = \{\, x \in \mathbf{R}^n \mid \|x\| = 1 \,\}$. We shall then have to consider sets $A \subset V$ that are open or closed relative to V; for example, we would like to call $A = \{\, x \in V \mid \|x - e_1\| < \frac{1}{2} \,\}$ an open set in V, where $e_1 \in V$ is the first standard basis vector in $\mathbf{R}^n$. Note that this set A is neither open nor closed in $\mathbf{R}^n$. In the following definition we recover the definitions given above in case $V = \mathbf{R}^n$.

Definition 1.2.16. Let V be any fixed subset in $\mathbf{R}^n$ and let A be a subset of V. Then A is said to be *open in* V if there exists an open set $O \subset \mathbf{R}^n$, such that

$$A = V \cap O.$$

This defines a topology on V that is called the *relative topology* on V.

An *open neighborhood* of A in V is an open set in V containing A, and a *neighborhood* of A in V is any set in V containing an open neighborhood of A in V. Furthermore, A is said to be *closed* in V if $V \setminus A$ is open in V. A point $a \in V$ is said to be a *cluster point* of A in V if $(V \cap B(a;\delta)) \cap A = B(a;\delta) \cap A \ne \emptyset$, for all $\delta > 0$. The set of all cluster points of A in V is called the *closure* of A in V and is denoted by $\overline{A}^V$. Finally we define $\partial_V A$, the *boundary* of A in V, by

$$\partial_V A = \overline{A}^V \cap \overline{V \setminus A}^V. \qquad ❍$$

Observe that the statement A is open in V is not equivalent to A is open and A is contained in V, except when V is open in $\mathbf{R}^n$, see Proposition 1.2.17.(i) below.

The following proposition asserts that all these new sets arise by taking intersections of corresponding sets in $\mathbf{R}^n$ with the fixed set V.

Proposition 1.2.17. *Let V be a fixed subset in $\mathbf{R}^n$ and let A be a subset of V.*

(i) *Suppose V is open in $\mathbf{R}^n$, then A is open in V if and only if it is open in $\mathbf{R}^n$.*

(ii) *A is closed in V if and only if $A = V \cap F$ for some closed F in $\mathbf{R}^n$. Suppose V is closed in $\mathbf{R}^n$, then A is closed in V if and only if it is closed in $\mathbf{R}^n$.*

(iii) $\overline{A}^V = V \cap \overline{A}$.

(iv) $\partial_V A = V \cap \partial A$.

Proof. (i). This is immediate from the definitions.
(ii). If A is closed in V then $A = V \setminus P$ with P open in V; and since $P = V \cap O$ with O open in $\mathbf{R}^n$, we find

$$A = V \cap P^c = V \cap (V \cap O)^c = V \cap (V^c \cup O^c) = V \cap O^c = V \cap F$$

where $F := O^c$ is closed in $\mathbf{R}^n$. Conversely, if $A = V \cap F$ with F closed in $\mathbf{R}^n$, then

$$V \setminus A = V \cap (V \cap F)^c = V \cap (V^c \cup F^c) = V \cap F^c = V \cap O$$

where $O := F^c$ is open in $\mathbf{R}^n$.
(iii). Suppose $a \in V \cap \overline{A}$. Then for every $\delta > 0$ we have $B(a;\delta) \cap A \neq \emptyset$; and since $A \subset V$ it follows that for every $\delta > 0$ we have $(V \cap B(a;\delta)) \cap A \neq \emptyset$, which shows $a \in \overline{A}^V$. The implications all reverse.
(iv). Using (iii) we see

$$\partial_V A = (V \cap \overline{A}) \cap (V \cap \overline{V \setminus A}) = (V \cap \overline{A}) \cap (V \cap \overline{\mathbf{R}^n \setminus A}) = V \cap (\overline{A} \cap \overline{A^c}) = V \cap \partial A. \square$$

1.3 Limits and continuous mappings

In what follows we consider mappings or functions f that are defined on a subset of $\mathbf{R}^n$ and take values in $\mathbf{R}^p$, with n and $p \in \mathbf{N}$; thus $f : \mathbf{R}^n \supset\!\to \mathbf{R}^p$. Such an f is a function of the vector variable $x \in \mathbf{R}^n$. Since $x = (x_1, \dots, x_n)$ with $x_j \in \mathbf{R}$, we also say that f is a function of the n real variables $x_1, \dots, x_n$. Therefore we say that $f : \mathbf{R}^n \supset\!\!\to \mathbf{R}^p$ is a *function of several real variables* if $n \geq 2$. Furthermore, a function $f : \mathbf{R}^n \supset\!\!\to \mathbf{R}$ is called a *scalar function* while $f : \mathbf{R}^n \supset\!\!\to \mathbf{R}^p$ with $p \geq 2$ is called a *vector-valued function* or *mapping*. If $1 \leq j \leq n$ and $a \in \text{dom}(f)$, we have the j-th *partial mapping* $\mathbf{R} \supset\!\!\to \mathbf{R}^p$ associated with f and a, which in a neighborhood of a_j is given by

$$t \mapsto f(a_1, \dots, a_{j-1}, t, a_{j+1}, \dots, a_n). \tag{1.5}$$

Suppose that f is defined on an open set $U \subset \mathbf{R}^n$ and that $a \in U$, then, by the definition of open set, there exists an open ball centered at a which is contained in U. Applying Lemma 1.1.7.(v) in combination with translation and scaling, we can find an open cube centered at a and contained in U. But this implies that there exists $\delta > 0$ such that the partial mappings in (1.5) are defined on open intervals in $\mathbf{R}$ of the form $\left]\, a_j - \delta, a_j + \delta \,\right[$.

Definition 1.3.1. Let $A \subset \mathbf{R}^n$ and let $a \in \overline{A}$; let $f : A \to \mathbf{R}^p$ and let $b \in \mathbf{R}^p$. Then the mapping f is said to have *limit b* at a, with notation $\lim_{x \to a} f(x) = b$, if for every $\epsilon > 0$ there exists $\delta > 0$ satisfying

$$x \in A \quad \text{and} \quad \|x - a\| < \delta \quad \Longrightarrow \quad \|f(x) - b\| < \epsilon. \qquad ❍$$

We may reformulate this definition in terms of open balls (see Definition 1.2.1): $\lim_{x \to a} f(x) = b$ if and only if for every $\epsilon > 0$ there exists $\delta > 0$ satisfying

$$f(A \cap B(a; \delta)) \subset B(b; \epsilon), \qquad \text{or equivalently} \qquad A \cap B(a; \delta) \subset f^{-1}(B(b; \epsilon)).$$

Here $f^{-1}(B)$, the *inverse image* under f of a set $B \subset \mathbf{R}^p$, is defined as

$$f^{-1}(B) = \{\, x \in A \mid f(x) \in B \,\}.$$

Furthermore, the definition of limit can be rephrased in terms of neighborhoods (see Definitions 1.2.6 and 1.2.16).

Proposition 1.3.2. *In the notation of Definition 1.3.1 we have* $\lim_{x \to a} f(x) = b$ *if and only if for every neighborhood* V *of* b *in* $\mathbf{R}^p$ *the inverse image* $f^{-1}(V)$ *is a neighborhood of* a *in* A.

Proof. $\Rightarrow$. Consider a neighborhood V of b in $\mathbf{R}^p$. By Definitions 1.2.6 and 1.2.2 we can find $\epsilon > 0$ with $B(b; \epsilon) \subset V$. Next select $\delta > 0$ as in Definition 1.3.1. Then, by Definition 1.2.16, $U := A \cap B(a; \delta)$ is an open neighborhood of a in A satisfying $U \subset f^{-1}(V)$; therefore $f^{-1}(V)$ is a neighborhood of a in A.
$\Leftarrow$. For every $\epsilon > 0$ the set $B(b; \epsilon) \subset \mathbf{R}^p$ is open, hence it is a neighborhood of b in $\mathbf{R}^p$. Thus $f^{-1}(B(b; \epsilon))$ is a neighborhood of a in A. By Definition 1.2.16 we can find $\delta > 0$ with $A \cap B(a; \delta) \subset f^{-1}(B(b; \epsilon))$, which shows that the requirement of Definition 1.3.1 is satisfied. ❑

There is also a criterion for the existence of a limit in terms of sequences of points.

Lemma 1.3.3. *In the notation of Definition 1.3.1 we have* $\lim_{x\to a} f(x) = b$ *if and only if, for every sequence* $(x_k)_{k\in\mathbf{N}}$ *with* $\lim_{k\to\infty} x_k = a$*, we have* $\lim_{k\to\infty} f(x_k) = b$.

Proof. The necessity is obvious. Conversely, suppose that the condition is satisfied and $\lim_{x\to a} f(x) \neq b$. Then there exists $\epsilon > 0$ such that for every $k \in \mathbf{N}$ we can find $x_k \in A$ satisfying $\|x_k - a\| < \frac{1}{k}$ and $\|f(x_k) - b\| \geq \epsilon$. The sequence $(x_k)_{k\in\mathbf{N}}$ then converges to a, but $(f(x_k))_{k\in\mathbf{N}}$ does not converge to b. ❑

We now come to the definition of continuity of a mapping.

Definition 1.3.4. Consider $a \in A \subset \mathbf{R}^n$ and a mapping $f : A \to \mathbf{R}^p$. Then f is said to be *continuous* at a if $\lim_{x\to a} f(x) = f(a)$, that is, for every $\epsilon > 0$ there exists $\delta > 0$ satisfying

$$x \in A \quad \text{and} \quad \|x - a\| < \delta \quad \Longrightarrow \quad \|f(x) - f(a)\| < \epsilon.$$

Or, equivalently,

$$A \cap B(a; \delta) \subset f^{-1}(B(f(a); \epsilon)). \tag{1.6}$$

Or, equivalently,

$$f^{-1}(V) \text{ is a neighborhood of } a \text{ in } A \text{ if } V \text{ is a neighborhood of } f(a) \text{ in } \mathbf{R}^p.$$

Let $B \subset A$; we say that f is *continuous* on B if f is continuous at every point $a \in B$. Finally, f is *continuous* if it is continuous on A. ❍

Example 1.3.5. An important class of continuous mappings is formed by the $f : \mathbf{R}^n \supset\!\to \mathbf{R}^p$ that are *Lipschitz continuous*, that is, for which there exists $k > 0$ with

$$\|f(x) - f(x')\| \leq k\|x - x'\| \qquad (x, x' \in \operatorname{dom}(f)).$$

Such a number k is called a *Lipschitz constant* for f. The norm function $x \mapsto \|x\|$ is Lipschitz continuous on $\mathbf{R}^n$ with Lipschitz constant 1.

For $(x_1, x_2) \in \mathbf{R}^n \times \mathbf{R}^n$ we have $\|(x_1, x_2)\|^2 = \|x_1\|^2 + \|x_2\|^2$, which implies

$$\max\{\|x_1\|, \|x_2\|\} \leq \|(x_1, x_2)\| \leq \|x_1\| + \|x_2\|. \tag{1.7}$$

Next, the mapping: $\mathbf{R}^n \times \mathbf{R}^n \to \mathbf{R}^n$ of addition, given by $(x_1, x_2) \mapsto x_1 + x_2$, is Lipschitz continuous with Lipschitz constant 2. In fact, using the triangle inequality and the estimate (1.7) we see

$$\begin{aligned} \|x_1 + x_2 - (x_1' + x_2')\| &\leq \|x_1 - x_1'\| + \|x_2 - x_2'\| \\ &\leq 2\|(x_1 - x_1', x_2 - x_2')\| = 2\|(x_1, x_2) - (x_1', x_2')\|. \end{aligned}$$

Furthermore, the mapping: $\mathbf{R}^n \times \mathbf{R}^n \to \mathbf{R}$ of taking the inner product, with $(x_1, x_2) \mapsto \langle x_1, x_2 \rangle$, is continuous. Using the Cauchy–Schwarz inequality, we obtain this result from

$$\begin{aligned}
|\langle x_1, x_2\rangle - \langle x_1', x_2'\rangle| &= |\langle x_1, x_2\rangle - \langle x_1', x_2\rangle + \langle x_1', x_2\rangle - \langle x_1', x_2'\rangle| \\
&\le |\langle x_1 - x_1', x_2\rangle| + |\langle x_1', x_2 - x_2'\rangle| \le \|x_1 - x_1'\|\,\|x_2\| + \|x_1'\|\,\|x_2 - x_2'\| \\
&\le (\|x_2\| + \|x_1'\|)\|(x_1, x_2) - (x_1', x_2')\| \\
&\le (\|x_1\| + \|x_2\| + 1)\|(x_1, x_2) - (x_1', x_2')\|,
\end{aligned}$$

if $\|x_1 - x_1'\| \le 1$.

Finally, suppose f_1 and $f_2 : \mathbf{R}^n \to \mathbf{R}^p$ are continuous. Then the mapping $f : \mathbf{R}^n \to \mathbf{R}^p \times \mathbf{R}^p$ with $f(x) = (f_1(x), f_2(x))$ is continuous too. Indeed,

$$\begin{aligned}
\|f(x) - f(x')\| &= \|(f_1(x) - f_1(x'), f_2(x) - f_2(x'))\| \\
&\le \|f_1(x) - f_1(x')\| + \|f_2(x) - f_2(x')\|.
\end{aligned}$$ ☆

Lemma 1.3.3 obviously implies the following:

Lemma 1.3.6. *Let $a \in A \subset \mathbf{R}^n$ and let $f : A \to \mathbf{R}^p$. Then f is continuous at a if and only if, for every sequence $(x_k)_{k\in\mathbf{N}}$ with $\lim_{k\to\infty} x_k = a$, we have $\lim_{k\to\infty} f(x_k) = f(a)$. If the latter condition is satisfied, we obtain*

$$\lim_{k\to\infty} f(x_k) = f(\lim_{k\to\infty} x_k).$$

We will show that for a continuous mapping f the inverse images under f of open sets in $\mathbf{R}^p$ are open sets in $\mathbf{R}^n$, and that a similar statement is valid for closed sets. The proof of the latter assertion requires a result from set theory, viz. if $f : A \to B$ is a mapping of sets, we have

$$f^{-1}(B \setminus F) = A \setminus f^{-1}(F) \qquad (F \subset B). \tag{1.8}$$

Indeed, for $a \in A$,

$$\begin{aligned}
a \in f^{-1}(B \setminus F) &\iff f(a) \in B \setminus F \iff f(a) \notin F \\
&\iff a \notin f^{-1}(F) \iff a \in A \setminus f^{-1}(F).
\end{aligned}$$

Theorem 1.3.7. *Consider $A \subset \mathbf{R}^n$ and a mapping $f : A \to \mathbf{R}^p$. Then the following are equivalent.*

(i) *f is continuous.*

(ii) *$f^{-1}(O)$ is open in A for every open set O in $\mathbf{R}^p$. In particular, if A is open in $\mathbf{R}^n$ then: $f^{-1}(O)$ is open in $\mathbf{R}^n$ for every open set O in $\mathbf{R}^p$.*

(iii) *$f^{-1}(F)$ is closed in A for every closed set F in $\mathbf{R}^p$. In particular, if A is closed in $\mathbf{R}^n$ then: $f^{-1}(F)$ is closed in $\mathbf{R}^n$ for every closed set F in $\mathbf{R}^p$.*

Furthermore, if f is continuous, the level set $N(c) := f^{-1}(\{c\})$ *is closed in A, for every $c \in \mathbf{R}^p$.*

Proof. (i) ⇒ (ii). Consider an open set $O \subset \mathbf{R}^p$. Let $a \in f^{-1}(O)$ be arbitrary, then $f(a) \in O$ and therefore O, being open, is a neighborhood of $f(a)$ in $\mathbf{R}^p$. Proposition 1.3.2 then implies that $f^{-1}(O)$ is a neighborhood of a in A. This shows that a is an interior point of $f^{-1}(O)$ in A, which proves that $f^{-1}(O)$ is open in A.
(ii) ⇒ (i). Let $a \in A$ and let V be an arbitrary neighborhood of $f(a)$ in $\mathbf{R}^p$. Then there exists an open set O in $\mathbf{R}^p$ with $f(a) \in O \subset V$. Thus $a \in f^{-1}(O) \subset f^{-1}(V)$ while $f^{-1}(O)$ is an open neighborhood of a in A. The result now follows from Proposition 1.3.2.
(ii) ⇔ (iii). This is obvious from Formula (1.8). ❑

Example 1.3.8. A continuous mapping does not necessarily take open sets into open sets (for instance, a constant mapping), nor closed ones into closed ones. Furthermore, consider $f : \mathbf{R} \to \mathbf{R}$ given by $f(x) = \frac{2x}{1+x^2}$. Then $f(\mathbf{R}) = [\,-1, 1\,]$ where $\mathbf{R}$ is open and $[\,-1, 1\,]$ is not. Furthermore, $\mathbf{N}$ is closed in $\mathbf{R}$ but $f(\mathbf{N}) = \{\frac{2n}{1+n^2} \mid n \in \mathbf{N}\}$ is not closed, 0 clearly lying in its closure but not in the set itself. ✩

As for limits of sequences, Lemma 1.1.7.(iv) implies the following:

Proposition 1.3.9. *Let $A \subset \mathbf{R}^n$ and let $a \in \overline{A}$ and $b \in \mathbf{R}^p$. Consider $f : A \to \mathbf{R}^p$ with corresponding component functions $f_i : A \to \mathbf{R}$. Then*

$$\lim_{x \to a} f(x) = b \qquad \Longleftrightarrow \qquad \lim_{x \to a} f_i(x) = b_i \qquad (1 \le i \le p).$$

Furthermore, if $a \in A$, then the mapping f is continuous at a if and only if all the component functions of f are continuous at a.

In other words, limits and continuity of a mapping $f : \mathbf{R}^n \supset\!\to \mathbf{R}^p$ can be reduced to limits and continuity of the component functions $f_i : \mathbf{R}^n \supset\!\to \mathbf{R}$, for $1 \le i \le p$. On the other hand, f need not be continuous at $a \in \mathbf{R}^n$ if all the j-th partial mappings as in Formula (1.5) are continuous at $a_j \in \mathbf{R}$, for $1 \le j \le n$. This follows from the example of $a = 0 \in \mathbf{R}^2$ and $f : \mathbf{R}^2 \to \mathbf{R}$ with $f(x) = 0$ if x satisfies $x_1 x_2 = 0$, and $f(x) = 1$ elsewhere. More interesting is the following:

Example 1.3.10. Consider $f : \mathbf{R}^2 \to \mathbf{R}$ given by

$$f(x) = \begin{cases} \dfrac{x_1 x_2}{\|x\|^2}, & x \neq 0; \\ 0, & x = 0. \end{cases}$$

Obviously $f(x) = f(tx)$ for $x \in \mathbf{R}^2$ and $0 \neq t \in \mathbf{R}$, which implies that the restriction of f to straight lines through the origin with exclusion of the origin is constant, with the constant continuously dependent on the direction x of the line. Were f continuous at 0, then $f(x) = \lim_{t\downarrow 0} f(tx) = f(0) = 0$, for all $x \in \mathbf{R}^2$. Thus f would be identically equal to 0, which clearly is not the case. Therefore, f cannot be continuous at 0. Nevertheless, both partial functions $x_1 \mapsto f(x_1, 0) = 0$ and $x_2 \mapsto f(0, x_2) = 0$, which are associated with 0, are continuous everywhere. ☆

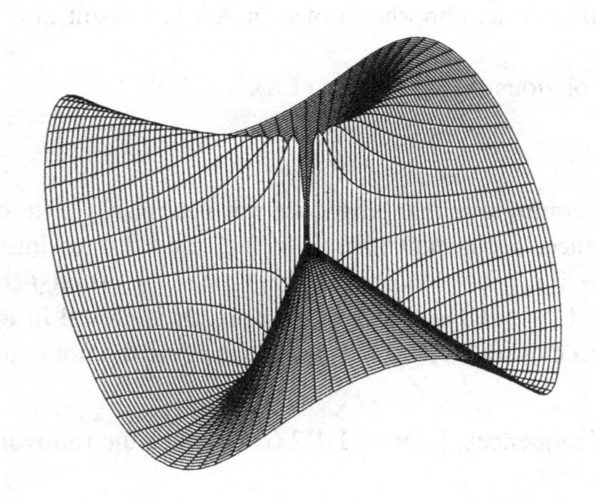

Illustration for Example 1.3.11

Example 1.3.11. Even the requirement that the restriction of $g : \mathbf{R}^2 \to \mathbf{R}$ to every straight line through 0 be continuous at 0 and attain the value $g(0)$ is not sufficient to ensure the continuity of g at 0. To see this, let f denote the function from the preceding example and consider g given by

$$g(x) = f(x_1, x_2^2) = \begin{cases} \dfrac{x_1 x_2^2}{x_1^2 + x_2^4}, & x \neq 0; \\ 0, & x = 0. \end{cases}$$

In this case

$$\lim_{t\downarrow 0} g(tx) = \lim_{t\downarrow 0} t \frac{x_1 x_2^2}{x_1^2 + t^2 x_2^4} = 0 \qquad (x_1 \neq 0); \qquad g(tx) = 0 \qquad (x_1 = 0).$$

Nevertheless, g still fails to be continuous at 0, as is obvious from

$$g(\lambda t^2, t) = g(\lambda, 1) = \frac{\lambda}{\lambda^2+1} \in \Big[-\frac{1}{2}, \frac{1}{2} \Big] \qquad (\lambda \in \mathbf{R},\ 0 \neq t \in \mathbf{R}).$$

Consequently, the function g is constant on the parabolas $\{ x \in \mathbf{R}^2 \mid x_1 = \lambda x_2^2 \}$, for $\lambda \in \mathbf{R}$ (whose union is $\mathbf{R}^2$ with the x_1-axis excluded), and in each neighborhood of 0 it assumes all values between $-\frac{1}{2}$ and $\frac{1}{2}$. ☆

1.4 Composition of mappings

Definition 1.4.1. Let $f : \mathbf{R}^n \supset\!\to \mathbf{R}^p$ and $g : \mathbf{R}^p \supset\!\to \mathbf{R}^q$. The *composition* $g \circ f : \mathbf{R}^n \supset\!\to \mathbf{R}^q$ has domain equal to $\mathrm{dom}(f) \cap f^{-1}(\mathrm{dom}(g))$ and is defined by

$$(g \circ f)(x) = g(f(x)).$$

In other words, it is the mapping

$$g \circ f : x \mapsto f(x) \mapsto g(f(x)) : \mathbf{R}^n \supset\!\to \mathbf{R}^p \supset\!\to \mathbf{R}^q.$$

In particular, let f_1, f_2 and $f : \mathbf{R}^n \supset\!\to \mathbf{R}^p$ and $\lambda \in \mathbf{R}$. Then we define the *sum* $f_1 + f_2 : \mathbf{R}^n \supset\!\to \mathbf{R}^p$ and the *scalar multiple* $\lambda f : \mathbf{R}^n \supset\!\to \mathbf{R}^p$ as the compositions

$$\begin{aligned} f_1 + f_2 :&\quad x \mapsto (f_1(x), f_2(x)) \mapsto f_1(x) + f_2(x) : &&\mathbf{R}^n \supset\!\to \mathbf{R}^{2p} \supset\!\to \mathbf{R}^p, \\ \lambda f :&\quad x \mapsto (\lambda, f(x)) \mapsto \lambda f(x) : &&\mathbf{R}^n \supset\!\to \mathbf{R}^{p+1} \supset\!\to \mathbf{R}^p, \end{aligned}$$

for $x \in \bigcap_{1\le k\le 2} \mathrm{dom}(f_k)$ and $x \in \mathrm{dom}(f)$, respectively. Next we define the *product* $\langle f_1, f_2 \rangle : \mathbf{R}^n \supset\!\to \mathbf{R}$ as the composition

$$\langle f_1, f_2 \rangle : x \mapsto (f_1(x), f_2(x)) \mapsto \langle f_1(x), f_2(x) \rangle : \mathbf{R}^n \supset\!\to \mathbf{R}^p \times \mathbf{R}^p \supset\!\to \mathbf{R},$$

for $x \in \bigcap_{1\le k\le 2} \mathrm{dom}(f_k)$. Finally, assume $p = 1$. Then we write $f_1 f_2$ for the product $\langle f_1, f_2 \rangle$ and define the *reciprocal function* $\frac{1}{f} : \mathbf{R}^n \supset\!\to \mathbf{R}$ by

$$\frac{1}{f} : x \mapsto f(x) \mapsto \frac{1}{f(x)} : \mathbf{R}^n \supset\!\to \mathbf{R} \supset\!\to \mathbf{R} \qquad (x \in \mathrm{dom}(f) \setminus f^{-1}(\{0\})).$$ ○

Exactly as in the theory of real functions of one variable one obtains corresponding results for the limits and the continuity of these new mappings.

Theorem 1.4.2 (Substitution Theorem). *Suppose $f : \mathbf{R}^n \supset\!\to \mathbf{R}^p$ and $g : \mathbf{R}^p \supset\!\to \mathbf{R}^q$; let $a \in \overline{\mathrm{dom}(f)}$, $b \in \overline{\mathrm{dom}(g)}$ and $c \in \mathbf{R}^q$, while $\lim_{x\to a} f(x) = b$ and $\lim_{y\to b} g(y) = c$. Then we have the following properties.*

(i) $\lim_{x\to a}(g\circ f)(x) = c$. *In particular, if* $b \in \mathrm{dom}(g)$ *and* g *is continuous at* b, *then*

$$\lim_{x\to a} g(f(x)) = g(\lim_{x\to a} f(x)).$$

(ii) *Let* $a \in \mathrm{dom}(f)$ *and* $f(a) \in \mathrm{dom}(g)$. *If* f *and* g *are continuous at* a *and* $f(a)$, *respectively, then* $g \circ f : \mathbf{R}^n \supset\!\!\to \mathbf{R}^q$ *is continuous at* a.

Proof. Ad (i). Let W be an arbitrary neighborhood of c in $\mathbf{R}^q$. A reformulation of the data by means of Proposition 1.3.2 gives the existence of a neighborhood V of b in $\mathrm{dom}(g)$, and U of a in $\mathrm{dom}(f) \cap f^{-1}(\mathrm{dom}(g))$, respectively, satisfying

$$V \subset g^{-1}(W), \qquad U \subset f^{-1}(V).$$

Combination of these inclusions proves the first equality in assertion (i), because

$$(g(f(U)) \subset g(V) \subset W, \qquad \text{thus} \qquad U \subset (g\circ f)^{-1}(W).$$

As for the interchange of the limit with the mapping g, note that

$$\lim_{x\to a} g(f(x)) = \lim_{x\to a}(g\circ f)(x) = c = g(b) = g(\lim_{x\to a} f(x)).$$

Assertion (ii) is a direct consequence of (i). ❑

The following corollary is immediate from the Substitution Theorem in conjunction with Example 1.3.5.

Corollary 1.4.3. *For* f_1 *and* $f_2 : \mathbf{R}^n \supset\!\!\to \mathbf{R}^p$ *and* $\lambda \in \mathbf{R}$, *we have the following.*

(i) *If* $a \in \overline{\bigcap_{1\le k\le 2} \mathrm{dom}(f_k)}$, *then* $\lim_{x\to a}(\lambda f_1 + f_2)(x) = \lambda \lim_{x\to a} f_1(x) + \lim_{x\to a} f_2(x)$.

(ii) *If* $a \in \bigcap_{1\le k\le 2} \mathrm{dom}(f_k)$ *and* f_1 *and* f_2 *are continuous at* a, *then* $\lambda f_1 + f_2$ *is continuous at* a.

(iii) *If* $a \in \overline{\bigcap_{1\le k\le 2} \mathrm{dom}(f_k)}$, *then we have* $\lim_{x\to a}\langle f_1, f_2\rangle(x) = \langle \lim_{x\to a} f_1(x), \lim_{x\to a} f_2(x)\rangle$.

(iv) *If* $a \in \bigcap_{1\le k\le 2} \mathrm{dom}(f_k)$ *and* f_1 *and* f_2 *are continuous at* a, *then* $\langle f_1, f_2\rangle$ *is continuous at* a.

Furthermore, assume $f : \mathbf{R}^n \supset\!\!\to \mathbf{R}$.

(v) *If* $\lim_{x\to a} f(x) \neq 0$, *then* $\lim_{x\to a} \frac{1}{f}(x) = \frac{1}{\lim_{x\to a} f(x)}$.

(vi) *If* $a \in \mathrm{dom}(f)$, $f(a) \neq 0$ *and* f *is continuous at* a, *then* $\frac{1}{f}$ *is continuous at* a.

Example 1.4.4. According to Lemma 1.1.7.(iv) the *coordinate functions* $\mathbf{R}^n \to \mathbf{R}$ with $x \mapsto x_j$, for $1 \le j \le n$, are continuous. It then follows from part (iv) in the corollary above and by mathematical induction that *monomial functions* $\mathbf{R}^n \to \mathbf{R}$, that is, functions of the form $x \mapsto x_1^{\alpha_1} \cdots x_n^{\alpha_n}$, with $\alpha_j \in \mathbf{N}_0$ for $1 \le j \le n$, are continuous. In turn, this fact and part (ii) of the corollary imply that *polynomial functions* on $\mathbf{R}^n$, which by definition are linear combinations of monomial functions on $\mathbf{R}^n$, are continuous. Furthermore, *rational functions* on $\mathbf{R}^n$ are defined as quotients of polynomial functions on that subset of $\mathbf{R}^n$ where the quotient is well-defined. In view of parts (vi) and (iv) rational functions on $\mathbf{R}^n$ are continuous too. As the composition $g \circ f$ of a rational function f on $\mathbf{R}^n$ by a continuous function g on $\mathbf{R}$ is also continuous by the Substitution Theorem 1.4.2.(ii), the continuity of functions on $\mathbf{R}^n$ that are given by formulae can often be decided by mere inspection. ☆

1.5 Homeomorphisms

Example 1.5.1. For functions $f : \mathbf{R} \supset\!\!\to \mathbf{R}$ we have the following result. Let $I \subset \mathbf{R}$ be an interval and let $f : I \to \mathbf{R}$ be a continuous injective function. Then f is strictly monotonic. The inverse function of f, which is defined on the interval $f(I)$, is continuous and strictly monotonic. For instance, from the construction of the trigonometric functions we know that $\tan : \left]-\frac{\pi}{2}, \frac{\pi}{2}\right[\to \mathbf{R}$ is a continuous bijection, hence it follows that the inverse function $\arctan : \mathbf{R} \to \left]-\frac{\pi}{2}, \frac{\pi}{2}\right[$ is continuous.

However, for continuous mappings $f : \mathbf{R}^n \supset\!\!\to \mathbf{R}^p$, with $n \ge 2$, that are bijective: $\mathrm{dom}(f) \to \mathrm{im}(f)$, the inverse is not automatically continuous. For example, consider $f : \left]-\pi, \pi\right] \to S^1 := \{ x \in \mathbf{R}^2 \mid \|x\| = 1 \}$ given by $f(t) = (\cos t, \sin t)$. Then f is a continuous bijection, while

$$\lim_{t \downarrow -\pi} f(t) = \lim_{t \downarrow -\pi} (\cos t, \sin t) = (-1, 0) = f(\pi).$$

If $f^{-1} : S^1 \to \left]-\pi, \pi\right]$ were continuous at $(-1, 0) \in S^1$, then we would have by the Substitution Theorem 1.4.2.(i)

$$\pi = f^{-1}(f(\pi)) = f^{-1}\Big(\lim_{t \downarrow -\pi} f(t)\Big) = \lim_{t \downarrow -\pi} t = -\pi. \qquad ☆$$

In order to describe phenomena like this we introduce the notion of a homeomorphism (ὁμοῖος = similar and ἡ μορφή = shape).

Definition 1.5.2. Let $A \subset \mathbf{R}^n$ and $B \subset \mathbf{R}^p$. A mapping $f : A \to B$ is said to be a *homeomorphism* if f is a continuous bijection and the inverse mapping $f^{-1} : B \to A$ is continuous, that is, $(f^{-1})^{-1}(O) = f(O)$ is open in B, for every open set O in A. If this is the case, A and B are called *homeomorphic* sets.

A mapping $f : A \to B$ is said to be *open* if the image under f of every open set in A is open in B, and f is said to be *closed* if the image under f of every closed set in A is closed in B. ❍

Example 1.5.3. In this terminology, $]-\frac{\pi}{2}, \frac{\pi}{2}[$ and $\mathbf{R}$ are homeomorphic sets.

Let $p < n$. Then the orthogonal projection $f : \mathbf{R}^n \to \mathbf{R}^p$ with $f(x) = (x_1, \dots, x_p)$ is a continuous open mapping. However, f is not closed; for example, consider $f : \mathbf{R}^2 \to \mathbf{R}$ and the closed subset $\{ x \in \mathbf{R}^2 \mid x_1 x_2 = 1 \}$, which has the nonclosed image $\mathbf{R} \setminus \{0\}$ in $\mathbf{R}$. ☆

Proposition 1.5.4. *Let $A \subset \mathbf{R}^n$ and $B \subset \mathbf{R}^p$ and let $f : A \to B$ be a bijection. Then the following are equivalent.*

(i) *f is a homeomorphism.*

(ii) *f is continuous and open.*

(iii) *f is continuous and closed.*

Proof. (i) $\Leftrightarrow$ (ii) follows from the definitions. For (i) $\Leftrightarrow$ (iii), use Formula (1.8).❑

At this stage the reader probably expects a theorem stating that, if $U \subset \mathbf{R}^n$ is open and $V \subset \mathbf{R}^n$ and if $f : U \to V$ is a homeomorphism, then V is open in $\mathbf{R}^n$. Indeed, under the further assumption of differentiability of f and of its inverse mapping f^{-1}, results of this kind will be established in this book, see Example 2.4.9 and Section 3.2. Moreover, *Brouwer's Theorem* states that a continuous and injective mapping $f : U \to \mathbf{R}^n$ defined on an open set $U \subset \mathbf{R}^n$ is an open mapping. Related to this result is the *invariance of dimension*: if a neighborhood of a point in $\mathbf{R}^n$ is mapped continuously and injectively onto a neighborhood of a point in $\mathbf{R}^p$, then $p = n$. As yet, however, no proofs of these latter results are known at the level of this text. The condition of injectivity of the mapping is necessary for the invariance of dimension, see Exercises 1.45 and 1.53, where it is demonstrated, surprisingly, that there exist continuous **surjective** mappings: $I \to I^2$ with $I = [\,0, 1\,]$ and that these never are injective.

1.6 Completeness

The more profound properties of continuous mappings turn out to be consequences of the following.

Theorem 1.6.1. *Every nonempty set $A \subset \mathbf{R}$ which is bounded from above has a* supremum *or* least upper bound $a \in \mathbf{R}$, *with notation* $a = \sup A$; *that is, a has the following properties:*

(i) $x \leq a$, *for all* $x \in A$;

(ii) *for every* $\delta > 0$, *there exists* $x \in A$ *with* $a - \delta < x$.

Note that assertion (i) says that a is an upper bound for A while (ii) asserts that no smaller number than a is an upper bound for A; therefore a is rightly called the least upper bound for A. Similarly, we have the notion of an *infimum* or *greatest lower bound*.

Depending on one's choice of the defining properties for the set $\mathbf{R}$ of real numbers, Theorem 1.6.1 is either an axiom or indeed a theorem if the set $\mathbf{R}$ has been constructed on the basis of other axioms. Here we take this theorem as the starting point for our further discussions. A direct consequence is the Theorem of Bolzano–Weierstrass. For this we need a further definition.

We say that a sequence $(x_k)_{k\in\mathbf{N}}$ in $\mathbf{R}^n$ is *bounded* if there exists $M > 0$ such that $\|x_k\| \leq M$, for all $k \in \mathbf{N}$.

Theorem 1.6.2 (Bolzano–Weierstrass on R). *Every bounded sequence in* $\mathbf{R}$ *possesses a convergent subsequence.*

Proof. We denote our sequence by $(x_k)_{k\in\mathbf{N}}$. Consider

$$a = \sup A \qquad \text{where} \qquad A = \{\, x \in \mathbf{R} \mid x < x_k \text{ for infinitely many } k \in \mathbf{N} \,\}.$$

Then $a \in \mathbf{R}$ is well-defined because A is nonempty and bounded from above, as the sequence is bounded. Next let $\delta > 0$ be arbitrary. By the definition of supremum, only a finite number of x_k satisfy $a + \delta < x_k$, while there are infinitely many x_k with $a - \delta < x_k$ in view of Theorem 1.6.1.(ii). Accordingly, we find infinitely many $k \in \mathbf{N}$ with $a - \delta < x_k < a + \delta$. Now successively select $\delta = \frac{1}{l}$ with $l \in \mathbf{N}$, and obtain a strictly increasing sequence of indices $(k_l)_{l\in\mathbf{N}}$ with $|x_{k_l} - a| < \frac{1}{l}$. The subsequence $(x_{k_l})_{l\in\mathbf{N}}$ obviously converges to a. ❑

There is no straightforward extension of Theorem 1.6.1 to $\mathbf{R}^n$ because a reasonable ordering on $\mathbf{R}^n$ that is compatible with the vector operations does not exist if $n \geq 2$. Nevertheless, the Theorem of Bolzano–Weierstrass is quite easy to generalize.

Theorem 1.6.3 (Bolzano–Weierstrass on $\mathbf{R}^n$). *Every bounded sequence in* $\mathbf{R}^n$ *possesses a convergent subsequence.*

Proof. Let $(x^{(k)})_{k\in\mathbf{N}}$ be our bounded sequence. We reduce to $\mathbf{R}$ by considering the sequence $(x_1^{(k)})_{k\in\mathbf{N}}$ of first components, which is a bounded sequence in $\mathbf{R}$. By the preceding theorem we then can extract a subsequence that is convergent in $\mathbf{R}$. In order to prevent overburdening the notation we now assume that $(x^{(k)})_{k\in\mathbf{N}}$ denotes the corresponding subsequence in $\mathbf{R}^n$. The sequence $(x_2^{(k)})_{k\in\mathbf{N}}$ of second components of that sequence is bounded in $\mathbf{R}$ too, and therefore we can extract a convergent subsequence. The corresponding subsequence in $\mathbf{R}^n$ now has the property that the sequences of both its first and second components converge. We go on extracting subsequences; after the n-th extraction there are still infinitely many terms left and we have a subsequence that converges in all components, which implies that it is convergent on the strength of Proposition 1.1.10. ❑

Definition 1.6.4. A sequence $(x_k)_{k\in\mathbf{N}}$ in $\mathbf{R}^n$ is said to be a *Cauchy sequence* if for every $\epsilon > 0$ there exists $N \in \mathbf{N}$ such that

$$k \geq N \quad \text{and} \quad l \geq N \quad \Longrightarrow \quad \|x_k - x_l\| < \epsilon.$$ ❍

A Cauchy sequence is bounded. In fact, take $\epsilon = 1$ in the definition above and let N be the corresponding element in $\mathbf{N}$. Then we obtain by the reverse triangle inequality from Lemma 1.1.7.(iii) that $\|x_k\| \leq \|x_N\| + 1$, for all $k \geq N$. Hence

$$\|x_k\| \leq M := \max\{\, \|x_1\|, \dots, \|x_N\|, \ \|x_N\| + 1\,\}.$$

Next, consider a sequence $(x_k)_{k\in\mathbf{N}}$ in $\mathbf{R}^n$ that converges to, say, $a \in \mathbf{R}^n$. This is a Cauchy sequence. Indeed, select $\epsilon > 0$ arbitrarily. Then we can find $N \in \mathbf{N}$ with $\|x_k - a\| < \frac{\epsilon}{2}$, for all $k \geq N$. Now the triangle inequality from Lemma 1.1.7.(i) implies, for all k and $l \geq N$,

$$\|x_k - x_l\| \leq \|x_k - a\| + \|x_l - a\| < \frac{\epsilon}{2} + \frac{\epsilon}{2} = \epsilon.$$

Note, however, that the definition of Cauchy sequence does not involve a limit, so that it is not immediately clear whether every Cauchy sequence has a limit. In fact, this is the case for $\mathbf{R}^n$, but it is not true for every Cauchy sequence with terms in $\mathbf{Q}^n$, for instance.

Theorem 1.6.5 (Completeness of $\mathbf{R}^n$). *Every Cauchy sequence in $\mathbf{R}^n$ is convergent in $\mathbf{R}^n$. This property of $\mathbf{R}^n$ is called its* completeness.

Proof. A Cauchy sequence is bounded and thus it follows from the Theorem of Bolzano–Weierstrass that it has a subsequence convergent to a limit in $\mathbf{R}^n$. But then the Cauchy property implies that the whole sequence converges to the same limit. ❑

Note that the notions of Cauchy sequence and of completeness can be generalized to arbitrary metric spaces in a straightforward fashion.

1.7 Contractions

In this section we treat a first consequence of completeness.

Definition 1.7.1. Let $F \subset \mathbf{R}^n$, let $f : F \to \mathbf{R}^n$ be a mapping that is Lipschitz continuous with a Lipschitz constant $\epsilon < 1$. Then f is said to be a *contraction* in F with *contraction factor* $\leq \epsilon$. In other words,

$$\|f(x) - f(x')\| \leq \epsilon \|x - x'\| < \|x - x'\| \qquad (x, x' \in F). \qquad ❍$$

Lemma 1.7.2 (Contraction Lemma). *Assume $F \subset \mathbf{R}^n$ closed and $x_0 \in F$. Let $f : F \to F$ be a contraction with contraction factor $\leq \epsilon$. Then there exists a unique point $x \in F$ with*

$$f(x) = x; \qquad \textit{furthermore} \qquad \|x - x_0\| \leq \frac{1}{1-\epsilon} \|f(x_0) - x_0\|.$$

Proof. Because f is a mapping of F in F, a sequence $(x_k)_{k \in \mathbf{N}_0}$ in F may be defined inductively by

$$x_{k+1} = f(x_k) \qquad (k \in \mathbf{N}_0).$$

By mathematical induction on $k \in \mathbf{N}_0$ it follows that

$$\|x_k - x_{k+1}\| = \|f(x_{k-1}) - f(x_k)\| \leq \epsilon \|x_{k-1} - x_k\| \leq \cdots \leq \epsilon^k \|x_1 - x_0\|.$$

Repeated application of the triangle inequality then gives, for all $m \in \mathbf{N}$,

$$\begin{aligned} \|x_k - x_{k+m}\| &\leq \|x_k - x_{k+1}\| + \cdots + \|x_{k+m-1} - x_{k+m}\| \\ &\leq (1 + \cdots + \epsilon^{m-1}) \epsilon^k \|f(x_0) - x_0\| \leq \frac{\epsilon^k}{1-\epsilon} \|f(x_0) - x_0\|. \end{aligned} \tag{1.9}$$

In other words, $(x_k)_{k \in \mathbf{N}_0}$ is a Cauchy sequence in F. Because of the completeness of $\mathbf{R}^n$ (see Theorem 1.6.5) there exists $x \in \mathbf{R}^n$ with $\lim_{k \to \infty} x_k = x$; and because F is closed, $x \in F$. Since f is continuous we therefore get from Lemma 1.3.6

$$f(x) = f(\lim_{k \to \infty} x_k) = \lim_{k \to \infty} f(x_k) = \lim_{k \to \infty} x_{k+1} = x,$$

i.e. x is a *fixed point* of f. Also, x is the unique fixed point in F of f. Indeed, suppose x' is another fixed point in F, then

$$0 < \|x - x'\| = \|f(x) - f(x')\| < \|x - x'\|,$$

which is a contradiction. Finally, in Inequality (1.9) we may take the limit for $m \to \infty$; in that way we obtain an *estimate for the rate of convergence* of the sequence $(x_k)_{k \in \mathbf{N}_0}$:

$$\|x_k - x\| \leq \frac{\epsilon^k}{1-\epsilon} \|f(x_0) - x_0\| \qquad (k \in \mathbf{N}_0). \tag{1.10}$$

The last assertion in the lemma then follows for $k = 0$. ❑

Example 1.7.3. Let $a \geq \frac{1}{2}$, and define $F = \big[\sqrt{\frac{a}{2}}, \infty\big[$ and $f : F \to \mathbf{R}$ by

$$f(x) = \frac{1}{2}\Big(x + \frac{a}{x}\Big).$$

Then $f(x) \in F$ for all $x \in F$, since $(\sqrt{x} - \sqrt{\frac{a}{x}})^2 \geq 0$ implies $f(x) \geq \sqrt{a}$. Furthermore,

$$-\frac{1}{2} \leq f'(x) = \frac{1}{2}\Big(1 - \frac{a}{x^2}\Big) \leq \frac{1}{2} \qquad (x \in F).$$

Accordingly, application of the Mean Value Theorem on $\mathbf{R}$ (see Formula (2.14)) yields that f is a contraction with contraction factor $\leq \frac{1}{2}$. The fixed point for f equals $\sqrt{a} \in F$. Using the estimate (1.10) we see that the sequence $(x_k)_{k\in\mathbf{N}_0}$ satisfies

$$|x_k - \sqrt{a}| \leq 2^{-k}|a - 1| \qquad \text{if} \qquad x_0 = a, \qquad x_{k+1} = \frac{1}{2}\Big(x_k + \frac{a}{x_k}\Big) \qquad (k \in \mathbf{N}).$$

In fact, the convergence is much stronger, as can be seen from

$$x_{k+1}^2 - a = \frac{1}{4x_k^2}\,(x_k^2 - a)^2 \qquad (k \in \mathbf{N}_0). \tag{1.11}$$

Furthermore, this implies $x_{k+1}^2 \geq a$, and from this it follows in turn that $x_{k+2} \leq x_{k+1}$, i.e. the sequence $(x_k)_{k\in\mathbf{N}_0}$ decreases monotonically towards its limit $\sqrt{a}$. ☆

For another proof of the Contraction Lemma based on Theorem 1.8.8 below, see Exercise 1.29.

If the appropriate estimates can be established, one may use the Contraction Lemma to prove surjectivity of mappings $f : \mathbf{R}^n \to \mathbf{R}^n$, for instance by considering $g : \mathbf{R}^n \to \mathbf{R}^n$ with $g(x) = x - f(x) + y$ for a given $y \in \mathbf{R}^n$. A fixed point x for g then satisfies $f(x) = y$ (this technique is applied in the proof of Proposition 3.2.3). In particular, taking $y = 0$ then leads to a zero x for f.

1.8 Compactness and uniform continuity

There is a class of infinite sets, called compact sets, that in certain limited aspects behave very much like finite sets. Consider the infinite pigeon-hole principle: if $(x_k)_{k\in\mathbf{N}}$ is a sequence in $\mathbf{R}$ all of whose terms belong to a finite set K, then at least one element of K must be equal to x_k for an infinite number of indices k. For infinite sets K this statement is obviously false. In this case, however, we could hope for a slightly weaker conclusion: that K contains a point that is arbitrarily closely approximated, and this infinitely often. For many purposes in analysis such an approximation is just as good as equality.

Definition 1.8.1. A set $K \subset \mathbf{R}^n$ is said to be *compact* (more precisely, *sequentially compact*) if every sequence of elements in K contains a subsequence which converges to a point in K. ○

Later, in Definition 1.8.16, we shall encounter another definition of compactness, which is the standard one in more general spaces than $\mathbf{R}^n$; in the latter, however, several different notions of compactness all coincide.

Note that the definition of compactness of K refers to the points of K only and to the distance function between the points in K, but does not refer to points outside K. Therefore, compactness is an absolute concept, unlike the properties of being open or being closed, which depend on the ambient space $\mathbf{R}^n$.

It is immediate from Definition 1.8.1 and Lemma 1.2.12 that we have the following:

Lemma 1.8.2. *A subset of a compact set $K \subset \mathbf{R}^n$ is compact if and only if it is closed in K.*

In Example 1.3.8 we saw that continuous mappings do not necessarily preserve closed sets; on the other hand, they do preserve compact sets. In this sense compact and finite sets behave similarly: the image of a finite set under a mapping is a finite set too.

Theorem 1.8.3 (Continuous image of compact is compact). *Let $K \subset \mathbf{R}^n$ be compact and $f : K \to \mathbf{R}^p$ a continuous mapping. Then $f(K) \subset \mathbf{R}^p$ is compact.*

Proof. Consider an arbitrary sequence $(y_k)_{k\in\mathbf{N}}$ of elements in $f(K)$. Then there exists a sequence $(x_k)_{k\in\mathbf{N}}$ of elements in K with $f(x_k) = y_k$. By the compactness of K the sequence $(x_k)_{k\in\mathbf{N}}$ contains a convergent subsequence, with limit $a \in K$, say. By going over to this subsequence, we have $a = \lim_{k\to\infty} x_k$ and from Lemma 1.3.6 we get $f(a) = \lim_{k\to\infty} f(x_k) = \lim_{k\to\infty} y_k$. But this says that $(y_k)_{k\in\mathbf{N}}$ has a convergent subsequence whose limit belongs to $f(K)$. ❑

Observe that there are two ingredients in the definition of compactness: one related to the Theorem of Bolzano–Weierstrass and thus to boundedness, viz., that a sequence has a convergent subsequence; and one related to closedness, viz., that the limit of a convergent sequence belongs to the set, see Lemma 1.2.12.(iii). The subsequent characterization of compact sets will be used throughout this book.

Theorem 1.8.4. *For a set $K \subset \mathbf{R}^n$ the following assertions are equivalent.*

(i) *K is bounded and closed.*

(ii) *K is compact.*

Proof. **(i) ⇒ (ii).** Consider an arbitrary sequence $(x_k)_{k\in\mathbf{N}}$ of elements contained in K. As this sequence is bounded it has, by the Theorem of Bolzano–Weierstrass, a subsequence convergent to an element $a \in \mathbf{R}^n$. Then $a \in K$ according to Lemma 1.2.12.(iii).
(ii) ⇒ (i). Assume K not bounded. Then we can find a sequence $(x_k)_{k\in\mathbf{N}}$ satisfying $x_k \in K$ and $\|x_k\| \geq k$, for $k \in \mathbf{N}$. Obviously, in this case the extraction of a convergent subsequence is impossible, so that K cannot be compact. Finally, K being closed follows from Lemma 1.2.12. Indeed, all subsequences of a convergent sequence converge to the same limit. ❑

It follows from the two preceding theorems that $[\,-1, 1\,]$ and $\mathbf{R}$ are not homeomorphic; indeed, the former set is compact while the latter is not. On the other hand, $]\,-1, 1\,[$ and $\mathbf{R}$ are homeomorphic, see Example 1.5.3.

Theorem 1.8.4 also implies that the inverse image of a compact set under a continuous mapping is closed (see Theorem 1.3.7.(iii)); nevertheless, the inverse image of a compact set might fail to be compact. For instance, consider $f : \mathbf{R}^n \to \{0\}$; or, more interestingly, $f : \mathbf{R} \to \mathbf{R}^2$ with $f(t) = (\cos t, \sin t)$. Then the image $\mathrm{im}(f) = S^1 := \{\, x \in \mathbf{R}^2 \mid \|x\| = 1\,\}$, which is compact in $\mathbf{R}^2$, but $f^{-1}(S^1) = \mathbf{R}$, which is noncompact.

Definition 1.8.5. A mapping $f : \mathbf{R}^n \to \mathbf{R}^p$ is said to be *proper* if the inverse image under f of every compact set in $\mathbf{R}^p$ is compact in $\mathbf{R}^n$, thus $f^{-1}(K) \subset \mathbf{R}^n$ compact for every compact $K \subset \mathbf{R}^p$. ❍

Intuitively, a proper mapping is one that maps points "near infinity" in $\mathbf{R}^n$ to points "near infinity" in $\mathbf{R}^p$.

Theorem 1.8.6. *Let $f : \mathbf{R}^n \to \mathbf{R}^p$ be proper and continuous. Then f is a closed mapping.*

Proof. Let F be a closed set in $\mathbf{R}^n$. We verify that $f(F)$ satisfies the condition in Lemma 1.2.12.(iii). Indeed, let $(x_k)_{k\in\mathbf{N}}$ be a sequence of points in F with the property that $(f(x_k))_{k\in\mathbf{N}}$ is convergent, with limit $b \in \mathbf{R}^p$. Thus, for k sufficiently large, we have $f(x_k) \in K = \{\, y \in \mathbf{R}^p \mid \|y - b\| \leq 1\,\}$, while K is compact in $\mathbf{R}^p$. This implies that $x_k \in f^{-1}(K) \cap F \subset \mathbf{R}^n$, which is compact too, f being proper and F being closed. Hence the sequence $(x_k)_{k\in\mathbf{N}}$ has a convergent subsequence, which we will also denote by $(x_k)_{k\in\mathbf{N}}$, with limit $a \in F$. But then Lemma 1.3.6 gives $f(a) = \lim_{k\to\infty} f(x_k) = b$, that is $b \in f(F)$. ❑

In Example 1.5.1 we saw that the inverse of a continuous bijection defined on a subset of $\mathbf{R}^n$ is not necessarily continuous. But for mappings with a compact domain in $\mathbf{R}^n$ we do have a positive result.

Theorem 1.8.7. *Suppose that $K \subset \mathbf{R}^n$ is compact, let $L \subset \mathbf{R}^p$, and let $f : K \to L$ be a continuous bijective mapping. Then L is a compact set and $f : K \to L$ is a homeomorphism.*

Proof. Let $F \subset K$ be closed in K. Proposition 1.2.17.(ii) then implies that F is closed in $\mathbf{R}^n$; and thus F is compact according to Lemma 1.8.2. Using Theorems 1.8.3 and 1.8.4 we see that $f(F)$ is closed in $\mathbf{R}^p$, and thus in L. The assertion of the theorem now follows from Proposition 1.5.4. ❑

Observe that this theorem also follows from Theorem 1.8.6.

Obviously, a scalar function on a finite set is bounded and attains its minimum and maximum values. In this sense compact sets resemble finite sets.

Theorem 1.8.8. *Let $K \subset \mathbf{R}^n$ be a compact nonempty set, and let $f : K \to \mathbf{R}$ be a continuous function. Then there exist a and $b \in K$ such that*

$$f(a) \leq f(x) \leq f(b) \qquad (x \in K).$$

Proof. $f(K) \subset \mathbf{R}$ is a compact set by Theorem 1.8.3, and Theorem 1.8.4 then implies that $\sup f(K)$ is well-defined. The definitions of supremum and of the compactness of $f(K)$ then give that $\sup f(K) \in f(K)$. But this yields the existence of the desired element $b \in K$. For $a \in K$ consider $\inf f(K)$. ❑

A useful application of this theorem is to show the equivalence of all norms on the finite-dimensional vector space $\mathbf{R}^n$. This implies that definitions formulated in terms of the Euclidean norm, like those of limit and continuity in Section 1.3, or of differentiability in Definition 2.2.2, are in fact independent of the particular choice of the norm.

Definition 1.8.9. A *norm* on $\mathbf{R}^n$ is a function $\nu : \mathbf{R}^n \to \mathbf{R}$ satisfying the following conditions, for $x, y \in \mathbf{R}^n$ and $\lambda \in \mathbf{R}$.

(i) *Positivity:* $\nu(x) \geq 0$, with equality if and only if $x = 0$.

(ii) *Homogeneity:* $\nu(\lambda x) = |\lambda|\, \nu(x)$.

(iii) *Triangle inequality:* $\nu(x + y) \leq \nu(x) + \nu(y)$. ○

Corollary 1.8.10 (Equivalence of norms on $\mathbf{R}^n$). *For every norm ν on $\mathbf{R}^n$ there exist numbers $c_1 > 0$ and $c_2 > 0$ such that*

$$c_1 \|x\| \leq \nu(x) \leq c_2 \|x\| \qquad (x \in \mathbf{R}^n).$$

In particular, ν is Lipschitz continuous, with c_2 being a Lipschitz constant for ν.

Proof. We begin by proving the second inequality. From Formula (1.1) we have $x = \sum_{1\le j\le n} x_j e_j \in \mathbf{R}^n$, and thus we obtain from the Cauchy–Schwarz inequality (see Proposition 1.1.6)

$$\nu(x) = \nu\Big(\sum_{1\le j\le n} x_j e_j\Big) \le \sum_{1\le j\le n} |x_j|\,\nu(e_j) \le \|x\|\,\|(\nu(e_1), \dots, \nu(e_n))\| =: c_2\|x\|.$$

Next we show that ν is Lipschitz continuous. Indeed, using $\nu(x) = \nu(x-a+a) \le \nu(x-a) + \nu(a)$ we obtain

$$|\nu(x) - \nu(a)| \le \nu(x-a) \le c_2\|x-a\| \qquad (x,\ a \in \mathbf{R}^n).$$

Finally, we consider the restriction of ν to the compact $K = \{x \in \mathbf{R}^n \mid \|x\| = 1\}$. By Theorem 1.8.8 there exists $a \in K$ with $\nu(a) \le \nu(x)$, for all $x \in K$. Set $c_1 = \nu(a)$, then $c_1 > 0$ by Definition 1.8.9.(i). For arbitrary $0 \ne x \in \mathbf{R}^n$ we have $\frac{1}{\|x\|}x \in K$, and hence

$$c_1 \le \nu\Big(\frac{1}{\|x\|}x\Big) = \frac{\nu(x)}{\|x\|}. \qquad \square$$

Definition 1.8.11. We denote by $\mathrm{Aut}(\mathbf{R}^n)$ the set of *bijective linear mappings* or *automorphisms* (αὐτός = of or by itself) of $\mathbf{R}^n$ into itself. ❍

Corollary 1.8.12. *Let $A \in \mathrm{Aut}(\mathbf{R}^n)$ and define $\nu : \mathbf{R}^n \to \mathbf{R}$ by $\nu(x) = \|Ax\|$. Then ν is a norm on $\mathbf{R}^n$ and hence there exist numbers $c_1 > 0$ and $c_2 > 0$ satisfying*

$$c_1\|x\| \le \|Ax\| \le c_2\|x\|, \qquad c_2^{-1}\|x\| \le \|A^{-1}x\| \le c_1^{-1}\|x\| \qquad (x \in \mathbf{R}^n).$$

In the analysis in several real variables the notion of uniform continuity is as important as in the analysis in one variable. Roughly speaking, uniform continuity of a mapping f means that it is continuous and that the δ in Definition 1.3.4 applies for all $a \in \mathrm{dom}(f)$ simultaneously.

Definition 1.8.13. Let $f : \mathbf{R}^n \supset\!\!\to \mathbf{R}^p$ be a mapping. Then f is said to be *uniformly continuous* if for every $\epsilon > 0$ there exists $\delta > 0$ satisfying

$$x,\ y \in \mathrm{dom}(f) \quad \text{and} \quad \|x-y\| < \delta \quad \Longrightarrow \quad \|f(x) - f(y)\| < \epsilon.$$

Equivalently, phrased in terms of neighborhoods: f is uniformly continuous if for every neighborhood V of 0 in $\mathbf{R}^p$ there exists a neighborhood U of 0 in $\mathbf{R}^n$ such that

$$x,\ y \in \mathrm{dom}(f) \quad \text{and} \quad x - y \in U \quad \Longrightarrow \quad f(x) - f(y) \in V. \qquad ❍$$

Example 1.8.14. Any Lipschitz continuous mapping is uniformly continuous. In particular, every $A \in \mathrm{Aut}(\mathbf{R}^n)$ is uniformly continuous; this follows directly from Corollary 1.8.12. In fact, any linear mapping $A : \mathbf{R}^n \to \mathbf{R}^p$, whether bijective or not, is Lipschitz continuous and therefore uniformly continuous. For a proof, observe that Formula (1.1) and Lemma 1.1.7.(iv) imply, for $x \in \mathbf{R}^n$,

$$\|Ax\| = \| \sum_{1\le j\le n} x_j Ae_j\| \le \sum_{1\le j\le n} |x_j|\, \|Ae_j\| \le \|x\| \sum_{1\le j\le n} \|Ae_j\| =: k\|x\|.$$

The function $f : \mathbf{R} \to \mathbf{R}$ with $f(x) = x^2$ is not uniformly continuous. ☆

Theorem 1.8.15. *Let $K \subset \mathbf{R}^n$ be compact and $f : K \to \mathbf{R}^p$ a continuous mapping. Then f is uniformly continuous.*

Proof. Suppose f is not uniformly continuous. Then there exists $\epsilon > 0$ with the following property. For every $\delta > 0$ we can find a pair of points x and $y \in K$ with $\|x - y\| < \delta$ and nevertheless $\|f(x) - f(y)\| \ge \epsilon$. In particular, for every $k \in \mathbf{N}$ there exist x_k and $y_k \in K$ with

$$\|x_k - y_k\| < \frac{1}{k} \qquad \text{and} \qquad \|f(x_k) - f(y_k)\| \ge \epsilon. \tag{1.12}$$

By the compactness of K the sequence $(x_k)_{k\in\mathbf{N}}$ has a convergent subsequence, with a limit in K. Go over to this subsequence. Next $(y_k)_{k\in\mathbf{N}}$ has a convergent subsequence, also with a limit in K. Again, we switch to this subsequence, to obtain $a := \lim_{k\to\infty} x_k$ and $b := \lim_{k\to\infty} y_k$. Furthermore, by taking the limit in the first inequality in (1.12) we see that $a = b \in K$. On the other hand, from the continuity of f at a we obtain $\lim_{k\to\infty} f(x_k) = f(a) = \lim_{k\to\infty} f(y_k)$. The continuity of the Euclidean norm then gives $\lim_{k\to\infty} \|f(x_k) - f(y_k)\| = 0$; this is in contradiction with the second inequality in (1.12). ❑

In many cases a set $K \subset \mathbf{R}^n$ and a function $f : K \to \mathbf{R}$ are known to possess a certain *local property*, that is, for every $x \in K$ there exists a neighborhood $U(x)$ of x in $\mathbf{R}^n$ such that the restriction of f to $U(x)$ has the property in question. For example, continuity of f on K by definition has the following meaning: for every $\epsilon > 0$ and every $x \in K$ there exists an open neighborhood $U(x)$ of x in $\mathbf{R}^n$ such that for all $x' \in U(x) \cap K$ one has $|f(x) - f(x')| < \epsilon$. It then follows that $K \subset \bigcup_{x\in K} U(x)$. Thus one finds collections of open sets in $\mathbf{R}^n$ with the property that K is contained in their union.

In such a case, one may wish to ascertain whether it follows that the set K, or the function f, possesses the corresponding *global property*. The question may be asked, for example, whether f is uniformly continuous on K, that is, whether for every $\epsilon > 0$ there exists one open set U in $\mathbf{R}^n$ such that for all $x, x' \in K$ with $x - x' \in U$ one has $|f(x) - f(x')| < \epsilon$. On the basis of Theorem 1.8.15 this does indeed follow if K is compact. On the other hand, consider $f(x) = \frac{1}{x}$, for $x \in K := \,]0,1[$. For every $x \in K$ there exists a neighborhood $U(x)$ of x in $\mathbf{R}$ such that $f|_{U(x)}$ is bounded and such that $K = \bigcup_{x \in K} U(x)$, that is, f is locally bounded on K. Yet it is not true that f is globally bounded on K. This motivates the following:

Definition 1.8.16. A collection $\mathcal{O} = \{ O_i \mid i \in I \}$ of open sets in $\mathbf{R}^n$ is said to be an *open covering* of a set $K \subset \mathbf{R}^n$ if

$$K \subset \bigcup_{O \in \mathcal{O}} O.$$

A subcollection $\mathcal{O}' \subset \mathcal{O}$ is said to be an *(open) subcovering* of K if $\mathcal{O}'$ is a covering of K; if in addition $\mathcal{O}'$ is a finite collection, it is said to be a *finite (open) subcovering*.

A subset $K \subset \mathbf{R}^n$ is said to be *compact* if **every** open covering of K contains a finite subcovering of K. ❍

In topology, this definition of compactness rather than the Definition 1.8.1 of (sequential) compactness is the proper one to use. In spaces like $\mathbf{R}^n$, however, the two definitions 1.8.1 and 1.8.16 of compactness coincide, which is a consequence of the following:

Theorem 1.8.17 (Heine–Borel). *For a set $K \subset \mathbf{R}^n$ the following assertions are equivalent.*

(i) *K is bounded and closed.*

(ii) *K is compact in the sense of Definition 1.8.16.*

The proof uses a technical concept.

Definition 1.8.18. An *n-dimensional rectangle B* parallel to the coordinate axes is a subset of $\mathbf{R}^n$ of the form

$$B = \{ x \in \mathbf{R}^n \mid a_j \le x_j \le b_j \ \ (1 \le j \le n) \}$$

where it is assumed that $a_j, b_j \in \mathbf{R}$ and $a_j \le b_j$, for $1 \le j \le n$. ❍

Proof. (i) ⇒ (ii). Choose a rectangle $B \subset \mathbf{R}^n$ such that $K \subset B$. Then define, by induction over $k \in \mathbf{N}_0$, a collection of rectangles $\mathcal{B}_k$ as follows. Let $\mathcal{B}_0 = \{B\}$ and let $\mathcal{B}_{k+1}$ be the collection of rectangles obtained by subdividing each rectangle $B \in \mathcal{B}_k$ into 2^n rectangles by "halving the edges of B".

Let $\mathcal{O}$ be an arbitrary open covering of K, and suppose that $\mathcal{O}$ does not contain a finite subcovering of K. Because $K \subset \bigcup_{B_1 \in \mathcal{B}_1} B_1$, there exists a rectangle $B_1 \in \mathcal{B}_1$

such that $\mathcal{O}$ does not contain a finite subcovering of $K \cap B_1 \neq \emptyset$. But this implies the existence of a rectangle B_2 with

$$B_2 \in \mathcal{B}_2, \qquad B_2 \subset B_1, \qquad \mathcal{O} \text{ contains no finite subcovering of } K \cap B_2 \neq \emptyset.$$

By induction over k we find a sequence of rectangles $(B_k)_{k \in \mathbf{N}}$ such that

$$B_k \in \mathcal{B}_k, \qquad B_k \subset B_{k-1}, \qquad \mathcal{O} \text{ contains no finite subcovering of } K \cap B_k \neq \emptyset. \tag{1.13}$$

Now let $x_k \in K \cap B_k$ be arbitrary. Since $B_k \subset B_l$ for $k > l$ and since diameter $B_l = 2^{-l}$ diameter B, we see that $(x_k)_{k \in \mathbf{N}}$ is a Cauchy sequence in $\mathbf{R}^n$. In view of the completeness of $\mathbf{R}^n$, see Theorem 1.6.5, this means that there exists $x \in \mathbf{R}^n$ such that $x = \lim_{k \to \infty} x_k$; since K is closed, we have $x \in K$. Thus there is $O \in \mathcal{O}$ with $x \in O$. Now O is open in $\mathbf{R}^n$, and so we can find $\delta > 0$ such that

$$B(x; \delta) \subset O. \tag{1.14}$$

Because $\lim_{k \to \infty}$ diameter $B_k = 0$ and $\lim_{k \to \infty} x_k - x = 0$, there is $k \in \mathbf{N}$ with

$$\text{diameter } B_k + \|x_k - x\| < \delta.$$

Let $y \in B_k$ be arbitrary. Then

$$\|y - x\| \leq \|y - x_k\| + \|x_k - x\| \leq \text{diameter } B_k + \|x_k - x\| < \delta;$$

that is, $y \in B(x; \delta)$. But now it follows from (1.14) that $B_k \subset O$, which is in contradiction with (1.13). Therefore $\mathcal{O}$ must contain a finite subcovering of K.
(ii) $\Rightarrow$ (i). The boundedness of K follows by covering K by open balls in $\mathbf{R}^n$ about the origin of radius k, for $k \in \mathbf{N}$. Now K admits of a finite subcovering by such balls, from which it follows that K is bounded.

In order to demonstrate that K is closed, we prove that $\mathbf{R}^n \setminus K$ is open. Indeed, choose $y \notin K$, and define $O_j = \{x \in \mathbf{R}^n \mid \|x - y\| > \frac{1}{j}\}$ for $j \in \mathbf{N}$. This gives

$$K \subset \mathbf{R}^n \setminus \{y\} = \bigcup_{j \in \mathbf{N}} O_j.$$

This open covering of the compact set K admits of a finite subcovering, and consequently

$$K \subset \bigcup_{1 \leq k \leq l} O_{j_k} = O_{j_0} \qquad \text{if} \qquad j_0 = \max\{\, j_k \mid 1 \leq k \leq l\,\}.$$

Therefore $\|x - y\| > \frac{1}{j_0}$ if $x \in K$; and so $y \in \{x \in \mathbf{R}^n \mid \|x - y\| < \frac{1}{j_0}\} \subset \mathbf{R}^n \setminus K$. ❑

The proof of the next theorem is a characteristic application of compactness.

Theorem 1.8.19 (Dini's Theorem). *Let $K \subset \mathbf{R}^n$ be compact and suppose that the sequence $(f_k)_{k \in \mathbf{N}}$ of continuous functions on K has the following properties.*

(i) $(f_k)_{k\in\mathbf{N}}$ *converges pointwise on* K *to a continuous function* f, *in other words,* $\lim_{k\to\infty} f_k(x) = f(x)$, *for all* $x \in K$.

(ii) $(f_k)_{k\in\mathbf{N}}$ *is monotonically decreasing, that is,* $f_k(x) \geq f_{k+1}(x)$, *for all* $k \in \mathbf{N}$ *and* $x \in K$.

Then $(f_k)_{k\in\mathbf{N}}$ *converges uniformly on* K *to the function* f. *More precisely, for every* $\epsilon > 0$ *there exists* $N \in \mathbf{N}$ *such that we have*

$$0 \leq f_k(x) - f(x) < \epsilon \qquad (k \geq N,\ x \in K).$$

Proof. For each $k \in \mathbf{N}$ let $g_k = f_k - f$. Then g_k is continuous on K, while $\lim_{k\to\infty} g_k(x) = 0$ and $g_k(x) \geq g_{k+1}(x) \geq 0$, for all $x \in K$ and $k \in \mathbf{N}$. Let $\epsilon > 0$ be arbitrary and define $O_k(\epsilon) = g_k^{-1}(\,[\,0, \epsilon\,[\,)$; note that $O_k(\epsilon) \subset O_l(\epsilon)$ if $k < l$. Then the collection of sets $\{\, O_k(\epsilon) \mid k \in \mathbf{N} \,\}$ is an open covering of K, because of the pointwise convergence of $(g_k)_{k\in\mathbf{N}}$. By the compactness of K there exists a finite subcovering. Let N be largest of the indices k labeling the sets in that subcovering, then $K \subset O_N(\epsilon)$, and the assertion follows. ❑

To conclude this section we give an alternative characterization of the concept of compactness.

Definition 1.8.20. Let $K \subset \mathbf{R}^n$ be a subset. A collection $\{F_i \mid i \in I\}$ of sets in $\mathbf{R}^n$ has the *finite intersection property relative to* K if for every finite subset $J \subset I$

$$K \cap \bigcap_{i\in J} F_i \neq \emptyset. \qquad \bigcirc$$

Proposition 1.8.21. *A set* $K \subset \mathbf{R}^n$ *is compact if and only if, for every collection* $\{F_i \mid i \in I\}$ *of closed subsets in* $\mathbf{R}^n$ *having the finite intersection property relative to* K, *we have*

$$K \cap \bigcap_{i\in I} F_i \neq \emptyset.$$

Proof. In this proof $\mathcal{O}$ consistently represents a collection $\{O_i \mid i \in I\}$ of open subsets in $\mathbf{R}^n$. For such $\mathcal{O}$ we define $\mathcal{F} = \{F_i \mid i \in I\}$ with $F_i := \mathbf{R}^n \setminus O_i$; then $\mathcal{F}$ is a collection of closed subsets in $\mathbf{R}^n$. Further, in this proof J invariably represents a finite subset of I. Now the following assertions (i) through (iv) are successively equivalent:

(i) K is not compact;

(ii) there exists $\mathcal{O}$ with $K \subset \bigcup_{i\in I} O_i$, and for every J one has $K \setminus \bigcup_{i\in J} O_i \neq \emptyset$;

(iii) there is $\mathcal{O}$ with $\mathbf{R}^n \setminus K \supset \bigcap_{i\in I}(\mathbf{R}^n \setminus O_i)$, and for every J one has $K \cap \bigcap_{i\in J}(\mathbf{R}^n \setminus O_i) \neq \emptyset$;

(iv) there exists $\mathcal{F}$ with $K \cap \bigcap_{i\in I} F_i = \emptyset$, and for every J one has $K \cap \bigcap_{i\in J} F_i \neq \emptyset$.

In (ii) $\Leftrightarrow$ (iii) we used DeMorgan's laws from (1.3). The assertion follows. ❑

1.9 Connectedness

Let $I \subset \mathbf{R}$ be an interval and $f : I \to \mathbf{R}$ a continuous function. Then f has the *intermediate value property*, that is, f assumes on I all values in between any two of the values it assumes; as a consequence $f(I)$ is an interval too. It is characteristic of open, or closed, intervals in $\mathbf{R}$ that these cannot be written as the union of two disjoint open, or closed, subintervals, respectively. We take this indecomposability as a clue for generalization to $\mathbf{R}^n$.

Definition 1.9.1. A set $A \subset \mathbf{R}^n$ is said to be *disconnected* if there exist open sets U and V in $\mathbf{R}^n$ such that

$$A\cap U \neq \emptyset, \qquad A\cap V \neq \emptyset, \qquad (A\cap U)\cap(A\cap V) = \emptyset, \qquad (A\cap U)\cup(A\cap V) = A.$$

In other words, A is the union of two disjoint nonempty subsets that are open in A. Furthermore, the set A is said to be *connected* if A is not disconnected. ○

We recall the definition of an *interval* I contained in $\mathbf{R}$: for every a and $b \in I$ and any $c \in \mathbf{R}$ the inequalities $a < c < b$ imply that $c \in I$.

Proposition 1.9.2 (Connectedness of interval). *The only connected subsets of* $\mathbf{R}$ *containing more than one point are* $\mathbf{R}$ *and the intervals (open, closed, or half-open).*

Proof. Let $I \subset \mathbf{R}$ be connected. If I were not an interval, then by definition there must be a and $b \in I$ and $c \notin I$ with $a < c < b$. Then $I \cap \{x \in \mathbf{R} \mid x < c\}$ and $I \cap \{x \in \mathbf{R} \mid x > c\}$ would be disjoint nonempty open subsets in I whose union is I.

Let $I \subset \mathbf{R}$ be an interval. If I were not connected, then $I = U \cup V$, where U and V are disjoint nonempty open sets in I. Then there would be $a \in U$ and $b \in V$ satisfying $a < b$ (rename if necessary). Define $c = \sup\{x \in \mathbf{R} \mid [a, x[\, \subset U\}$; then $c \leq b$ and thus $c \in I$, because I is an interval. Clearly $c \in \overline{U}^I$; noting that $U = I \setminus V$ is closed in I, we must have $c \in U$. However, U is also open in I, and since I is an interval there exists $\delta > 0$ with $]c - \delta, c + \delta[\, \subset U$, which violates the definition of c. ❑

Lemma 1.9.3. *The following assertions are equivalent for a set $A \subset \mathbf{R}^n$.*

(i) *A is disconnected.*

(ii) *There exists a surjective continuous function sending A to the two-point set $\{0, 1\}$.*

We recall the definition of the *characteristic function* 1_A of a set $A \subset \mathbf{R}^n$, with

$$1_A(x) = 1 \quad \text{if} \quad x \in A, \qquad 1_A(x) = 0 \quad \text{if} \quad x \notin A.$$

Proof. For (i) $\Rightarrow$ (ii) use the characteristic function of $A \cap U$ with U as in Definition 1.9.1. It is immediate that (ii) $\Rightarrow$ (i). ❑

Theorem 1.9.4 (Continuous image of connected is connected). *Let $A \subset \mathbf{R}^n$ be connected and let $f : A \to \mathbf{R}^p$ be a continuous mapping. Then $f(A)$ is connected in $\mathbf{R}^p$.*

Proof. If $f(A)$ were disconnected, there would be a continuous surjection $g : f(A) \to \{0, 1\}$, and then $g \circ f : A \to \{0, 1\}$ would also be a continuous surjection, violating the connectedness of A. ❑

Theorem 1.9.5 (Intermediate Value Theorem). *Let $A \subset \mathbf{R}^n$ be connected and let $f : A \to \mathbf{R}$ be a continuous function. Then $f(A)$ is an interval in $\mathbf{R}$; in particular, f takes on all values between any two that it assumes.*

Proof. $f(A)$ is connected in $\mathbf{R}$ according to Theorem 1.9.4, so by Proposition 1.9.2 it is an interval. ❑

Lemma 1.9.6. *The union of any family of connected subsets of $\mathbf{R}^n$ having at least one point in common is also connected. Furthermore, the closure of a connected set is connected.*

Proof. Let $x \in \mathbf{R}^n$ and $A = \cup_{\alpha \in \mathcal{A}} A_\alpha$ with $x \in A_\alpha$ and $A_\alpha \subset \mathbf{R}^n$ connected. Suppose $f : A \to \{0, 1\}$ is continuous. Since each A_α is connected and $x \in A_\alpha$, we have $f(x') = f(x)$, for all $x' \in A_\alpha$ and $\alpha \in \mathcal{A}$. Thus f cannot be surjective. For the second assertion, let $A \subset \mathbf{R}^n$ be connected and $f : \overline{A} \to \{0, 1\}$ be a continuous function. The continuity of f then implies that it cannot be surjective. ❑

Definition 1.9.7. Let $A \subset \mathbf{R}^n$ and $x \in A$. The *connected component* of x in A is the union of all connected subsets of A containing x. ❍

Using the Lemmata 1.9.6 and 1.9.3 one easily proves the following:

Proposition 1.9.8. *Let $A \subset \mathbf{R}^n$ and $x \in A$. Then we have the following assertions.*

(i) *The connected component of x in A is connected and is a maximal connected set in A.*

(ii) *The connected component of x in A is closed in A.*

(iii) *The set of connected components of the points of A forms a partition of A, that is, every point of A lies in a connected component, and different connected components are disjoint.*

(iv) *A continuous function $f : A \to \{0, 1\}$ is constant on the connected components of A.*

Note that connected components of A need not be open subsets of A. For instance, consider the subset of the rationals $\mathbf{Q}$ in $\mathbf{R}$. The connected component of each point $x \in \mathbf{Q}$ equals $\{x\}$.

Chapter 2

Differentiation

Locally, in shrinking neighborhoods of a point, a differentiable mapping can be approximated by an affine mapping in such a way that the difference between the two mappings vanishes faster than the distance to the point in question. This condition forces the graphs of the original mapping and of the affine mapping to be tangent at their point of intersection. The behavior of a differentiable mapping is substantially better than that of a merely continuous mapping.

At this stage linear algebra starts to play an important role. We reformulate the definition of differentiability so that our earlier results on continuity can be used to the full, thereby minimizing the need for additional estimates. The relationship between being differentiable in this sense and possessing derivatives with respect to each of the variables individually is discussed next. We develop rules for computing derivatives; among these, the chain rule for the derivative of a composition, in particular, has many applications in subsequent parts of the book. A differentiable function has good properties as long as its derivative does not vanish; critical points, where the derivative vanishes, therefore deserve to be studied in greater detail. Higher-order derivatives and Taylor's formula come next, because of their role in determining the behavior of functions near critical points, as well as in many other applications. The chapter closes with a study of the interaction between the operations of taking a limit, of differentiation and of integration; in particular, we investigate conditions under which they may be interchanged.

2.1 Linear mappings

We begin by fixing our notation for linear mappings and matrices because no universally accepted conventions exist. We denote by

$$\mathrm{Lin}(\mathbf{R}^n, \mathbf{R}^p)$$

the *linear space of all linear mappings* from $\mathbf{R}^n$ to $\mathbf{R}^p$, thus $A \in \mathrm{Lin}(\mathbf{R}^n, \mathbf{R}^p)$ if and only if

$$A : \mathbf{R}^n \to \mathbf{R}^p, \qquad A(\lambda x + y) = \lambda Ax + Ay \qquad (\lambda \in \mathbf{R},\ x, y \in \mathbf{R}^n).$$

From linear algebra we know that, having chosen the standard basis $(e_1, \dots, e_n)$ in $\mathbf{R}^n$, we obtain the *matrix* of A as the rectangular array consisting of pn numbers in $\mathbf{R}$, whose j-th column equals the element $Ae_j \in \mathbf{R}^p$, for $1 \le j \le n$; that is

$$(Ae_1\, Ae_2\, \cdots\, Ae_j\, \cdots\, Ae_n).$$

In other words, if $(e'_1, \dots, e'_p)$ denotes the standard basis in $\mathbf{R}^p$ and if we write, see Formula (1.1),

$$Ae_j = \sum_{1 \le i \le p} a_{ij} e'_i, \qquad \text{then } A \text{ has the matrix} \qquad \begin{pmatrix} a_{11} & \dots & a_{1n} \\ \vdots & & \vdots \\ a_{p1} & \dots & a_{pn} \end{pmatrix},$$

with p rows and n columns. The number $a_{ij} \in \mathbf{R}$ is called the (i, j)-th *entry* or *coefficient* of the matrix of A, with i labeling rows and j columns. In this book we will sparingly use the matrix representation of a linear operator with respect to bases other than the standard bases. Nevertheless, we will keep the concepts of a linear transformation and of its matrix representations separate, even though we will not denote them in different ways, in order to avoid overburdening the notation. Write

$$\mathrm{Mat}(p \times n, \mathbf{R})$$

for the *linear space of $p \times n$ matrices with coefficients in* $\mathbf{R}$. In the special case where $n = p$, we set

$$\mathrm{End}(\mathbf{R}^n) := \mathrm{Lin}(\mathbf{R}^n, \mathbf{R}^n), \qquad \mathrm{Mat}(n, \mathbf{R}) := \mathrm{Mat}(n \times n, \mathbf{R}).$$

Here End comes from *endomorphism* (ἔνδον = within).

In Definition 1.8.11 we already encountered the special case of the subset of the bijective elements in $\mathrm{End}(\mathbf{R}^n)$, which are also called automorphisms of $\mathbf{R}^n$. We denote this set, and the set of corresponding matrices by, respectively,

$$\mathrm{Aut}(\mathbf{R}^n) \subset \mathrm{End}(\mathbf{R}^n), \qquad \mathbf{GL}(n, \mathbf{R}) \subset \mathrm{Mat}(n, \mathbf{R}).$$

$\mathbf{GL}(n, \mathbf{R})$ is the *general linear group of invertible matrices* in $\mathrm{Mat}(n, \mathbf{R})$. In fact, composition of bijective linear mappings corresponds to multiplication of the associated invertible matrices, and both operations satisfy the group axioms: thus $\mathrm{Aut}(\mathbf{R}^n)$ and $\mathbf{GL}(n, \mathbf{R})$ are groups, and they are **noncommutative** except in the case of $n = 1$.

For practical purposes we identify the linear space $\mathrm{Mat}(p \times n, \mathbf{R})$ with the linear space $\mathbf{R}^{pn}$ by means of the bijective linear mapping $c : \mathrm{Mat}(p \times n, \mathbf{R}) \to \mathbf{R}^{pn}$, defined by

$$A = \begin{pmatrix} a_{11} & \dots & a_{1n} \\ \vdots & & \vdots \\ a_{p1} & \dots & a_{pn} \end{pmatrix} \mapsto c(A) := (a_{11}, \dots, a_{p1}, a_{12}, \dots, a_{p2}, \dots, a_{pn}). \tag{2.1}$$

Observe that this map is obtained by putting the columns Ae_j of A below each other, and then writing the resulting vector as a row for typographical reasons. We summarize the results above in the following linear isomorphisms of vector spaces

$$\mathrm{Lin}(\mathbf{R}^n, \mathbf{R}^p) \simeq \mathrm{Mat}(p \times n, \mathbf{R}) \simeq \mathbf{R}^{pn}. \tag{2.2}$$

We now endow $\mathrm{Lin}(\mathbf{R}^n, \mathbf{R}^p)$ with a norm $\|\cdot\|_{\mathrm{Eucl}}$, the *Euclidean norm* (also called the *Frobenius norm* or *Hilbert–Schmidt norm*), as follows. For $A \in \mathrm{Lin}(\mathbf{R}^n, \mathbf{R}^p)$ we set

$$\|A\|_{\mathrm{Eucl}} := \|c(A)\| = \sqrt{\sum_{1\le j\le n} \|Ae_j\|^2} = \sqrt{\sum_{1\le i\le p,\, 1\le j\le n} a_{ij}^2}. \tag{2.3}$$

Intermezzo on linear algebra. There exists a more natural definition of the Euclidean norm. That definition, however, requires some notions from linear algebra. Since these will be needed in this book anyway, both in the theory and in the exercises, we recall them here.

Given $A \in \mathrm{Lin}(\mathbf{R}^n, \mathbf{R}^p)$, we have $A^t \in \mathrm{Lin}(\mathbf{R}^p, \mathbf{R}^n)$, the *adjoint linear operator* of A with respect to the standard inner products on $\mathbf{R}^n$ and $\mathbf{R}^p$. It is characterized by the property

$$\langle Ax, y\rangle = \langle x, A^t y\rangle \qquad (x \in \mathbf{R}^n,\ y \in \mathbf{R}^p).$$

It is immediate from $a_{ij} = \langle Ae_j, e_i\rangle = \langle e_j, A^t e_i\rangle = a^t_{ji}$ that the matrix of A^t with respect to the standard basis of $\mathbf{R}^p$, and that of $\mathbf{R}^n$, is the *transpose* of the matrix of A with respect to the standard basis of $\mathbf{R}^n$, and that of $\mathbf{R}^p$, respectively; in other words, it is obtained from the matrix of A by interchanging the rows with the columns.

For the standard basis $(e_1, \dots, e_n)$ in $\mathbf{R}^n$ we obtain

$$\langle Ae_i, Ae_j\rangle = \langle e_i, (A^t A)e_j\rangle.$$

Consequently, the composition $A^t A \in \mathrm{End}(\mathbf{R}^n)$ has the symmetric matrix of inner products

$$(\langle a_i, a_j\rangle)_{1\le i,\, j\le n} \qquad \text{with } a_j = Ae_j \in \mathbf{R}^p, \text{ the } j\text{-th column of the matrix of } A. \tag{2.4}$$

This matrix is known as *Gram's matrix* associated with the vectors $a_1, \dots, a_n \in \mathbf{R}^p$.

Next, we recall the definition of $\mathrm{tr}\, A \in \mathbf{R}$, the *trace* of $A \in \mathrm{End}(\mathbf{R}^n)$, as the coefficient of $-\lambda^{n-1}$ in the following *characteristic polynomial* of A:

$$\det(\lambda I - A) = \lambda^n - \lambda^{n-1}\,\mathrm{tr}\, A + \cdots + (-1)^n \det A.$$

Now the multiplicative property of the determinant gives

$$\det(\lambda I - A) = \det B(\lambda I - A)B^{-1} = \det(\lambda I - BAB^{-1}) \qquad (B \in \mathrm{Aut}(\mathbf{R}^n)),$$

and hence $\operatorname{tr} A = \operatorname{tr} BAB^{-1}$. This makes it clear that $\operatorname{tr} A$ does not depend on a particular matrix representation chosen for the operator A. Therefore, the trace can also be computed by

$$\operatorname{tr} A = \sum_{1 \le j \le n} a_{jj} \qquad \text{with} \qquad (a_{ij}) \in \mathrm{Mat}(n, \mathbf{R}) \text{ a corresponding matrix.}$$

With these definitions available it is obvious from (2.3) that

$$\|A\|^2_{\mathrm{Eucl}} = \sum_{1 \le j \le n} \|Ae_j\|^2 = \operatorname{tr}(A^t A) \qquad (A \in \mathrm{Lin}(\mathbf{R}^n, \mathbf{R}^p)).$$

Lemma 2.1.1. *For all* $A \in \mathrm{Lin}(\mathbf{R}^n, \mathbf{R}^p)$ *and* $h \in \mathbf{R}^n$,

$$\|A\,h\| \le \|A\|_{\mathrm{Eucl}}\, \|h\|. \tag{2.5}$$

Proof. The Cauchy–Schwarz inequality from Proposition 1.1.6 immediately yields

$$\|A\,h\|^2 = \sum_{1 \le i \le p} \Big(\sum_{1 \le j \le n} a_{ij} h_j \Big)^2 \le \sum_{1 \le i \le p} \Big(\sum_{1 \le j \le n} a_{ij}^2 \sum_{1 \le k \le n} h_k^2 \Big) = \|A\|^2_{\mathrm{Eucl}}\, \|h\|^2. \quad \square$$

The Euclidean norm of a linear operator A is easy to compute, but has the disadvantage of not giving the best possible estimate in Inequality (2.5), which would be

$$\inf\{\, C \ge 0 \mid \|Ah\| \le C\|h\| \text{ for all } h \in \mathbf{R}^n \,\}.$$

It is easy to verify that this number is equal to the *operator norm* $\|A\|$ of A, given by

$$\|A\| = \sup\{\, \|Ah\| \mid h \in \mathbf{R}^n,\ \|h\| = 1 \,\}.$$

In the general case the determination of $\|A\|$ turns out to be a complicated problem. From Corollary 1.8.10 we know that the two norms are equivalent. More explicitly, using $\|Ae_j\| \le \|A\|$, for $1 \le j \le n$, one sees

$$\|A\| \le \|A\|_{\mathrm{Eucl}} \le \sqrt{n}\, \|A\| \qquad (A \in \mathrm{Lin}(\mathbf{R}^n, \mathbf{R}^p)).$$

See Exercise 2.66 for a proof by linear algebra of this estimate and of Lemma 2.1.1.

In the linear space $\mathrm{Lin}(\mathbf{R}^n, \mathbf{R}^p)$ we can now introduce all concepts from Chapter 1 with which we are already familiar for the Euclidean space $\mathbf{R}^{pn}$. In particular we now know what open and closed sets in $\mathrm{Lin}(\mathbf{R}^n, \mathbf{R}^p)$ are, and we know what is meant by continuity of a mapping $\mathbf{R}^k \supset\!\to \mathrm{Lin}(\mathbf{R}^n, \mathbf{R}^p)$ or $\mathrm{Lin}(\mathbf{R}^n, \mathbf{R}^p) \to \mathbf{R}^k$. In particular, the mapping $A \mapsto \|A\|_{\mathrm{Eucl}}$ is continuous from $\mathrm{Lin}(\mathbf{R}^n, \mathbf{R}^p)$ to $\mathbf{R}$.

Finally, we treat two results that are needed only later on, but for which we have all the necessary concepts available at this stage.

Lemma 2.1.2. $\mathrm{Aut}(\mathbf{R}^n)$ *is an open set of the linear space* $\mathrm{End}(\mathbf{R}^n)$*, or equivalently,* $\mathbf{GL}(n, \mathbf{R})$ *is open in* $\mathrm{Mat}(n, \mathbf{R})$.

Proof. We note that a matrix A belongs to $\mathbf{GL}(n, \mathbf{R})$ if and only if $\det A \neq 0$, and that $A \mapsto \det A$ is a continuous function on $\mathrm{Mat}(n, \mathbf{R})$, because it is a polynomial function in the coefficients of A. If therefore $A_0 \in \mathbf{GL}(n, \mathbf{R})$, there exists a neighborhood U of A_0 in $\mathrm{Mat}(n, \mathbf{R})$ with $\det A \neq 0$ if $A \in U$; that is, with the property that $U \subset \mathbf{GL}(n, \mathbf{R})$. ❑

Remark. Some more linear algebra leads to the following additional result. We have *Cramer's rule*

$$\det A \cdot I = A\, A^\sharp = A^\sharp A \qquad (A \in \mathrm{Mat}(n, \mathbf{R})). \tag{2.6}$$

Here is $A^\sharp$ is the *complementary matrix*, given by

$$A^\sharp = (a^\sharp_{ij}) = ((-1)^{i-j} \det A_{ji}) \in \mathrm{Mat}(n, \mathbf{R}),$$

where $A_{ij} \in \mathrm{Mat}(n-1, \mathbf{R})$ is the matrix obtained from A by deleting the i-th row and the j-th column. ($\det A_{ij}$ is called the (i, j)-th *minor* of A.) In other words, with $e_j \in \mathbf{R}^n$ denoting the j-th standard basis vector,

$$a^\sharp_{ij} = \det(a_1 \cdots a_{i-1}\, e_j\, a_{i+1} \cdots a_n).$$

In fact, Cramer's rule follows by expanding $\det A$ by rows and columns, respectively, and using the antisymmetric properties of the determinant. In particular, for $A \in \mathbf{GL}(n, \mathbf{R})$, the inverse $A^{-1} \in \mathbf{GL}(n, \mathbf{R})$ is given by

$$A^{-1} = \frac{1}{\det A} A^\sharp. \tag{2.7}$$

The coefficients of A^{-1} are therefore rational functions of those of A, which implies the assertion in Lemma 2.1.2.

Definition 2.1.3. $A \in \mathrm{End}(\mathbf{R}^n)$ is said to be *self-adjoint* if its adjoint A^t equals A itself, that is, if

$$\langle Ax, y\rangle = \langle x, Ay\rangle \qquad (x, y \in \mathbf{R}^n).$$

We denote by $\mathrm{End}^+(\mathbf{R}^n)$ the linear subspace of $\mathrm{End}(\mathbf{R}^n)$ consisting of the self-adjoint operators. For $A \in \mathrm{End}^+(\mathbf{R}^n)$, the corresponding matrix $(a_{ij})_{1\le i,j\le n} \in \mathrm{Mat}(n, \mathbf{R})$ with respect to any orthonormal basis for $\mathbf{R}^n$ is *symmetric*, that is, it satisfies $a_{ij} = a_{ji}$, for $1 \le i, j \le n$.

$A \in \mathrm{End}(\mathbf{R}^n)$ is said to be *anti-adjoint* if A satisfies $A^t = -A$. We denote by $\mathrm{End}^-(\mathbf{R}^n)$ the linear subspace of $\mathrm{End}(\mathbf{R}^n)$ consisting of the anti-adjoint operators. The corresponding matrices are *antisymmetric*, that is, they satisfy $a_{ij} = -a_{ji}$, for $1 \le i, j \le n$. ❍

In the following lemma we show that $\mathrm{End}(\mathbf{R}^n)$ splits into the direct sum of the ± 1 eigenspaces of the operation of taking the adjoint acting in this linear space.

Lemma 2.1.4. *We have the direct sum decomposition*

$$\mathrm{End}(\mathbf{R}^n) = \mathrm{End}^+(\mathbf{R}^n) \oplus \mathrm{End}^-(\mathbf{R}^n).$$

Proof. For every $A \in \mathrm{End}(\mathbf{R}^n)$ we have $A = \frac{1}{2}(A+A^t)+\frac{1}{2}(A-A^t) \in \mathrm{End}^+(\mathbf{R}^n)+\mathrm{End}^-(\mathbf{R}^n)$. On the other hand, taking the adjoint of $0 = A_+ + A_- \in \mathrm{End}^+(\mathbf{R}^n) + \mathrm{End}^-(\mathbf{R}^n)$ gives $0 = A_+ - A_-$, and thus $A_+ = A_- = 0$. ❑

2.2 Differentiable mappings

We briefly review the definition of differentiability of functions $f : I \to \mathbf{R}$ defined on an interval $I \subset \mathbf{R}$. Remember that f is said to be differentiable at $a \in I$ if the limit

$$\lim_{x\to a} \frac{f(x) - f(a)}{x - a} =: f'(a) \in \mathbf{R} \tag{2.8}$$

exists. Then $f'(a)$ is called the derivative of f at a, also denoted by $Df(a)$ or $\frac{df}{dx}(a)$.

Note that the definition of differentiability in (2.8) remains meaningful for a mapping $f : \mathbf{R} \to \mathbf{R}^p$; nevertheless, a direct generalization to $f : \mathbf{R}^n \to \mathbf{R}^p$, for $n \geq 2$, of this definition is not feasible: $x - a$ is a vector in $\mathbf{R}^n$, and it is impossible to divide the vector $f(x) - f(a) \in \mathbf{R}^p$ by it. On the other hand, (2.8) is equivalent to

$$\lim_{x\to a} \frac{f(x) - f(a) - f'(a)(x - a)}{x - a} = 0,$$

which by the definition of limit is equivalent to

$$\lim_{x\to a} \frac{|f(x) - f(a) - f'(a)(x - a)|}{|x - a|} = 0.$$

This, however, involves a quotient of numbers in $\mathbf{R}$, and hence the generalization to mappings $f : \mathbf{R}^n \supset\!\!\to \mathbf{R}^p$ is straightforward.

As a motivation for results to come we formulate:

Proposition 2.2.1. *Let $f : I \to \mathbf{R}$ and $a \in I$. Then the following assertions are equivalent.*

(i) *f is differentiable at a.*

(ii) *There exists a function $\phi = \phi_a : I \to \mathbf{R}$ continuous at a, such that*

$$f(x) = f(a) + \phi_a(x)(x - a).$$

(iii) *There exist a number* $L(a) \in \mathbf{R}$ *and a function* $\epsilon_a : \mathbf{R} \supset\!\!\to \mathbf{R}$, *such that, for* $h \in \mathbf{R}$ *with* $a + h \in I$,

$$f(a+h) = f(a) + L(a)\,h + \epsilon_a(h) \qquad \textit{with} \qquad \lim_{h\to 0} \frac{\epsilon_a(h)}{h} = 0.$$

If one of these conditions is satisfied, we have

$$f'(a) = \phi_a(a) = L(a), \qquad \phi_a(x) = f'(a) + \frac{\epsilon_a(x-a)}{x-a} \qquad (x \neq a). \qquad (2.9)$$

Proof. (i) ⇒ (ii). Define

$$\phi_a(x) = \begin{cases} \dfrac{f(x) - f(a)}{x-a}, & x \in I \setminus \{a\}; \\ f'(a), & x = a. \end{cases}$$

Then, indeed, we have $f(x) = f(a) + (x-a)\phi_a(x)$ and ϕ_a is continuous at a, because

$$\phi_a(a) = f'(a) = \lim_{x\to a} \frac{f(x)-f(a)}{x-a} = \lim_{x\to a,\, x\neq a} \phi_a(x).$$

(ii) ⇒ (iii). We can take $L(a) = \phi_a(a)$ and $\epsilon_a(h) = (\phi_a(a+h) - \phi_a(a))h$.
(iii) ⇒ (i). Straightforward. ❑

In the theory of differentiation of functions depending on several variables it is essential to consider limits for $x \in \mathbf{R}^n$ freely approaching $a \in \mathbf{R}^n$, from all possible directions; therefore, in what follows, we require our mappings to be defined on open sets. We now introduce:

Definition 2.2.2. Let $U \subset \mathbf{R}^n$ be an open set, let $a \in U$ and $f : U \to \mathbf{R}^p$. Then f is said to be a *differentiable mapping* at a if there exists $Df(a) \in \mathrm{Lin}(\mathbf{R}^n, \mathbf{R}^p)$ such that

$$\lim_{x\to a} \frac{\|f(x) - f(a) - Df(a)(x-a)\|}{\|x-a\|} = 0.$$

$Df(a)$ is said to be **the** *(total) derivative* of f at a. Indeed, in Lemma 2.2.3 below, we will verify that $Df(a)$ is uniquely determined once it exists. We say that f is *differentiable* if it is differentiable at every point of its domain U. In that case the *derivative* or *derived mapping* Df of f is the operator-valued mapping defined by

$$Df : U \to \mathrm{Lin}(\mathbf{R}^n, \mathbf{R}^p) \qquad \text{with} \qquad a \mapsto Df(a). \qquad ❍$$

In other words, in this definition we require the existence of a mapping $\epsilon_a : \mathbf{R}^n \supset\!\!\to \mathbf{R}^p$ satisfying, for all h with $a + h \in U$,

$$f(a+h) = f(a) + Df(a)h + \epsilon_a(h) \qquad \text{with} \qquad \lim_{h\to 0} \frac{\|\epsilon_a(h)\|}{\|h\|} = 0. \qquad (2.10)$$

In the case where $n = p = 1$, the mapping $L \mapsto L(1)$ gives a linear isomorphism $\mathrm{End}(\mathbf{R}) \xrightarrow{\sim} \mathbf{R}$. This new definition of the derivative $Df(a) \leftrightarrow Df(a)1 \in \mathbf{R}$ therefore coincides (up to isomorphism) with the original $f'(a) \in \mathbf{R}$.

Lemma 2.2.3. *$Df(a) \in \mathrm{Lin}(\mathbf{R}^n, \mathbf{R}^p)$ is uniquely determined if f as above is differentiable at a.*

Proof. Consider any $h \in \mathbf{R}^n$. For $t \in \mathbf{R}$ with $|t|$ sufficiently small we have $a + th \in U$ and hence, by Formula (2.10),

$$Df(a)h = \frac{1}{t}(f(a + th) - f(a)) - \frac{1}{t}\epsilon_a(th).$$

It then follows, again by (2.10), that $Df(a)h = \lim_{t\to 0} \frac{1}{t}(f(a + th) - f(a))$; in particular $Df(a)h$ is completely determined by f. ❑

Definition 2.2.4. If f as above is differentiable at a, then we say that the mapping

$$\mathbf{R}^n \to \mathbf{R}^p \qquad \text{given by} \qquad x \mapsto f(a) + Df(a)(x - a)$$

is the *best affine approximation* to f at a. It is the unique affine approximation for which the difference mapping ϵ_a satisfies the estimate in (2.10). ❍

In Chapter 5 we will define the notion of a geometric tangent space and in Theorem 5.1.2 we will see that the geometric tangent space of $\mathrm{graph}(f) = \{\, (x, f(x)) \in \mathbf{R}^{n+p} \mid x \in \mathrm{dom}(f) \,\}$ at $(a, f(a))$ equals the graph $\{\, (x, f(a) + Df(a)(x - a)) \in \mathbf{R}^{n+p} \mid x \in \mathbf{R}^n \,\}$ of the best affine approximation to f at a. For $n = p = 1$, this gives the familiar concept of the geometric tangent line at a point to the graph of f.

Example 2.2.5. (Compare with Example 1.3.5.) Every $A \in \mathrm{Lin}(\mathbf{R}^n, \mathbf{R}^p)$ is differentiable and we have $DA(a) = A$, for all $a \in \mathbf{R}^n$. Indeed, $A(a+h) - A(a) = A(h)$, for every $h \in \mathbf{R}^n$; and there is no remainder term. (Of course, the best affine approximation of a linear mapping is the mapping itself.) The derivative of A is the constant mapping $DA : \mathbf{R}^n \to \mathrm{Lin}(\mathbf{R}^n, \mathbf{R}^p)$ with $DA(a) = A$. For example, the derivative of the *identity mapping* $I \in \mathrm{Aut}(\mathbf{R}^n)$ with $I(x) = x$ satisfies $DI = I$. Furthermore, as the mapping: $\mathbf{R}^n \times \mathbf{R}^n \to \mathbf{R}^n$ of addition, given by $(x_1, x_2) \mapsto x_1 + x_2$, is linear, it is its own derivative.

Next, define $g : \mathbf{R}^n \times \mathbf{R}^n \to \mathbf{R}$ by $g(x_1, x_2) = \langle x_1, x_2 \rangle$. Then g is differentiable while $Dg(a_1, a_2) \in \mathrm{Lin}(\mathbf{R}^n \times \mathbf{R}^n, \mathbf{R})$ is the mapping satisfying

$$Dg(a_1, a_2)(h_1, h_2) = \langle a_1, h_2 \rangle + \langle a_2, h_1 \rangle \qquad (a_1, a_2, h_1, h_2 \in \mathbf{R}^n).$$

Indeed,

$$g(a_1 + h_1, a_2 + h_2) - g(a_1, a_2) = \langle h_1, a_2\rangle + \langle a_1, h_2\rangle + \langle h_1, h_2\rangle \qquad \text{with}$$

$$0 \le \lim_{(h_1,h_2)\to 0} \frac{|\langle h_1, h_2\rangle|}{\|(h_1, h_2)\|} \le \lim_{(h_1,h_2)\to 0} \frac{\|h_1\|\,\|h_2\|}{\|(h_1, h_2)\|} \le \lim_{(h_1,h_2)\to 0} \|h_2\| = 0.$$

Here we used the Cauchy–Schwarz inequality from Proposition 1.1.6 and the estimate (1.7).

Finally, let f_1 and $f_2 : \mathbf{R}^n \to \mathbf{R}^p$ be differentiable and define $f : \mathbf{R}^n \to \mathbf{R}^p \times \mathbf{R}^p$ by $f(x) = (f_1(x), f_2(x))$. Then f is differentiable and $Df(a) \in \mathrm{Lin}(\mathbf{R}^n, \mathbf{R}^p \times \mathbf{R}^p)$ is given by

$$Df(a)h = (Df_1(a)h, Df_2(a)h) \qquad (a, h \in \mathbf{R}^n). \qquad ☆$$

Example 2.2.6. Let $U \subset \mathbf{R}^n$ be open, let $a \in U$ and let $f : U \to \mathbf{R}^n$ be differentiable at a. Suppose $Df(a) \in \mathrm{Aut}(\mathbf{R}^n)$. Then a is an *isolated* zero for $f - f(a)$, which means that there exists a neighborhood U_1 of a in U such that $x \in U_1$ and $f(x) = f(a)$ imply $x = a$. Indeed, applying part (i) of the Substitution Theorem 1.4.2 with the linear and thus continuous $Df(a)^{-1} : \mathbf{R}^n \to \mathbf{R}^n$ (see Example 1.8.14), we also have

$$\lim_{h\to 0} \frac{1}{\|h\|} Df(a)^{-1}\epsilon_a(h) = 0.$$

Therefore we can select $\delta > 0$ such that $\|Df(a)^{-1}\epsilon_a(h)\| < \|h\|$, for $0 < \|h\| < \delta$. This said, consider $a + h \in U$ with $0 < \|h\| < \delta$ and $f(a + h) = f(a)$. Then Formula (2.10) gives

$$0 = Df(a)h + \epsilon_a(h), \qquad \text{and thus} \qquad \|h\| = \|Df(a)^{-1}\epsilon_a(h)\| < \|h\|,$$

and this contradiction proves the assertion. In Proposition 3.2.3 we shall return to this theme. ☆

As before we obtain:

Lemma 2.2.7 (Hadamard). *Let $U \subset \mathbf{R}^n$ be an open set, let $a \in U$ and $f : U \to \mathbf{R}^p$. Then the following assertions are equivalent.*

(i) *The mapping f is differentiable at a.*

(ii) *There exists an operator-valued mapping $\phi = \phi_a : U \to \mathrm{Lin}(\mathbf{R}^n, \mathbf{R}^p)$ continuous at a, such that*

$$f(x) = f(a) + \phi_a(x)(x - a).$$

If one of these conditions is satisfied we have $\phi_a(a) = Df(a)$.

Proof. **(i) ⇒ (ii).** In the notation of Formula (2.10) we define (compare with Formula (2.9))

$$\phi_a(x) = \begin{cases} Df(a) + \dfrac{1}{\|x-a\|^2}\epsilon_a(x-a)(x-a)^t, & x \in U \setminus \{a\}; \\ Df(a), & x = a. \end{cases}$$

Observe that the matrix product of the column vector $\epsilon_a(x-a) \in \mathbf{R}^p$ with the row vector $(x-a)^t \in \mathbf{R}^n$ yields a $p \times n$ matrix, which is associated with a mapping in $\mathrm{Lin}(\mathbf{R}^n, \mathbf{R}^p)$. Or, in other words, since $(x-a)^t y = \langle x-a, y\rangle \in \mathbf{R}$ for $y \in \mathbf{R}^n$,

$$\phi_a(x)y = Df(a)y + \frac{\langle x-a, y\rangle}{\|x-a\|^2}\epsilon_a(x-a) \in \mathbf{R}^p \qquad (x \in U \setminus \{a\},\ y \in \mathbf{R}^n).$$

Now, indeed, we have $f(x) = f(a) + \phi_a(x)(x-a)$. A direct computation gives $\|\epsilon_a(h)\, h^t\|_{\mathrm{Eucl}} = \|\epsilon_a(h)\|\,\|h\|$, hence

$$\lim_{h\to 0}\frac{\|\epsilon_a(h)\, h^t\|_{\mathrm{Eucl}}}{\|h\|^2} = \lim_{h\to 0}\frac{\|\epsilon_a(h)\|}{\|h\|} = 0.$$

This shows that ϕ_a is continuous at a.
(ii) ⇒ (i). Straightforward upon using Lemma 2.1.1. ❑

From the representation of f in Hadamard's Lemma 2.2.7.(ii) we obtain at once:

Corollary 2.2.8. *f is continuous at a if f is differentiable at a.*

Once more Lemma 1.1.7.(iv) implies:

Proposition 2.2.9. *Let $U \subset \mathbf{R}^n$ be an open set, let $a \in U$ and $f : U \to \mathbf{R}^p$. Then the following assertions are equivalent.*

(i) *The mapping f is differentiable at a.*

(ii) *Every component function $f_i : U \to \mathbf{R}$ of f, for $1 \le i \le p$, is differentiable at a.*

2.3 Directional and partial derivatives

Suppose f to be differentiable at a and let $0 \neq v \in \mathbf{R}^n$ be fixed. Then we can apply Formula (2.10) with $h = tv$ for $t \in \mathbf{R}$. Since $Df(a) \in \mathrm{Lin}(\mathbf{R}^n, \mathbf{R}^p)$, we find

$$f(a + tv) - f(a) - tDf(a)v = \epsilon_a(tv),$$
$$\lim_{t\to 0} \frac{\|\epsilon_a(tv)\|}{|t|} = \|v\| \lim_{tv\to 0} \frac{\|\epsilon_a(tv)\|}{\|tv\|} = 0. \tag{2.11}$$

This says that the mapping $g : \mathbf{R} \supset\!\!\to \mathbf{R}^p$ given by $g(t) = f(a+tv)$ is differentiable at 0 with derivative $g'(0) = Df(a)v \in \mathbf{R}^p$.

Definition 2.3.1. Let $U \subset \mathbf{R}^n$ be an open set, let $a \in U$ and $f : U \to \mathbf{R}^p$. We say that f has a *directional derivative* at a in the direction of $v \in \mathbf{R}^n$ if the function $\mathbf{R} \supset\!\!\to \mathbf{R}^p$ with $t \mapsto f(a + tv)$ is differentiable at 0. In that case

$$\frac{d}{dt}\Big|_{t=0} f(a + tv) = \lim_{t\to 0} \frac{1}{t}(f(a + tv) - f(a)) =: D_v f(a) \in \mathbf{R}^p$$

is called the *directional derivative* of f at a in the direction of v.

In the particular case where v is equal to the standard basis vector $e_j \in \mathbf{R}^n$, for $1 \leq j \leq n$, we call

$$D_{e_j} f(a) =: D_j f(a) \in \mathbf{R}^p$$

the j-th *partial derivative* of f at a. If all n partial derivatives of f at a exist we say that f is *partially differentiable* at a. Finally, f is called *partially differentiable* if it is so at every point of U. ❍

Note that

$$D_j f(a) = \frac{d}{dt}\Big|_{t=a_j} f(a_1, \dots, a_{j-1}, t, a_{j+1}, \dots, a_n).$$

This is the derivative of the j-th partial mapping associated with f and a, for which the remaining variables are kept fixed.

Proposition 2.3.2. *Let f be as in Definition 2.3.1. If f is differentiable at a it has the following properties.*

(i) *f has directional derivatives at a in all directions $v \in \mathbf{R}^n$ and $D_v f(a) = Df(a)v$. In particular, the mapping $\mathbf{R}^n \to \mathbf{R}^p$ given by $v \mapsto D_v f(a)$ is linear.*

(ii) *f is partially differentiable at a and*

$$Df(a)v = \sum_{1\leq j\leq n} v_j D_j f(a) \qquad (v \in \mathbf{R}^n).$$

Proof. Assertion (i) follows from Formula (2.11). The partial differentiability in (ii) is a consequence of (i); the formula follows from $Df(a) \in \mathrm{Lin}(\mathbf{R}^n, \mathbf{R}^p)$ and $v = \sum_{1 \le j \le n} v_j e_j$ (see (1.1)). ❑

Example 2.3.3. Consider the function g from Example 1.3.11. Then we have, for $v \in \mathbf{R}^2$,

$$D_v g(0) = \lim_{t \to 0} \frac{g(tv) - g(0)}{t} = \lim_{t \to 0} \frac{v_1 v_2^2}{v_1^2 + t^2 v_2^4} = \begin{cases} \dfrac{v_2^2}{v_1}, & v_1 \neq 0; \\ 0, & v_1 = 0. \end{cases}$$

In other words, directional derivatives of g at 0 in all directions do exist. However, g is not continuous at 0, and a fortiori therefore not differentiable at 0. Moreover, $v \mapsto D_v g(0)$ is not a linear function on $\mathbf{R}^2$; indeed, $D_{e_1+e_2} g(0) = 1 \neq 0 = D_{e_1} g(0) + D_{e_2} g(0)$ where e_1 and e_2 are the standard basis vectors in $\mathbf{R}^2$. ✩

Remark on notation. Another current notation for the j-th partial derivative of f at a is $\partial_j f(a)$; in this book, however, we will stick to $D_j f(a)$. For $f : \mathbf{R}^n \supset\!\to \mathbf{R}^p$ differentiable at a, the matrix of $Df(a) \in \mathrm{Lin}(\mathbf{R}^n, \mathbf{R}^p)$ with respect to the standard bases in $\mathbf{R}^n$, and $\mathbf{R}^p$, respectively, now takes the form

$$Df(a) = (D_1 f(a) \cdots D_j f(a) \cdots D_n f(a)) \in \mathrm{Mat}(p \times n, \mathbf{R});$$

its column vectors $D_j f(a) = Df(a) e_j \in \mathbf{R}^p$ being precisely the images under $Df(a)$ of the standard basis vectors e_j, for $1 \le j \le n$. And if one wants to emphasize the component functions $f_1, \dots, f_p$ of f one writes the matrix as

$$Df(a) = \big(D_j f_i(a)\big)_{1 \le i \le p,\, 1 \le j \le n} = \begin{pmatrix} D_1 f_1(a) & \cdots & D_n f_1(a) \\ \vdots & & \vdots \\ D_1 f_p(a) & \cdots & D_n f_p(a) \end{pmatrix}.$$

This matrix is called the *Jacobi matrix* of f at a. Note that this notation justifies our choice for the roles of the indices i, which customarily label f_i, and j, which label x_j, in order to be compatible with the convention that i labels the rows of a matrix and j the columns.

Classically, *Jacobi's notation* for the partial derivatives has become customary, viz.

$$\frac{\partial f(x)}{\partial x_j} := D_j f(x) \qquad \text{and} \qquad \frac{\partial f_i(x)}{\partial x_j} := D_j f_i(x).$$

A problem with Jacobi's notation is that one commits oneself to a specific designation of the variables in $\mathbf{R}^n$, viz. x in this case. As a consequence, substitution of variables can lead to absurd formulae or, at least, to formulae that require careful interpretation. Indeed, consider $\mathbf{R}^2$ with the new basis vectors $e_1' = e_1 + e_2$ and $e_2' = e_2$; then $x_1e_1 + x_2e_2 = y_1e_1' + y_2e_2'$ implies $x = (y_1, y_1 + y_2)$. This gives

$$x_1 = y_1, \qquad \frac{\partial}{\partial x_1} = D_{e_1} \neq D_{e_1'} = \frac{\partial}{\partial y_1},$$

$$x_2 \neq y_2, \qquad \frac{\partial}{\partial x_2} = D_{e_2} = D_{e_2'} = \frac{\partial}{\partial y_2}.$$

Thus $\frac{\partial}{\partial x_j}$ also depends on the choice of the remaining x_k with $k \neq j$. Furthermore, $\frac{\partial y_2}{\partial y_1}$ is either 0 (if we consider y_1 and y_2 as independent variables) or -1 (if we use $y = (x_1, x_2 - x_1)$); the meaning of this notation, therefore, seriously depends on the context.

On the other hand, a disadvantage of our notation is that the formulae for matrix multiplication look less natural. This, in turn, could be avoided with the notation $D_j f^i$ or f^i_j, where rows and columns, are labeled with upper and lower indices, respectively. Such notation, however, is not customary. We will not be dogmatic about the use of our convention; in fact, in the case of special coordinate systems, like spherical coordinates, the notation with the partials ∂ is the one of preference.

As further complications we mention that $D_j f(a)$ is sometimes used, especially in Fourier theory, for $\frac{1}{\sqrt{-1}} \frac{\partial f}{\partial x_j}(a)$, while for scalar functions $f : \mathbf{R}^n \supset\!\rightarrow \mathbf{R}$ one encounters $f_j(a)$ for $D_j f(a)$. In the latter case caution is needed as $D_i D_j f(a)$ corresponds to $f_{ji}(a)$, the operations in this case being from the right–hand side. Furthermore, for scalar functions $f : \mathbf{R}^n \supset\!\rightarrow \mathbf{R}$, the customary notation is $df(a)$ instead of $Df(a)$; despite this we will write $Df(a)$, for the sake of unity in notation. Finally, in Chapter 8, we will introduce the operation d of exterior differentiation acting, among other things, on functions.

Up to now we have been studying consequences of the hypothesis of differentiability of a mapping. In Example 2.3.3 we saw that even the existence of all directional derivatives does not guarantee differentiability. The next theorem gives a sufficient condition for the differentiability of a mapping, namely, the continuity of all its partial derivatives.

Theorem 2.3.4. *Let $U \subset \mathbf{R}^n$ be an open set, let $a \in U$ and $f : U \to \mathbf{R}^p$. Then f is differentiable at a if f is partially differentiable in a neighborhood of a and all its partial derivatives are continuous at a.*

Proof. On the strength of Proposition 2.2.9 we may assume that $p = 1$. In order to interpolate between $h \in \mathbf{R}^n$ and 0 stepwise and parallel to the coordinate axes, we write $h^{(j)} = \sum_{1 \leq k \leq j} h_k e_k \in \mathbf{R}^n$ for $n \geq j \geq 0$ (see (1.1)). Note that $h^{(j)} = h^{(j-1)} + h_j e_j$. Then, for h sufficiently small,

$$f(a+h) - f(a) = \sum_{n \geq j \geq 1} (f(a+h^{(j)}) - f(a+h^{(j-1)})) = \sum_{n \geq j \geq 1} (g_j(h_j) - g_j(0)),$$

where the $g_j : \mathbf{R} \supset\!\!\to \mathbf{R}$ are defined by $g_j(t) = f(a + h^{(j-1)} + te_j)$. Since f is partially differentiable in a neighborhood of a there exists for every $1 \le j \le n$ an open interval in $\mathbf{R}$ containing 0 on which g_j is differentiable. Now apply the Mean Value Theorem on $\mathbf{R}$ successively to each of the g_j. This gives the existence of $\tau_j = \tau_j(h) \in \mathbf{R}$ between 0 and h_j, for $1 \le j \le n$, satisfying

$$f(a+h)-f(a) = \sum_{1\le j\le n} D_j f(a+h^{(j-1)}+\tau_j e_j)\, h_j =: \phi_a(a+h)\, h \qquad (a+h \in U).$$

The continuity at a of the $D_j f$ implies the continuity at a of the mapping $\phi_a : U \to \mathrm{Lin}(\mathbf{R}^n, \mathbf{R})$, where the matrix of $\phi_a(a+h)$ with respect to the standard bases is given by

$$\phi_a(a+h) = (D_1 f(a+\tau_1 e_1), \dots, D_n f(a + h^{(n-1)} + \tau_n e_n)).$$

Thus we have shown that f satisfies condition (ii) in Hadamard's Lemma 2.2.7, which implies that f is differentiable at a. ❑

The continuity of the partial derivatives of f is not necessary for the differentiability of f, see Exercises 2.12 and 2.13.

Example 2.3.5. Consider the function g from Example 2.3.3. From that example we know already that g is not differentiable at 0 and that $D_1 g(0) = D_2 g(0) = 0$. By direct computation we find

$$D_1 g(x) = -\frac{x_2^2(x_1^2 - x_2^4)}{(x_1^2 + x_2^4)^2}, \qquad D_2 g(x) = \frac{2x_1 x_2(x_1^2 - x_2^4)}{(x_1^2 + x_2^4)^2} \qquad (x \in \mathbf{R}^2 \setminus \{0\}).$$

In particular, both partial derivatives of g are well-defined on all of $\mathbf{R}^2$. Then it follows from Theorem 2.3.4 that at least one of these partial derivatives has to be discontinuous at 0. To see this, note that

$$\lim_{t\downarrow 0} D_1 g(te_2) = \lim_{t\downarrow 0} \frac{t^6}{t^8} = \infty \ne 0 = D_1 g(0).$$

In fact, in this case both partial derivatives are discontinuous at 0, because

$$\lim_{t\downarrow 0} D_2 g(t(e_1+e_2)) = \lim_{t\downarrow 0} \frac{2t^2(t^2-t^4)}{t^4(1+t^2)^2} = \lim_{t\downarrow 0} \frac{2(1-t^2)}{(1+t^2)^2} = 2 \ne 0 = D_2 g(0). \quad ☆$$

We recall from (2.2) the linear isomorphism of vector spaces $\mathrm{Lin}(\mathbf{R}^n, \mathbf{R}^p) \simeq \mathbf{R}^{pn}$, having chosen the standard bases in $\mathbf{R}^n$ and $\mathbf{R}^p$. Using this we know what is meant by the continuity of the derivative $Df : U \to \mathrm{Lin}(\mathbf{R}^n, \mathbf{R}^p)$ for a differentiable mapping $f : U \to \mathbf{R}^p$.

Definition 2.3.6. Let $U \subset \mathbf{R}^n$ be open and $f : U \to \mathbf{R}^p$. If f is continuous, then we write $f \in C^0(U, \mathbf{R}^p) = C(U, \mathbf{R}^p)$. Furthermore, f is said to be *continuously differentiable* or a C^1 *mapping* if f is differentiable with continuous derivative $Df : U \to \text{Lin}(\mathbf{R}^n, \mathbf{R}^p)$; we write $f \in C^1(U, \mathbf{R}^p)$. ○

The definition above is in terms of the (total) derivative of f; alternatively, it could be phrased in terms of the partial derivatives of f. Condition (ii) below is a useful criterion for differentiability if f is given by explicit formulae.

Theorem 2.3.7. *Let $U \subset \mathbf{R}^n$ be open and $f : U \to \mathbf{R}^p$. Then the following are equivalent.*

(i) $f \in C^1(U, \mathbf{R}^p)$.

(ii) *f is partially differentiable and all of its partial derivatives $U \to \mathbf{R}^p$ are continuous.*

Proof. **(i) ⇒ (ii).** Proposition 2.3.2.(ii) implies the partial differentiability of f and Proposition 1.3.9 gives the continuity of the partial derivatives of f.
(ii) ⇒ (i). The differentiability of f follows from Theorem 2.3.4, while the continuity of Df follows from Proposition 1.3.9. ❑

2.4 Chain rule

The following result is one of the cornerstones of analysis in several variables.

Theorem 2.4.1 (Chain rule). *Let $U \subset \mathbf{R}^n$ and $V \subset \mathbf{R}^p$ be open and consider mappings $f : U \to \mathbf{R}^p$ and $g : V \to \mathbf{R}^q$. Let $a \in U$ be such that $f(a) \in V$. Suppose f is differentiable at a and g at $f(a)$. Then $g \circ f : U \to \mathbf{R}^q$, the composition of g and f, is differentiable at a, and we have*

$$D(g \circ f)(a) = Dg(f(a)) \circ Df(a) \in \text{Lin}(\mathbf{R}^n, \mathbf{R}^q).$$

Or, if $f(U) \subset V$,

$$D(g \circ f) = ((Dg) \circ f) \circ Df : U \to \text{Lin}(\mathbf{R}^n, \mathbf{R}^q).$$

Proof. Note that $\text{dom}(g \circ f) = \text{dom}(f) \cap f^{-1}(\text{dom}(g)) = U \cap f^{-1}(V)$ is open in $\mathbf{R}^n$ in view of Corollary 2.2.8, Theorem 1.3.7.(ii) and Lemma 1.2.5.(ii). We formulate the assumptions according to Hadamard's Lemma 2.2.7.(ii). Thus:

$$f(x) - f(a) = \phi(x)(x - a), \qquad \lim_{x \to a} \phi(x) = Df(a),$$

$$g(y) - g(f(a)) = \psi(y)(y - f(a)), \qquad \lim_{y \to f(a)} \psi(y) = Dg(f(a)).$$

Putting $y = f(x)$ and inserting the first equation into the second one, we find

$$(g \circ f)(x) - (g \circ f)(a) = \psi(f(x))\phi(x)(x - a).$$

Furthermore, $\lim_{x \to a} f(x) = f(a)$ and $\lim_{y \to f(a)} \psi(y) = Dg(f(a))$ give, in view of the Substitution Theorem 1.4.2.(i)

$$\lim_{x \to a} \psi(f(x)) = Dg(f(a)).$$

Therefore Corollary 1.4.3 implies

$$\lim_{x \to a} \psi(f(x))\phi(x) = \lim_{x \to a} \psi(f(x)) \lim_{x \to a} \phi(x) = Dg(f(a))Df(a).$$

But then the implication (ii) $\Rightarrow$ (i) from Hadamard's Lemma 2.2.7 says that $g \circ f$ is differentiable at a and that $D(g \circ f)(a) = Dg(f(a))Df(a)$. ❑

Corollary 2.4.2 (Chain rule for Jacobi matrices). *Let f and g be as in Theorem 2.4.1. In terms of the coefficients of the corresponding Jacobi matrices the chain rule takes the form, because $(g \circ f)_i = g_i \circ f$,*

$$D_j(g_i \circ f) = \sum_{1 \le k \le p} ((D_k g_i) \circ f) D_j f_k \qquad (1 \le j \le n,\ 1 \le i \le q).$$

Denoting the variable in $\mathbf{R}^n$ by x, and the one in $\mathbf{R}^p$ by y, we obtain in Jacobi's notation for partial derivatives

$$\frac{\partial(g_i \circ f)}{\partial x_j} = \sum_{1 \le k \le p} \left(\frac{\partial g_i}{\partial y_k} \circ f\right) \frac{\partial f_k}{\partial x_j} \qquad (1 \le i \le q,\ 1 \le j \le n).$$

In this context, it is customary to write $y_k = f_k(x_1, \dots, x_n)$ and $z_i = g_i(y_1, \dots, y_p)$, and write the chain rule (suppressing the points at which functions are to be evaluated) in the form

$$\frac{\partial z_i}{\partial x_j} = \sum_{1 \le k \le p} \frac{\partial z_i}{\partial y_k} \frac{\partial y_k}{\partial x_j} \qquad (1 \le i \le q,\ 1 \le j \le n).$$

Proof. The first formula is immediate from the expression

$$h_{ij} = \sum_{1 \le k \le p} g_{ik} f_{kj} \qquad (1 \le i \le q,\ 1 \le j \le n)$$

for the coefficients of a matrix $H = GF \in \mathrm{Mat}(q \times n, \mathbf{R})$ which is the product of $G \in \mathrm{Mat}(q \times p, \mathbf{R})$ and $F \in \mathrm{Mat}(p \times n, \mathbf{R})$. ❑

Corollary 2.4.3. *Consider the mappings f_1 and $f_2 : \mathbf{R}^n \supset\!\rightarrow \mathbf{R}^p$ and $\lambda \in \mathbf{R}$. Suppose f_1 and f_2 are differentiable at $a \in \mathbf{R}^n$. Then we have the following.*

(i) $\lambda f_1 + f_2 : \mathbf{R}^n \supset\!\rightarrow \mathbf{R}^p$ *is differentiable at a, with derivative*

$$D(\lambda f_1 + f_2)(a) = \lambda\, Df_1(a) + Df_2(a) \in \mathrm{Lin}(\mathbf{R}^n, \mathbf{R}^p).$$

(ii) $\langle f_1, f_2 \rangle : \mathbf{R}^n \supset\!\rightarrow \mathbf{R}$ *is differentiable at a, with derivative*

$$D(\langle f_1, f_2 \rangle)(a) = \langle Df_1(a), f_2(a) \rangle + \langle f_1(a), Df_2(a) \rangle \in \mathrm{Lin}(\mathbf{R}^n, \mathbf{R}).$$

Suppose $f : \mathbf{R}^n \supset\!\rightarrow \mathbf{R}$ is a differentiable function at $a \in \mathbf{R}^n$.

(iii) *If $f(a) \neq 0$, then $\frac{1}{f}$ is differentiable at a, with derivative*

$$D\left(\frac{1}{f}\right)(a) = -\frac{1}{f(a)^2} Df(a) \in \mathrm{Lin}(\mathbf{R}^n, \mathbf{R}).$$

Proof. From Example 2.2.5 we know that $f : \mathbf{R}^n \to \mathbf{R}^p \times \mathbf{R}^p$ is differentiable at a if $f(x) = (f_1(x), f_2(x))$, while $Df(a)h = (Df_1(a)h, Df_2(a)h$, for $(a, h \in \mathbf{R}^n)$.
(i). Define $g : \mathbf{R}^p \times \mathbf{R}^p \to \mathbf{R}^p$ by $g(y_1, y_2) = \lambda y_1 + y_2$. By Definition 1.4.1 we have $\lambda f_1 + f_2 = g \circ f$. The differentiability of $\lambda f_1 + f_2$ as well as the formula for its derivative now follow from Example 2.2.5 and the chain rule.
(ii). Define $g : \mathbf{R}^p \times \mathbf{R}^p \to \mathbf{R}$ by $g(y_1, y_2) = \langle y_1, y_2 \rangle$. By Definition 1.4.1 we have $\langle f_1, f_2 \rangle = g \circ f$. The differentiability of $\langle f_1, f_2 \rangle$ now follows from Example 2.2.5 and the chain rule. Furthermore, using the formulae from Example 2.2.5 and the chain rule, we find, for $a, h \in \mathbf{R}^n$,

$$\begin{aligned} D(\langle f_1, f_2 \rangle)(a)h &= D(g \circ f)(a)h = Dg((f_1(a), f_2(a))(Df_1(a)h, Df_2(a)h)) \\ &= \langle f_1(a), Df_2(a)h \rangle + \langle f_2(a), Df_1(a)h \rangle \in \mathbf{R}. \end{aligned}$$

(iii). Define $g : \mathbf{R} \setminus \{0\} \to \mathbf{R}$ by $g(y) = \frac{1}{y}$ and observe $Dg(b)k = -\frac{1}{b^2}k \in \mathbf{R}$ for $b \neq 0$ and $k \in \mathbf{R}$. Now apply the chain rule. ❑

Example 2.4.4. For arbitrary n but $p = 1$, write the formula in Corollary 2.4.3.(ii) as $D(f_1 f_2) = f_1 Df_2 + f_2 Df_1$. Note that Leibniz' rule $(f_1 f_2)' = f_1' f_2 + f_1 f_2'$ for the derivative of the product $f_1 f_2$ of two functions f_1 and $f_2 : \mathbf{R} \to \mathbf{R}$ now follows upon taking $n = 1$. ☆

Example 2.4.5. Suppose $f : \mathbf{R} \to \mathbf{R}^n$ and $g : \mathbf{R}^n \to \mathbf{R}$ are differentiable. Then the derivative $D(g \circ f) : \mathbf{R} \to \mathrm{End}(\mathbf{R}) \simeq \mathbf{R}$ of $g \circ f : \mathbf{R} \to \mathbf{R}$ is given by

$$(g \circ f)' = D(g \circ f) = ((D_1 g, \dots, D_n g) \circ f) \begin{pmatrix} Df_1 \\ \vdots \\ Df_n \end{pmatrix} = \sum_{1 \le i \le n} ((D_i g) \circ f) Df_i. \tag{2.12}$$

Or, in Jacobi's notation

$$\frac{d(g \circ f)}{dt}(t) = \sum_{1 \le i \le n} \frac{\partial g}{\partial x_i}(f(t)) \frac{df_i}{dt}(t) \qquad (t \in \mathbf{R}).$$ ☆

Example 2.4.6. Suppose $f : \mathbf{R}^n \to \mathbf{R}^n$ and $g : \mathbf{R}^n \to \mathbf{R}$ are differentiable. Then $D(g \circ f) : \mathbf{R}^n \to \mathrm{Lin}(\mathbf{R}^n, \mathbf{R})$ is given by

$$D_j(g \circ f) = ((Dg) \circ f) D_j f = \sum_{1 \le k \le n} ((D_k g) \circ f) D_j f_k \qquad (1 \le j \le n).$$

Or, in Jacobi's notation

$$\frac{\partial (g \circ f)}{\partial x_j}(x) = \sum_{1 \le k \le n} \frac{\partial g}{\partial y_k}(f(x)) \frac{\partial f_k}{\partial x_j}(x) \qquad (1 \le j \le n,\ y = f(x)).$$ ☆

As a direct application of the chain rule we derive the following lemma on the interchange of differentiation and a linear mapping.

Lemma 2.4.7. *Let $L \in \mathrm{Lin}(\mathbf{R}^p, \mathbf{R}^q)$. Then $D \circ L = L \circ D$, that is, for every mapping $f : U \to \mathbf{R}^p$ differentiable at $a \in U$ we have, if U is open in $\mathbf{R}^n$,*

$$D(Lf) = L(Df).$$

Proof. Using the chain rule and the linearity of L (see Example 2.2.5) we find

$$D(L \circ f)(a) = DL(f(a)) \circ Df(a) = L \circ (Df)(a).$$ ❑

Example 2.4.8 (Derivative of norm). The norm $\| \cdot \| : x \mapsto \|x\|$ is differentiable on $\mathbf{R}^n \setminus \{0\}$, and $D\| \cdot \|(x) \in \mathrm{Lin}(\mathbf{R}^n, \mathbf{R})$, its derivative at any x in this set, satisfies

$$D\| \cdot \|(x)h = \frac{\langle x, h \rangle}{\|x\|}, \qquad \text{in other words} \qquad D_j \| \cdot \|(x) = \frac{x_j}{\|x\|} \qquad (1 \le j \le n).$$

Indeed, by the chain rule the derivative of $x \mapsto \|x\|^2$ is $h \mapsto 2\|x\|\, D\| \cdot \|(x)h$ on the one hand, and on the other it is $h \mapsto 2\langle x, h \rangle$, by Example 2.2.5. Using the chain rule once more we now see, for every $p \in \mathbf{R}$,

$$D\| \cdot \|^p(x)h = p\|x\|^{p-1} \frac{\langle x, h \rangle}{\|x\|} = p\langle x, h \rangle \|x\|^{p-2},$$

that is, $D_j \| \cdot \|^p(x) = p x_j \|x\|^{p-2}$. The identity for the partial derivative also can be verified by direct computation. ☆

Example 2.4.9. Suppose that $U \subset \mathbf{R}^n$ and $V \subset \mathbf{R}^p$ are open subsets, that $f : U \to V$ is a differentiable bijection, and that the *inverse mapping* $f^{-1} : V \to U$ is differentiable too. (Note that the last condition is not automatically satisfied, as can be seen from the example of $f : \mathbf{R} \to \mathbf{R}$ with $f(x) = x^3$.) Then $Df(x) \in \mathrm{Lin}(\mathbf{R}^n, \mathbf{R}^p)$ is invertible, which implies $n = p$ and therefore $Df(x) \in \mathrm{Aut}(\mathbf{R}^n)$, and additionally

$$D(f^{-1})(f(x)) = Df(x)^{-1} \qquad (x \in U).$$

In fact, $f^{-1} \circ f = I$ on U and thus $D(f^{-1})(f(x))Df(x) = DI(x) = I$ by the chain rule. ☆

Example 2.4.10 (Exponential of linear mapping). Let $\|\cdot\|$ be the Euclidean norm on $\mathrm{End}(\mathbf{R}^n)$. If A and $B \in \mathrm{End}(\mathbf{R}^n)$, then we have $\|AB\| \le \|A\|\,\|B\|$. This follows by recognizing the matrix coefficients of AB as inner products and by applying the Cauchy–Schwarz inequality from Proposition 1.1.6. In particular, $\|A^k\| \le \|A\|^k$, for all $k \in \mathbf{N}$. Now fix $A \in \mathrm{End}(\mathbf{R}^n)$. We consider $e^{\|A\|} \in \mathbf{R}$ as the limit of the Cauchy sequence $\big(\sum_{0 \le k \le l} \frac{1}{k!}\|A\|^k\big)_{l \in \mathbf{N}}$ in $\mathbf{R}$. We obtain that $\big(\sum_{0 \le k \le l} \frac{1}{k!} A^k\big)_{l \in \mathbf{N}}$ is a Cauchy sequence in $\mathrm{End}(\mathbf{R}^n)$, and in view of Theorem 1.6.5 $\exp A$, the *exponential* of A, is well-defined in the complete space $\mathrm{End}(\mathbf{R}^n)$ by

$$\exp A := e^A := \sum_{k \in \mathbf{N}_0} \frac{1}{k!} A^k := \lim_{l \to \infty} \sum_{0 \le k \le l} \frac{1}{k!} A^k \in \mathrm{End}(\mathbf{R}^n).$$

Note that $\|e^A\| \le e^{\|A\|}$.

Next we define $\gamma : \mathbf{R} \to \mathrm{End}(\mathbf{R}^n)$ by $\gamma(t) = e^{tA}$. Then γ is a differentiable mapping, with the derivative

$$D\gamma(t) = \gamma'(t) : h \mapsto h\,A\,e^{tA} = h\,e^{tA}\,A$$

belonging to $\mathrm{Lin}\,(\mathbf{R}, \mathrm{End}(\mathbf{R}^n)) \simeq \mathrm{End}(\mathbf{R}^n)$. Indeed, from the result above we know that the power series $\sum_{k \in \mathbf{N}_0} t^k a_k$ in t converges, for all $t \in \mathbf{R}$, where the coefficients a_k are given by $\frac{1}{k!} A^k \in \mathrm{End}(\mathbf{R}^n)$. Furthermore, it is straightforward to verify that the theorem on the termwise differentiation of a convergent power series for obtaining the derivative of the series remains valid in the case where the coefficients a_k belong to $\mathrm{End}(\mathbf{R}^n)$ instead of to $\mathbf{R}$ or $\mathbf{C}$. Accordingly we conclude that $\gamma : \mathbf{R} \to \mathrm{End}(\mathbf{R}^n)$ satisfies the following *ordinary differential equation* on $\mathbf{R}$ with *initial condition*:

$$\gamma'(t) = A\,\gamma(t) \quad (t \in \mathbf{R}), \qquad \gamma(0) = I. \tag{2.13}$$

In particular we have $A = \gamma'(0)$. The foregoing asserts the existence of a solution of Equation (2.13). We now prove that the solutions of (2.13) are unique, in other words that each solution $\widetilde{\gamma}$ of (2.13) equals γ. In fact,

$$\frac{d(e^{-tA}\widetilde{\gamma}(t))}{dt} = e^{-tA}(\,-A\,\widetilde{\gamma}(t) + \widetilde{\gamma}'(t)) = 0;$$

and from this we deduce $e^{-tA}\widetilde{\gamma}(t) = I$, for all $t \in \mathbf{R}$, by substituting $t = 0$. Conclude that e^{-tA} is invertible, and that a solution $\widetilde{\gamma}$ is uniquely determined, namely by $\widetilde{\gamma}(t) = (e^{-tA})^{-1}$. But γ also is a solution; therefore $\widetilde{\gamma}(t) = e^{tA}$, and so $(e^{-tA})^{-1} = e^{tA}$, for all $t \in \mathbf{R}$. Deduce $e^{tA} \in \mathrm{Aut}(\mathbf{R}^n)$, for all $t \in \mathbf{R}$. Finally, using (2.13), we see that $t \mapsto e^{(t+t')A}e^{-t'A}$, for $t' \in \mathbf{R}$, satisfies (2.13), and we conclude

$$e^{tA}e^{t'A} = e^{(t+t')A} \qquad (t,\ t' \in \mathbf{R}).$$

The family $(e^{tA})_{t\in\mathbf{R}}$ of automorphisms is called a *one-parameter group* in $\mathrm{Aut}(\mathbf{R}^n)$ with *infinitesimal generator* $A \in \mathrm{End}(\mathbf{R}^n)$. In $\mathrm{Aut}(\mathbf{R}^n)$ the usual multiplication rule for arbitrary exponents is not necessarily true, in other words, for A and B in $\mathrm{End}(\mathbf{R}^n)$ one need not have $e^A e^B = e^{A+B}$, particularly so when $AB \neq BA$. Try to find such A and B as prove this point. ☆

2.5 Mean Value Theorem

For differentiable functions $f : \mathbf{R}^n \to \mathbf{R}^p$ a direct analog of the Mean Value Theorem on $\mathbf{R}$

$$f(x) - f(x') = f'(\xi)(x - x') \qquad \text{with } \xi \in \mathbf{R} \text{ between } x \text{ and } x', \tag{2.14}$$

as it applies to differentiable functions $f : \mathbf{R} \to \mathbf{R}$, need not exist if $p > 1$. This is apparent from the example of $f : \mathbf{R} \to \mathbf{R}^2$ with $f(t) = (\cos t,\ \sin t)$. Indeed, $Df(t) = (-\sin t,\ \cos t)$, and now the formula from the Mean Value Theorem on $\mathbf{R}$

$$f(x) - f(x') = (x - x')Df(\xi)$$

cannot hold, because for $x = 2\pi$ and $x' = 0$ the left–hand side equals the vector 0, whereas the right–hand side is a vector of length 2π.

Still, a variant of the Mean Value Theorem exists in which the equality sign is replaced by an inequality sign. To formulate this, we introduce the line segment $L(x', x)$ in $\mathbf{R}^n$ with endpoints x' and $x \in \mathbf{R}^n$:

$$L(x', x) = \{\, x_t := x' + t(x - x') \in \mathbf{R}^n \mid 0 \le t \le 1 \,\}. \tag{2.15}$$

Lemma 2.5.1. *Let $U \subset \mathbf{R}^n$ be open and let $f : U \to \mathbf{R}^p$ be differentiable. Assume that the line segment $L(x', x)$ is completely contained in U. Then for all $a \in \mathbf{R}^p$ there exists $\xi = \xi(a) \in L(x', x)$ such that*

$$\langle\, a,\ f(x) - f(x') \,\rangle = \langle\, a,\ Df(\xi)(x - x') \,\rangle. \tag{2.16}$$

Proof. Because U is open, there exists a number $\delta > 0$ such that $x_t \in U$ if $t \in I := \,]-\delta,\ 1+\delta\,[\, \subset \mathbf{R}$. Let $a \in \mathbf{R}^p$ and define $g : I \to \mathbf{R}$ by $g(t) = \langle\, a,\ f(x_t) \,\rangle$.

From Example 2.4.5 and Corollary 2.4.3 it follows that g is differentiable on I, with derivative

$$g'(t) = \langle\, a,\ Df(x_t)(x - x')\,\rangle.$$

On account of the Mean Value Theorem applied with g and I there exists $0 < \tau < 1$ with $g(1) - g(0) = g'(\tau)$, i.e.

$$\langle\, a,\ f(x) - f(x')\,\rangle = \langle\, a,\ Df(x_\tau)(x - x')\,\rangle.$$

This proves Formula (2.16) with $\xi = x_\tau \in L(x', x) \subset U$. Note that ξ depends on g, and therefore on $a \in \mathbf{R}^p$. ❑

For a function f such as in Lemma 2.5.1 we may now apply Lemma 2.5.1 with the choice $a = f(x) - f(x')$. We then use the Cauchy–Schwarz inequality from Proposition 1.1.6 and divide by $\|f(x) - f(x')\|$ to obtain, after application of Inequality (2.5), the following estimate

$$\|f(x) - f(x')\| \leq \sup_{\xi \in L(x',x)} \|Df(\xi)\|_{\text{Eucl}} \, \|x - x'\|.$$

Definition 2.5.2. A set $A \subset \mathbf{R}^n$ is said to be *convex* if for every x' and $x \in A$ we have $L(x', x) \subset A$. ❍

From the preceding estimate we immediately obtain:

Theorem 2.5.3 (Mean Value Theorem). *Let U be a convex open subset of $\mathbf{R}^n$ and let $f : U \to \mathbf{R}^p$ be a differentiable mapping. Suppose that the derivative $Df : U \to \mathrm{Lin}(\mathbf{R}^n, \mathbf{R}^p)$ is bounded on U, that is, there exists $k > 0$ with $\|Df(\xi)h\| \leq k\|h\|$, for all $\xi \in U$ and $h \in \mathbf{R}^n$, which is the case if $\|Df(\xi)\|_{\text{Eucl}} \leq k$. Then f is Lipschitz continuous on U with Lipschitz constant k, in other words*

$$\|f(x) - f(x')\| \leq k\,\|x - x'\| \qquad (x,\ x' \in U). \tag{2.17}$$

Example 2.5.4. Let $U \subset \mathbf{R}^n$ be an open set and consider a mapping $f : U \to \mathbf{R}^p$. Suppose $f : U \to \mathbf{R}^p$ is differentiable on U and $Df : U \to \mathrm{Lin}(\mathbf{R}^n, \mathbf{R}^p)$ is continuous at $a \in U$. Then, for every $\epsilon > 0$, we can find a $\delta > 0$ such that the mapping ϵ_a from Formula (2.10), satisfying

$$\epsilon_a(h) = f(a + h) - f(a) - Df(a)h \qquad (a + h \in U),$$

is Lipschitz continuous on $B(0; \delta)$ with Lipschitz constant ϵ. Indeed, ϵ_a is differentiable at h, with derivative $Df(a + h) - Df(a)$. Therefore the continuity of Df at a guarantees the existence of $\delta > 0$ such that

$$\|Df(a + h) - Df(a)\|_{\text{Eucl}} \leq \epsilon \qquad (\|h\| < \delta).$$

The Mean Value Theorem 2.5.3 now gives the Lipschitz continuity of ϵ_a on $B(0; \delta)$, with Lipschitz constant ϵ. ☆

Corollary 2.5.5. *Let $f : \mathbf{R}^n \to \mathbf{R}^p$ be a C^1 mapping and let $K \subset \mathbf{R}^n$ be compact. Then the restriction $f|_K$ of f to K is Lipschitz continuous.*

Proof. Define $g : \mathbf{R}^{2n} \to [\,0, \infty\,[$ by

$$g(x, x') = \begin{cases} \dfrac{\|f(x) - f(x')\|}{\|x - x'\|}, & x \neq x'; \\ 0, & x = x'. \end{cases}$$

We claim that g is locally bounded on $\mathbf{R}^{2n}$, that is, every $(x, x') \in \mathbf{R}^{2n}$ belongs to an open set $O(x, x')$ in $\mathbf{R}^{2n}$ such that the restriction of g to $O(x, x')$ is bounded. Indeed, given $(x, x') \in \mathbf{R}^{2n}$, select an open rectangle $O \subset \mathbf{R}^n$ (see Definition 1.8.18) containing both x and x', and set $O(x, x') := O \times O \subset \mathbf{R}^{2n}$. Now Theorem 1.8.8, applied in the case of the closed rectangle $\overline{O}$ and the continuous function $\|Df\|_{\text{Eucl}}$, gives the boundedness of Df on O. A rectangle being convex, the Mean Value Theorem 2.5.3 then implies the existence of $k = k(x, x') > 0$ with

$$g(y, y') \leq k(x, x') \qquad ((y, y') \in O(x, x')).$$

From the Theorem of Heine–Borel 1.8.17 it follows that $L := K \times K$ is compact in $\mathbf{R}^{2n}$, while $\{\, O(x, x') \mid (x, x') \in L \,\}$ is an open covering of L. According to Definition 1.8.16 we can extract a finite subcovering $\{\, O(x_l, x_l') \mid 1 \leq l \leq m \,\}$ of L. Finally, put

$$k = \max\{\, k(x_l, x_l') \mid 1 \leq l \leq m \,\}.$$

Then $g(y, y') \leq k$ for all $(y, y') \in L$, which proves the assertion of the corollary, with k being a Lipschitz constant for $f|_K$. ❑

2.6 Gradient

Lemma 2.6.1. *There exists a linear isomorphism between* $\text{Lin}(\mathbf{R}^n, \mathbf{R})$ *and* $\mathbf{R}^n$. *In fact, for every $L \in \text{Lin}(\mathbf{R}^n, \mathbf{R})$ there is a unique $l \in \mathbf{R}^n$ such that*

$$L(h) = \langle l, h \rangle \qquad (h \in \mathbf{R}^n).$$

Proof. The equality $h = \sum_{1 \leq j \leq n} h_j e_j \in \mathbf{R}^n$ from Formula (1.1) and the linearity of L imply

$$L(h) = \sum_{1 \leq j \leq n} h_j L(e_j) = \sum_{1 \leq j \leq n} l_j h_j = \langle l, h \rangle \qquad \text{with} \qquad l_j = L(e_j).$$ ❑

The isomorphism in the lemma is not *canonical*, which means that it is not determined by the structure of $\mathbf{R}^n$ as a linear space alone, but that some choice should be made for producing it; and in our proof the inner product on $\mathbf{R}^n$ was this extra ingredient.

Definition 2.6.2. Let $U \subset \mathbf{R}^n$ be an open set, let $a \in U$ and let $f : U \to \mathbf{R}$ be a function. Suppose f is differentiable at a. Then $Df(a) \in \mathrm{Lin}(\mathbf{R}^n, \mathbf{R})$ and therefore application of the lemma above gives the existence of the (column) vector $\mathrm{grad}\, f(a) \in \mathbf{R}^n$, the *gradient* of f at a, such that

$$Df(a)h = \langle\, \mathrm{grad}\, f(a), h \rangle \qquad (h \in \mathbf{R}^n).$$

More explicitly,

$$Df(a) = (D_1 f(a) \cdots D_n f(a)) : \mathbf{R}^n \to \mathbf{R},$$

$$\nabla f(a) := \mathrm{grad}\, f(a) = \begin{pmatrix} D_1 f(a) \\ \vdots \\ D_n f(a) \end{pmatrix} \in \mathbf{R}^n.$$

The symbol ∇ is pronounced as *nabla,* or *del*; the name nabla originates from an ancient stringed instrument in the form of a harp. The preceding identities imply

$$\mathrm{grad}\, f(a) = Df(a)^t \in \mathrm{Lin}(\mathbf{R}, \mathbf{R}^n) \simeq \mathbf{R}^n.$$

If f is differentiable on U we obtain in this way the mapping $\mathrm{grad}\, f : U \to \mathbf{R}^n$ with $x \mapsto \mathrm{grad}\, f(x)$, which is said to be the *gradient vector field* of f on U.

If $f \in C^1(U)$, the gradient vector field satisfies $\mathrm{grad}\, f \in C^0(U, \mathbf{R}^n)$. Here $\mathrm{grad} : C^1(U) \to C^0(U, \mathbf{R}^n)$ is a linear operator of function spaces; it is called the *gradient operator.* ○

With this notation Formula (2.12) takes the form

$$(g \circ f)' = \langle\, (\mathrm{grad}\, g) \circ f, Df \rangle. \tag{2.18}$$

Next we discuss the geometric meaning of the gradient vector in the following:

Theorem 2.6.3. *Let the function $f : \mathbf{R}^n \supset\!\!\to \mathbf{R}$ be differentiable at $a \in \mathbf{R}^n$.*

(i) *Near a the function f increases fastest in the direction of* $\mathrm{grad}\, f(a) \in \mathbf{R}^n$.

(ii) *The rate of increase in f is measured by the length of* $\mathrm{grad}\, f(a)$.

(iii) $\mathrm{grad}\, f(a)$ *is orthogonal to the level set $N(f(a)) = f^{-1}(\{f(a)\})$, in the sense that* $\mathrm{grad}\, f(a)$ *is orthogonal to every vector in $\mathbf{R}^n$ that is tangent at a to a differentiable curve in $\mathbf{R}^n$ through a that is locally contained in $N(f(a))$.*

(iv) *If f has a local extremum at a then* $\mathrm{grad}\, f(a) = 0$.

Proof. The rate of increase of the function f at the point a in an arbitrary direction $v \in \mathbf{R}^n$ is given by the directional derivative $D_v f(a)$ (see Proposition 2.3.2.(i)), which satisfies

$$|D_v f(a)| = |Df(a)v| = |\langle \operatorname{grad} f(a), v\rangle| \le \|\operatorname{grad} f(a)\| \, \|v\|,$$

in view of the Cauchy–Schwarz inequality from Proposition 1.1.6. Furthermore, it follows from the latter proposition that the rate of increase is maximal if v is a positive scalar multiple of grad $f(a)$, and this proves assertions (i) and (ii).

For (iii), consider a mapping $c : \mathbf{R} \to \mathbf{R}^n$ satisfying $c(0) = a$ and $f(c(t)) = f(a)$ for t in a neighborhood I of 0 in $\mathbf{R}$. As $f \circ c$ is locally constant near 0, Formula (2.18) implies

$$0 = (f \circ c)'(0) = \langle (\operatorname{grad} f)(c(0)), Dc(0)\rangle = \langle \operatorname{grad} f(a), Dc(0)\rangle.$$

The assertion follows as the vector $Dc(0) \in \mathbf{R}^n$ is tangent at a to the curve c through a (see Definition 5.1.1 for more details) while $c(t) \in N(f(a))$ for $t \in I$.

For (iv), consider the partial functions

$$g_j : t \mapsto f(a_1, \dots, a_{j-1}, t, a_{j+1}, \dots, a_n)$$

as defined in Formula (1.5); these are differentiable at a_j with derivative $D_j f(a)$. As f has a local extremum at a, so have the g_j at a_j, which implies $0 = g_j'(a_j) = D_j f(a)$, for $1 \le j \le n$. ❑

This description shows that the gradient of a function is in fact independent of the choice of the inner product on $\mathbf{R}^n$.

Definition 2.6.4. Let $f : \mathbf{R}^n \supset\!\to \mathbf{R}$ be differentiable at $a \in \mathbf{R}^n$. Then a is said to be a *critical* or *stationary point* for f if $Df(a) = 0$, or, equivalently, if grad $f(a) = 0$. Furthermore, $f(a) \in \mathbf{R}$ is called a *critical value* of f. ❍

Remark. Let D be a subset of $\mathbf{R}^n$ and $f : \mathbf{R}^n \to \mathbf{R}$ a differentiable function. Note that in the case where D is open, the restriction of f to D does not necessarily attain any extremum at all on D. On the other hand, if D is compact, Theorem 1.8.8 guarantees the existence of points where f restricted to D reaches an extremum. There are various possibilities for the location of these points. If the interior $\operatorname{int}(D) \neq \emptyset$, one uses Theorem 2.6.3.(iv) to find the critical points of f in $\operatorname{int}(D)$ as points where extrema may occur. But such points might also belong to the boundary ∂D of D and then a separate investigation is required. Finally, when D is unbounded, one also has to know the asymptotic behavior of $f(x)$ for $x \in \mathbf{R}^n$ going to infinity in order to decide whether the relative extrema are absolute. In this context, when determining minima, it might be helpful to select x^0 in D and to consider only the points in $D' = \{ x \in D \mid f(x) \le f(x^0) \}$. This is effective, in particular, when D' is bounded.

Since the first-order derivative of a function vanishes at a critical point, the properties of that function (whether it assumes a local maximum or minimum, or does not have any local extremum) depend on the higher-order derivatives, which we will study in the next section.

2.7 Higher-order derivatives

For a mapping $f : \mathbf{R}^n \supset\!\!\to \mathbf{R}^p$ the partial derivatives $D_i f$ are themselves mappings $\mathbf{R}^n \supset\!\!\to \mathbf{R}^p$ and they, in turn, can have partial derivatives. These are called the *second-order* partial derivatives, denoted by

$$D_j D_i f, \qquad \text{or in Jacobi's notation} \qquad \frac{\partial^2 f}{\partial x_j \partial x_i} \qquad (1 \le i, j \le n).$$

Example 2.7.1. $D_j D_i f$ is not necessarily the same as $D_i D_j f$. A counterexample is obtained by considering $f : \mathbf{R}^2 \to \mathbf{R}$ given by

$$f(x) = x_1 x_2 g(x) \qquad \text{with} \qquad g : \mathbf{R}^2 \to \mathbf{R} \qquad \text{bounded.}$$

Then

$$D_1 f(0, x_2) = \lim_{x_1 \to 0} \frac{f(x) - f(0, x_2)}{x_1} = x_2 \lim_{x_1 \to 0} g(x),$$

$$\text{and} \qquad D_2 D_1 f(0) = \lim_{x_2 \to 0} \big(\lim_{x_1 \to 0} g(x) \big),$$

provided this limit exists. Similarly, we have

$$D_1 D_2 f(0) = \lim_{x_1 \to 0} \big(\lim_{x_2 \to 0} g(x) \big).$$

For a counterexample we only have to choose a function g for which both limits exist but are different, for example

$$g(x) = \frac{x_1^2 - x_2^2}{\|x\|^2} \qquad (x \ne 0).$$

Then $|g(x)| \le 1$ for $x \in \mathbf{R}^2 \setminus \{0\}$, and

$$\lim_{x_1 \to 0} g(x) = -1 \qquad (x_2 \ne 0), \qquad \lim_{x_2 \to 0} g(x) = 1 \qquad (x_1 \ne 0),$$

and this implies $D_2 D_1 f(0) = -1$ and $D_1 D_2 f(0) = 1$. ✩

In the next theorem we formulate a sufficient condition for the equality of the mixed second-order partial derivatives.

Theorem 2.7.2 (Equality of mixed partial derivatives). *Let $U \subset \mathbf{R}^n$ be an open set, let $a \in U$ and suppose $f : U \to \mathbf{R}^p$ is a mapping for which $D_j D_i f$ and $D_i D_j f$ exist in a neighborhood of a and are continuous at a. Then*

$$D_j D_i f(a) = D_i D_j f(a) \qquad (1 \leq i, j \leq n).$$

Proof. Using a permutation of the indices if necessary, we may assume that $i = 1$ and $j = 2$ and, as a consequence, that $n = 2$. Furthermore, in view of Proposition 2.2.9 it is sufficient to verify the theorem for $p = 1$. We shall show that both sides in the identity are equal to

$$\lim_{x \to a} r(x) \qquad \text{where} \qquad r(x) = \frac{f(x) - f(a_1, x_2) - f(x_1, a_2) + f(a)}{(x_1 - a_1)(x_2 - a_2)}.$$

Note that the denominator is the area of the rectangle with vertices x, (a_1, x_2), a and (x_1, a_2) all in $\mathbf{R}^2$, while the numerator is the alternating sum of the values of f in these vertices.

Let U be a neighborhood of a where the first-order and the mixed second-order partial derivatives of f exist, and let $x \in U$. Define $g : \mathbf{R} \supset\!\!\to \mathbf{R}$ in a neighborhood of a_1 by

$$g(t) = f(t, x_2) - f(t, a_2), \qquad \text{then} \qquad r(x) = \frac{g(x_1) - g(a_1)}{(x_1 - a_1)(x_2 - a_2)}.$$

As g is differentiable the Mean Value Theorem for $\mathbf{R}$ now implies the existence of $\xi_1 = \xi_1(x) \in \mathbf{R}$ between a_1 and x_1 such that

$$r(x) = \frac{g'(\xi_1)}{x_2 - a_2} = \frac{D_1 f(\xi_1, x_2) - D_1 f(\xi_1, a_2)}{x_2 - a_2}.$$

If the function $h : \mathbf{R} \supset\!\!\to \mathbf{R}$ is defined in a sufficiently small neighborhood of a_2 by $h(t) = D_1 f(\xi_1, t)$, then h is differentiable and once again it follows by the Mean Value Theorem that there is $\xi_2 = \xi_2(x)$ between a_2 and x_2 such that

$$r(x) = h'(\xi_2) = D_2 D_1 f(\xi).$$

Now $\xi = \xi(x) \in \mathbf{R}^2$ satisfies $\|\xi - a\| \leq \|x - a\|$, for all $x \in \mathbf{R}^2$ as above, and the continuity of $D_2 D_1 f$ at a implies

$$\lim_{x \to a} r(x) = \lim_{\xi \to a} D_2 D_1 f(\xi) = D_2 D_1 f(a). \qquad (2.19)$$

Finally, repeat the preceding arguments with the roles of x_1 and x_2 interchanged; in general, one finds $\xi \in \mathbf{R}^2$ different from the element obtained above. Yet, this leads to

$$\lim_{x \to a} r(x) = D_1 D_2 f(a).$$

❏

Next we give the natural extension of the theory of differentiability to higher-order derivatives. Organizing the notations turns out to be the main part of the work.

Let U be an open subset of $\mathbf{R}^n$ and consider a differentiable mapping $f : U \to \mathbf{R}^p$. Then we have for the derivative $Df : U \to \mathrm{Lin}(\mathbf{R}^n, \mathbf{R}^p) \simeq \mathbf{R}^{pn}$ in view of Formula (2.2). If, in turn, Df is differentiable, then its derivative $D^2 f := D(Df)$, the *second-order derivative* of f, is a mapping $U \to \mathrm{Lin}\,(\mathbf{R}^n, \mathrm{Lin}(\mathbf{R}^n, \mathbf{R}^p)) \simeq \mathbf{R}^{pn^2}$. When considering higher-order derivatives, we quickly get a rather involved notation. This becomes more manageable if we recall some results in linear algebra.

Definition 2.7.3. We denote by $\mathrm{Lin}^k(\mathbf{R}^n, \mathbf{R}^p)$ the linear space of all *k-linear mappings* from the k-fold Cartesian product $\mathbf{R}^n \times \cdots \times \mathbf{R}^n$ taking values in $\mathbf{R}^p$. Thus $T \in \mathrm{Lin}^k(\mathbf{R}^n, \mathbf{R}^p)$ if and only if $T : \mathbf{R}^n \times \cdots \times \mathbf{R}^n \to \mathbf{R}^p$ and T is linear in each of the k variables varying in $\mathbf{R}^n$, when all the other variables are held fixed. ❍

Lemma 2.7.4. *There exists a natural isomorphism of linear spaces*

$$\mathrm{Lin}\,(\mathbf{R}^n,\ \mathrm{Lin}(\mathbf{R}^n, \mathbf{R}^p)) \xrightarrow{\sim} \mathrm{Lin}^2(\mathbf{R}^n, \mathbf{R}^p),$$

given by $\mathrm{Lin}\,(\mathbf{R}^n,\ \mathrm{Lin}(\mathbf{R}^n, \mathbf{R}^p)) \ni S \leftrightarrow T \in \mathrm{Lin}^2(\mathbf{R}^n, \mathbf{R}^p)$ *if and only if* $(Sh_1)h_2 = T(h_1, h_2)$, *for* h_1 *and* $h_2 \in \mathbf{R}^n$.

More generally, there exists a natural isomorphism of linear spaces

$$\mathrm{Lin}\,(\mathbf{R}^n, \mathrm{Lin}(\mathbf{R}^n, \dots, \mathrm{Lin}(\mathbf{R}^n, \mathbf{R}^p)\dots)) \xrightarrow{\sim} \mathrm{Lin}^k(\mathbf{R}^n, \mathbf{R}^p) \qquad (k \in \mathbf{N} \setminus \{1\}),$$

with $S \leftrightarrow T$ *if and only if* $(\cdots((Sh_1)h_2)\cdots)h_k = T(h_1, h_2, \dots, h_k)$, *for* h_1, ..., $h_k \in \mathbf{R}^n$.

Proof. The proof is by mathematical induction over $k \in \mathbf{N} \setminus \{1\}$.

First we treat the case where $k = 2$. For $S \in \mathrm{Lin}\,(\mathbf{R}^n,\ \mathrm{Lin}(\mathbf{R}^n, \mathbf{R}^p))$ define

$$\mu(S) : \mathbf{R}^n \times \mathbf{R}^n \to \mathbf{R}^p \qquad \text{by} \qquad \mu(S)(h_1, h_2) = (Sh_1)h_2 \qquad (h_1, h_2 \in \mathbf{R}^n).$$

It is obvious that $\mu(S)$ is linear in both of its variables separately, in other words, $\mu(S) \in \mathrm{Lin}^2(\mathbf{R}^n, \mathbf{R}^p)$. Conversely, for $T \in \mathrm{Lin}^2(\mathbf{R}^n, \mathbf{R}^p)$ define

$$\nu(T) : \mathbf{R}^n \to \{\,\text{mappings} : \mathbf{R}^n \to \mathbf{R}^p\,\} \qquad \text{by} \qquad (\nu(T)h_1)h_2 = T(h_1, h_2).$$

Then, from the linearity of T in the second variable it is straightforward that $\nu(T)h_1 \in \mathrm{Lin}(\mathbf{R}^n, \mathbf{R}^p)$, while from the linearity of T in the first variable it follows that $h_1 \mapsto \nu(T)h_1$ is a linear mapping on $\mathbf{R}^n$. Furthermore, $\nu \circ \mu = I$ since

$$(\nu \circ \mu(S)h_1)h_2 = \mu(S)(h_1, h_2) = (Sh_1)h_2 \qquad (h_1, h_2 \in \mathbf{R}^n).$$

Similarly, one proves $\mu \circ \nu = I$.

Now assume the assertion to be true for $k \geq 2$ and show in the same fashion as above that

$$\mathrm{Lin}\,(\mathbf{R}^n, \mathrm{Lin}^k(\mathbf{R}^n, \mathbf{R}^p)) \simeq \mathrm{Lin}^{k+1}(\mathbf{R}^n, \mathbf{R}^p). \qquad \square$$

Corollary 2.7.5. *For every $T \in \mathrm{Lin}^k(\mathbf{R}^n, \mathbf{R}^p)$ there exists $c > 0$ such that, for all $h_1, \dots, h_k \in \mathbf{R}^n$,*

$$\|T(h_1, \dots, h_k)\| \leq c\|h_1\| \cdots \|h_k\|.$$

Proof. Use mathematical induction over $k \in \mathbf{N}$ and the Lemmata 2.7.4 and 2.1.1. ❑

We now compute the derivative of a k-linear mapping, thereby generalizing the results in Example 2.2.5.

Proposition 2.7.6. *If $T \in \mathrm{Lin}^k(\mathbf{R}^n, \mathbf{R}^p)$ then T is differentiable. Furthermore, let $(a_1, \dots, a_k)$ and $(h_1, \dots, h_k) \in \mathbf{R}^n \times \cdots \times \mathbf{R}^n$, then $DT(a_1, \dots, a_k) \in \mathrm{Lin}(\mathbf{R}^n \times \cdots \times \mathbf{R}^n, \mathbf{R}^p)$ is given by*

$$DT(a_1, \dots, a_k)(h_1, \dots, h_k) = \sum_{1 \leq i \leq k} T(a_1, \dots, a_{i-1}, h_i, a_{i+1}, \dots, a_k).$$

Proof. The differentiability of T follows from the fact that the components of $T(a_1, \dots, a_k) \in \mathbf{R}^p$ are polynomials in the coordinates of the vectors a_1, ..., $a_k \in \mathbf{R}^n$, see Proposition 2.2.9 and Corollary 2.4.3. With the notation $h^{(i)} = (0, \dots, 0, h_i, 0, \dots, 0) \in \mathbf{R}^n \times \cdots \times \mathbf{R}^n$ we have

$$(h_1, \dots, h_k) = \sum_{1 \leq i \leq k} h^{(i)} \in \mathbf{R}^n \times \cdots \times \mathbf{R}^n.$$

Next, using the linearity of $DT(a_1, \dots, a_k)$ we obtain

$$DT(a_1, \dots, a_k)(h_1, \dots, h_k) = \sum_{1 \leq i \leq k} DT(a_1, \dots, a_k)h^{(i)}.$$

According to the chain rule the mapping

$$\mathbf{R}^n \to \mathbf{R}^p \qquad \text{with} \qquad x_i \mapsto T(a_1, \dots, a_{i-1}, x_i, a_{i+1}, \dots, a_k) \tag{2.20}$$

has derivative at $a_i \in \mathbf{R}^n$ in the direction $h_i \in \mathbf{R}^n$ equal to $DT(a_1, \dots, a_k)h^{(i)}$. On the other hand, the mapping in (2.20) is linear because T is k-linear, and therefore Example 2.2.5 implies that $T(a_1, \dots, a_{i-1}, h_i, a_{i+1}, \dots, a_k)$ is another expression for this derivative. ❑

Example 2.7.7. Let $k \in \mathbf{N}$ and define $f : \mathbf{R} \to \mathbf{R}$ by $f(t) = t^k$. Observe that $f(t) = T(t, \dots, t)$ where $T \in \mathrm{Lin}^k(\mathbf{R}, \mathbf{R})$ is given by $T(x_1, \dots, x_k) = x_1 \cdots x_k$. The proposition then implies that $f'(t) = kt^{k-1}$. ☆

With a slight abuse of notation, we will, in subsequent parts of this book, mostly consider $D^2 f(a)$ as an element of $\mathrm{Lin}^2(\mathbf{R}^n, \mathbf{R}^p)$, if $a \in U$, that is,

$$D^2 f(a)(h_1, h_2) = (D^2 f(a)h_1)h_2 \qquad (h_1, h_2 \in \mathbf{R}^n).$$

Similarly, we have $D^k f(a) \in \mathrm{Lin}^k(\mathbf{R}^n, \mathbf{R}^p)$, for $k \in \mathbf{N}$.

Definition 2.7.8. Let $f : U \to \mathbf{R}^p$, with U open in $\mathbf{R}^n$. Then the notation $f \in C^0(U, \mathbf{R}^p)$ means that f is continuous. By induction over $k \in \mathbf{N}$ we say that f is a k *times continuously differentiable mapping*, notation $f \in C^k(U, \mathbf{R}^p)$, if $f \in C^{k-1}(U, \mathbf{R}^p)$ and if the derivative of order $k - 1$

$$D^{k-1} f : U \to \mathrm{Lin}^{k-1}(\mathbf{R}^n, \mathbf{R}^p)$$

is a continuously differentiable mapping. We write $f \in C^\infty(U, \mathbf{R}^p)$ if $f \in C^k(U, \mathbf{R}^p)$, for all $k \in \mathbf{N}$. If we want to consider $f \in C^k$ for $k \in \mathbf{N}$ or $k = \infty$, we write $f \in C^k$ with $k \in \mathbf{N}_\infty$. In the case where $p = 1$, we write $C^k(U)$ instead of $C^k(U, \mathbf{R}^p)$. ○

Rephrasing the definition above, we obtain at once that $f \in C^k(U, \mathbf{R}^p)$ if $f \in C^{k-1}(U, \mathbf{R}^p)$ and if, for all $i_1, \dots, i_{k-1} \in \{1, \dots, n\}$, the partial derivative of order $k - 1$

$$D_{i_{k-1}} \cdots D_{i_1} f$$

belongs to $C^1(U, \mathbf{R}^p)$. In view of Theorem 2.3.7 the latter condition is satisfied if all the partial derivatives of order k satisfy

$$D_{i_k} D_{i_{k-1}} \cdots D_{i_1} f := D_{i_k}(D_{i_{k-1}} \cdots D_{i_1} f) \in C^0(U, \mathbf{R}^p).$$

Clearly, $f \in C^k(U, \mathbf{R}^p)$ if and only if

$$Df \in C^{k-1}(U, \mathrm{Lin}(\mathbf{R}^n, \mathbf{R}^p)).$$

This can subsequently be used to prove, by induction over k, that the composition $g \circ f$ is a C^k mapping if f and g are C^k mappings. Indeed, we have $D(g \circ f) = ((Dg) \circ f) \circ Df$ on account of the chain rule. Now $(Dg) \circ f$ is a C^{k-1} mapping, being a composition of the C^k mapping f with the C^{k-1} mapping Dg, where we use the induction hypothesis. Furthermore, Df is a C^{k-1} mapping, and composition of linear mappings is infinitely differentiable. Hence we obtain that $D(g \circ f)$ is a C^{k-1} mapping.

Theorem 2.7.9. *If $U \subset \mathbf{R}^n$ is open and $f \in C^2(U, \mathbf{R}^p)$, then the bilinear mapping $D^2 f(a) \in \mathrm{Lin}^2(\mathbf{R}^n, \mathbf{R}^p)$ satisfies*

$$D^2 f(a)(h_1, h_2) = (D_{h_1} D_{h_2} f)(a) \qquad (a, h_1, h_2 \in \mathbf{R}^n).$$

Furthermore, $D^2 f(a)$ is symmetric, that is,

$$D^2 f(a)(h_1, h_2) = D^2 f(a)(h_2, h_1) \qquad (a, h_1, h_2 \in \mathbf{R}^n).$$

More generally, let $f \in C^k(U, \mathbf{R}^p)$, for $k \in \mathbf{N}$. Then $D^k f(a) \in \mathrm{Lin}^k(\mathbf{R}^n, \mathbf{R}^p)$ satisfies

$$D^k f(a)(h_1, \dots, h_k) = (D_{h_1} \dots D_{h_k} f)(a) \qquad (a, h_i \in \mathbf{R}^n, 1 \le i \le k).$$

Moreover, $D^k f(a)$ is symmetric, that is, for $a, h_i \in \mathbf{R}^n$ where $1 \le i \le k$,

$$D^k f(a)(h_1, h_2, \dots, h_k) = D^k f(a)(h_{\sigma(1)}, h_{\sigma(2)}, \dots, h_{\sigma(k)}),$$

for every $\sigma \in S_k$, the permutation group *on k elements, which consists of bijections of the set $\{1, 2, \dots, k\}$.*

Proof. We have the linear mapping $L : \mathrm{Lin}(\mathbf{R}^n, \mathbf{R}^p) \to \mathbf{R}^p$ of evaluation at the fixed element $h_2 \in \mathbf{R}^n$, which is given by $L(S) = Sh_2$. Using this we can write

$$D_{h_2} f = L \circ Df : U \to \mathbf{R}^p, \qquad \text{since} \qquad D_{h_2} f(a) = Df(a)h_2 \qquad (a \in U).$$

From Lemma 2.4.7 we now obtain

$$D(D_{h_2} f)(a) = D(L \circ Df)(a) = (L \circ D^2 f)(a) \in \mathrm{Lin}(\mathbf{R}^n, \mathbf{R}^p) \qquad (a \in U).$$

Applying this identity to $h_1 \in \mathbf{R}^n$ and recalling that $D^2 f(a)h_1 \in \mathrm{Lin}(\mathbf{R}^n, \mathbf{R}^p)$, we get

$$\begin{aligned}(D_{h_1} D_{h_2} f)(a) &= D_{h_1}(D_{h_2} f)(a) = L(D^2 f(a)h_1) \\ &= (D^2 f(a)h_1)h_2 = D^2 f(a)(h_1, h_2),\end{aligned}$$

where we made use of Lemma 2.7.4 in the last step. Theorem 2.7.2 on the equality of mixed partial derivatives now implies the symmetry of $D^2 f(a)$.

The general result follows by mathematical induction over $k \in \mathbf{N}$. ❑

2.8 Taylor's formula

In this section we discuss Taylor's formula for mappings in $C^k(U, \mathbf{R}^p)$ with U open in $\mathbf{R}^n$. We begin by recalling the definition, and a proof of the formula for functions depending on one real variable.

Definition 2.8.1. Let I be an open interval in $\mathbf{R}$ with $a \in I$, and let $\gamma \in C^k(I, \mathbf{R}^p)$ for $k \in \mathbf{N}$. Then

$$\sum_{0\le j\le k} \frac{h^j}{j!}\gamma^{(j)}(a) \qquad (h \in \mathbf{R})$$

is the *Taylor polynomial* of order k of γ at a. The *k-th remainder* $h \mapsto R_k(a, h)$ of γ at a is defined by

$$\gamma(a+h) = \sum_{0\le j\le k} \frac{h^j}{j!}\gamma^{(j)}(a) + R_k(a, h) \qquad (a+h \in I). \qquad \bigcirc$$

Under the assumption of one extra degree of differentiability for γ we can give a formula for the k-th remainder, which can be quite useful for obtaining explicit estimates.

Lemma 2.8.2 (Integral formula for the k-th remainder). *Let I be an open interval in $\mathbf{R}$ with $a \in I$, and let $\gamma \in C^{k+1}(I, \mathbf{R}^p)$ for $k \in \mathbf{N}_0$. Then we have*

$$\gamma(a+h) = \sum_{0\le j\le k} \frac{h^j}{j!}\gamma^{(j)}(a) + \frac{h^{k+1}}{k!}\int_0^1 (1-t)^k\gamma^{(k+1)}(a+th)\,dt \qquad (a+h \in I).$$

Here the integration of the vector-valued function at the right–hand side is per component function. Furthermore, there exists a constant $c > 0$ such that

$$\|R_k(a, h)\| \le c|h|^{k+1} \quad \textit{when} \quad h \to 0 \quad \textit{in} \quad \mathbf{R};$$

thus $\|R_k(a, h)\| = \mathcal{O}(|h|^{k+1}), \quad h \to 0.$

Proof. By going over to $\widetilde{\gamma}(t) = \gamma(a+th)$ and noting that $\widetilde{\gamma}^{(j)}(t) = h^j\gamma^{(j)}(a+th)$ according to the chain rule, it is sufficient to consider the case where $a = 0$ and $h = 1$. Now use mathematical induction over $k \in \mathbf{N}_0$. Indeed, from the Fundamental Theorem of Integral Calculus on $\mathbf{R}$ and the continuity of $\gamma^{(1)}$ we obtain

$$\gamma(1) - \gamma(0) = \int_0^1 \gamma^{(1)}(t)\,dt.$$

Furthermore, for $k \in \mathbf{N}$,

$$\frac{1}{(k-1)!}\int_0^1 (1-t)^{k-1}\gamma^{(k)}(t)\,dt = \frac{1}{k!}\gamma^{(k)}(0) + \frac{1}{k!}\int_0^1 (1-t)^k\gamma^{(k+1)}(t)\,dt,$$

based on $\frac{(1-t)^{k-1}}{(k-1)!} = -\frac{d}{dt}\frac{(1-t)^k}{k!}$ and an integration by parts.

The estimate for $R_k(a, h)$ is obvious from the continuity of $\gamma^{(k+1)}$ on $[\,0, 1\,]$; in fact,

$$\left\| \frac{1}{k!} \int_0^1 (1-t)^k \gamma^{(k+1)}(t)\,dt \right\| \le \frac{1}{(k+1)!} \max_{t\in[0,1]} \|\gamma^{(k+1)}(t)\|. \qquad \square$$

In the weaker case where $\gamma \in C^k(I, \mathbf{R}^p)$, for $k \in \mathbf{N}$, the formula above (applied with k replaced by $k-1$) can still be used to find an integral expression as well as an estimate for the k-th remainder term. Indeed, note that $R_{k-1}(a,h) = \frac{h^k}{k!}\gamma^{(k)}(a) + R_k(a,h)$, which implies

$$\begin{aligned} R_k(a,h) &= R_{k-1}(a,h) - \frac{h^k}{k!}\gamma^{(k)}(a) \\ &= \frac{h^k}{(k-1)!}\int_0^1 (1-t)^{k-1}(\gamma^{(k)}(a+th) - \gamma^{(k)}(a))\,dt. \end{aligned} \tag{2.21}$$

Now the continuity of $\gamma^{(k)}$ on I implies that for every $\epsilon > 0$ there is a $\delta > 0$ such that

$$\|\gamma^{(k)}(a+h) - \gamma^{(k)}(a)\| < \epsilon \qquad \text{whenever} \qquad |h| < \delta.$$

Using this estimate in Formula (2.21) we see that for every $\epsilon > 0$ there exists a $\delta > 0$ such that for every $h \in \mathbf{R}$ with $|h| < \delta$

$$\|R_k(a,h)\| < \epsilon\,\frac{1}{k!}\,|h|^k; \qquad \text{in other words} \qquad \|R_k(a,h)\| = o(|h|^k), \quad h \to 0. \tag{2.22}$$

Therefore, in this case the k-th remainder is of order lower than $|h|^k$ when $h \to 0$ in $\mathbf{R}$.

At this stage we are sufficiently prepared to handle the case of mappings defined on open sets in $\mathbf{R}^n$.

Theorem 2.8.3 (Taylor's formula: first version). *Assume U to be a convex open subset of $\mathbf{R}^n$ (see Definition 2.5.2) and let $f \in C^{k+1}(U, \mathbf{R}^p)$, for $k \in \mathbf{N}_0$. Then we have, for all a and $a+h \in U$,* Taylor's formula *with the* integral formula for the *k-th remainder $R_k(a,h)$,*

$$f(a+h) = \sum_{0\le j\le k} \frac{1}{j!} D^j f(a)(h^j) + \frac{1}{k!}\int_0^1 (1-t)^k D^{k+1} f(a+th)(h^{k+1})\,dt.$$

Here $D^j f(a)(h^j)$ means $D^j f(a)(h, \dots, h)$. The sum at the right–hand side is called the Taylor polynomial *of order k of f at the point a. Moreover,*

$$\|R_k(a,h)\| = \mathcal{O}(\|h\|^{k+1}), \quad h \to 0.$$

Proof. Apply Lemma 2.8.2 to $\gamma(t) = f(a+th)$, where $t \in [0,1]$. From the chain rule we get

$$\frac{d}{dt} f(a+th) = Df(a+th)h = D_h f(a+th).$$

Hence, by mathematical induction over $j \in \mathbf{N}$ and Theorem 2.7.9,

$$\gamma^{(j)}(t) = \Big(\frac{d}{dt}\Big)^j f(a+th) = D^j f(a+th)(\underbrace{h, \dots, h}_{j}) = D^j f(a+th)(h^j).$$

In the same manner as in the proof of Lemma 2.8.2 the estimate for $R(a, h)$ follows once we use Corollary 2.7.5. ❑

Substituting $h = \sum_{1\le i\le n} h_i e_i$ and applying Theorem 2.7.9, we see

$$D^j f(a)(h^j) = \sum_{(i_1,\dots,i_j)} h_{i_1} \cdots h_{i_j} \, (D_{i_1} \cdots D_{i_j} f)(a). \tag{2.23}$$

Here the summation is over all ordered j-tuples $(i_1, \dots, i_j)$ of indices i_s belonging to $\{1, \dots, n\}$. However, the right–hand side of Formula (2.23) contains a number of double counts because the order of differentiation is irrelevant according to Theorem 2.7.9. In other words, all that matters is the number of times an index i_s occurs. For a choice $(i_1, \dots, i_j)$ of indices we therefore write the number of times an index i_s equals i as α_i, and this for $1 \le i \le n$. In this way we assign to $(i_1, \dots, i_j)$ the *multi-index*

$$\alpha = (\alpha_1, \dots, \alpha_n) \in \mathbf{N}_0^n, \qquad \text{with} \qquad |\alpha| := \sum_{1\le i\le n} \alpha_i = j.$$

Next write, for $\alpha \in \mathbf{N}_0^n$,

$$\alpha! = \alpha_1! \cdots \alpha_n! \in \mathbf{N}, \qquad h^\alpha = h_1^{\alpha_1} \cdots h_n^{\alpha_n} \in \mathbf{R}, \qquad D^\alpha = D_1^{\alpha_1} \cdots D_n^{\alpha_n}. \tag{2.24}$$

Denote the number of choices of different $(i_1, \dots, i_j)$ all leading to the same element α by $m(\alpha) \in \mathbf{N}$. We claim

$$m(\alpha) = \frac{j!}{\alpha!} \qquad (|\alpha| = j).$$

Indeed, $m(\alpha)$ is completely determined by

$$\sum_{|\alpha|=j} m(\alpha) h^\alpha = \sum_{(i_1,\dots,i_j)} h_{i_1} \cdots h_{i_j} = (h_1 + \cdots + h_n)^j.$$

One finds that $m(\alpha)$ equals the product of the number of combinations of j objects, taken α_1 at a time, times the number of combinations of $j - \alpha_1$ objects, taken α_2 at a time, and so on, times the number of combinations of $j - (\alpha_1 + \cdots + \alpha_{n-1})$ objects, taken α_n at a time. Therefore

$$m(\alpha) = \binom{j}{\alpha_1}\binom{j-\alpha_1}{\alpha_2} \cdots \binom{j - (\alpha_1 + \cdots + \alpha_{n-1})}{\alpha_n} = \frac{j!}{\alpha_1! \cdots \alpha_n!} = \frac{j!}{\alpha!}.$$

Using this and Theorem 2.7.9 we can rewrite Formula (2.23) as

$$\frac{1}{j!} D^j f(a)(h^j) = \sum_{|\alpha|=j} \frac{h^\alpha}{\alpha!} D^\alpha f(a).$$

Now Theorem 2.8.3 leads immediately to

Theorem 2.8.4 (Taylor's formula: second version with multi-indices). *Let U be a convex open subset of $\mathbf{R}^n$ and $f \in C^{k+1}(U, \mathbf{R}^p)$, for $k \in \mathbf{N}_0$. Then we have, for all a and $a + h \in U$*

$$f(a+h) = \sum_{|\alpha| \leq k} \frac{h^\alpha}{\alpha!} D^\alpha f(a) + (k+1) \sum_{|\alpha|=k+1} \frac{h^\alpha}{\alpha!} \int_0^1 (1-t)^k D^\alpha f(a+th)\, dt.$$

We now derive an estimate for the k-th remainder $R_k(a, h)$ under the (weaker) assumption that $f \in C^k(U, \mathbf{R}^p)$. As in Formula (2.21) we get

$$R_k(a, h) = \frac{1}{(k-1)!} \int_0^1 (1-t)^{k-1} (D^k f(a+th)(h^k) - D^k f(a)(h^k))\, dt.$$

If we copy the proof of the estimate (2.22) using Corollary 2.7.5, we find

$$\|R_k(a, h)\| = o(\|h\|^k), \qquad h \to 0. \tag{2.25}$$

Therefore, in this case the k-th remainder is of order lower than $\|h\|^k$ when $h \to 0$ in $\mathbf{R}^n$.

Remark. Under the assumption that $f \in C^\infty(U, \mathbf{R}^p)$, that is, f can be differentiated an arbitrary number of times, one may ask whether the *Taylor series* of f at a, which is called the *MacLaurin series* of f if $a = 0$,

$$\sum_{\alpha \in \mathbf{N}_0^n} \frac{h^\alpha}{\alpha!} D^\alpha f(a)$$

converges to $f(a+h)$. This comes down to finding suitable estimates for $R_k(a, h)$ in terms of k, for $k \to \infty$. By analogy with the case of functions of a single real variable, this leads to a theory of *real-analytic functions* of n real variables, which we shall not go into (however, see Exercise 8.12). In the case of a single real variable one may settle the convergence of the series by means of convergence criteria and try to prove the convergence to $f(a+h)$ by studying the differential equation satisfied by the series; see Exercise 0.11 for an example of this alternative technique.

2.9 Critical points

We need some more concepts from linear algebra for our investigation of the nature of critical points. These results will also be needed elsewhere in this book.

Lemma 2.9.1. *There is a linear isomorphism between* $\mathrm{Lin}^2(\mathbf{R}^n, \mathbf{R})$ *and* $\mathrm{End}(\mathbf{R}^n)$. *In fact, for every* $T \in \mathrm{Lin}^2(\mathbf{R}^n, \mathbf{R})$ *there is a unique* $\lambda(T) \in \mathrm{End}(\mathbf{R}^n)$ *satisfying*

$$T(h, k) = \langle \lambda(T)h, k \rangle \qquad (h, k \in \mathbf{R}^n).$$

Proof. From Lemmata 2.7.4 and 2.6.1 we obtain the linear isomorphisms

$$\mathrm{Lin}^2(\mathbf{R}^n, \mathbf{R}) \xrightarrow{\sim} \mathrm{Lin}\,(\mathbf{R}^n,\ \mathrm{Lin}(\mathbf{R}^n, \mathbf{R})) \xrightarrow{\sim} \mathrm{Lin}(\mathbf{R}^n, \mathbf{R}^n) = \mathrm{End}(\mathbf{R}^n),$$

and following up the isomorphisms we get the identity. ❑

Now let U be an open subset of $\mathbf{R}^n$, let $a \in U$ and $f \in C^2(U)$. In this case, the bilinear form $D^2 f(a) \in \mathrm{Lin}^2(\mathbf{R}^n, \mathbf{R})$ is called the *Hessian* of f at a. If we apply Lemma 2.9.1 with $T = D^2 f(a)$, we find $Hf(a) \in \mathrm{End}(\mathbf{R}^n)$, satisfying

$$D^2 f(a)(h, k) = \langle\, Hf(a)h, k\rangle \qquad (h, k \in \mathbf{R}^n).$$

According to Theorem 2.7.9 we have $\langle\, Hf(a)h, k\rangle = \langle h, Hf(a)k\rangle$, for all h and $k \in \mathbf{R}^n$; hence, $Hf(a)$ is self-adjoint, see Definition 2.1.3. With respect to the standard basis in $\mathbf{R}^n$ the symmetric matrix of $Hf(a)$, which is called the *Hessian matrix* of f at a, is given by

$$(D_j D_i f(a))_{1\le i,j\le n}.$$

If in addition U is convex, we can rewrite Taylor's formula from Theorem 2.8.3, for $a + h \in U$, as

$$f(a+h) = f(a) + \langle\,\mathrm{grad}\, f(a), h\rangle + \frac{1}{2}\langle\, Hf(a)h, h\rangle + R_2(a, h)$$

$$\text{with} \quad \lim_{h\to 0}\frac{\|R_2(a, h)\|}{\|h\|^2} = 0.$$

Accordingly, at a critical point a for f (see Definition 2.6.4),

$$f(a+h) = f(a) + \frac{1}{2}\langle\, Hf(a)h, h\rangle + R_2(a, h) \quad \text{with} \quad \lim_{h\to 0}\frac{\|R_2(a, h)\|}{\|h\|^2} = 0. \tag{2.26}$$

The quadratic term dominates the remainder term. Therefore we now first turn our attention to the quadratic term. In particular, we are interested in the case where the quadratic term does not change sign, or in other words, where $Hf(a)$ satisfies the following:

Definition 2.9.2. Let $A \in \mathrm{End}^+(\mathbf{R}^n)$, thus A is self-adjoint. Then A is said to be *positive (semi)definite*, and *negative (semi)definite*, if, for all $x \in \mathbf{R}^n \setminus \{0\}$,

$$\langle Ax, x\rangle \ge 0 \quad \text{or} \quad \langle Ax, x\rangle > 0, \qquad \text{and} \qquad \langle Ax, x\rangle \le 0 \quad \text{or} \quad \langle Ax, x\rangle < 0,$$

respectively. Moreover, A is said to be *indefinite* if it is neither positive nor negative semidefinite. ❍

The following theorem, which is well-known from linear algebra, clarifies the meaning of the condition of being (semi)definite or indefinite. The theorem and its corollary will be required in Section 7.3. In this text we prove the theorem by analytical means, using the theory we have been developing. This proof of the diagonalizability of a symmetric matrix does not make use of the characteristic polynomial $\lambda \mapsto \det(\lambda I - A)$, nor of the Fundamental Theorem of Algebra. Furthermore, it provides an algorithm for the actual computation of the eigenvalues of a self-adjoint operator, see Exercise 2.67. See Exercises 2.65 and 5.47 for other proofs of the theorem.

Theorem 2.9.3 (Spectral Theorem for Self-adjoint Operator). *For any* $A \in \mathrm{End}^+(\mathbf{R}^n)$ *there exist an orthonormal basis* $(a_1, \dots, a_n)$ *for* $\mathbf{R}^n$ *and a vector* $(\lambda_1, \dots, \lambda_n) \in \mathbf{R}^n$ *such that*

$$Aa_j = \lambda_j a_j \qquad (1 \le j \le n).$$

In particular, if we introduce the Rayleigh quotient $R : \mathbf{R}^n \setminus \{0\} \to \mathbf{R}$ *for* A *by*

$$R(x) = \frac{\langle Ax, x\rangle}{\langle x, x\rangle}, \qquad \textit{then} \quad \max\{\, \lambda_j \mid 1 \le j \le n \,\} = \max\{\, R(x) \mid x \in \mathbf{R}^n \setminus \{0\} \,\}. \tag{2.27}$$

A similar statement is valid if we replace max *by* min *everywhere in (2.27).*

Proof. As the Rayleigh quotient for A satisfies $R(tx) = R(x)$, for all $t \in \mathbf{R} \setminus \{0\}$ and $x \in \mathbf{R}^n \setminus \{0\}$, the values of R are the same as those of the restriction of R to the unit sphere $S^{n-1} := \{\, x \in \mathbf{R}^n \mid \|x\| = 1 \,\}$. But because S^{n-1} is compact and R is continuous, R restricted to S^{n-1} attains its maximum at some point $a \in S^{n-1}$ because of Theorem 1.8.8. Passing back to $\mathbf{R}^n \setminus \{0\}$, we have shown the existence of a critical point a for R defined on the open set $\mathbf{R}^n \setminus \{0\}$. Theorem 2.6.3.(iv) then says that $\operatorname{grad} R(a) = 0$. Because A is self-adjoint, and also considering Example 2.2.5, we see that the derivative of $g : x \mapsto \langle Ax, x\rangle$ is given by $Dg(x)h = 2\langle Ax, h\rangle$, for all x and $h \in \mathbf{R}^n$. Hence, as $\|a\| = 1$,

$$0 = \operatorname{grad} R(a) = 2\langle a, a\rangle\, Aa - 2\langle Aa, a\rangle\, a$$

implies $Aa = R(a)\, a$. In other words, a is an eigenvector for A with eigenvalue as in the right–hand side of Formula (2.27).

Next, set $V = \{\, x \in \mathbf{R}^n \mid \langle x, a\rangle = 0 \,\}$. Then V is a linear subspace of dimension $n - 1$, while

$$\langle Ax, a\rangle = \langle x, Aa\rangle = \lambda\langle x, a\rangle = 0 \qquad (x \in V)$$

implies that $A(V) \subset V$. Furthermore, A acting as a linear operator in V is again self-adjoint. Therefore the assertion of the theorem follows by mathematical induction over $n \in \mathbf{N}$. ❑

Phrased differently, there exists an orthonormal basis consisting of *eigenvectors* associated with the real *eigenvalues* of the operator A. In particular, writing any $x \in \mathbf{R}^n$ as a linear combination of the basis vectors, we get

$$x = \sum_{1 \le j \le n} \langle x, a_j\rangle a_j, \qquad \text{and thus} \qquad Ax = \sum_{1 \le j \le n} \lambda_j \langle x, a_j\rangle a_j \qquad (x \in \mathbf{R}^n).$$

From this formula we see that the general quadratic form is a weighted sum of squares; indeed,

$$\langle Ax, x\rangle = \sum_{1\le j\le n} \lambda_j \langle x, a_j\rangle^2 \qquad (x \in \mathbf{R}^n). \tag{2.28}$$

For yet another reformulation we need the following:

Definition 2.9.4. We define $\mathbf{O}(\mathbf{R}^n)$, the *orthogonal group*, consisting of the *orthogonal* operators in $\mathrm{End}(\mathbf{R}^n)$ by

$$\mathbf{O}(\mathbf{R}^n) = \{\, O \in \mathrm{End}(\mathbf{R}^n) \mid O^t O = I \,\}. \qquad \bigcirc$$

Observe that $O \in \mathbf{O}(\mathbf{R}^n)$ implies $\langle x, y\rangle = \langle O^t Ox, y\rangle = \langle Ox, Oy\rangle$, for all x and $y \in \mathbf{R}^n$. In other words, $O \in \mathbf{O}(\mathbf{R}^n)$ if and only if O preserves the inner product on $\mathbf{R}^n$. Therefore $(Oe_1, \dots, Oe_n)$ is an orthonormal basis for $\mathbf{R}^n$ if $(e_1, \dots, e_n)$ denotes the standard basis in $\mathbf{R}^n$; and conversely, every orthonormal basis arises in this manner.

Denote by $O \in \mathbf{O}(\mathbf{R}^n)$ the operator that sends e_j to the basis vector a_j from Theorem 2.9.3, then $O^{-1}AOe_j = O^tAOe_j = \lambda_j e_j$, for $1 \le j \le n$. Thus, conjugation of $A \in \mathrm{End}^+(\mathbf{R}^n)$ by $O^{-1} = O^t \in \mathbf{O}(\mathbf{R}^n)$ is seen to turn A into the operator O^tAO with diagonal matrix, having the eigenvalues λ_j of A on the diagonal. At this stage, we can derive Formula (2.28) in the following way, for all $x \in \mathbf{R}^n$,

$$\langle Ax, x\rangle = \langle O(O^tAO)O^tx, x\rangle = \langle (O^tAO)O^tx, O^tx\rangle = \sum_{1\le j\le n} \lambda_j (O^tx)_j^2. \tag{2.29}$$

In this context, the following terminology is current. The number in $\mathbf{N}_0$ of negative eigenvalues of $A \in \mathrm{End}^+(\mathbf{R}^n)$, counted with multiplicities, is called the *index* of A. Also, the number in $\mathbf{Z}$ of positive minus the number of negative eigenvalues of A, all counted with multiplicities, is called the *signature* of A. Finally, the multiplicity in $\mathbf{N}_0$ of 0 as an eigenvalue of A is called the *nullity* of A. The result that index, signature and nullity are invariants of A, in particular, that they are independent of the construction in the proof of Theorem 2.9.3, is known as *Sylvester's law of inertia.* For a proof, observe that Formula (2.29) determines a decomposition of $\mathbf{R}^n$ into a direct sum of linear subspaces $V_+ \oplus V_0 \oplus V_-$, where A is positive definite, identically zero, and negative definite on V_+, V_0, and V_-, respectively. Let $W_+ \oplus W_0 \oplus W_-$ be another such decomposition. Then $V_+ \cap (W_0 \oplus W_-) = (0)$ and, therefore, $\dim V_+ + \dim(W_0 \oplus W_-) \le n$, i.e., $\dim V_+ \le \dim W_+$. Similarly, $\dim W_+ \le \dim V_+$. In fact, it follows from the theory of the characteristic polynomial of A that the eigenvalues counted with multiplicities are invariants of A.

Taking $A = Hf(a)$, we find the *index*, the *signature* and the *nullity* of f at the point a.

Corollary 2.9.5. *Let* $A \in \mathrm{End}^+(\mathbf{R}^n)$.

(i) *If A is positive definite there exists $c > 0$ such that*

$$\langle Ax, x \rangle \geq c\|x\|^2 \qquad (x \in \mathbf{R}^n).$$

(ii) *If A is positive semidefinite, then all of its eigenvalues are ≥ 0, and in particular,* $\det A \geq 0$ *and* $\operatorname{tr} A \geq 0$.

(iii) *If A is indefinite, then it has at least one positive as well as one negative eigenvalue (in particular, $n \geq 2$).*

Proof. Assertion (i) is clear from Formula (2.28) with $c := \min\{\, \lambda_j \mid 1 \leq j \leq n \,\} > 0$; a different proof uses Theorem 1.8.8 and the fact that the Rayleigh quotient for A is positive on S^{n-1}. Assertions (ii) and (iii) follow directly from the Spectral Theorem. ❑

Example 2.9.6 (Gram's matrix). Given $A \in \mathrm{Lin}(\mathbf{R}^n, \mathbf{R}^p)$, we have $A^tA \in \mathrm{End}(\mathbf{R}^n)$ with the symmetric matrix of inner products $(\langle a_i, a_j \rangle)_{1 \leq i,\, j \leq n}$, or, as in Section 2.1, Gram's matrix associated with the vectors $a_1, \dots, a_n \in \mathbf{R}^p$. Now $(A^tA)^t = A^t(A^t)^t = A^tA$ gives that A^tA is self-adjoint, while $\langle A^tAx, x \rangle = \langle Ax, Ax \rangle \geq 0$, for all $x \in \mathbf{R}^n$, gives that it is positive semidefinite. Accordingly $\det(A^tA) \geq 0$ and $\operatorname{tr}(A^tA) \geq 0$. ☆

Now we are ready to formulate a sufficient condition for a local extremum at a critical point.

Theorem 2.9.7 (Second-derivative test). *Let U be a convex open subset of $\mathbf{R}^n$ and let $a \in U$ be a critical point for $f \in C^2(U)$. Then we have the following assertions.*

(i) *If $Hf(a) \in \mathrm{End}(\mathbf{R}^n)$ is positive definite, then f has a local strict minimum at a.*

(ii) *If $Hf(a)$ is negative definite, then f has a local strict maximum at a.*

(iii) *If $Hf(a)$ is indefinite, then f has no local extremum at a. In this case a is called a* saddle point.

Proof. **(i).** Select $c > 0$ as in Corollary 2.9.5.(i) for $Hf(a)$ and $\delta > 0$ such that $\frac{\|R_2(a,h)\|}{\|h\|^2} < \frac{c}{4}$ for $\|h\| < \delta$. Then we obtain from Formula (2.26) and the reverse triangle inequality

$$f(a+h) \geq f(a) + \frac{c}{2}\|h\|^2 - \frac{c}{4}\|h\|^2 = f(a) + \frac{c}{4}\|h\|^2 \qquad (h \in B(a; \delta)).$$

(ii). Apply (i) to $-f$.
(iii). Since $D^2 f(a)$ is indefinite, there exist h_1 and $h_2 \in \mathbf{R}^n$, with

$$\mu_1 := \frac{1}{2}\langle Hf(a)h_1, h_1\rangle > 0 \qquad \text{and} \qquad \mu_2 := \frac{1}{2}\langle Hf(a)h_2, h_2\rangle < 0,$$

respectively. Furthermore, we can find $T > 0$ such that, for all $0 < t < T$,

$$\mu_1 - \frac{\|R_2(a, th_1)\|}{\|th_1\|^2}\|h_1\|^2 > 0 \qquad \text{and} \qquad \mu_2 + \frac{\|R_2(a, th_2)\|}{\|th_2\|^2}\|h_2\|^2 < 0.$$

But this implies, for all $0 < t < T$,

$$f(a + th_1) = f(a) + \mu_1 t^2 + R_2(a, th_1) > f(a);$$
$$f(a + th_2) = f(a) + \mu_2 t^2 + R_2(a, th_2) < f(a).$$

This proves that in this case f does not attain an extremum at a. ❑

Definition 2.9.8. In the notation of Theorem 2.9.7 we say that a is a *nondegenerate critical point* for f if $Hf(a) \in \mathrm{Aut}(\mathbf{R}^n)$. ❍

Remark. Observe that the second-derivative test is conclusive if the critical point is nondegenerate, because in that case all eigenvalues are nonzero, which implies that one of the three cases listed in the test must apply. On the other hand, examples like $f : \mathbf{R}^2 \to \mathbf{R}$ with $f(x) = \pm(x_1^4 + x_2^4)$, and $= x_1^4 - x_2^4$, show that f may have a minimum, a maximum, and a saddle point, respectively, at 0 while the second-order derivative vanishes at 0.

Applying Example 2.2.6 with Df instead of f, we see that a nondegenerate critical point is isolated, that is, has a neighborhood free from other critical points.

Example 2.9.9. Consider $f : \mathbf{R}^2 \to \mathbf{R}$ given by $f(x) = x_1x_2 - x_1^2 - x_2^2 - 2x_1 - 2x_2 + 4$. Then $D_1 f(x) = x_2 - 2x_1 - 2 = 0$ and $D_2 f(x) = x_1 - 2x_2 - 2 = 0$, implying that $a = -(2, 2)$ is the only critical point of f. Furthermore, $D_1^2 f(a) = -2$, $D_1D_2 f(a) = D_2D_1 f(a) = 1$ and $D_2^2 f(a) = -2$; thus $Hf(a)$, and its characteristic equation, become, in that order

$$\begin{pmatrix} -2 & 1 \\ 1 & -2 \end{pmatrix}, \qquad \begin{vmatrix} \lambda + 2 & -1 \\ -1 & \lambda + 2 \end{vmatrix} = \lambda^2 + 4\lambda + 3 = (\lambda + 1)(\lambda + 3) = 0.$$

Hence, -1 and -3 are eigenvalues of $Hf(a)$, which implies that $Hf(a)$ is negative definite; and accordingly f has a local strict maximum at a. Furthermore, we see that a is a nondegenerate critical point. The Taylor polynomial of order 2 of f at a equals f, the latter being a quadratic polynomial itself; thus

$$\begin{aligned} f(x) &= 8 + \tfrac{1}{2}(x - a)^t Hf(a)(x - a) \\ &= 8 + \frac{1}{2}(x_1 + 2, x_2 + 2)\begin{pmatrix} -2 & 1 \\ 1 & -2 \end{pmatrix}\begin{pmatrix} x_1 + 2 \\ x_2 + 2 \end{pmatrix}. \end{aligned}$$

In order to put $Hf(a)$ in normal form, we observe that

$$O = \frac{1}{\sqrt{2}}\begin{pmatrix} 1 & -1 \\ 1 & 1 \end{pmatrix}$$

is the matrix of $O \in \mathbf{O}(\mathbf{R}^n)$ that sends the standard basis vectors e_1 and e_2 to the normalized eigenvectors $\frac{1}{\sqrt{2}}(1, 1)^t$ and $\frac{1}{\sqrt{2}}(-1, 1)^t$, corresponding to the eigenvalues -1 and -3, respectively. Note that

$$O^t(x - a) = \frac{1}{\sqrt{2}}\begin{pmatrix} 1 & 1 \\ -1 & 1 \end{pmatrix}\begin{pmatrix} x_1 + 2 \\ x_2 + 2 \end{pmatrix} = \frac{1}{\sqrt{2}}\begin{pmatrix} x_1 + x_2 + 4 \\ x_2 - x_1 \end{pmatrix}.$$

From Formula (2.29) we therefore obtain

$$f(x) = 8 - \frac{1}{4}(x_1 + x_2 + 4)^2 - \frac{3}{4}(x_1 - x_2)^2.$$

Again, we see that f attains a local strict maximum at a; furthermore, this is seen to be an absolute maximum. For every $c < 8$ the level curve $\{ x \in \mathbf{R}^2 \mid f(x) = c \}$ is an ellipse with its major axis along $\mathbf{R}(1, 1)$ and minor axis along $a + \mathbf{R}(-1, 1)$. ✫

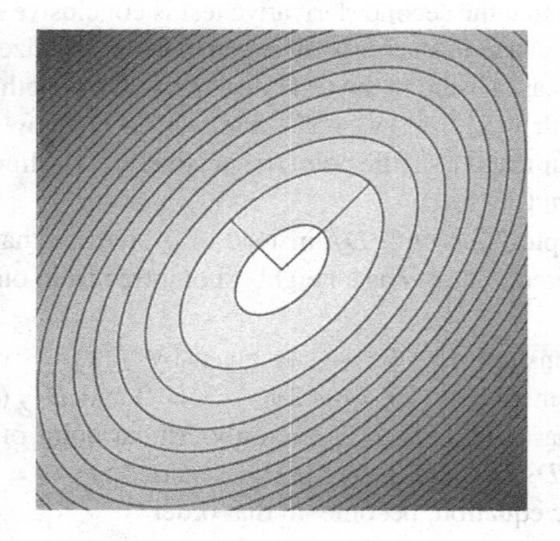

Illustration for Example 2.9.9

2.10 Commuting limit operations

We turn our attention to the commuting properties of certain limit operations. Earlier, in Theorem 2.7.2, we encountered the result that mixed partial derivatives may be taken in either order: a pair of differentiation operations may be interchanged. The Fundamental Theorem of Integral Calculus on $\mathbf{R}$ asserts the following interchange of a differentiation and an integration.

Theorem 2.10.1 (Fundamental Theorem of Integral Calculus on R). *Let $I = [\,a, b\,]$ and assume $f : I \to \mathbf{R}$ is continuous. If $f' : I \to \mathbf{R}$ is also continuous then*

$$\frac{d}{dx}\int_a^x f(\xi)\,d\xi = f(x) = f(a) + \int_a^x f'(\xi)\,d\xi \qquad (x \in I).$$

We recall that the first identity, which may be established using estimates, is the essential one. The second identity then follows from the fact that both f and $x \mapsto \int_a^x f'(\xi)\,d\xi$ have the same derivative.

The theorem is of great importance in this section, where we study the differentiability or integrability of integrals depending on a parameter. As a first step we consider the continuity of such integrals.

Theorem 2.10.2 (Continuity Theorem). *Suppose K is a compact subset of $\mathbf{R}^n$ and $J = [\,c, d\,]$ a compact interval in $\mathbf{R}$. Let $f : K \times J \to \mathbf{R}^p$ be a continuous mapping. Then the mapping $F : K \to \mathbf{R}^p$, given by*

$$F(x) = \int_J f(x, t)\,dt,$$

is continuous. Here the integration is by components. In particular we obtain, for $a \in K$,

$$\lim_{x\to a}\int_J f(x, t)\,dt = \int_J \lim_{x\to a} f(x, t)\,dt.$$

Proof. We have, for x and $a \in K$,

$$\|F(x) - F(a)\| = \Big\|\int_J (f(x, t) - f(a, t))\,dt\Big\| \le \int_J \|f(x, t) - f(a, t)\|\,dt.$$

On account of Theorem 1.8.15 the mapping f is uniformly continuous on the compact subset $K \times J$ of $\mathbf{R}^n \times \mathbf{R}$. Hence, given any $\epsilon > 0$ we can find $\delta > 0$ such that

$$\|f(x, t) - f(a, t_0)\| < \frac{\epsilon}{d - c} \qquad \text{if} \qquad \|(x, t) - (a, t_0)\| < \delta.$$

In particular, we obtain this estimate for f if $\|x - a\| < \delta$ and $t = t_0$. Therefore, for such x and a

$$\|F(x) - F(a)\| \le \int_J \frac{\epsilon}{d - c}\,dt = \epsilon.$$

This proves the continuity of F at every $a \in K$. ❑

Example 2.10.3. Let $f(x, t) = \cos xt$. Then the conditions of the theorem are satisfied for every compact set $K \subset \mathbf{R}$ and $J = [\,0, 1\,]$. In particular, for x varying in compacta containing 0,

$$F(x) = \int_0^1 \cos xt\,dt = \begin{cases} \dfrac{\sin x}{x}, & x \neq 0; \\ 1, & x = 0. \end{cases}$$

Accordingly, the continuity of F at 0 yields the well-known limit $\lim_{x\to 0}\frac{\sin x}{x} = 1$. ✩

Theorem 2.10.4 (Differentiation Theorem). *Suppose K is a compact subset of $\mathbf{R}^n$ with nonempty interior U and $J = [\,c, d\,]$ a compact interval in $\mathbf{R}$. Let $f : K \times J \to \mathbf{R}^p$ be a mapping with the following properties:*

(i) *f is continuous;*

(ii) *the total derivative $D_1 f : U \times J \to \mathrm{Lin}(\mathbf{R}^n, \mathbf{R}^p)$ with respect to the variable in U is continuous.*

Then $F : U \to \mathbf{R}^p$, given by $F(x) = \int_J f(x, t)\, dt$, is a differentiable mapping satisfying

$$D \int_J f(x, t)\, dt = \int_J D_1 f(x, t)\, dt \qquad (x \in U).$$

Proof. Let $a \in U$ and suppose $h \in \mathbf{R}^n$ is sufficiently small. The derivative of the mapping: $\mathbf{R} \to \mathbf{R}^p$ with $s \mapsto f(a + sh, t)$ is given by $s \mapsto D_1 f(a + sh, t)\, h$; hence we have, according to the Fundamental Theorem 2.10.1,

$$f(a + h, t) - f(a, t) = \int_0^1 D_1 f(a + sh, t)\, ds\, h.$$

Furthermore, the right–hand side depends continuously on $t \in J$. Therefore

$$\begin{aligned} F(a + h) - F(a) &= \int_J (f(a + h, t) - f(a, t))\, dt \\ &= \int_J \int_0^1 D_1 f(a + sh, t)\, ds\, dt\, h =: \phi(a + h)h, \end{aligned}$$

where $\phi : U \to \mathrm{Lin}(\mathbf{R}^n, \mathbf{R}^p)$. Applying the Continuity Theorem 2.10.2 twice we obtain

$$\lim_{h \to 0} \phi(a + h) = \int_J \int_0^1 \lim_{h \to 0} D_1 f(a + sh, t)\, ds\, dt = \int_J D_1 f(a, t)\, dt = \phi(a).$$

On the strength of Hadamard's Lemma 2.2.7 this implies the differentiability of F at a, with derivative $\int_J D_1 f(a, t)\, dt$. ❑

Example 2.10.5. If we use the Differentiation Theorem 2.10.4 for computing F' and F'' where F denotes the function from Example 2.10.3, we obtain

$$\int_0^1 t \sin xt\, dt = \frac{1}{x^2}(\sin x - x \cos x),$$

$$\int_0^1 t^2 \cos xt\, dt = \frac{1}{x^3}((x^2 - 2) \sin x + 2x \cos x).$$

Application of the Continuity Theorem 2.10.2 now gives

$$\lim_{x \to 0} \frac{1}{x^3}((x^2 - 2) \sin x + 2x \cos x) = \frac{1}{3}.$$ ☆

Example 2.10.6. Let $a > 0$ and define $f : [0, a] \times [0, 1] \to \mathbf{R}$ by

$$f(x,t) = \begin{cases} \arctan \dfrac{t}{x}, & 0 < x \le a; \\ \dfrac{\pi}{2}, & x = 0. \end{cases}$$

Then f is bounded everywhere but discontinuous at $(0, 0)$ and $D_1 f(x,t) = -\frac{t}{x^2+t^2}$, which implies that $D_1 f$ is unbounded on $[0, a] \times [0, 1]$ near $(0, 0)$. Therefore the conclusions of the Theorems 2.10.2 and 2.10.4 above do not follow automatically in this case. Indeed, $F(x) = \int_0^1 f(x,t)\,dt$ is given by $F(0) = \frac{\pi}{2}$ and

$$F(x) = \int_0^1 \arctan \frac{t}{x}\,dt = \arctan \frac{1}{x} + \frac{1}{2} x \log \frac{x^2}{1+x^2} \qquad (x \in \mathbf{R}_+).$$

We see that F is continuous at 0 but that F is not differentiable at 0 (use Exercise 0.3 if necessary). Still, one can compute that $F'(x) = \frac{1}{2} \log \frac{x^2}{1+x^2}$, for all $x \in \mathbf{R}_+$, which may be done in two different ways.

Similar problems arise with $G : [0, \infty[\to \mathbf{R}$ given by $G(0) = -2$ and

$$G(x) = \int_0^1 \log(x^2 + t^2)\,dt = \log(1 + x^2) - 2 + 2x \arctan \frac{1}{x} \qquad (x \in \mathbf{R}_+).$$

Then $G'(x) = 2 \arctan \frac{1}{x}$, thus $\lim_{x \downarrow 0} G'(x) = \pi$ while the partial derivative of the integrand with respect to x vanishes for $x = 0$. ☆

In the Differentiation Theorem 2.10.4 we considered the operations of differentiation with respect to a variable x and integration with respect to a variable t, and we showed that under suitable conditions these commute. There are also the operations of differentiation with respect to t and integration with respect to x. The following theorem states the remarkable result that the commutativity of any pair of these operations associated with distinct variables implies the commutativity of the other two pairs.

Theorem 2.10.7. *Let $I = [a, b]$ and $J = [c, d]$ be intervals in $\mathbf{R}$. Then the following assertions are equivalent.*

(i) *Let f and $D_1 f$ be continuous on $I \times J$. Then $x \mapsto \int_J f(x, y)\,dy$ is differentiable, and*

$$D \int_J f(x,y)\,dy = \int_J D_1 f(x,y)\,dy \qquad (x \in \operatorname{int}(I)).$$

(ii) *Let f, $D_2 f$ and $D_1 D_2 f$ be continuous on $I \times J$ and assume $D_1 f(x, c)$ exists for $x \in I$. Then $D_1 f$ and $D_2 D_1 f$ exist on the interior of $I \times J$, and on that interior*

$$D_1 D_2 f = D_2 D_1 f.$$

(iii) **(Interchanging the order of integration).** *Let f be continuous on $I \times J$. Then the functions $x \mapsto \int_J f(x, y)\,dy$ and $y \mapsto \int_I f(x, y)\,dx$ are continuous, and*

$$\int_I \int_J f(x, y)\,dy\,dx = \int_J \int_I f(x, y)\,dx\,dy.$$

Furthermore, all three of these statements are valid.

Proof. For $1 \le i \le 2$, we define I_i and E_i acting on $C(I \times J)$ by

$$I_1 f(x, y) = \int_a^x f(\xi, y)\,d\xi, \qquad I_2 f(x, y) = \int_c^y f(x, \eta)\,d\eta;$$

$$E_1 f(x, y) = f(a, y), \qquad E_2 f(x, y) = f(x, c).$$

The following identities (1) and (2) (where well-defined) are mere reformulations of the Fundamental Theorem 2.10.1, while the remaining ones are straightforward, for $1 \le i \le 2$,

(1) $D_i I_i = I$, (2) $I_i D_i = I - E_i$,

(3) $D_i E_i = 0$, (4) $E_i I_i = 0$,

(5) $D_j E_i = E_i D_j \quad (j \ne i)$, (6) $I_j E_i = E_i I_j \quad (j \ne i)$.

Now we are prepared for the proof of the equivalence; in this proof we write 'ass' for 'assumption'.
(i) ⇒ (ii), that is, $D_1 I_2 = I_2 D_1$ implies $D_1 D_2 = D_2 D_1$.
We have

$$D_1 \underset{(2)}{=} D_1 I_2 D_2 + D_1 E_2 \underset{\text{ass}}{=} I_2 D_1 D_2 + D_1 E_2.$$

This shows that $D_1 f$ exists. Furthermore, $D_1 f$ is differentiable with respect to y and

$$D_2 D_1 \underset{(5)}{=} D_2 I_2 D_1 D_2 + D_2 E_2 D_1 \underset{(1)+(3)}{=} D_1 D_2.$$

(ii) ⇒ (iii), that is, $D_1 D_2 = D_2 D_1$ implies $I_1 I_2 = I_2 I_1$.
The continuity of the integrals follows from Theorem 2.10.2. Using (1) we see that $I_2 I_1 f$ satisfies the conditions imposed on f in (ii) (note that $I_2 I_1 f(x, c) = 0$). Therefore

$$I_1 I_2 \underset{(1)}{=} I_1 I_2 D_1 I_1 \underset{(1)}{=} I_1 I_2 D_1 D_2 I_2 I_1 \underset{\text{ass}}{=} I_1 I_2 D_2 D_1 I_2 I_1 \underset{(2)}{=} I_1 D_1 I_2 I_1 - I_1 E_2 D_1 I_2 I_1$$

$$\underset{(2)+(5)}{=} I_2 I_1 - E_1 I_2 I_1 - I_1 D_1 E_2 I_2 I_1 \underset{(6)+(4)}{=} I_2 I_1 - I_2 E_1 I_1 \underset{(4)}{=} I_2 I_1.$$

(iii) ⇒ (i), that is, $I_1 I_2 = I_2 I_1$ implies $D_1 I_2 = I_2 D_1$.
Observe that $D_1 f$ satisfies the conditions imposed on f in (iii), hence

$$D_1 I_2 \underset{(2)}{=} D_1 I_2 I_1 D_1 + D_1 I_2 E_1 \underset{\text{ass}+(6)}{=} D_1 I_1 I_2 D_1 + D_1 E_1 I_2 \underset{(1)+(3)}{=} I_2 D_1.$$

Apply this identity to f and evaluate at $y = d$.

The Differentiation Theorem 2.10.4 implies the validity of assertion (i), and therefore the assertions (ii) and (iii) also hold. ❑

In assertion (ii) the condition on $D_1 f(x, c)$ is necessary for the following reason. If $f(x, y) = g(x)$, then $D_2 f$ and $D_1 D_2 f$ both equal 0, but $D_1 f$ exists only if g is differentiable. Note that assertion (ii) is a stronger form of Theorem 2.7.2, as the conditions in (ii) are weaker.

In Corollary 6.4.3 we will give a different proof for assertion (iii) that is based on the theory of Riemann integration.

In more intuitive terms the assertions in the theorem above can be put as follows.

(i) A person slices a cake from west to east; then the rate of change of the area of a cake slice equals the average over the slice of the rate of change of the height of the cake.

(ii) A person walks on a hillside and points a flashlight along a tangent to the hill; then the rate at which the beam's slope changes when walking north and pointing east equals its rate of change when walking east and pointing north.

(iii) A person gets just as much cake to eat if he slices it from west to east or from south to north.

Example 2.10.8 (Frullani's integral). Let $0 < a < b$ and define $f : [0, 1] \times [a, b]$ by $f(x, y) = x^y$. Then f is continuous and satisfies

$$\int_0^1 \int_a^b f(x, y)\, dy\, dx = \int_0^1 \int_a^b e^{y \log x}\, dy\, dx = \int_0^1 \frac{x^b - x^a}{\log x}\, dx.$$

On the other hand,

$$\int_a^b \int_0^1 f(x, y)\, dx\, dy = \int_a^b \frac{1}{y+1}\, dy = \log \frac{b+1}{a+1}.$$

Theorem 2.10.7.(iii) now implies

$$\int_0^1 \frac{x^b - x^a}{\log x}\, dx = \log \frac{b+1}{a+1}.$$

Using the substitution $x = e^{-t}$ one deduces *Frullani's integral*

$$\int_{\mathbf{R}_+} \frac{e^{-at} - e^{-bt}}{t}\, dt = \log \frac{b}{a}. \qquad ☆$$

Define $f : \mathbf{R}^2 \to \mathbf{R}$ by $f(x, t) = x e^{-xt}$. Then f is continuous and the improper integral $F(x) = \int_{\mathbf{R}_+} f(x, t)\, dt$ exists, for every $x \geq 0$. Nevertheless, $F : [\,0, \infty\,[\; \to \mathbf{R}$ is discontinuous, since it only assumes the values 0, at 0, and 1, elsewhere on $\mathbf{R}_+$. Without extra conditions the Continuity Theorem 2.10.2 apparently is not valid for improper integrals. In order to clarify the situation consider the rate of convergence of the improper integral. In fact, for $d \in \mathbf{R}_+$ and $x \in \mathbf{R}_+$ we have

$$1 - \int_0^d x e^{-xt}\, dt = e^{-xd},$$

and this only tends to 0 if xd tends to ∞. Hence the convergence becomes slower and slower for x approaching 0. The purpose of the next definition is to rule out this sort of behavior.

Definition 2.10.9. Suppose A is a subset of $\mathbf{R}^n$ and $J = [\, c, \infty\, [$ an unbounded interval in $\mathbf{R}$. Let $f : A \times J \to \mathbf{R}^p$ be a continuous mapping and define $F : A \to \mathbf{R}^p$ by $F(x) = \int_J f(x, t)\, dt$. We say that $\int_J f(x, t)\, dt$ is *uniformly convergent* for $x \in A$ if for every $\epsilon > 0$ there exists $D > c$ such that

$$x \in A \quad \text{and} \quad d \geq D \quad \Longrightarrow \quad \left\| F(x) - \int_c^d f(x, t)\, dt \right\| < \epsilon. \qquad ❍$$

The following lemma gives a useful criterion for uniform convergence; its proof is straightforward.

Lemma 2.10.10 (De la Vallée-Poussin's test). *Suppose A is a subset of $\mathbf{R}^n$ and $J = [\, c, \infty\, [$ an unbounded interval in $\mathbf{R}$. Let $f : A \times J \to \mathbf{R}^p$ be a continuous mapping. Suppose there exists a function $g : J \to [\, 0, \infty\, [$ such that $\| f(x, t) \| \leq g(t)$, for all $(x, t) \in A \times J$, and $\int_J g(t)\, dt$ is convergent. Then $\int_J f(x, t)\, dt$ is uniformly convergent for $x \in A$.*

The test above is only applicable to integrands that are absolutely convergent. In the case of conditional convergence of an integral one might try to transform it into an absolutely convergent integral through integration by parts.

Example 2.10.11. For every $0 < p \leq 1$, we prove the uniform convergence for $x \in I := [\, 0, \infty\, [$ of

$$F_p(x) := \int_{\mathbf{R}_+} f_p(x, t)\, dt := \int_{\mathbf{R}_+} e^{-xt} \frac{\sin t}{t^p}\, dt.$$

Note that the integrand is continuous at 0, hence our concern here is its behavior at ∞. To this end, observe that $\int e^{-xt} \sin t\, dt = -\frac{e^{-xt}}{1+x^2}(\cos t + x \sin t) =: g(x, t)$. In particular, for all $(x, t) \in I \times I$,

$$|g(x, t)| \leq \frac{1 + x}{1 + x^2} \leq \frac{2 + 2x^2}{1 + x^2} = 2.$$

Integration by parts now yields, for all $d < d' \in \mathbf{R}_+$,

$$\int_d^{d'} e^{-xt} \frac{\sin t}{t^p}\, dt = \left[\frac{1}{t^p} g(x,t) \right]_d^{d'} + p \int_d^{d'} \frac{1}{t^{p+1}} g(x,t)\, dt.$$

For $d' \to \infty$, the boundary term converges and the second integral is absolutely convergent. Therefore $F_p(x)$ converges, even for $x = 0$. Furthermore, the uniform convergence now follows from the estimate, valid for all $x \in I$ and $d \in \mathbf{R}_+$,

$$\left| \int_d^{\infty} e^{-xt} \frac{\sin t}{t^p}\, dt \right| \le \frac{2}{d^p} + 2p \int_d^{\infty} \frac{1}{t^{p+1}}\, dt = \frac{4}{d^p}.$$

Note that, in particular, we have proved the convergence of

$$\int_{\mathbf{R}_+} \frac{\sin t}{t^p}\, dt \qquad (0 < p \le 1).$$

Nevertheless, this convergence is only conditional. Indeed, $\int_{\mathbf{R}_+} \frac{\sin t}{t}\, dt$ is not absolutely convergent as follows from

$$\begin{aligned} \int_{\mathbf{R}_+} \frac{|\sin t|}{t}\, dt &= \sum_{k \in \mathbf{N}_0} \int_{k\pi}^{(k+1)\pi} \frac{|\sin t|}{t}\, dt = \sum_{k \in \mathbf{N}_0} \int_0^{\pi} \frac{\sin t}{t + k\pi}\, dt \\ &\ge \sum_{k \in \mathbf{N}_0} \frac{1}{(k+1)\pi} \int_0^{\pi} \sin t\, dt. \end{aligned}$$

☆

Theorem 2.10.12 (Continuity Theorem). *Suppose K is a compact subset of $\mathbf{R}^n$ and $J = [\, c, \infty\, [$ an unbounded interval in $\mathbf{R}$. Let $f : K \times J \to \mathbf{R}^p$ be a continuous mapping. Suppose that the mapping $F : K \to \mathbf{R}^p$ given by $F(x) = \int_J f(x,t)\, dt$ is well-defined and that the integral converges uniformly for $x \in K$. Then F is continuous. In particular we obtain, for $a \in K$,*

$$\lim_{x \to a} \int_J f(x,t)\, dt = \int_J \lim_{x \to a} f(x,t)\, dt.$$

Proof. Let $\epsilon > 0$ be arbitrary and select $d > c$ such that, for every $x \in K$,

$$\left\| F(x) - \int_c^d f(x,t)\, dt \right\| < \frac{\epsilon}{3}.$$

Now apply Theorem 2.10.2 with $f : K \times [\, c, d\,] \to \mathbf{R}^p$ and $a \in K$. This gives the existence of $\delta > 0$ such that we have, for all $x \in K$ with $\|x - a\| < \delta$,

$$\Delta := \left\| \int_c^d f(x,t)\, dt - \int_c^d f(a,t)\, dt \right\| < \frac{\epsilon}{3}.$$

Hence we find, for such x,

$$\begin{aligned}\|F(x) - F(a)\| &\leq \Big\|F(x) - \int_c^d f(x,t)\,dt\Big\| + \Delta \\ &+ \Big\|\int_c^d f(a,t)\,dt - F(a)\Big\| < 3\frac{\epsilon}{3} = \epsilon. \end{aligned}$$ ❑

Theorem 2.10.13 (Differentiation Theorem). *Suppose K is a compact subset of $\mathbf{R}^n$ with nonempty interior U and $J = [\,c, \infty\,[$ an unbounded interval in $\mathbf{R}$. Let $f : K \times J \to \mathbf{R}^p$ be a mapping with the following properties.*

(i) *f is continuous and $\int_J f(x,t)\,dt$ converges.*

(ii) *The total derivative $D_1 f : U \times J \to \mathrm{Lin}(\mathbf{R}^n, \mathbf{R}^p)$ with respect to the variable in U is continuous.*

(iii) *There exists a function $g : J \to [\,0, \infty\,[$ such that $\|D_1 f(x,t)\|_{\mathrm{Eucl}} \leq g(t)$, for all $(x,t) \in U \times J$, and such that $\int_J g(t)\,dt$ is convergent.*

Then $F : U \to \mathbf{R}^p$, given by $F(x) = \int_J f(x,t)\,dt$, is a differentiable mapping satisfying

$$D\int_J f(x,t)\,dt = \int_J D_1 f(x,t)\,dt.$$

Proof. Verify that the proof of the Differentiation Theorem 2.10.4 can be adapted to the current situation. ❑

See Theorem 6.12.4 for a related version of the Differentiation Theorem.

Example 2.10.14. We have the following integral (see Exercises 0.14, 6.60 and 8.19 for other proofs)

$$\int_{\mathbf{R}_+} \frac{\sin t}{t}\,dt = \frac{\pi}{2}.$$

In fact, from Example 2.10.11 we know that $F(x) := F_1(x) = \int_{\mathbf{R}_+} e^{-xt}\frac{\sin t}{t}\,dt$ is uniformly convergent for $x \in I$, and thus the Continuity Theorem 2.10.12 gives the continuity of $F : I \to \mathbf{R}$. Furthermore, let $a > 0$ be arbitrary. Then we have $|D_1 f(x,t)| = |e^{-xt}\sin t| \leq e^{-at}$, for every $x > a$. Accordingly, it follows from De la Vallée-Poussin's test and the Differentiation Theorem 2.10.13 that $F : \,]\,a, \infty\,[\, \to \mathbf{R}$ is differentiable with derivative (see Exercise 0.8 if necessary)

$$F'(x) = -\int_{\mathbf{R}_+} e^{-xt}\sin t\,dt = -\frac{1}{1+x^2}.$$

Since a is arbitrary this implies $F(x) = c - \arctan x$, for all $x \in \mathbf{R}_+$. But a substitution of variables gives $F(\frac{x}{p}) = \int_{\mathbf{R}_+} e^{-xt} \frac{\sin pt}{t}\, dt$, for every $p \in \mathbf{R}_+$. Applying the Continuity Theorem 2.10.12 once again we therefore obtain $c - \frac{\pi}{2} = \lim_{p \downarrow 0} F(\frac{x}{p}) = 0$, which gives $F(x) = \frac{\pi}{2} - \arctan x$, for $x \in \mathbf{R}_+$. Taking $x = 0$ and using the continuity of F we find the equality. ☆

Theorem 2.10.15. *Let $I = [\,a, b\,]$ and $J = [\,c, \infty\,[$ be intervals in $\mathbf{R}$. Let f be continuous on $I \times J$ and let $\int_J f(x, y)\, dy$ be uniformly convergent for $x \in I$. Then*

$$\int_I \int_J f(x, y)\, dy dx = \int_J \int_I f(x, y)\, dx dy.$$

Proof. Let $\epsilon > 0$ be arbitrary. On the strength of the uniform convergence of $\int_J f(x, t) dt$ for $x \in I$ we can find $d > c$ such that, for all $x \in I$,

$$\Big| \int_d^\infty f(x, y)\, dy \Big| < \frac{\epsilon}{b - a}.$$

According to Theorem 2.10.7.(iii) we have

$$\int_c^d \int_I f(x, y)\, dx\, dy = \int_I \int_c^d f(x, y)\, dy\, dx.$$

Therefore

$$\Big| \int_I \int_J f(x, y)\, dy\, dx - \int_c^d \int_I f(x, y)\, dx\, dy \Big|$$

$$= \Big| \int_I \int_d^\infty f(x, y)\, dy\, dx \Big| \le \int_I \frac{\epsilon}{b - a}\, dx = \epsilon. \qquad \square$$

Example 2.10.16. For all $p \in \mathbf{R}_+$ we have according to De la Vallée-Poussin's test that the improper integral $\int_{\mathbf{R}_+} e^{-py} \sin xy\, dy$ is uniformly convergent for $x \in \mathbf{R}$, because $|e^{-py} \sin xy| \le e^{-py}$. Hence for all $0 < a < b$ in $\mathbf{R}$ (see Exercise 0.8 if necessary)

$$\int_{\mathbf{R}_+} e^{-py} \frac{\cos ay - \cos by}{y}\, dy = \int_{\mathbf{R}_+} e^{-py} \int_a^b \sin xy\, dx\, dy$$

$$= \int_a^b \int_{\mathbf{R}_+} e^{-py} \sin xy\, dy\, dx = \int_a^b \frac{x}{p^2 + x^2}\, dx = \frac{1}{2} \log\Big(\frac{p^2 + b^2}{p^2 + a^2}\Big).$$

Application of the Continuity Theorem 2.10.12 now gives

$$\int_{\mathbf{R}_+} \frac{\cos ay - \cos by}{y}\, dy = \log\Big(\frac{b}{a}\Big). \qquad ☆$$

Results like the preceding three theorems can also be formulated for integrands f having a singularity at one of the endpoints of the interval $J = [\,c, d\,]$.

Chapter 3

Inverse Function and Implicit Function Theorems

The main theme in this chapter is that the local behavior of a differentiable mapping, near a point, is qualitatively determined by that of its derivative at the point in question. Diffeomorphisms are differentiable substitutions of variables appearing, for example, in the description of geometrical objects in Chapter 4 and in the Change of Variables Theorem in Chapter 6. Useful criteria, in terms of derivatives, for deciding whether a mapping is a diffeomorphism are given in the Inverse Function Theorems. Next comes the Implicit Function Theorem, which is a fundamental result concerning existence and uniqueness of solutions for a system of n equations in n unknowns in the presence of parameters. The theorem will be applied in the study of geometrical objects; it also plays an important part in the Change of Variables Theorem for integrals.

3.1 Diffeomorphisms

A suitable change of variables, or diffeomorphism (this term is a contraction of differentiable and homeomorphism), can sometimes solve a problem that looks intractable otherwise. Changes of variables, which were already encountered in the integral calculus on **R**, will reappear later in the Change of Variables Theorem 6.6.1.

Example 3.1.1 (Polar coordinates). (See also Example 6.6.4.) Let

$$V = \{ (r, \alpha) \in \mathbf{R}^2 \mid r \in \mathbf{R}_+, \ -\pi < \alpha < \pi \},$$

and let

$$U = \mathbf{R}^2 \setminus \{ (x_1, 0) \in \mathbf{R}^2 \mid x_1 \le 0 \}$$

be the plane excluding the nonpositive part of the x_1-axis. Define $\Psi : V \to U$ by

$$\Psi(r, \alpha) = r\,(\cos\alpha,\ \sin\alpha).$$

Then Ψ is infinitely differentiable. Moreover Ψ is injective, because $\Psi(r, \alpha) = \Psi(r', \alpha')$ implies $r = \|\Psi(r, \alpha)\| = \|\Psi(r', \alpha')\| = r'$; therefore $\alpha \equiv \alpha' \bmod 2\pi$, and so $\alpha = \alpha'$. And Ψ is also surjective; indeed, the inverse mapping $\Phi : U \to V$ assigns to the point $x \in U$ its *polar coordinates* $(r, \alpha) \in V$ and is given by

$$\Phi(x) = \Big(\|x\|,\ \arg\Big(\frac{1}{\|x\|}x\Big)\Big) = \Big(\|x\|,\ 2\arctan\Big(\frac{x_2}{\|x\| + x_1}\Big)\Big).$$

Here $\arg : S^1 \setminus \{(-1, 0)\} \to\ ]-\pi, \pi[$, with $S^1 = \{\, u \in \mathbf{R}^2 \mid \|u\| = 1 \,\}$ the unit circle in $\mathbf{R}^2$, is the *argument function*, i.e. the C^∞ inverse of $\alpha \mapsto (\cos\alpha,\ \sin\alpha)$, given by

$$\arg(u) = 2\arctan\Big(\frac{u_2}{1 + u_1}\Big) \qquad (u \in S^1 \setminus \{(-1, 0)\}).$$

Indeed, $\arg u = \alpha$ if and only if $u = (\cos\alpha,\ \sin\alpha)$, and so (compare with Exercise 0.1)

$$\tan\frac{\alpha}{2} = \frac{\sin\frac{\alpha}{2}}{\cos\frac{\alpha}{2}} = \frac{2\sin\frac{\alpha}{2}\cos\frac{\alpha}{2}}{2\cos^2\frac{\alpha}{2}} = \frac{\sin\alpha}{1 + \cos\alpha} = \frac{u_2}{1 + u_1}.$$

On U, Φ is then a C^∞ mapping. Consequently Ψ is a bijective C^∞ mapping, and so is its inverse Φ. ☆

Definition 3.1.2. Let U and V be open subsets in $\mathbf{R}^n$ and $k \in \mathbf{N}_\infty$. A bijective mapping $\Phi : U \to V$ is said to be a C^k *diffeomorphism* from U to V if $\Phi \in C^k(U, \mathbf{R}^n)$ and $\Psi := \Phi^{-1} \in C^k(V, \mathbf{R}^n)$. If such a Φ exists, U and V are called C^k diffeomorphic. Alternatively, one speaks of the *(regular)* C^k *change of coordinates* $\Phi : U \to V$, where $y = \Phi(x)$ stands for the new coordinates of the point $x \in U$. Obviously, $\Psi : V \to U$ is also a regular C^k change of coordinates.

Now let $\Psi : V \to U$ be a C^k diffeomorphism and $f : U \to \mathbf{R}^p$ a mapping. Then

$$\Psi^* f = f \circ \Psi : V \to \mathbf{R}, \qquad \text{that is} \qquad \Psi^* f(y) = f(\Psi(y)) \qquad (y \in V),$$

is called the mapping obtained from f by *pullback under the diffeomorphism* Ψ. Because $\Psi \in C^k$, the chain rule implies that $\Psi^* f \in C^k(V, \mathbf{R}^p)$ if and only if $f \in C^k(U, \mathbf{R}^p)$ (for "only if", note that $f = (\Psi^* f) \circ \Psi^{-1}$, where $\Psi^{-1} \in C^k$). ❍

In the definition above we assume that U and V are open subsets of the same space $\mathbf{R}^n$. From Example 2.4.9 we see that this is no restriction.

Note that a diffeomorphism Φ is, in particular, a homeomorphism; and therefore Φ is an open mapping, by Proposition 1.5.4. This implies that for every open set

$O \subset U$ the restriction $\Phi|_O$ of Φ to O is a C^k diffeomorphism from O to an open subset of V.

When performing a substitution of variables Ψ one should always clearly distinguish the mappings under consideration, that is, replace $f \in C^k(U, \mathbf{R}^p)$ by $f \circ \Psi \in C^k(V, \mathbf{R}^p)$, even though f and $f \circ \Psi$ assume the same values. Otherwise, absurd results may arise. As an example, consider the substitution $x = \Psi(y) = (y_1, y_1 + y_2)$ in $\mathbf{R}^2$ from the Remark on notation in Section 2.3. If we write $g = f \circ \Psi \in C^1(\mathbf{R}^2, \mathbf{R})$, given $f \in C^1(\mathbf{R}^2, \mathbf{R})$, then $g(y) = f(y_1, y_1 + y_2) = f(x)$. Therefore

$$\frac{\partial g}{\partial y_1}(y) = \frac{\partial f}{\partial x_1}(x) + \frac{\partial f}{\partial x_2}(x), \qquad \frac{\partial g}{\partial y_2}(y) = \frac{\partial f}{\partial x_2}(x),$$

and these formulae become cumbersome if one writes $g = f$.

It is evident that even in the simple case of the substitution of polar coordinates $x = \Psi(r, \alpha)$ in $\mathbf{R}^2$, showing Ψ to be a diffeomorphism is rather laborious, especially if we want to prove the existence (which requires Ψ to be both injective and surjective) and the differentiability of the inverse mapping Φ by explicit calculation of Φ (that is, by solving $x = \Psi(y)$ for y in terms of x). In the following we discuss how a study of the inverse mapping may be replaced by the analysis of the total derivative of the mapping, which is a matter of linear algebra.

3.2 Inverse Function Theorems

Recall Example 2.4.9, which says that the total derivative of a diffeomorphism is always an invertible linear mapping. Our next aim is to find a partial converse of this result. To establish whether a C^1 mapping Φ is locally, near a point a, a C^1 diffeomorphism, it is sufficient to verify that $D\Phi(a)$ is invertible. Thus there is no need to explicitly determine the inverse of Φ (this is often very difficult, if not impossible). Note that this result is a generalization to $\mathbf{R}^n$ of the Inverse Function Theorem on $\mathbf{R}$, which has the following formulation. Let $I \subset \mathbf{R}$ be an open interval and let $f : I \to \mathbf{R}$ be differentiable, with $f'(x) \neq 0$, for every $x \in I$. Then f is a bijection from I onto an open interval J in $\mathbf{R}$, and the inverse function $g : J \to I$ is differentiable with $g'(y) = f'(g(y))^{-1}$, for all $y \in J$.

Lemma 3.2.1. *Suppose that U and V are open subsets of $\mathbf{R}^n$ and that $\Phi : U \to V$ is a bijective mapping with inverse $\Psi := \Phi^{-1} : V \to U$. Assume that Φ is differentiable at $a \in U$ and that Ψ is continuous at $b = \Phi(a) \in V$. Then Ψ is differentiable at b if and only if $D\Phi(a) \in \mathrm{Aut}(\mathbf{R}^n)$, and if this is the case, then $D\Psi(b) = D\Phi(a)^{-1}$.*

Proof. If Ψ is differentiable at b the chain rule immediately gives the invertibility of $D\Phi(a)$ as well as the formula for $D\Psi(b)$.

Let us now assume $D\Phi(a) \in \mathrm{Aut}(\mathbf{R}^n)$. For x near a we put $y = \Phi(x)$, which means $x = \Psi(y)$. Applying Hadamard's Lemma 2.2.7 to Φ we then see

$$y - b = \Phi(x) - \Phi(a) = \phi(x)(x - a),$$

where ϕ is continuous at a and $\det \phi(a) = \det D\Phi(a) \neq 0$ because $D\Phi(a) \in \mathrm{Aut}(\mathbf{R}^n)$. Hence there exists a neighborhood U_1 of a in U such that for $x \in U_1$ we have $\det \phi(x) \neq 0$, and thus $\phi(x) \in \mathrm{Aut}(\mathbf{R}^n)$. From the continuity of Ψ at b we deduce the existence of a neighborhood V_1 of b in V such that $y \in V_1$ gives $\Psi(y) \in U_1$, which implies $\phi(\Psi(y)) \in \mathrm{Aut}(\mathbf{R}^n)$. Because substitution of $x = \Psi(y)$ yields

$$y - b = \Phi(\Psi(y)) - \Phi(a) = (\phi \circ \Psi)(y)(\Psi(y) - a),$$

we know at this stage

$$\Psi(y) - a = (\phi \circ \Psi)(y)^{-1}(y - b) \qquad (y \in V_1).$$

Furthermore, $(\phi \circ \Psi)^{-1} : V_1 \to \mathrm{Aut}(\mathbf{R}^n)$ is continuous at b as this mapping is the composition of the following three maps: $y \mapsto \Psi(y)$ from V_1 to U_1, which is continuous at b; then $x \mapsto \phi(x)$ from $U_1 \to \mathrm{Aut}(\mathbf{R}^n)$, which is continuous at a; and $A \mapsto A^{-1}$ from $\mathrm{Aut}(\mathbf{R}^n)$ into itself, which is continuous according to Formula (2.7). We also obtain from the Substitution Theorem 1.4.2.(i)

$$\lim_{y \to b} (\phi \circ \Psi)(y)^{-1} = \lim_{x \to a} \phi(x)^{-1} = D\Phi(a)^{-1}.$$

But then Hadamard's Lemma 2.2.7 says that Ψ is differentiable at b, with derivative $D\Psi(b) = D\Phi(a)^{-1}$. ❑

Next we formulate a global version of the lemma above.

Proposition 3.2.2. *Suppose that U and V are open subsets of $\mathbf{R}^n$ and that $\Phi : U \to V$ is a homeomorphism of class C^1. Then Φ is a C^1 diffeomorphism if and only if $D\Phi(a) \in \mathrm{Aut}(\mathbf{R}^n)$, for all $a \in U$.*

Proof. The condition on $D\Phi$ is obviously necessary. On the other hand, if the condition is satisfied, it follows from Lemma 3.2.1 that $\Psi := \Phi^{-1} : V \to U$ is differentiable at every $b \in V$. It remains to be shown that $\Psi \in C^1(V, U)$, that is, that the following mapping is continuous (cf. Lemma 2.1.2)

$$D\Psi : V \to \mathrm{Aut}(\mathbf{R}^n) \qquad \text{with} \qquad D\Psi(b) = D\Phi(\Psi(b))^{-1}.$$

This continuity follows as the map is the composition of the following three maps: $b \mapsto \Psi(b)$ from V to U, which is continuous as Φ is a homeomorphism; $a \mapsto D\Phi(a)$ from U to $\mathrm{Aut}(\mathbf{R}^n)$, which is continuous as Φ is a C^1 mapping; and the continuous mapping $A \mapsto A^{-1}$ from $\mathrm{Aut}(\mathbf{R}^n)$ into itself. ❑

It is remarkable that the conditions in Proposition 3.2.2 can be weakened, at least locally. The requirement that Φ be a homeomorphism can be replaced by that of Φ being a C^1 mapping. Under that assumption we still can show that Φ is **locally** bijective with a differentiable local inverse. The main problem is to establish surjectivity of Φ onto a neighborhood of $\Phi(a)$, which may be achieved by means of the Contraction Lemma 1.7.2.

Proposition 3.2.3. *Suppose that U_0 and V_0 are open subsets of $\mathbf{R}^n$ and that $\Phi \in C^1(U_0, V_0)$. Assume that $D\Phi(a) \in \mathrm{Aut}(\mathbf{R}^n)$ for some $a \in U_0$. Then there exist open neighborhoods U of a in U_0 and V of $b = \Phi(a)$ in V_0 such that $\Phi|_U : U \to V$ is a homeomorphism.*

Proof. By going over to the mapping $D\Phi(a)^{-1} \circ \Phi$ instead of Φ, we may assume that $D\Phi(a) = I$, the identity. And by going over to $x \mapsto \Phi(x + a) - b$ we may also suppose $a = 0$ and $b = \Phi(a) = 0$.

We define $\Xi : U_0 \to \mathbf{R}^n$ by $\Xi(x) = x - \Phi(x)$; then $\Xi(0) = 0$ and $D\Xi(0) = 0$. So, by the continuity of $D\Xi$ at 0 there exists $\delta > 0$ such that

$$B = \{ x \in \mathbf{R}^n \mid \|x\| \leq \delta \} \qquad \text{satisfies} \qquad B \subset U_0, \qquad \|D\Xi(x)\|_{\mathrm{Eucl}} \leq \frac{1}{2},$$

for all $x \in B$. From the Mean Value Theorem 2.5.3, we see that $\|\Xi(x)\| \leq \frac{1}{2}\|x\|$, for $x \in B$; and thus Ξ maps B into $\frac{1}{2}B$. We now claim: Φ is surjective onto $\frac{1}{2}B$. More precisely, given $y \in \frac{1}{2}B$, there exists a unique $x \in B$ satisfying $\Phi(x) = y$. For showing this, consider the mapping

$$\Xi_y : U_0 \to \mathbf{R}^n \qquad \text{with} \qquad \Xi_y(x) = y + \Xi(x) = x + y - \Phi(x) \qquad (y \in \mathbf{R}^n).$$

Obviously, x is a fixed point of Ξ_y if and only if x satisfies $\Phi(x) = y$. For $y \in \frac{1}{2}B$ and $x \in B$ we have $\Xi(x) \in \frac{1}{2}B$ and thus $\Xi_y(x) \in B$; hence, Ξ_y may be viewed as a mapping of the closed set B into itself. The bound of $\frac{1}{2}$ on its derivative together with the Mean Value Theorem shows that $\Xi_y : B \to B$ is a contraction with contraction factor $\leq \frac{1}{2}$. By the Contraction Lemma 1.7.2, it follows that Ξ_y has a unique fixed point $x \in B$, that is, $\Phi(x) = y$. This proves the claim. Moreover, the estimate in the Contraction Lemma gives $\|x\| \leq 2\|y\|$.

Let $V = \mathrm{int}(\frac{1}{2}B)$, then V is an open neighborhood of $0 = \Phi(a)$ in V_0. Next, define the local inverse $\Psi : V \to B$ by $\Psi(y) = x$ if $y = \Phi(x)$. This inverse is continuous, because $x = \Phi(x) + \Xi(x)$ implies

$$\|x - x'\| \leq \|\Phi(x) - \Phi(x')\| + \|\Xi(x) - \Xi(x')\| \leq \|\Phi(x) - \Phi(x')\| + \frac{1}{2}\|x - x'\|,$$

and hence, for y and $y' \in V$,

$$\|\Psi(y) - \Psi(y')\| \leq 2\|y - y'\|.$$

Set $U = \Psi(V)$. Then U equals the inverse image $\Phi^{-1}(V)$. As Φ is continuous and V is open, it follows that U is an open neighborhood of $0 = a \in U_0$. It is clear now that the mappings $\Phi : U \to V$ and $\Psi : V \to U$ are bijective inverses of each other; as they are continuous, they are homeomorphisms. ❑

A proof of the proposition above that does not require the Contraction Lemma, but uses Theorem 1.8.8 instead, can be found in Exercise 3.23.

Theorem 3.2.4 (Local Inverse Function Theorem). *Let $U_0 \subset \mathbf{R}^n$ be open and $a \in U_0$. Let $\Phi : U_0 \to \mathbf{R}^n$ be a C^1 mapping and $D\Phi(a) \in \mathrm{Aut}(\mathbf{R}^n)$. Then there exists an open neighborhood U of a contained in U_0 such that $V := \Phi(U)$ is an open subset of $\mathbf{R}^n$, and*

$$\Phi|_U : U \to V \quad \textit{is a } C^1 \textit{ diffeomorphism.}$$

Proof. Apply Proposition 3.2.3 to find open sets U and V as in that proposition. By shrinking U and V further if necessary, we may assume that $D\Phi(x) \in \mathrm{Aut}(\mathbf{R}^n)$ for all $x \in U$. Indeed, $\mathrm{Aut}(\mathbf{R}^n)$ is open in $\mathrm{End}(\mathbf{R}^n)$ according to Lemma 2.1.2, and therefore its inverse image under the continuous mapping $D\Phi$ is open. The conclusion now follows from Proposition 3.2.2. ❑

Example 3.2.5. Let $\Phi : \mathbf{R} \to \mathbf{R}$ be given by $\Phi(x) = x^2$. Then $\Phi(0) = 0$ and $D\Phi(0) = 0 \notin \mathrm{Aut}(\mathbf{R})$. For every open subset U of $\mathbf{R}$ containing 0 one has $\Phi(U) \subset [\,0, \infty[$; and this proves that $\Phi(U)$ cannot be an open neighborhood of $\Phi(0)$. In addition, there does not exist an open neighborhood U of 0 such that $\Phi|_U$ is injective. ☆

Definition 3.2.6. Let $U \subset \mathbf{R}^n$ be an open subset, $x \in U$ and $\Phi \in C^1(U, \mathbf{R}^n)$. Then Φ is called *regular* or *singular*, respectively, at x, and x is called a *regular* or a *singular* or *critical* point of Φ, respectively, depending on whether

$$D\Phi(x) \in \mathrm{Aut}(\mathbf{R}^n), \qquad \text{or} \qquad \notin \mathrm{Aut}(\mathbf{R}^n). \qquad ○$$

Note that in the present case the three following statements are equivalent.

(i) Φ is regular at x.

(ii) $\det D\Phi(x) \neq 0$.

(iii) $\operatorname{rank} D\Phi(x) := \dim \operatorname{im} D\Phi(x)$ is maximal.

Furthermore, $U_{\text{reg}} := \{ x \in U \mid \Phi$ regular at $x \}$ is an open subset of U.

Corollary 3.2.7. *Let U be open in $\mathbf{R}^n$ and let $f \in C^1(U, \mathbf{R}^n)$ be regular on U, then f is an open mapping (see Definition 1.5.2).*

Note that a mapping f as in the corollary is locally injective, that is, every point in U has a neighborhood in which f is injective. But f need not be injective on U under these circumstances.

Theorem 3.2.8 (Global Inverse Function Theorem). *Let U be open in $\mathbf{R}^n$ and $\Phi \in C^1(U, \mathbf{R}^n)$. Then*

$$V = \Phi(U) \quad \text{is open in } \mathbf{R}^n \qquad \text{and} \qquad \Phi : U \to V \quad \text{is a } C^1 \text{ diffeomorphism}$$

if and only if

$$\Phi \quad \text{is injective and regular on} \quad U. \tag{3.1}$$

Proof. Condition (3.1) is obviously necessary. Conversely assume the condition is met. This implies that Φ is an open mapping; in particular, $V = \Phi(U)$ is open in $\mathbf{R}^n$. Moreover, Φ is a bijection which is both continuous and open; thus Φ is a homeomorphism according to Proposition 1.5.4. Therefore Proposition 3.2.2 gives that Φ is a C^1 diffeomorphism. ❑

We now come to the Inverse Function Theorems for a mapping Φ that is of class C^k, for $k \in \mathbf{N}_\infty$, instead of class C^1. In that case we have the conclusion that Φ is a C^k diffeomorphism, that is, the inverse mapping is a C^k mapping too. This is an immediate consequence of the following:

Proposition 3.2.9. *Let U and V be open subsets of $\mathbf{R}^n$ and let $\Phi : U \to V$ be a C^1 diffeomorphism. Let $k \in \mathbf{N}_\infty$ and suppose that Φ is a C^k mapping. Then the inverse of Φ is a C^k mapping and therefore Φ itself is a C^k diffeomorphism.*

Proof. Use mathematical induction over $k \in \mathbf{N}$. For $k = 1$ the assertion is a tautology; therefore, assume it holds for $k - 1$. Now, from the chain rule (see Example 2.4.9) we know, writing $\Psi = \Phi^{-1}$,

$$D\Psi : V \to \operatorname{Aut}(\mathbf{R}^n) \qquad \text{satisfies} \qquad D\Psi(y) = D\Phi(\Psi(y))^{-1} \qquad (y \in V).$$

This shows that $D\Psi$ is the composition of the C^{k-1} mapping $\Psi : V \to U$, the C^{k-1} mapping $D\Phi : U \to \operatorname{Aut}(\mathbf{R}^n)$, and the mapping $A \mapsto A^{-1}$ from $\operatorname{Aut}(\mathbf{R}^n)$ into itself, which is even C^∞, as follows from Cramer's rule (2.6) (see also Exercise 2.45). Hence, as the composition of C^{k-1} mappings $D\Psi$ is of class C^{k-1}, which implies that Ψ is a C^k mapping. ❑

Note the similarity of this proof to that of Proposition 3.2.2.

In subsequent parts of the book, in the theory as well as in the exercises, Φ and Ψ will often be used on an equal basis to denote a diffeomorphism. Nevertheless, the choice usually is dictated by the final application: is the old variable x linked to the new variable y by means of the substitution Ψ, that is $x = \Psi(y)$; or does y arise as the image of x under the mapping Φ, i.e. $y = \Phi(x)$?

3.3 Applications of Inverse Function Theorems

Application A. (See also Example 6.6.6.) Define

$$\Phi : \mathbf{R}^2_+ \to \mathbf{R}^2 \qquad \text{by} \qquad \Phi(x) = (x_1^2 - x_2^2,\ 2x_1x_2).$$

Then Φ is an injective C^∞ mapping. Indeed, for $x \in \mathbf{R}^2_+$ and $\widetilde{x} \in \mathbf{R}^2_+$ with $\Phi(x) = \Phi(\widetilde{x})$ one has

$$x_1^2 - x_2^2 = \widetilde{x}_1^2 - \widetilde{x}_2^2 \qquad \text{and} \qquad x_1x_2 = \widetilde{x}_1\widetilde{x}_2.$$

Therefore

$$(\widetilde{x}_1^2 + x_2^2)(\widetilde{x}_2^2 - x_2^2) = \widetilde{x}_1^2\widetilde{x}_2^2 - x_2^2(\widetilde{x}_1^2 - \widetilde{x}_2^2) - x_2^4 = x_1^2x_2^2 - x_2^2(x_1^2 - x_2^2) - x_2^4 = 0.$$

Because $\widetilde{x}_1^2 + x_2^2 \neq 0$, it follows that $x_2^2 = \widetilde{x}_2^2$, and so $x_2 = \widetilde{x}_2$; hence also $x_1 = \widetilde{x}_1$. Now choose

$$\begin{aligned} U &= \{\, x \in \mathbf{R}^2_+ \mid 1 < x_1^2 - x_2^2 < 9,\ 1 < 2x_1x_2 < 4 \,\}, \\ V &= \{\, y \in \mathbf{R}^2 \mid 1 < y_1 < 9,\ 1 < y_2 < 4 \,\}. \end{aligned}$$

It then follows that $\Phi : U \to V$ is a bijective C^∞ mapping. Additionally, for $x \in U$,

$$\det D\Phi(x) = \det\begin{pmatrix} 2x_1 & -2x_2 \\ 2x_2 & 2x_1 \end{pmatrix} = 4\|x\|^2 \neq 0.$$

According to the Global Inverse Function Theorem $\Phi : U \to V$ is a C^∞ diffeomorphism; but the same then is true of $\Psi := \Phi^{-1} : V \to U$. Here Ψ is the regular C^∞ coordinate transformation with $x = \Psi(y)$; in other words, Ψ assigns to a given $y \in V$ the unique solution $x \in U$ of the equations

$$x_1^2 - x_2^2 = y_1, \qquad 2x_1x_2 = y_2.$$

Note that

$$\|x\|^2 = \sqrt{x_1^4 + 2x_1^2x_2^2 + x_2^4} = \sqrt{x_1^4 - 2x_1^2x_2^2 + x_2^4 + 4x_1^2x_2^2} = \sqrt{y_1^2 + y_2^2} = \|y\|.$$

This also implies

$$\det D\Psi(y) = (\det D\Phi(x))^{-1} = \frac{1}{4\|x\|^2} = \frac{1}{4\|y\|}.$$

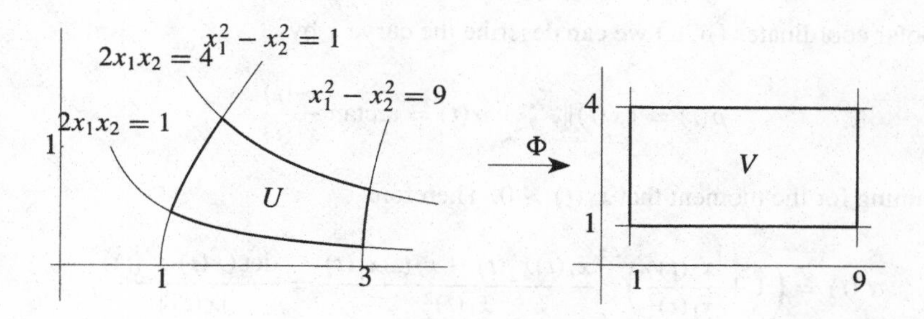

Illustration for Application 3.3.A

Application B. (See also Example 6.6.8.) Verification of the conditions for the Global Inverse Function Theorem may be complicated by difficulties in proving the injectivity. This is the case in the following problem.

Let $x : I \to \mathbf{R}^2$ be a C^k curve, for $k \in \mathbf{N}$, in the plane such that

$$x(t) = (x_1(t), x_2(t)) \qquad \text{and} \qquad x'(t) = (x_1'(t), x_2'(t))$$

are linearly independent vectors in $\mathbf{R}^2$, for every $t \in I$. Prove that for every $s \in I$ there is an open interval J around s in I such that

$$\Psi :]0, 1[\times J \to \mathbf{R}^2 \qquad \text{with} \qquad \Psi(r, t) = r \cdot x(t)$$

is a C^k diffeomorphism onto the image P_J. The image P_J under Ψ is called the *area swept out* during the interval of time J.

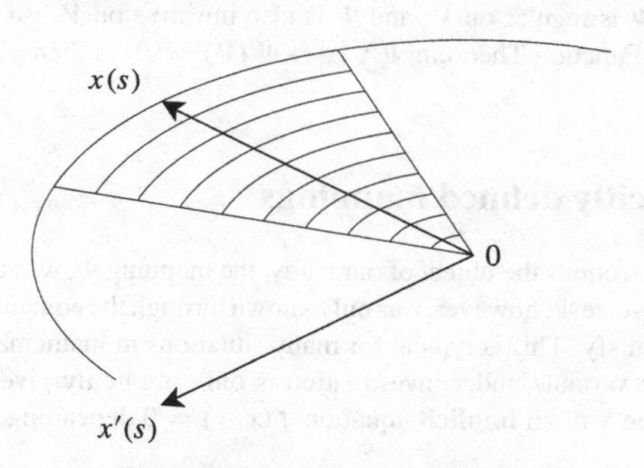

Illustration for Application 3.3.B

Indeed, because $x(t)$ and $x'(t)$ are linearly independent, it follows in particular that $x(t) \neq 0$ and

$$\det\,(x(t)\;\, x'(t)) = x_1(t)x_2'(t) - x_2(t)x_1'(t) \neq 0.$$

In polar coordinates (ρ, α) we can describe the curve x by

$$\rho(t) = \|x(t)\|, \qquad \alpha(t) = \arctan \frac{x_2(t)}{x_1(t)},$$

assuming for the moment that $x_1(t) > 0$. Therefore

$$\alpha'(t) = \left(1 + \frac{x_2(t)^2}{x_1(t)^2}\right)^{-1} \frac{x_1(t)x_2'(t) - x_2(t)x_1'(t)}{x_1(t)^2} = \frac{\det\,(x(t)\ x'(t))}{\|x(t)\|^2}.$$

And this equality is also obtained if we take for $\alpha(t)$ the expression from Example 3.1.1. As a consequence $\alpha'(t) \neq 0$ always, which means that the mapping $t \mapsto \alpha(t)$ is strictly monotonic, and therefore injective. Next, consider the mapping $\Psi :]\,0, 1\,[\times I \to \mathbf{R}^2$ with $\Psi(r, t) = r \cdot x(t)$. Let $s \in I$ be fixed; consider t_1, t_2 sufficiently close to s and such that, for certain r_1, r_2

$$\Psi(r_1, t_1) = \Psi(r_2, t_2), \qquad \text{or} \qquad r_1 \cdot x(t_1) = r_2 \cdot x(t_2).$$

It follows that $\alpha(t_1) = \alpha(t_2)$, initially modulo a multiple of 2π, but since t_1 and t_2 are near each other we conclude $\alpha(t_1) = \alpha(t_2)$. From the injectivity of α follows $t_1 = t_2$, and therefore also $r_1 = r_2$. In other words, there exists an open interval J around s in I such that $\Psi :]\,0, 1\,[\times J \to \mathbf{R}^2$ is injective. Also, for all $(r, t) \in]\,0, 1\,[\times J =: V$,

$$\det D\Psi(r, t) = \det \begin{pmatrix} x_1(t) & r x_1'(t) \\ x_2(t) & r x_2'(t) \end{pmatrix} = r \det\,(x(t)\ x'(t)) \neq 0.$$

Consequently Ψ is regular on V, and Ψ is also injective on V. According to the Global Inverse Function Theorem $\Psi : V \to \Psi(V) = P_J$ is then a C^k diffeomorphism.

3.4 Implicitly defined mappings

In the previous sections the object of our study, the mapping Φ, was usually **explicitly** given; its inverse Ψ, however, was only known through the equation $\Phi\Psi - I = 0$ that it should satisfy. This is typical for many situations in mathematics, when the mapping or the variable under investigation is only implicitly given, as a (hypothetical) solution x of an **implicit** equation $f(x, y) = 0$ depending on parameters y.

A representative example is the polynomial equation $p(x) = \sum_{0 \le i \le n} a_i x^i = 0$. Here we want to know the dependence of the solution x on the coefficients a_i. Many complications arise: the problem usually is underdetermined, it involves more variables than there are equations; is there any solution at all in $\mathbf{R}$; and what about the different solutions? Our next goal is the development of the standard tool for handling such situations: the Implicit Function Theorem.

(A) The problem

Consider a system of n *equations* that depend on p *parameters* $y_1, \dots, y_p$ in $\mathbf{R}$, for n *unknowns* $x_1, \dots, x_n$ in $\mathbf{R}$:

$$\begin{aligned} f_1(x_1, \dots, x_n; y_1, \dots, y_p) &= 0, \\ \vdots \qquad\qquad & \ \vdots \ \ \vdots \\ f_n(x_1, \dots, x_n; y_1, \dots, y_p) &= 0. \end{aligned}$$

A condensed notation for this is

$$f(x; y) = 0, \tag{3.2}$$

where

$$f : \mathbf{R}^n \times \mathbf{R}^p \supset\!\to \mathbf{R}^n, \qquad x = (x_1, \dots, x_n) \in \mathbf{R}^n, \qquad y = (y_1, \dots, y_p) \in \mathbf{R}^p.$$

Assume that for a special value $y^0 = (y_1^0, \dots, y_p^0)$ of the parameter there exists a solution $x^0 = (x_1^0, \dots, x_n^0)$ of $f(x; y) = 0$; i.e.

$$f(x^0; y^0) = 0.$$

It would then be desirable to establish that, for y near y^0, there also exists a solution x near x^0 of $f(x; y) = 0$. If this is the case, we obtain a mapping

$$\psi : \mathbf{R}^p \supset\!\to \mathbf{R}^n \qquad \text{with} \qquad \psi(y) = x \qquad \text{and} \qquad f(x; y) = 0.$$

This mapping ψ, which assigns to parameters y near y^0 the unique solution $x = \psi(y)$ of $f(x; y) = 0$, is called the *function implicitly defined by* $f(x; y) = 0$. If f depends on x and y differentiably, one also expects ψ to depend on y differentiably.

The Implicit Function Theorem 3.5.1 decides the validity of these assertions as regards:

- existence,
- uniqueness,
- differentiable dependence on parameters,

of the solutions $x = \psi(y)$ of $f(x; y) = 0$, for y near y^0.

(B) An idea about the solution

Assume f to be a differentiable function of (x, y), and in particular therefore f to be defined on an open set. From the definition of differentiability:

$$\begin{aligned} f(x; y) - f(x^0; y^0) &= Df(x^0; y^0)\begin{pmatrix} x - x^0 \\ y - y^0 \end{pmatrix} + R(x, y) \\ &= \big(D_x f(x^0; y^0) \ \ D_y f(x^0; y^0) \big)\begin{pmatrix} x - x^0 \\ y - y^0 \end{pmatrix} + R(x, y) \\ &= D_x f(x^0; y^0)(x - x^0) + D_y f(x^0; y^0)(y - y^0) + R(x, y). \end{aligned} \tag{3.3}$$

Here

$$D_x f(x^0; y^0) \in \mathrm{End}(\mathbf{R}^n), \qquad \text{and} \qquad D_y f(x^0; y^0) \in \mathrm{Lin}(\mathbf{R}^p, \mathbf{R}^n),$$

denote the derivatives of f with respect to the variable $x \in \mathbf{R}^n$, and $y \in \mathbf{R}^p$; in Jacobi's notation their matrices with respect to the standard bases are, respectively,

$$\begin{pmatrix} \dfrac{\partial f_1}{\partial x_1}(x^0; y^0) & \cdots & \dfrac{\partial f_1}{\partial x_n}(x^0; y^0) \\ \vdots & & \vdots \\ \dfrac{\partial f_n}{\partial x_1}(x^0; y^0) & \cdots & \dfrac{\partial f_n}{\partial x_n}(x^0; y^0) \end{pmatrix},$$

$$\begin{pmatrix} \dfrac{\partial f_1}{\partial y_1}(x^0; y^0) & \cdots & \dfrac{\partial f_1}{\partial y_p}(x^0; y^0) \\ \vdots & & \vdots \\ \dfrac{\partial f_n}{\partial y_1}(x^0; y^0) & \cdots & \dfrac{\partial f_n}{\partial y_p}(x^0; y^0) \end{pmatrix}.$$

For R the usual estimates from Formula (2.10) hold:

$$\lim_{(x,y)\to(x^0,y^0)} \frac{\|R(x, y)\|}{\|(x - x^0, y - y^0)\|} = 0,$$

or alternatively, for every $\epsilon > 0$ there exists a $\delta > 0$ such that

$$\|R(x, y)\| \le \epsilon(\|x - x^0\|^2 + \|y - y^0\|^2)^{1/2},$$

provided $\|x - x^0\|^2 + \|y - y^0\|^2 < \delta^2$. Therefore, given y near y^0, the existence of a solution x near x^0 requires

$$D_x f(x^0; y^0)(x - x^0) + D_y f(x^0; y^0)(y - y^0) + R(x, y) = 0. \tag{3.4}$$

The Implicit Function Theorem then asserts that a unique solution $x = \psi(y)$ of Equation (3.4), and therefore of Equation (3.2), exists if the *linearized problem*

$$D_x f(x^0; y^0)(x - x^0) + D_y f(x^0; y^0)(y - y^0) = 0$$

can be solved for x, in other words if $D_x f(x^0; y^0) \in \mathrm{End}(\mathbf{R}^n)$ is an invertible mapping. Or, rephrasing yet again, if

$$D_x f(x^0; y^0) \in \mathrm{Aut}(\mathbf{R}^n). \tag{3.5}$$

(C) Formula for the derivative of the solution

If we now also know that $\psi : \mathbf{R}^p \supset\!\!\to \mathbf{R}^n$ is differentiable, we have the composition of differentiable mappings

$$\mathbf{R}^p \supset\!\!\to \mathbf{R}^n \times \mathbf{R}^p \to \mathbf{R}^n \qquad \text{given by} \qquad y \overset{\psi \times I}{\longmapsto} (\psi(y), y) \overset{f}{\mapsto} f(\psi(y), y) = 0.$$

In this case application of the chain rule leads to

$$Df(\psi(y), y) \circ D(\psi \times I)(y) = 0, \tag{3.6}$$

or, more explicitly,

$$\begin{aligned} &(\, D_x f(\psi(y), y) \;\; D_y f(\psi(y), y)) \begin{pmatrix} D\psi(y) \\ I \end{pmatrix} \\ &\quad = D_x f(\psi(y), y)\, D\psi(y) + D_y f(\psi(y), y) = 0. \end{aligned}$$

In other words, we have the identity of mappings in $\mathrm{Lin}(\mathbf{R}^p, \mathbf{R}^n)$

$$D_x f(\psi(y), y)\, D\psi(y) = -D_y f(\psi(y), y).$$

Remarkably, under the same condition (3.5) as before we obtain a formula for the derivative $D\psi(y^0) \in \mathrm{Lin}(\mathbf{R}^p, \mathbf{R}^n)$ at y^0 of the implicitly defined function ψ:

$$D\psi(y^0) = -D_x f(\psi(y^0); y^0)^{-1} \circ D_y f(\psi(y^0); y^0).$$

(D) The conditions are necessary

To show that problems may arise if condition (3.5) is not met, we consider the equation

$$f(x; y) = x^2 - y = 0.$$

Let $y^0 > 0$ and let x^0 be a solution > 0 (e.g. $y^0 = 1, x^0 = 1$). For $y > 0$ sufficiently close to y^0, there exists precisely one solution x near x^0, and this solution $x = \sqrt{y}$ depends differentiably on y. Of course, if $x^0 < 0$, then we have the unique solution $x = -\sqrt{y}$.

This should be compared with the situation for $y^0 = 0$, with the solution $x^0 = 0$. Note that

$$D_x f(0; 0) = 2x \;|_{x=0} = 0.$$

For y near $y^0 = 0$ with $y > 0$ there are **two** solutions x near $x^0 = 0$ (i.e. $x = \pm\sqrt{y}$), whereas for y near y^0 with $y < 0$ there is **no** solution $x \in \mathbf{R}$. Furthermore, the positive and negative solutions both depend nondifferentiably on $y \geq 0$ at $y = y^0$, since

$$\lim_{y \downarrow 0} \frac{\pm\sqrt{y}}{y} = \pm\infty.$$

In addition

$$\lim_{y \downarrow 0} \psi'(y) = \lim_{y \downarrow 0} \pm \frac{1}{2\sqrt{y}} = \pm\infty.$$

In other words, the two solutions obtained for positive y collide, their speed increasing without bound as $y \downarrow 0$; next, after y has passed through 0 to become negative, they have vanished.

3.5 Implicit Function Theorem

The proof of the following theorem is by reduction to a situation that can be treated by means of the Local Inverse Function Theorem. This is done by adding (dummy) equations so that the number of equations equals the number of variables. For another proof of the Implicit Function Theorem see Exercise 3.28.

Theorem 3.5.1 (Implicit Function Theorem). *Let $k \in \mathbf{N}_\infty$, let W be open in $\mathbf{R}^n \times \mathbf{R}^p$ and $f \in C^k(W, \mathbf{R}^n)$. Assume*

$$(x^0, y^0) \in W, \qquad f(x^0; y^0) = 0, \qquad D_x f(x^0; y^0) \in \mathrm{Aut}(\mathbf{R}^n).$$

Then there exist open neighborhoods U of x^0 in $\mathbf{R}^n$ and V of y^0 in $\mathbf{R}^p$ with the following properties:

$$\textit{for every} \quad y \in V \quad \textit{there exists a unique} \quad x \in U \qquad \textit{with} \qquad f(x; y) = 0.$$

In this way we obtain a C^k mapping: $\mathbf{R}^p \supset\!\!\to \mathbf{R}^n$ satisfying

$$\psi : V \to U \qquad \text{with} \qquad \psi(y) = x \qquad \text{and} \qquad f(x; y) = 0,$$

which is uniquely determined by these properties. Furthermore, the derivative $D\psi(y) \in \mathrm{Lin}(\mathbf{R}^p, \mathbf{R}^n)$ of ψ at y is given by

$$D\psi(y) = -D_x f(\psi(y); y)^{-1} \circ D_y f(\psi(y); y) \qquad (y \in V).$$

Proof. Define $\Phi \in C^k(W, \mathbf{R}^n \times \mathbf{R}^p)$ by

$$\Phi(x, y) = (f(x, y), y).$$

Then $\Phi(x^0, y^0) = (0, y^0)$, and we have the matrix representation

$$D\Phi(x, y) = \begin{pmatrix} D_x f(x, y) & D_y f(x, y) \\ 0 & I \end{pmatrix}.$$

From $D_x f(x^0, y^0) \in \mathrm{Aut}(\mathbf{R}^n)$ we obtain $D\Phi(x^0, y^0) \in \mathrm{Aut}(\mathbf{R}^n \times \mathbf{R}^p)$. According to the Local Inverse Function Theorem 3.2.4 there exist an open neighborhood W_0 of (x^0, y^0) contained in W and an open neighborhood V_0 of $(0, y^0)$ in $\mathbf{R}^n \times \mathbf{R}^p$ such that $\Phi : W_0 \to V_0$ is a C^k diffeomorphism. Let U and V_1 be open neighborhoods of x^0 and y^0 in $\mathbf{R}^n$ and $\mathbf{R}^p$, respectively, such that $U \times V_1 \subset W_0$. Then Φ is a diffeomorphism of $U \times V_1$ onto an open subset of V_0, so that by restriction of Φ we may assume that $U \times V_1$ is W_0. Denote by $\Psi : V_0 \to W_0$ the inverse C^k diffeomorphism, which is of the form

$$\Psi(z, y) = (\psi(z, y), y) \qquad ((z, y) \in V_0),$$

for a C^k mapping $\psi : V_0 \to \mathbf{R}^n$. Note that we have the equivalence

$$(x, y) \in W_0 \quad \text{and} \quad f(x, y) = z \qquad \Longleftrightarrow \qquad (z, y) \in V_0 \quad \text{and} \quad \psi(z, y) = x.$$

In these relations, take $z = 0$. Now define the subset V of $\mathbf{R}^p$ containing y^0 by $V = \{ y \in V_1 \mid (0, y) \in V_0 \}$. Then V is open as it is the inverse image of the open set V_0 under the continuous mapping $y \mapsto (0, y)$. If, with a slight abuse of notation, we define $\psi : V \to \mathbf{R}^n$ by $\psi(y) = \psi(0, y)$, then ψ is a C^k mapping satisfying $f(x, y) = 0$ if and only if $x = \psi(y)$. Now, obviously,

$$y \in V \quad \text{and} \quad x = \psi(y) \quad \Longrightarrow \quad x \in U, \quad (x, y) \in W \quad \text{and} \quad f(x, y) = 0. \ \square$$

3.6 Applications of the Implicit Function Theorem

Application A (Simple zeros are C^∞ functions of coefficients). We now investigate how the zeros of a polynomial function $p : \mathbf{R} \to \mathbf{R}$ with

$$p(x) = \sum_{0 \le i \le n} a_i x^i,$$

depend on the parameter $a = (a_0, \dots, a_n) \in \mathbf{R}^{n+1}$ formed by the coefficients of p.

Let c be a zero of p. Then there exists a polynomial function q on $\mathbf{R}$ such that

$$p(x) = p(x) - p(c) = \sum_{0 \le i \le n} a_i (x^i - c^i) = (x - c)\, q(x) \qquad (x \in \mathbf{R}).$$

We say that c is a *simple zero* of p if $q(c) \neq 0$. (Indeed, if $q(c) = 0$, the same argument gives $q(x) = (x - c)\, r(x)$; and so $p(x) = (x - c)^2\, r(x)$, with r another polynomial function.) Using the identity

$$p'(x) = q(x) + (x - c) q'(x)$$

we then see that c is a simple zero of p if and only if $p'(c) \neq 0$.

We now define

$$f : \mathbf{R} \times \mathbf{R}^{n+1} \to \mathbf{R} \qquad \text{by} \qquad f(x; a) = f(x; a_0, \dots, a_n) = \sum_{0 \le i \le n} a_i x^i.$$

Then f is a C^∞ function, while $x \in \mathbf{R}$ is the unknown and $a = (a_0, \dots, a_n) \in \mathbf{R}^{n+1}$ the parameter. Further assume that c^0 is a simple zero of the polynomial function p^0 corresponding to the special value $a^0 = (a_0^0, \dots, a_n^0) \in \mathbf{R}^{n+1}$; we then have

$$f(c^0; a^0) = 0, \qquad D_x f(c^0; a^0) = (p^0)'(c^0) \neq 0.$$

By the Implicit Function Theorem there exist numbers $\eta > 0$ and $\delta > 0$ such that a polynomial function p corresponding to values $a = (a_0, \dots, a_n) \in \mathbf{R}^{n+1}$ of the parameter near a^0, that is, satisfying $|a_i - a_i^0| < \eta$ with $0 \le i \le n$, also has precisely one simple zero $c \in \mathbf{R}$ with $|c - c^0| < \delta$. It further follows that this c is a C^∞ function of the coefficients $a_0, \dots, a_n$ of p.

This positive result should be contrasted with the *Abel–Ruffini Theorem.* That theorem in algebra asserts that there does not exist a formula which gives the zeros of a general polynomial function of degree n in terms of the coefficients of that function by means of addition, subtraction, multiplication, division and extraction of roots, if $n \geq 5$.

Application B. For a suitably chosen neighborhood U of 0 in $\mathbf{R}$ the equation in the unknown $x \in \mathbf{R}$

$$x^2 y_1 + e^{2x} + y_2 = 0$$

has a unique solution $x \in U$ if $y = (y_1, y_2)$ varies in a suitable neighborhood V of $(1, -1)$ in $\mathbf{R}^2$.

Indeed, first define $f : \mathbf{R} \times \mathbf{R}^2 \to \mathbf{R}$ by

$$f(x; y) = x^2 y_1 + e^{2x} + y_2.$$

Then $f(0; 1, -1) = 0$, and $D_x f(0; 1, -1) \in \mathrm{End}(\mathbf{R})$ is given by

$$D_x f(0; 1, -1) = 2xy_1 + 2e^{2x} \,|_{(0;1,-1)} = 2 \neq 0.$$

According to the Implicit Function Theorem there exist neighborhoods U of 0 in $\mathbf{R}$ and V of $(1, -1)$ in $\mathbf{R}^2$, and a differentiable mapping $\psi : V \to U$ such that

$$\psi(1, -1) = 0 \qquad \text{and} \qquad x = \psi(y) \qquad \text{satisfies} \qquad x^2 y_1 + e^{2x} + y_2 = 0.$$

Furthermore,

$$D_y f(x; y) \in \mathrm{Lin}(\mathbf{R}^2, \mathbf{R}) \qquad \text{is given by} \qquad D_y f(x; y) = (x^2, 1).$$

It follows that we have the following equality of elements in $\mathrm{Lin}(\mathbf{R}^2, \mathbf{R})$:

$$D\psi(1, -1) = -D_x f(0; 1, -1)^{-1} \circ D_y f(0; 1, -1) = -\frac{1}{2}\,(0, 1) = (0, -\frac{1}{2}).$$

Having shown $x : y \mapsto x(y)$ to be a well-defined differentiable function on V of y, we can also take the equations

$$D_j(x(y)^2 y_1 + e^{2x(y)} + y_2) = 0 \qquad (j = 1, 2,\ y \in V)$$

as the starting point for the calculation of $D_j x(1, -1)$, for $j = 1, 2$. (Here D_j denotes partial differentiation with respect to y_j.) In what follows we shall write $x(y)$ as x, and $D_j x(y)$ as $D_j x$. For $j = 1$ and 2, respectively, this gives (compare with Formula (3.6))

$$2x\,(D_1 x)\,y_1 + x^2 + 2e^{2x}\,D_1 x = 0, \qquad \text{and} \qquad 2x\,(D_2 x)\,y_1 + 2e^{2x}\,D_2 x + 1 = 0.$$

Therefore, at $(x; y) = (0; 1, -1)$,

$$2D_1x(1, -1) = 0, \qquad \text{and} \qquad 2D_2x(1, -1) + 1 = 0.$$

This way of determining the partial derivatives is known as the *method of implicit differentiation.*

Higher-order partial derivatives at $(1, -1)$ of x with respect to y_1 and y_2 can also be determined in this way. To calculate $D_2^2x(1, -1)$, for example, we use

$$D_2(2x\,(D_2x)\,y_1 + 2e^{2x}\,D_2x + 1) = 0.$$

Hence

$$2(D_2x)^2 y_1 + 2x\,(D_2^2x)\,y_1 + 4e^{2x}(D_2x)^2 + 2e^{2x}\,D_2^2x = 0,$$

and thus

$$2\frac{1}{4} + 4\frac{1}{4} + 2D_2^2x(1, -1) = 0, \qquad \text{i.e.} \qquad D_2^2x(1, -1) = -\frac{3}{4}.$$

Application C. Let $f \in C^1(\mathbf{R})$ satisfy $f(0) \neq 0$. Consider the following equation for $x \in \mathbf{R}$ dependent on $y \in \mathbf{R}$

$$y\,f(x) = \int_0^x f(yt)\,dt.$$

Then there are neighborhoods U and V of 0 in $\mathbf{R}$ such that for every $y \in V$ there exists a unique solution $x = x(y) \in \mathbf{R}$ with $x \in U$. Moreover $y \mapsto x(y)$ is a C^1 mapping on V and $x'(0) = 1$.

Indeed, define $F : \mathbf{R} \times \mathbf{R} \to \mathbf{R}$ by

$$F(x; y) = y\,f(x) - \int_0^x f(yt)\,dt.$$

Then

$$D_xF(x; y) = D_1F(x; y) = yf'(x) - f(yx), \qquad \text{so} \qquad D_1F(0; 0) = -f(0);$$

and using the Differentiation Theorem 2.10.4 we find

$$D_yF(x; y) = D_2F(x; y) = f(x) - \int_0^x t\,f'(yt)\,dt, \qquad \text{so} \qquad D_2F(0; 0) = f(0).$$

Because D_1F and $D_2F : \mathbf{R} \times \mathbf{R} \to \mathrm{End}(\mathbf{R})$ exist and because both are continuous, partly on account of the Continuity Theorem 2.10.2, it follows from Theorem 2.3.4 that F is (totally) differentiable; and we also find that F is a C^1 function. In addition

$$F(0; 0) = 0, \qquad D_1F(0; 0) = -f(0) \neq 0.$$

Therefore $D_1F(0; 0) \in \mathrm{Aut}(\mathbf{R})$ is the mapping $t \mapsto -f(0)\,t$. The first assertion then follows because the conditions of the Implicit Function Theorem are satisfied. Finally

$$x'(0) = -D_1F(0; 0)^{-1}\,D_2F(0; 0) = \frac{1}{f(0)}\,f(0) = 1.$$

Application D. Considering that the proof of the Implicit Function Theorem made intensive use of the theory of linear mappings, we hardly expect the theorem to add much to this theory. However, for the sake of completeness we now discuss its implications for the theory of square systems of linear equations:

$$\begin{array}{cccc} a_{11}x_1+ & \cdots & +a_{1n}x_n - b_1 & = & 0, \\ \vdots & & \vdots & & \vdots \\ a_{n1}x_1+ & \cdots & +a_{nn}x_n - b_n & = & 0. \end{array} \tag{3.7}$$

We write this as $f(x; y) = 0$, where the unknown x and the parameter y, respectively, are given by

$$x = (x_1, \dots, x_n) \in \mathbf{R}^n,$$

$$y = (a_{11}, \dots, a_{n1}, a_{12}, \dots, a_{n2}, \dots, a_{nn}, b_1, \dots, b_n) = (A, b) \in \mathbf{R}^{n^2+n}.$$

Now, for all $(x; y) \in \mathbf{R}^n \times \mathbf{R}^{n^2+n}$,

$$D_j f_i(x; y) = a_{ij}.$$

The condition $D_x f(x^0; y^0) \in \mathrm{Aut}(\mathbf{R}^n)$ therefore implies $A^0 \in \mathbf{GL}(n, \mathbf{R})$, and this is independent of the vector b^0. In fact, for every $b \in \mathbf{R}^n$ there exists a unique solution x^0 of $f(x^0; (A^0, b)) = 0$, if $A^0 \in \mathbf{GL}(n, \mathbf{R})$. Let $y^0 = (A^0, b^0)$ and x^0 be such that

$$f(x^0; y^0) = 0 \qquad \text{and} \qquad D_x f(x^0; y^0) \in \mathrm{Aut}(\mathbf{R}^n).$$

The Implicit Function Theorem now leads to the slightly stronger result that for neighboring matrices A and vectors b, respectively, there exists a unique solution x of the system (3.7). This was indeed to be expected because neighboring matrices A are also contained in $\mathbf{GL}(n, \mathbf{R})$, according to Lemma 2.1.2. A further conclusion is that the solution x of the system (3.7) depends in a C^∞-manner on the coefficients of the matrix A and of the inhomogeneous term b. Of course, inspection of the formulae from linear algebra for the solution of the system (3.7), Cramer's rule in particular, also leads to this result.

Application E. Let M be the linear space $\mathrm{Mat}(2, \mathbf{R})$ of 2×2 matrices with coefficients in $\mathbf{R}$, and let $A \in M$ be arbitrary. Then there exist an open neighborhood V of 0 in $\mathbf{R}$ and a C^∞ mapping $\psi : V \to M$ such that

$$\psi(0) = I \qquad \text{and} \qquad \psi(t)^2 + tA\psi(t) = I \qquad (t \in V).$$

Furthermore,

$$\psi'(0) \in \mathrm{Lin}(\mathbf{R}, M) \qquad \text{is given by} \qquad s \mapsto -\frac{1}{2} s\, A.$$

Indeed, consider $f : M \times \mathbf{R} \to M$ with $f(X; t) = X^2 + tAX - I$. Then $f(I; 0) = 0$. For the computation of $D_X f(X; t)$, observe that $X \mapsto X^2$ equals the composition $X \mapsto (X, X) \mapsto g(X, X)$ where the mapping g with $g(X_1, X_2) = X_1 X_2$ belongs to $\mathrm{Lin}^2(M, M)$. Furthermore, $X \mapsto tAX$ belongs to $\mathrm{Lin}(M, M)$. In view of Proposition 2.7.6 we now find

$$D_X f(X; t)H = HX + XH + tAH \qquad (H \in M).$$

In particular

$$f(I; 0) = 0, \qquad D_X f(I; 0) : H \mapsto 2H.$$

According to the Implicit Function Theorem there exist a neighborhood V of 0 in $\mathbf{R}$ and a neighborhood U of I in M such that for every $t \in V$ there is a unique $\psi(t) \in M$ with

$$\psi(t) \in U \qquad \text{and} \qquad f(\psi(t), t) = \psi(t)^2 + tA\psi(t) - I = 0.$$

In addition, ψ is a C^∞ mapping. Since $t \mapsto tAX$ is linear we have

$$D_t f(X; t)s = sAX.$$

In particular, therefore, $D_t f(I; 0)s = sA$, and hence the result

$$\psi'(0) = -D_X f(I; 0)^{-1} \circ D_t f(I; 0) : s \mapsto -\frac{1}{2} sA.$$

This conclusion can also be arrived at by means of implicit differentiation, as follows. Differentiating $\psi(t)^2 + tA\psi(t) = I$ with respect to t one finds

$$\psi(t)\psi'(t) + \psi'(t)\psi(t) + A\psi(t) + tA\psi'(t) = 0.$$

Substitution of $t = 0$ gives $2\psi'(0) + A = 0$.

3.7 Implicit and Inverse Function Theorems on C

We identify $\mathbf{C}^n$ with $\mathbf{R}^{2n}$ as in Formula (1.2). A mapping $f : U \to \mathbf{C}^p$, with U open in $\mathbf{C}^n$, is called *holomorphic* or *complex-differentiable* if f is continuously differentiable over the field $\mathbf{R}$ and if for every $x \in U$ the real-linear mapping $Df(x) : \mathbf{R}^{2n} \to \mathbf{R}^{2p}$ is in fact complex-linear from $\mathbf{C}^n$ to $\mathbf{C}^p$. This is equivalent to the requirement that $Df(x)$ commutes with multiplication by i:

$$Df(x)(iz) = i\, Df(x)z, \qquad (z \in \mathbf{C}^n).$$

Here the "multiplication by i" may be regarded as a special real-linear transformation in $\mathbf{C}^n$ or $\mathbf{C}^p$, respectively. In this way we obtain for $n = p = 1$ the holomorphic functions of one complex variable, see Definition 8.3.9 and Lemma 8.3.10.

Theorem 3.7.1 (Local Inverse Function Theorem: complex-differentiable case). *Let $U_0 \subset \mathbf{C}^n$ be open and $a \in U_0$. Further assume $\Phi : U_0 \to \mathbf{C}^n$ complex-differentiable and $D\Phi(a) \in \mathrm{Aut}(\mathbf{R}^{2n})$. Then there exists an open neighborhood U of a contained in U_0 such that $\Phi : U \to \Phi(U) = V$ is bijective, V open in $\mathbf{C}^n$, and the inverse mapping: $V \to U$ complex-differentiable.*

Remark. For $n = 1$, the derivative $D\Phi(x^0) : \mathbf{R}^2 \to \mathbf{R}^2$ is invertible as a real-linear mapping if and only if the complex derivative $\Phi'(x^0)$ is different from zero.

Theorem 3.7.2 (Implicit Function Theorem: complex-differentiable case). *Assume W to be open in $\mathbf{C}^n \times \mathbf{C}^p$ and $f : W \to \mathbf{C}^n$ complex-differentiable. Let*

$$(x^0, y^0) \in W, \qquad f(x^0; y^0) = 0, \qquad D_x f(x^0; y^0) \in \mathrm{Aut}(\mathbf{C}^n).$$

Then there exist open neighborhoods U of x^0 in C^n and V of y^0 in $\mathbf{R}^p$ with the following properties:

$$\textit{for every} \quad y \in V \quad \textit{there exists a unique} \quad x \in U \qquad \text{with} \quad f(x; y) = 0.$$

In this way we obtain a complex-differentiable mapping: $\mathbf{C}^p \supset\!\to \mathbf{C}^n$ satisfying

$$\psi : V \to U \qquad \text{with} \qquad \psi(y) = x \qquad \text{and} \qquad f(x; y) = 0,$$

which is uniquely determined by these properties. Furthermore, the derivative $D\psi(y) \in \mathrm{Lin}(\mathbf{C}^p, \mathbf{C}^n)$ of ψ at y is given by

$$D\psi(y) = -D_x f(\psi(y); y)^{-1} \circ D_y f(\psi(y); y) \qquad (y \in V).$$

Proof. This follows from the Implicit Function Theorem 3.5.1; only the penultimate assertion requires some explanation. Because $D_x f(x^0; y^0)$ is complex-linear, its inverse is too. In addition, $D_y f(x^0; y^0)$ is complex-linear, which makes $D\psi(y) : \mathbf{C}^p \to \mathbf{C}^n$ complex-linear. ❑

Chapter 4

Manifolds

Manifolds are common geometrical objects which are intensively studied in many areas of mathematics, such as algebra, analysis and geometry. In the present chapter we discuss the definition of manifolds and some of their properties while their infinitesimal structure, the tangent spaces, are studied in Chapter 5. In Volume II, in Chapter 6 we calculate integrals on sets bounded by manifolds, and in Chapters 7 and 8 we study integrals on the manifolds themselves. Although it is natural to define a manifold by means of equations in the ambient space, we often work on the manifold itself via (local) parametrizations. In the former case manifolds arise as null sets or as inverse images, and then submersions are useful in describing them; in the latter case manifolds are described as images under mappings, and then immersions are used.

4.1 Introductory remarks

In analysis, a subset V of $\mathbf{R}^n$ can often be described as the *graph*

$$V = \text{graph}(f) = \{\,(w,\ f(w)) \in \mathbf{R}^n \mid w \in \text{dom}(f)\,\}$$

of a mapping $f : \mathbf{R}^d \supset\!\!\rightarrow \mathbf{R}^{n-d}$ suitably chosen with an open set as its domain of definition. Examples are straight lines in $\mathbf{R}^2$, planes in $\mathbf{R}^3$, the *spiral* or *helix* (ἡ ἕλιξ = curl of hair) V in $\mathbf{R}^3$ where

$$V = \{\,(w,\ \cos w,\ \sin w) \in \mathbf{R}^3 \mid w \in \mathbf{R}\,\} \qquad \text{with} \qquad f(w) = (\cos w,\ \sin w).$$

Certain other sets V, like the nondegenerate conics in $\mathbf{R}^2$ (ellipse, hyperbola, parabola) and the nondegenerate quadrics in $\mathbf{R}^3$, present a different case. Although locally (that is, for every $x \in V$ in a neighborhood of x in $\mathbf{R}^n$) they can be written as graphs, they may fail to have this property globally. In other words, there does not always exist a single f such that all of V is the graph of that f. Nevertheless, sets of this kind are of such importance that they have been given a name of their own.

Definition 4.1.1. A set V is called a *manifold* if V can locally be written as the graph of a mapping. ❍

Sets $\widetilde{V}$ of a different type may then be defined by requiring $\widetilde{V}$ to be bounded by a number of manifolds V; here one may think of cubes, balls, spherical shells, etc. in $\mathbf{R}^3$.

There are two other common ways of specifying sets V: in Definitions 4.1.2 and 4.1.3 the set V is described as an image, or inverse image, respectively, under a mapping.

Definition 4.1.2. A subset V of $\mathbf{R}^n$ is said to be a *parametrized set* when V occurs as an image under a mapping $\phi : \mathbf{R}^d \supset\!\!\to \mathbf{R}^n$ defined on an open set, that is to say

$$V = \mathrm{im}(\phi) = \{\, \phi(y) \in \mathbf{R}^n \mid y \in \mathrm{dom}(\phi)\,\}. \qquad ❍$$

Definition 4.1.3. A subset V of $\mathbf{R}^n$ is said to be a *zero-set* when there exists a mapping $g : \mathbf{R}^n \supset\!\!\to \mathbf{R}^{n-d}$ defined on an open set, such that

$$V = g^{-1}(\{0\}) = \{\, x \in \mathrm{dom}(g) \mid g(x) = 0\,\}. \qquad ❍$$

Obviously, the local variants of Definitions 4.1.2 and 4.1.3 also exist, in which V satisfies the definition locally.

The unit circle V in the (x_1, x_3)-plane in $\mathbf{R}^3$ satisfies Definitions 4.1.1 – 4.1.3. In fact, with $f(w) = \sqrt{1-w^2}$,

$$\begin{aligned} V &= \{\, (w,\, 0,\, \pm f(w)) \in \mathbf{R}^3 \mid |w| < 1\,\} \cup \{\, (\pm f(w),\, 0,\, w) \in \mathbf{R}^3 \mid |w| < 1\,\}, \\ V &= \{\, (\cos y,\, 0,\, \sin y) \in \mathbf{R}^3 \mid y \in \mathbf{R}\,\}, \\ V &= \{\, x \in \mathbf{R}^3 \mid g(x) = (x_2,\, x_1^2 + x_3^2 - 1) = (0, 0)\,\}. \end{aligned}$$

By contrast, the coordinate axes V in $\mathbf{R}^2$, while constituting a zero-set $V = \{\, x \in \mathbf{R}^2 \mid x_1 x_2 = 0\,\}$, do not form a manifold everywhere, because near the point $(0, 0)$ it is not possible to write V as the graph of any function.

In the following we shall be more precise about Definitions 4.1.1 – 4.1.3, and investigate the various relationships between them. Note that the mapping belonging to the description as a graph

$$w \mapsto (w,\, f(w))$$

is ipso facto injective, whereas a parametrization

$$y \mapsto \phi(y)$$

is not necessarily. Furthermore, a graph V can always be written as a zero-set

$$V = \{ (w, t) \mid t - f(w) = 0 \}.$$

For this reason graphs are regarded as rather "elementary" objects.

Note that the dimensions of the domain and range spaces of the mappings are always chosen such that a point in V has d "degrees of freedom", leaving exceptions aside. If, for example, $g(x) = 0$ then $x \in \mathbf{R}^n$ satisfies $n - d$ equations $g_1(x) = 0, \dots, g_{n-d}(x) = 0$, and therefore x retains $n - (n - d) = d$ degrees of freedom.

4.2 Manifolds

Definition 4.2.1. Let $\emptyset \neq V \subset \mathbf{R}^n$ be a subset and $x \in V$, and let also $k \in \mathbf{N}_\infty$ and $0 \leq d \leq n$. Then V is said to be a C^k *submanifold of* $\mathbf{R}^n$ *at* x *of dimension* d if there exists an open neighborhood U of x in $\mathbf{R}^n$ such that $V \cap U$ equals the graph of a C^k mapping f of an open subset W of $\mathbf{R}^d$ into $\mathbf{R}^{n-d}$, that is

$$V \cap U = \{ (w, f(w)) \in \mathbf{R}^n \mid w \in W \subset \mathbf{R}^d \}.$$

Here a choice has been made as to which $n - d$ coordinates of a point in $V \cap U$ are functions of the other d coordinates.

If V has this property for every $x \in V$, with constant k and d, but possibly a different choice of the d independent coordinates, V is called a C^k*submanifold in* $\mathbf{R}^n$ *of dimension* d. (For the dimensions n and 0 see the following example.) ❍

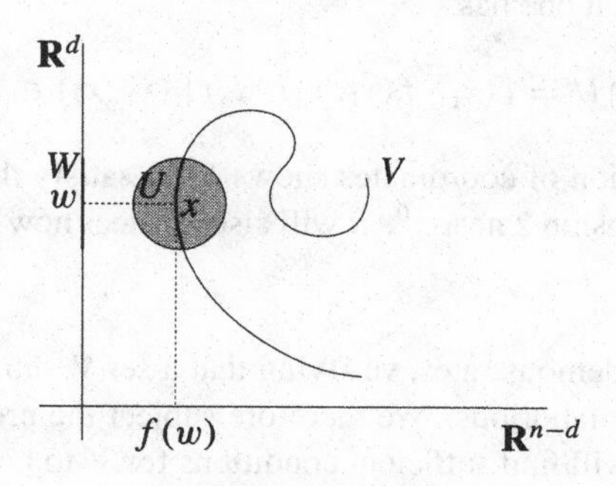

Illustration for Definition 4.2.1

Clearly, if V has dimension d at x then it also has dimension d at nearby points. Thus the conditions that the dimension of V at x be $d \in \mathbf{N}_0$ define a partition of V

into open subsets. Therefore, if V is connected, it has the same dimension at each of its points.

If $d = 1$, we speak of a C^k *curve*; if $d = 2$, we speak of a C^k *surface*; and if $d = n - 1$, we speak of a C^k *hypersurface*. According to these definitions curves and (hyper)surfaces are sets; in subsequent parts of this book, however, the words curve or (hyper)surface may refer to mappings defining the sets.

Example 4.2.2. Every open subset $V \subset \mathbf{R}^n$ satisfies the definition of a C^∞ submanifold in $\mathbf{R}^n$ of dimension n. Indeed, take $W = V$ and define $f : V \to \mathbf{R}^0 = \{0\}$ by $f(x) = 0$, for all $x \in V$. Then $x \in V$ can be identified with $(x, f(x)) \in \mathbf{R}^n \times \mathbf{R}^0 \simeq \mathbf{R}^n$. And likewise, every point in $\mathbf{R}^n$ is a C^∞ submanifold in $\mathbf{R}^n$ of dimension 0.☆

Example 4.2.3. The unit sphere $S^2 \subset \mathbf{R}^3$, with $S^2 = \{ x \in \mathbf{R}^3 \mid \|x\| = 1 \}$, is a C^∞ submanifold in $\mathbf{R}^3$ of dimension 2. To see this, let $x^0 \in S^2$. The coordinates of x^0 cannot vanish all at the same time; so, for example, we may assume $x_2^0 \neq 0$, say $x_2^0 < 0$. One then has $1 - (x_1^0)^2 - (x_3^0)^2 = (x_2^0)^2 > 0$, and therefore

$$x^0 = (x_1^0,\ -\sqrt{1 - (x_1^0)^2 - (x_3^0)^2},\ x_3^0).$$

A similar description now is obtained for all points $x \in S^2 \cap U$, where U is the neighborhood $\{x \in \mathbf{R}^3 \mid x_2 < 0\}$ of x^0 in $\mathbf{R}^3$. Indeed, let W be the open neighborhood $\{ (x_1, x_3) \in \mathbf{R}^2 \mid x_1^2 + x_3^2 < 1 \}$ of (x_1^0, x_3^0) in $\mathbf{R}^2$, and further define

$$f : W \to \mathbf{R} \qquad \text{by} \qquad f(x_1, x_3) = -\sqrt{1 - x_1^2 - x_3^2}.$$

Then f is a C^∞ mapping because the polynomial under the root sign cannot attain the value 0. In addition one has

$$S^2 \cap U = \{ (x_1,\ f(x_1, x_3),\ x_3) \mid (x_1, x_3) \in W \}.$$

Subsequent permutation of coordinates shows S^2 to satisfy the definition of a C^∞ submanifold of dimension 2 near x^0. It will also be clear how to deal with the other possibilities for x^0. ☆

As this example demonstrates, verifying that a set V satisfies the definition of a submanifold can be laborious. We therefore subject the properties of manifolds to further study; we will find sufficient conditions for V to be a manifold.

Remark 1. Let V be a C^k submanifold, for $k \in \mathbf{N}_\infty$, in $\mathbf{R}^n$ of dimension d. Then define, in the notation of Definition 4.2.1, the mapping $\phi : W \to \mathbf{R}^n$ by

$$\phi(w) = (w,\ f(w)).$$

Then ϕ is an injective C^k mapping, and therefore $\phi : W \to \phi(W)$ is bijective. The inverse mapping $\phi^{-1} : \phi(W) \to W$, with

$$(w, f(w)) \mapsto w$$

is continuous, because it is the projection mapping onto the first factor. In addition $D\phi(w)$, for $w \in W$, is injective because

$$D\phi(w) = \begin{array}{c} d \\ \\ n-d \end{array} \overset{d}{\left(\begin{array}{c} I_d \\ \hline Df(w) \end{array} \right)}.$$

Definition 4.2.4. Let $D \subset \mathbf{R}^d$ be a nonempty open subset, $d \le n$ and $k \in \mathbf{N}_\infty$, and $\phi \in C^k(D, \mathbf{R}^n)$. Let $y \in D$, then ϕ is said to be a C^k *immersion at* y if

$$D\phi(y) \in \mathrm{Lin}(\mathbf{R}^d, \mathbf{R}^n) \quad \text{is injective.}$$

Further, ϕ is called a C^k *immersion* if ϕ is a C^k immersion at y, for all $y \in D$.

A C^k immersion ϕ is said to be a C^k *embedding* if in addition ϕ is injective (note that the induced mapping $\phi : D \to \phi(D)$ is bijective in that case), and if $\phi^{-1} : \phi(D) \to D$ is continuous. In particular, therefore, the mapping $\phi : D \to \phi(D)$ is bijective and continuous, and possesses a continuous inverse; in other words, it is a homeomorphism, see Definition 1.5.2. Accordingly, we can now say that a C^k embedding $\phi : D \to \phi(D)$ is a C^k immersion which is also a homeomorphism onto its image.

Another way of saying this is that ϕ gives a C^k *parametrization of* $\phi(D)$ *by* D. Conversely, the mapping

$$\kappa := \phi^{-1} : \phi(D) \to D$$

is called a *coordinatization* or a *chart* for $\phi(D)$. ❍

Using Definition 4.2.4 we can reformulate Remark 1 above as follows. If V is a C^k submanifold for $k \in \mathbf{N}_\infty$ in $\mathbf{R}^n$ of dimension d, there exists, for every $x \in V$, a neighborhood U of x in $\mathbf{R}^n$ such that $V \cap U$ possesses a C^k parametrization by a d-dimensional set. From the Immersion Theorem 4.3.1, which we shall presently prove, a converse result can be obtained. If $\phi : D \to \mathbf{R}^n$ is a C^k immersion at a point $y \in D$, then $\phi(D)$ near $x = \phi(y)$ is a C^k submanifold in $\mathbf{R}^n$ of dimension $d = \dim D$. In other words, the **image** under a mapping ϕ that is an immersion at y is a submanifold near $\phi(y)$.

Remark 2. Assume that V is a C^k submanifold of $\mathbf{R}^n$ at the point $x^0 \in V$ of dimension d. If U and f are the neighborhood of x^0 in $\mathbf{R}^n$ and the mapping from Definition 4.2.1, respectively, then every $x \in U$ can be written as $x = (w, t)$ with $w \in W \subset \mathbf{R}^d$, $t \in \mathbf{R}^{n-d}$; we can further define the C^k mapping g by

$$g : U \to \mathbf{R}^{n-d}, \qquad g(x) = g(w, t) = t - f(w).$$

Then

$$V \cap U = \{ x \in U \mid g(x) = 0 \}$$

and

$$Dg(x) \in \operatorname{Lin}(\mathbf{R}^n, \mathbf{R}^{n-d}) \quad \text{is surjective, for all } x \in U.$$

Indeed,

$$Dg(x) = \Big(D_w g(w, t) \;\Big|\; D_t g(w, t) \Big) = {\scriptstyle n-d} \overset{\qquad d \qquad\qquad n-d}{\Big(-Df(w) \;\Big|\; I_{n-d} \Big)}.$$

Note that the surjectivity of $Dg(x)$ implies that the column rank of $Dg(x)$ equals $n - d$. But then, in view of the Rank Lemma 4.2.7 below, the row rank also equals $n-d$, which implies that the rows, $Dg_1(x), \ldots, Dg_{n-d}(x)$, are linearly independent vectors in $\mathbf{R}^n$. Formulated in a different way: $V \cap U$ can be described as the solution set of $n-d$ equations $g_1(x) = 0, \ldots, g_{n-d}(x) = 0$ that are independent in the sense that the system $Dg_1(x), \ldots, Dg_{n-d}(x)$ of vectors in $\mathbf{R}^n$ is linearly independent.

Definition 4.2.5. The number $n - d$ of equations required for a local description of V is called the *codimension of V in $\mathbf{R}^n$*, notation

$$\operatorname{codim}_{\mathbf{R}^n} V = n - d. \qquad ❍$$

Definition 4.2.6. Let $U \subset \mathbf{R}^n$ be a nonempty open subset, $d \in \mathbf{N}_0$ and $k \in \mathbf{N}_\infty$, and $g \in C^k(U, \mathbf{R}^{n-d})$. Let $x \in U$, then g is said to be a C^k *submersion at* x if

$$Dg(x) \in \operatorname{Lin}(\mathbf{R}^n, \mathbf{R}^{n-d}) \quad \text{is surjective.}$$

Further, g is called a C^k *submersion* if g is a C^k submersion at x, for every $x \in U$. ❍

Making use of Definition 4.2.6 we can reformulate the above Remark 2 as follows. If V is a C^k manifold in $\mathbf{R}^n$, there exists for every $x \in V$ an open neighborhood U of x in $\mathbf{R}^n$ such that $V \cap U$ is the zero-set of a C^k submersion, with values in a $\operatorname{codim}_{\mathbf{R}^n} V$-dimensional space. From the Submersion Theorem 4.5.2, which we shall prove later on, a converse result can be obtained. If $g : U \to \mathbf{R}^{n-d}$

is a C^k submersion at a point $x \in U$ with $g(x) = 0$, then $U \cap g^{-1}(\{0\})$ near x is a C^k submanifold in $\mathbf{R}^n$ of dimension d. In other words, the **inverse image** under a mapping g that is a submersion at x, of $g(x)$, is a submanifold near x.

From linear algebra we recall the definition of the *rank* of $A \in \mathrm{Lin}(\mathbf{R}^n, \mathbf{R}^p)$ as the dimension r of the image of A, which then satisfies $r \leq \min(n, p)$. We need the result that A and its adjoint $A^t \in \mathrm{Lin}(\mathbf{R}^p, \mathbf{R}^n)$ have equal rank r. For the matrix of A with respect to the standard bases $(e_1, \dots, e_n)$ and $(e'_1, \dots, e'_p)$ in $\mathbf{R}^n$ and $\mathbf{R}^p$, respectively, this means that the maximal number of linearly independent columns equals the maximal number of linearly independent rows. For the sake of completeness we give an efficient proof of this result. Furthermore, in view of the principle that, locally near a point, a mapping behaves similar to its derivative at the point in question, it suggests normal forms for mappings, as we will see in the Immersion and Submersion Theorems (and the Rank Theorem, see Exercise 4.35).

Lemma 4.2.7 (Rank Lemma). *For $A \in \mathrm{Lin}(\mathbf{R}^n, \mathbf{R}^p)$ the following conditions are equivalent.*

(i) *A has rank r.*

(ii) *There exist $\Phi \in \mathrm{Aut}(\mathbf{R}^p)$ and $\Psi \in \mathrm{Aut}(\mathbf{R}^n)$ such that*

$$\Phi \circ A \circ \Psi(x) = (x_1, \dots, x_r, 0, \dots, 0) \in \mathbf{R}^p \qquad (x \in \mathbf{R}^n).$$

In particular, A and A^t have equal rank. Furthermore, if A is injective, then we may arrange that $\Psi = I$; therefore in this case

$$\Phi \circ A(x) = (x_1, \dots, x_n, 0, \dots, 0) \qquad (x \in \mathbf{R}^n).$$

On the other hand, suppose A is surjective. Then we can choose $\Phi = I$; therefore

$$A \circ \Psi(x) = (x_1, \dots, x_p) \qquad (x \in \mathbf{R}^n).$$

Finally, if $A \in \mathrm{Aut}(\mathbf{R}^n)$, then we take either Φ or Ψ equal to A^{-1}, and the remaining operator equal to I.

Proof. Only (i) $\Rightarrow$ (ii) needs verification. In view of the equality $\dim(\ker A) + r = n$, we can find a basis $(a_{r+1}, \dots, a_n)$ of $\ker A \subset \mathbf{R}^n$ and vectors $a_1, \dots, a_r$ complementing this basis to a basis of $\mathbf{R}^n$. Define $\Psi \in \mathrm{Aut}(\mathbf{R}^n)$ setting $\Psi e_j = a_j$, for $1 \leq j \leq n$. Then $A\Psi(e_j) = Aa_j$, for $1 \leq j \leq r$, and $A\Psi(e_j) = 0$, for $r < j \leq n$. The vectors $b_j = Aa_j$, for $1 \leq j \leq r$, form a basis of $\mathrm{im}\, A \subset \mathbf{R}^p$. Let us complement them by vectors $b_{r+1}, \dots, b_p$ to a basis of $\mathbf{R}^p$. Define $\Phi \in \mathrm{Aut}(\mathbf{R}^p)$ by $\Phi b_i = e'_i$, for $1 \leq i \leq p$. Then the operators Φ and Ψ are the required ones, since

$$\Phi \circ A \circ \Psi(e_j) = \begin{cases} e'_j, & 1 \leq j \leq r; \\ 0, & r < j \leq n. \end{cases}$$

The equality of ranks follows by means of the equality

$$\Psi^t \circ A^t \circ \Phi^t(x_1, \dots, x_p) = (x_1, \dots, x_r, 0, \dots, 0) \in \mathbf{R}^n.$$

If A is injective, then $r = n$ and $\ker A = (0)$, and thus we choose $\Psi = I$. If A is surjective, then $r = p$; we then select the complementary vectors $a_1, \dots, a_p \in \mathbf{R}^n$ such that $Ae_i = e'_i$, for $1 \le i \le p$. Then $\Phi = I$. ❑

4.3 Immersion Theorem

The following theorem says that, at least locally near a point y^0, an immersion can be given the same form as its derivative at y^0 in the Rank Lemma 4.2.7.

Locally, an immersion $\phi : D_0 \to \mathbf{R}^n$ with D_0 an open subset of $\mathbf{R}^d$, equals the restriction of a diffeomorphism of $\mathbf{R}^n$ to a linear subspace of dimension d. In fact, the Immersion Theorem asserts that there exist an open neighborhood D of y^0 in D_0 and a diffeomorphism Ψ acting in the **image space** $\mathbf{R}^n$, such that on D

$$\phi = \Psi \circ \iota,$$

where $\iota : \mathbf{R}^d \to \mathbf{R}^n$ denotes the *standard embedding* of $\mathbf{R}^d$ into $\mathbf{R}^n$, defined by

$$\iota(y_1, \dots, y_d) = (y_1, \dots, y_d, 0, \dots, 0).$$

Theorem 4.3.1 (Immersion Theorem). *Let $D_0 \subset \mathbf{R}^d$ be an open subset, let $d < n$ and $k \in \mathbf{N}_\infty$, and let $\phi : D_0 \to \mathbf{R}^n$ be a C^k mapping. Let $y^0 \in D_0$, let $\phi(y^0) = x^0$, and assume ϕ to be an immersion at y^0, that is $D\phi(y^0) \in \mathrm{Lin}(\mathbf{R}^d, \mathbf{R}^n)$ is injective. Then there exists an open neighborhood D of y^0 in D_0 such that the following assertions hold.*

(i) *$\phi(D)$ is a C^k submanifold in $\mathbf{R}^n$ of dimension d.*

(ii) *There exist an open neighborhood U of x^0 in $\mathbf{R}^n$ and a C^k diffeomorphism $\Phi : U \to \Phi(U) \subset \mathbf{R}^n$ such that for all $y \in D$*

$$\Phi \circ \phi(y) = (y, 0) \in \mathbf{R}^d \times \mathbf{R}^{n-d}.$$

In particular, the restriction of ϕ to D is an injection.

Proof. (i). Because $D\phi(y^0)$ is injective, the rank of $D\phi(y^0)$ equals d. Consequently, among the rows of the matrix of $D\phi(y^0)$ there are d which are linearly independent. We assume that these are the top d rows (this may require prior permutation of the coordinates of $\mathbf{R}^n$). As a result we find C^k mappings

$$g : D_0 \to \mathbf{R}^d \qquad \text{and} \qquad h : D_0 \to \mathbf{R}^{n-d}, \qquad \text{respectively, with}$$

$$g = (\phi_1, \dots, \phi_d), \qquad h = (\phi_{d+1}, \dots, \phi_n),$$

$$\phi(D_0) = \{\, (g(y),\ h(y)) \mid y \in D_0 \,\},$$

$$D\phi(y) = \begin{matrix} d \\ n-d \end{matrix} \left(\frac{Dg(y)}{Dh(y)} \right)^{d},$$

while $Dg(y^0) \in \mathrm{End}(\mathbf{R}^d)$ is surjective, and therefore invertible. And so, by the Local Inverse Function Theorem 3.2.4, there exist an open neighborhood D of y^0 in D_0 and an open neighborhood W of $g(y^0)$ in $\mathbf{R}^d$ such that $g : D \to W$ is a C^k diffeomorphism. Therefore we may now introduce $w := g(y) \in W$ as a new (independent) variable instead of y. The substitution of variables $y = g^{-1}(w)$ for $y \in D$ then implies

$$\phi(D) = \{ (w, f(w)) \mid w \in W \},$$

with W open in $\mathbf{R}^d$ and $f := h \circ g^{-1} \in C^k(W, \mathbf{R}^{n-d})$; this proves (i).

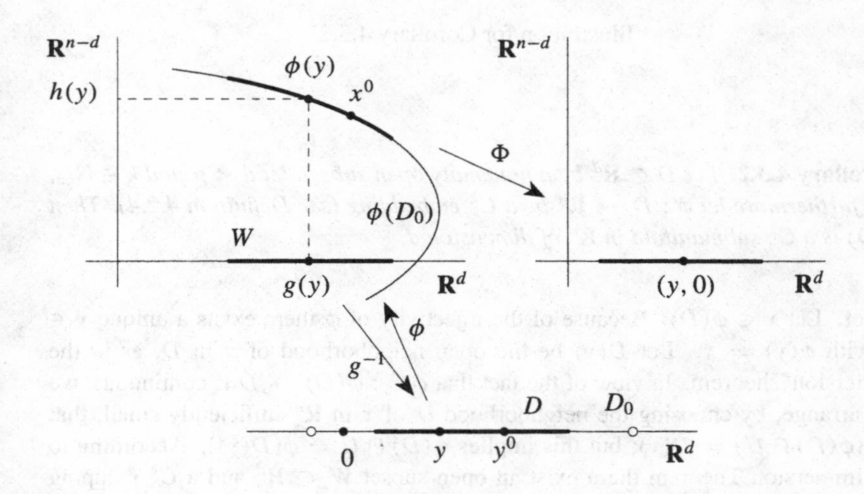

Immersion Theorem: near x^0 the coordinate transformation Φ straightens out the curved image set $\phi(D_0)$

(ii). To show this we define

$$\Psi : D \times \mathbf{R}^{n-d} \to W \times \mathbf{R}^{n-d} \quad \text{by} \quad \Psi(y, z) = \phi(y) + (0, z) = (g(y), h(y) + z).$$

From the proof of part (i) it follows that Ψ is invertible, with a C^k inverse, because

$$\Psi(y, z) = (w, t) \quad \Longleftrightarrow \quad y = g^{-1}(w) \quad \text{and} \quad z = t - h(y) = t - f(w).$$

Now let $\Phi := \Psi^{-1}$ and let U be the open neighborhood $W \times \mathbf{R}^{n-d}$ of x^0 in $\mathbf{R}^n$. Then $\Phi : U \to \Phi(U) \subset \mathbf{R}^n$ is a C^k diffeomorphism with the property

$$\Phi \circ \phi(y) = \Psi^{-1}(\phi(y) + (0, 0)) = (y, 0).$$ ❑

Remark. A mapping $\phi : \mathbf{R}^d \supset\!\!\rightarrow \mathbf{R}^n$ is also said to be *regular at* y^0 if it is immersive at y^0. Note that this is the case if and only if the rank of $D\phi(y^0)$ is maximal (compare with the definition of regularity in Definition 3.2.6). Accordingly, mappings $\mathbf{R}^d \supset\!\!\rightarrow \mathbf{R}^n$ with $d \le n$ in general may be expected to be immersive.

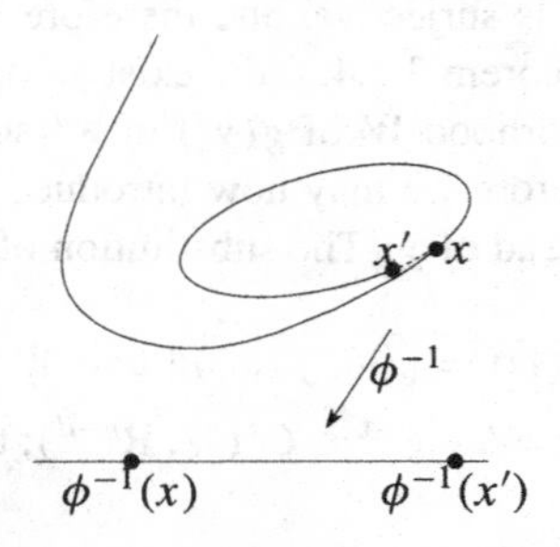

Illustration for Corollary 4.3.2

Corollary 4.3.2. *Let $D \subset \mathbf{R}^d$ be a nonempty open subset, let $d < n$ and $k \in \mathbf{N}_\infty$, and furthermore let $\phi : D \to \mathbf{R}^n$ be a C^k embedding (see Definition 4.2.4). Then $\phi(D)$ is a C^k submanifold in $\mathbf{R}^n$ of dimension d.*

Proof. Let $x \in \phi(D)$. Because of the injectivity of ϕ there exists a unique $y \in D$ with $\phi(y) = x$. Let $D(y)$ be the open neighborhood of y in D, as in the Immersion Theorem. In view of the fact that $\phi^{-1} : \phi(D) \to D$ is continuous, we can arrange, by choosing the neighborhood U of x in $\mathbf{R}^n$ sufficiently small, that $\phi^{-1}(\phi(D) \cap U) = D(y)$; but this implies $\phi(D) \cap U = \phi(D(y))$. According to the Immersion Theorem there exist an open subset $W \subset \mathbf{R}^d$ and a C^k mapping $f : W \to \mathbf{R}^{n-d}$ such that

$$\phi(D) \cap U = \phi(D(y)) = \{\, (w, f(w)) \mid w \in W \,\}.$$

Therefore $\phi(D)$ satisfies Definition 4.2.1 at the point x, but x was arbitrarily chosen in $\phi(D)$. ❑

With a view to later applications, in the theory of integration in Section 7.1, we formulate the following:

Lemma 4.3.3. *Let V be a C^k submanifold, for $k \in \mathbf{N}_\infty$, in $\mathbf{R}^n$ of dimension d. Suppose that D is an open subset of $\mathbf{R}^d$ and that $\phi : D \to \mathbf{R}^n$ is a C^k immersion satisfying $\phi(D) \subset V$. Then we have the following.*

(i) *$\phi(D)$ is an open subset of V (see Definition 1.2.16).*

(ii) *Furthermore, assume that ϕ is a C^k embedding. Let $\widetilde{D}$ be an open subset of $\mathbf{R}^{\widetilde{d}}$ and let $\widetilde{\phi}: \widetilde{D} \to \mathbf{R}^n$ be a C^k mapping satisfying $\widetilde{\phi}(\widetilde{D}) \subset V$. Then*

$$D_{\phi,\widetilde{\phi}} := \widetilde{\phi}^{-1}(\phi(D)) \subset \widetilde{D}$$

is an open subset of $\mathbf{R}^{\widetilde{d}}$ and $\phi^{-1} \circ \widetilde{\phi} : D_{\phi,\widetilde{\phi}} \to D$ is a C^k mapping.

(iii) *If, in addition, $\widetilde{d} = d$ and $\widetilde{\phi}$ is a C^k embedding too, then we have a C^k diffeomorphism*

$$\phi^{-1} \circ \widetilde{\phi} : D_{\phi,\widetilde{\phi}} \to D_{\widetilde{\phi},\phi}.$$

Proof. **(i)**. Select $x^0 \in \phi(D)$ arbitrarily and write $x^0 = \phi(y^0)$. Since $x^0 \in V$, by Definition 4.2.1 there exist an open neighborhood U_0 of x^0 in $\mathbf{R}^n$, an open subset W of $\mathbf{R}^d$ and a C^k mapping $f : W \to \mathbf{R}^{n-d}$ such that $V \cap U_0 = \{(w, f(w)) \in \mathbf{R}^n \mid w \in W\}$ (this may require prior permutation of the coordinates of $\mathbf{R}^n$). Then $V_0 := \phi^{-1}(U_0)$ is an open neighborhood of y^0 in D in view of the continuity of ϕ. As in the proof of the Immersion Theorem 4.3.1, we write $\phi = (g, h)$. Because $\phi(y) \in V \cap U_0$ for every $y \in V_0$, we can find $w \in W$ satisfying

$$(g(y), h(y)) = \phi(y) = (w, f(w)),$$

$$\text{thus} \qquad w = g(y) \quad \text{and} \quad h(y) = f(w) = (f \circ g)(y).$$

Set $w^0 = g(y^0)$. Because $h = f \circ g$ on the open neighborhood V_0 of y^0, the chain rule implies

$$Dh(y^0) = Df(w^0)\, Dg(y^0).$$

This said, suppose $v \in \mathbf{R}^d$ satisfies $Dg(y^0)v = 0$, then also $Dh(y^0)v = 0$, which implies $D\phi(y^0)v = 0$. In turn, this gives $v = 0$ because ϕ is an immersion at y^0. Accordingly $Dg(y^0) \in \mathrm{End}(\mathbf{R}^d)$ is injective and so belongs to $\mathrm{Aut}(\mathbf{R}^d)$. On the strength of the Local Inverse Function Theorem 3.2.4 there exists an open neighborhood D_0 of y^0 in V_0, such that the restriction of g to D_0 is a C^k diffeomorphism onto an open neighborhood W_0 of w^0 in $\mathbf{R}^d$. With these notations,

$$U(x^0) := U_0 \cap (W_0 \times \mathbf{R}^{n-d})$$

is an open subset of $\mathbf{R}^n$ and $\phi(D_0) = V \cap U(x^0)$. The union U of the open sets $U(x^0)$, for x^0 varying in $\phi(D)$, is open in $\mathbf{R}^n$ while $\phi(D) = V \cap U$. This proves assertion (i).

(ii). Assertion (i) now implies that

$$D_{\phi,\widetilde{\phi}} = \widetilde{\phi}^{-1}(\phi(D)) = \widetilde{\phi}^{-1}(V \cap U) = \widetilde{\phi}^{-1}(U) \subset \widetilde{D}$$

is an open subset of $\mathbf{R}^{\widetilde{d}}$, as $\widetilde{\phi} : \widetilde{D} \to \mathbf{R}^n$ is continuous and $U \subset \mathbf{R}^n$ is open. Let $\widetilde{y}^0 \in D_{\phi,\widetilde{\phi}}$, write $y^0 = \phi^{-1} \circ \widetilde{\phi}(\widetilde{y}^0)$ and let D_0 be the open neighborhood of y^0 in D

as above. As we did for ϕ, consider the decomposition $\widetilde{\phi} = (\widetilde{g}, \widetilde{h})$ with $\widetilde{g} : \widetilde{D} \to \mathbf{R}^d$ and $\widetilde{h} : \widetilde{D} \to \mathbf{R}^{n-d}$. If $y = \phi^{-1} \circ \widetilde{\phi}(\widetilde{y}) \in D_0$, then we have $\phi(y) = \widetilde{\phi}(\widetilde{y})$, therefore $g(y) = \widetilde{g}(\widetilde{y}) \in W_0$. Hence $y = g^{-1} \circ \widetilde{g}(\widetilde{y})$, because $g : D_0 \to W_0$ is a C^k diffeomorphism. Furthermore, we obtain that $\phi^{-1} \circ \widetilde{\phi} = g^{-1} \circ \widetilde{g}$ is a C^k mapping defined on an open neighborhood (viz. $\widetilde{g}^{-1}(W_0)$) of $\widetilde{y}^0$ in $D_{\phi,\widetilde{\phi}}$. As $\widetilde{y}^0$ is arbitrary, assertion (ii) now follows.

(iii). Apply part (ii), as is and also with the roles of ϕ and $\widetilde{\phi}$ interchanged. ❑

In particular, the preceding result shows that locally the dimension of a submanifold V of $\mathbf{R}^n$ is uniquely determined.

Remark. Following Definition 4.2.4, the C^k diffeomorphism

$$\kappa \circ \widetilde{\kappa}^{-1} = \phi^{-1} \circ \widetilde{\phi} : \widetilde{D} \cap (\widetilde{\kappa} \circ \kappa^{-1})(D) \to D \cap (\kappa \circ \widetilde{\kappa}^{-1})(\widetilde{D})$$

is said to be a *transition mapping* between the charts κ and $\widetilde{\kappa}$.

4.4 Examples of immersions

Example 4.4.1 (Circle). For every $r > 0$,

$$\phi : \mathbf{R} \to \mathbf{R}^2 \qquad \text{with} \qquad \phi(\alpha) = r(\cos\alpha,\ \sin\alpha)$$

is a C^∞ immersion. Its image is the circle

$$C_r = \{ x \in \mathbf{R}^2 \mid \|x\| = r \}.$$

We note that ϕ is by no means injective. However, we can make it so, by restricting ϕ to $]\,\alpha^0, \alpha^0 + 2\pi\,[$, for $\alpha^0 \in \mathbf{R}$. These are the maximal open intervals $I \subset \mathbf{R}$ for which $\phi|_I$ is injective. On these intervals $\phi(I)$ is a circle with one point left out. As in Example 3.1.1, we see that $\phi^{-1} : \phi(I) \to I$ is continuous. Consequently, $\phi : I \to \mathbf{R}^2$ is a C^∞ embedding. ☆

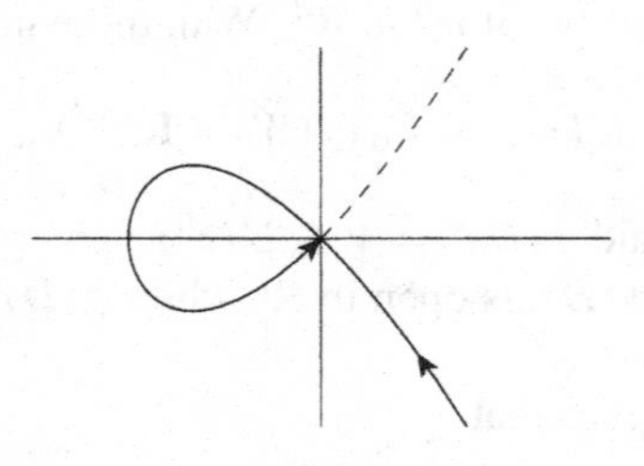

Illustration for Example 4.4.2

Example 4.4.2. The mapping $\widetilde{\phi} : \mathbf{R} \to \mathbf{R}^2$ with

$$\widetilde{\phi}(t) = (t^2 - 1,\ t^3 - t)$$

is a C^∞ immersion (even a polynomial immersion). One has

$$\widetilde{\phi}(s) = \widetilde{\phi}(t) \qquad \Longleftrightarrow \qquad s = t \quad \text{or} \quad s,\ t \in \{-1, 1\}.$$

This makes $\phi := \widetilde{\phi}|_{]-\infty,\ 1\,[}$ injective. But it does not make ϕ an embedding, because $\phi^{-1} : \phi(\,]-\infty,\ 1\,[\,) \to\ ]-\infty,\ 1\,[$ is not continuous at the point $(0, 0)$. Indeed, suppose this is the case. Then there exists a $\delta > 0$ such that

$$|t + 1| = |t - \phi^{-1}(0, 0)| < \frac{1}{2}, \tag{4.1}$$

whenever $t \in\ ]-\infty,\ 1\,[$ satisfies

$$\|\phi(t) - (0, 0)\| < \delta. \tag{4.2}$$

But because $\lim_{t\uparrow 1} \phi(t) = \phi(-1)$, there exists an element t satisfying inequality (4.2), and for which $0 < t < 1$. But that is in contradiction with inequality (4.1). ✩

Example 4.4.3 (Torus). (torus = protuberance, bulge.) Let D be the open square $]-\pi, \pi\,[\, \times\,]-\pi, \pi\,[\, \subset \mathbf{R}^2$ and define $\phi : D \to \mathbf{R}^3$ by

$$\phi(\alpha, \theta) = ((2 + \cos\theta)\, \cos\alpha,\ (2 + \cos\theta)\, \sin\alpha,\ \sin\theta).$$

Then ϕ is a C^∞ mapping, with derivative $D\phi(\alpha, \theta) \in \mathrm{Lin}(\mathbf{R}^2, \mathbf{R}^3)$ satisfying

$$D\phi(\alpha, \theta) = \begin{pmatrix} -(2 + \cos\theta)\, \sin\alpha & -\sin\theta\, \cos\alpha \\ (2 + \cos\theta)\, \cos\alpha & -\sin\theta\, \sin\alpha \\ 0 & \cos\theta \end{pmatrix}.$$

Next, ϕ is also an immersion on D. To verify this, first assume that $\theta \notin \{-\frac{\pi}{2}, \frac{\pi}{2}\}$; from this, $\cos\theta \neq 0$. Considering the last coordinates, we see that the column vectors in $D\phi(\alpha, \theta)$ are linearly independent. This conclusion also follows in the special case where $\theta \in \{-\frac{\pi}{2}, \frac{\pi}{2}\}$.

Moreover, ϕ is injective. Indeed, let $x = \phi(\alpha, \theta) = \phi(\widetilde{\alpha}, \widetilde{\theta}) = \widetilde{x}$. It then follows from $x_1^2 + x_2^2 = \widetilde{x}_1^2 + \widetilde{x}_2^2$ and $x_3 = \widetilde{x}_3$ that

$$(2 + \cos\theta)^2 = (2 + \cos\widetilde{\theta})^2 \qquad \text{and} \qquad \sin\theta = \sin\widetilde{\theta}.$$

Because $2 + \cos\theta$ and $2 + \cos\widetilde{\theta}$ are both > 0, we now have $\cos\theta = \cos\widetilde{\theta}$, $\sin\theta = \sin\widetilde{\theta}$; and therefore $\theta = \widetilde{\theta}$. But this implies $\cos\alpha = \cos\widetilde{\alpha}$ and $\sin\alpha = \sin\widetilde{\alpha}$; hence $\alpha = \widetilde{\alpha}$. Consequently, $\phi^{-1} : \phi(D) \to D$ is well-defined.

Finally, we prove the continuity of $\phi^{-1} : \phi(D) \to D$. If $\phi(\alpha, \theta) = x$, then

$$\begin{aligned}(\cos\alpha, \sin\alpha) &= (x_1^2 + x_2^2)^{-1/2}(x_1, x_2) &&=: h(x),\\ (\cos\theta, \sin\theta) &= ((x_1^2 + x_2^2)^{1/2} - 2, x_3) &&=: k(x).\end{aligned}$$

Here $h, k : \phi(D) \to \mathbf{R}^2$ are continuous mappings. The inverse of the mapping $\beta \mapsto (\cos\beta, \sin\beta)$ ($\beta \in\,]-\pi, \pi\,[$) is given by $g(u) := 2\arctan\left(\frac{u_2}{1+u_1}\right)$; and this mapping g also is continuous. Hence

$$(\alpha, \theta) = (g \circ h(x), g \circ k(x)),$$

which shows $\phi^{-1} : x \mapsto (\alpha, \theta)$ to be a continuous mapping $\phi(D) \to D$.

It follows that $\phi : D \to \mathbf{R}^3$ is an embedding, and according to Corollary 4.3.2 the image $\phi(D)$ is a C^∞ submanifold in $\mathbf{R}^3$ of dimension 2. Verify that $\phi(D)$ looks like the surface of a torus (donut) minus two intersecting circles. ☆

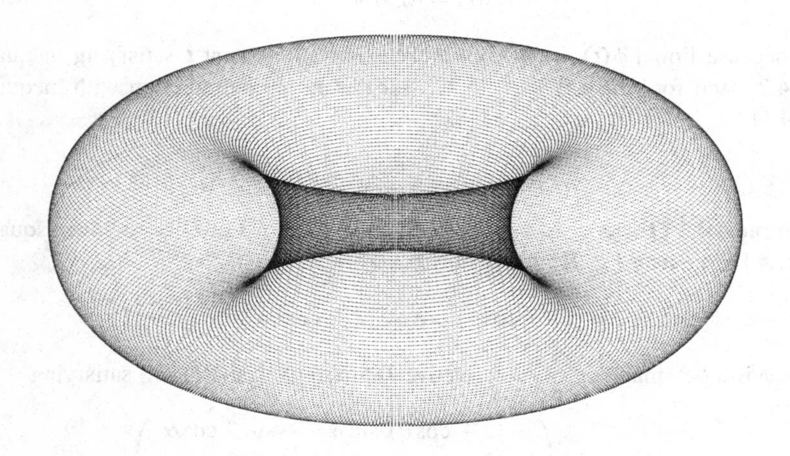

Illustration for Example 4.4.3: Transparent torus

4.5 Submersion Theorem

The following theorem says that, at least locally near a point x^0, a submersion can be given the same form as its derivative at x^0 in the Rank Lemma 4.2.7.

Locally, a submersion $g : U_0 \to \mathbf{R}^{n-d}$ with U_0 an open subset in $\mathbf{R}^n$, equals a diffeomorphism of $\mathbf{R}^n$ followed by projection onto a linear subspace of dimension $n-d$. In fact, the Submersion Theorem asserts that there exist an open neighborhood U of x^0 in U_0 and a diffeomorphism Φ acting in the **domain space** $\mathbf{R}^n$, such that on U,

$$g = \pi \circ \Phi,$$

where $\pi : \mathbf{R}^n \to \mathbf{R}^{n-d}$ denotes the *standard projection* onto the last $n-d$ coordinates, defined by

$$\pi(x_1, \dots, x_n) = (x_{d+1}, \dots, x_n).$$

Example 4.5.1. Let $g_1 : \mathbf{R}^3 \to \mathbf{R}$ with $g_1(x) = \|x\|^2$, and $g_2 : \mathbf{R}^3 \to \mathbf{R}$ with $g_2(x) = x_3$ be given functions. Assume, for (c_1, c_2) in a neighborhood of $(1, 0)$, that a point $x \in \mathbf{R}^3$ satisfies $g_1(x) = c_1$ and $g_2(x) = c_2$; that point x then lies on the intersection of a sphere and a plane, in other words, on a circle. Locally, such a point x is then uniquely determined by prescribing the coordinate x_1. Instead of the Cartesian coordinates (x_1, x_2, x_3), one may therefore also use the coordinates (y, c) on $\mathbf{R}^3$ to characterize x, where in the present case $y = y_1 = x_1$ and $c = (c_1, c_2)$. Accordingly, this leads to a substitution of variables $x = \Psi(y, c)$ in $\mathbf{R}^3$. Note that in these (y, c)-coordinates a circle is locally described as the **linear** submanifold of $\mathbf{R}^3$ given by c equals a constant vector.

The notion that, locally at least, a point $x \in \mathbf{R}^n$ may uniquely be determined by the values $(c_1, \dots, c_{n-d})$ taken by $n-d$ functions $g_1, \dots, g_{n-d}$ at x, plus a suitable choice of d Cartesian coordinates $(y_1, \dots, y_d)$ of x, is fundamental to the following Submersion Theorem 4.5.2. ☆

We recall the following definition from Theorem 1.3.7. Let $U \subset \mathbf{R}^n$ and let $g : U \to \mathbf{R}^p$. For all $c \in \mathbf{R}^p$ we define level set $N(c)$ by

$$N(c) = N(g, c) = \{\, x \in U \mid g(x) = c \,\}.$$

Theorem 4.5.2 (Submersion Theorem). *Let $U_0 \subset \mathbf{R}^n$ be an open subset, let $1 \le d < n$ and $k \in \mathbf{N}_\infty$, and let $g : U_0 \to \mathbf{R}^{n-d}$ be a C^k mapping. Let $x^0 \in U_0$, let $g(x^0) = c^0$, and assume g to be a submersion in x^0, that is $Dg(x^0) \in \mathrm{Lin}(\mathbf{R}^n, \mathbf{R}^{n-d})$ is surjective. Then there exist an open neighborhood U of x^0 in U_0 and an open neighborhood C of c^0 in $\mathbf{R}^{n-d}$ such that the following hold.*

(i) *The restriction of g to U is an open surjection.*

(ii) *The set $N(c) \cap U$ is a C^k submanifold in $\mathbf{R}^n$ of dimension d, for all $c \in C$.*

(iii) *There exists a C^k diffeomorphism $\Phi : U \to \mathbf{R}^n$ such that Φ maps the manifold $N(c) \cap U$ in $\mathbf{R}^n$ into the linear submanifold of $\mathbf{R}^n$ given by*

$$\{\, (x_1, \dots, x_n) \in \mathbf{R}^n \mid (x_{d+1}, \dots, x_n) = c \,\}.$$

(iv) *There exists a C^k diffeomorphism $\Psi : \Phi(U) \to U$ such that*

$$g \circ \Psi(y) = (y_{d+1}, \dots, y_n).$$

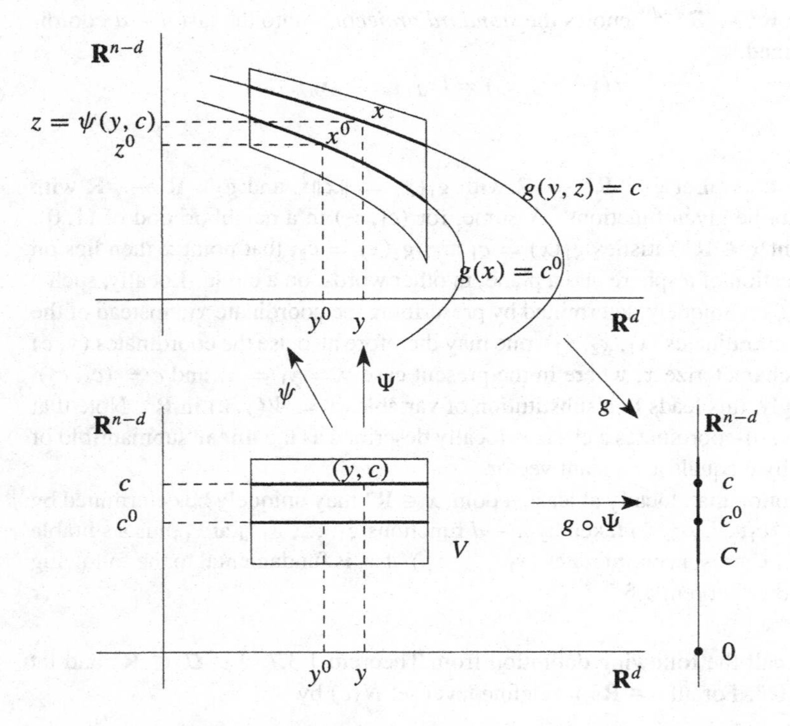

Submersion Theorem: near x^0 the coordinate transformation $\Phi := \Psi^{-1}$ straightens out the curved level sets $g(x) = c$

Proof. **(i).** Because $Dg(x^0)$ has rank $n - d$, among the columns of the matrix of $Dg(x^0)$ there are $n - d$ which are linearly independent. We assume that these are the $n - d$ columns at the right (this may require prior permutation of the coordinates of $\mathbf{R}^n$). Hence we can write $x \in \mathbf{R}^n$ as

$$x = (y, z) \qquad \text{with} \qquad y = (x_1, \dots, x_d) \in \mathbf{R}^d, \qquad z = (x_{d+1}, \dots, x_n) \in \mathbf{R}^{n-d},$$

while also

$$Dg(x) = {}_{n-d}\overset{\qquad\; d \qquad\qquad\quad n-d}{\Big(D_y g(y; z) \;\Big|\; D_z g(y; z)\Big)},$$

$$g(y^0; z^0) = c^0, \qquad D_z g(y^0; z^0) \in \mathrm{Aut}(\mathbf{R}^{n-d}).$$

The mapping $\mathbf{R}^{n-d} \supset\!\!\to \mathbf{R}^{n-d}$, defined by $z \mapsto g(y^0, z)$ satisfies the conditions of the Local Inverse Function Theorem 3.2.4. This implies assertion (i).

(ii). It also follows from the arguments above that we may apply the Implicit Function Theorem 3.5.1 to the system of equations

$$g(y; z) - c = 0 \qquad (y \in \mathbf{R}^d, \; z \in \mathbf{R}^{n-d}, \; c \in \mathbf{R}^{n-d})$$

in the unknown z with y and c as parameters. Accordingly, there exist η, γ and $\zeta > 0$ such that for every

$$(y, c) \in V := \{ (y, c) \in \mathbf{R}^d \times \mathbf{R}^{n-d} \mid \|y - y^0\| < \eta, \ \|c - c^0\| < \gamma \}$$

there is a unique $z = \psi(y, c) \in \mathbf{R}^{n-d}$ satisfying

$$(y, z) \in U_0, \qquad \|z - z^0\| < \zeta, \qquad g(y; z) = c.$$

Moreover, $\psi : V \to \mathbf{R}^{n-d}$, defined by $(y, c) \mapsto \psi(y, c)$, is a C^k mapping. If we now define $C := \{ c \in \mathbf{R}^{n-d} \mid \|c - c^0\| < \gamma \}$, we have, for $c \in C$

$$N(c) \cap \{ (y, z) \in \mathbf{R}^n \mid \|y - y^0\| < \eta, \ \|z - z^0\| < \zeta \}$$
$$= \{ (y, \ \psi(y, c)) \mid \ \|y - y^0\| < \eta \}.$$

Locally, therefore, and with c constant, $N(c)$ is the graph of the C^k function $y \mapsto \psi(y, c)$ with an open domain in $\mathbf{R}^d$; but this proves (ii).
(iii). Define

$$\Phi : U_0 \to \mathbf{R}^n \qquad \text{by} \qquad \Phi(y, z) = (y, \ g(y; z)).$$

From the proof of part (ii) we then have $\Phi(y, z) = (y, c) \in V$ if and only if $(y, z) = (y, \ \psi(y, \ c))$. It follows that, locally in a neighborhood of x^0, the mapping Φ is invertible with a C^k inverse. In other words, there exists an open neighborhood U of x^0 in U_0 such that $\Phi : U \to V$ is a C^k diffeomorphism of open sets in $\mathbf{R}^n$. For $(y, z) \in N(c) \cap U$ we have $g(y; z) = c$; and so $\Phi(y, z) = (y, c)$. Therefore we have now proved (iii).
(iv). Finally, let $\Psi := \Phi^{-1} : V \to U$. Because $\Phi(y, z) = (y, c)$ if and only if $g(y; z) = c$, one has for $(y, c) \in V$

$$g \circ \Psi(y, c) = g(y, z) = c.$$

Finally we note that in Exercise 7.36 we make use of the fact that one may assume, if $d = n - 1$ (and for smaller values of η and $\gamma > 0$ if necessary), that V is the open set in $\mathbf{R}^n$ with

$$V := \{ (y, c) \in \mathbf{R}^{n-1} \times \mathbf{R} \mid \ \|y - y^0\| < \eta, \ |c - c^0| < \gamma \},$$

and that U is the open neighborhood of x^0 in $\mathbf{R}^n$ defined by

$$U := \{ (y, z) \in \mathbf{R}^{n-1} \times \mathbf{R} \mid \ \|y - y^0\| < \eta, \ |g(y; z) - c^0| < \gamma \}.$$

To prove this, show by means of a first-order Taylor expansion that the estimate $\|z - z^0\| < \zeta$ follows from an estimate of the type $|g(y; z) - c^0| < \gamma$, because $D_z g(y^0; z^0) \neq 0$. ❑

Remark. A mapping $g : \mathbf{R}^n \supset\!\!\to \mathbf{R}^{n-d}$ is also said to be *regular at* x^0 if it is submersive at x^0. Note that this is the case if and only if rank $Dg(x^0)$ is maximal (compare with the definitions of regularity in Sections 3.2 and 4.3). Accordingly, mappings $\mathbf{R}^n \supset\!\!\to \mathbf{R}^{n-d}$ with $d \geq 0$ in general may be expected to be submersive.

If $g(x^0) = c^0$, then $N(c^0)$ is also said to be the *fiber of the mapping g* belonging to the value c^0. For regular x^0 this therefore looks like a manifold. The fibers $N(c)$ together form a *fiber bundle*: under the diffeomorphism Φ from the Submersion Theorem they are locally transferred into the linear submanifolds of $\mathbf{R}^n$ given by the last $n - d$ coordinates being constant.

4.6 Examples of submersions

Example 4.6.1 (Unit sphere S^{n-1}). (See Example 4.2.3). The *unit sphere*

$$S^{n-1} = \{ x \in \mathbf{R}^n \mid \|x\| = 1 \}$$

is a C^∞ submanifold in $\mathbf{R}^n$ of dimension $n - 1$. Indeed, defining the C^∞ mapping $g : \mathbf{R}^n \to \mathbf{R}$ by

$$g(x) = \|x\|^2,$$

one has $S^{n-1} = N(1)$. Let $x^0 \in S^{n-1}$, then

$$Dg(x^0) \in \operatorname{Lin}(\mathbf{R}^n, \mathbf{R}) \qquad \text{given by} \qquad Dg(x^0) = 2x^0$$

is surjective, x^0 being different from 0. But then, according to the Submersion Theorem there exists a neighborhood U of x^0 in $\mathbf{R}^n$ such that $S^{n-1} \cap U$ is a C^∞ submanifold in $\mathbf{R}^n$ of dimension $n - 1$.

More generally, the Submersion Theorem asserts that, locally near x^0, the fiber $N(g(x^0))$ can be described by writing a suitably chosen x_i as a C^∞ function of the x_j with $j \neq i$. ☆

Example 4.6.2 (Orthogonal matrices). The set $\mathbf{O}(n, \mathbf{R})$, the *orthogonal group*, consisting of the *orthogonal* matrices in $\operatorname{Mat}(n, \mathbf{R}) \simeq \mathbf{R}^{n^2}$, with

$$\mathbf{O}(n, \mathbf{R}) = \{ A \in \operatorname{Mat}(n, \mathbf{R}) \mid A^t A = I \}$$

is a C^∞ manifold of dimension $\frac{1}{2}n(n-1)$ in $\operatorname{Mat}(n, \mathbf{R})$. Let $\operatorname{Mat}^+(n, \mathbf{R})$ be the linear subspace in $\operatorname{Mat}(n, \mathbf{R})$ consisting of the symmetric matrices. Note that $A = (a_{ij}) \in \operatorname{Mat}^+(n, \mathbf{R})$ if and only if $a_{ij} = a_{ji}$. This means that A is completely determined by the a_{ij} with $i \geq j$, of which there are $1 + 2 + \cdots + n = \frac{1}{2}n(n+1)$. Consequently, $\operatorname{Mat}^+(n, \mathbf{R})$ is a linear subspace of $\operatorname{Mat}(n, \mathbf{R})$ of dimension $\frac{1}{2}n(n+1)$. We have

$$\mathbf{O}(n, \mathbf{R}) = g^{-1}(\{I\}),$$

where $g : \mathrm{Mat}(n, \mathbf{R}) \to \mathrm{Mat}^+(n, \mathbf{R})$ is the mapping with

$$g(A) = A^t A.$$

Note that indeed $(A^t A)^t = A^t A^{tt} = A^t A$. We shall now prove that g is a submersion at every $A \in \mathbf{O}(n, \mathbf{R})$. Since $\mathrm{Mat}(n, \mathbf{R})$ and $\mathrm{Mat}^+(n, \mathbf{R})$ are linear spaces, we may write, for every $A \in \mathbf{O}(n, \mathbf{R})$

$$Dg(A) : \mathrm{Mat}(n, \mathbf{R}) \to \mathrm{Mat}^+(n, \mathbf{R}).$$

Observe that g can be written as the composition $A \mapsto (A, A) \mapsto A^t A = \widetilde{g}(A, A)$, where $\widetilde{g} : (A_1, A_2) \mapsto A_1^t A_2$ is bilinear, that is $\widetilde{g} \in \mathrm{Lin}^2(\mathrm{Mat}(n, \mathbf{R}), \mathrm{Mat}(n, \mathbf{R}))$. Using Proposition 2.7.6 we now obtain, with $H \in \mathrm{Mat}(n, \mathbf{R})$,

$$Dg(A) : H \mapsto (H, H) \mapsto \widetilde{g}(H, A) + \widetilde{g}(A, H) = H^t A + A^t H.$$

Therefore, the surjectivity of $Dg(A)$ follows if, for every $C \in \mathrm{Mat}^+(n, \mathbf{R})$, we can find a solution $H \in \mathrm{Mat}(n, \mathbf{R})$ of $H^t A + A^t H = C$. Now $C = \frac{1}{2}C^t + \frac{1}{2}C$. So we try to solve H from

$$H^t A = \frac{1}{2}C^t \quad \text{and} \quad A^t H = \frac{1}{2}C; \qquad \text{hence} \qquad H = \frac{1}{2}(A^{-1})^t C = \frac{1}{2}AC.$$

We now see that the conditions of the Submersion Theorem are satisfied; therefore, near $A \in \mathbf{O}(n, \mathbf{R})$ the subset $\mathbf{O}(n, \mathbf{R}) = g^{-1}(\{I\})$ of $\mathrm{Mat}(n, \mathbf{R})$ is a C^∞ manifold of dimension equal to

$$\dim \mathrm{Mat}(n, \mathbf{R}) - \dim \mathrm{Mat}^+(n, \mathbf{R}) = n^2 - \frac{1}{2}n(n+1) = \frac{1}{2}n(n-1).$$

Because this is true for every $A \in \mathbf{O}(n, \mathbf{R})$ the assertion follows. ☆

Example 4.6.3 (Torus). We come back to Example 4.4.3. There an open subset V of a toroidal surface T was given as a parametrized set, that is, as the image $V = \phi(D)$ under $\phi : D = \,]-\pi, \pi\,[\times]-\pi, \pi\,[\to \mathbf{R}^3$ with

$$\phi(\alpha, \theta) = x = ((2 + \cos\theta)\cos\alpha,\ (2 + \cos\theta)\sin\alpha,\ \sin\theta). \tag{4.3}$$

We now try to write the set V, or, if we have to, a somewhat larger set, as the zero-set of a function $g : \mathbf{R}^3 \to \mathbf{R}$ to be determined. To achieve this, we *eliminate* α and θ from Formula (4.3), that is to say, we try to find a relation between the coordinates x_1, x_2 and x_3 of x which follows from the fact that $x \in V$. We have, for $x \in V$,

$$x_1^2 = (2 + \cos\theta)^2 \cos^2\alpha, \qquad x_2^2 = (2 + \cos\theta)^2 \sin^2\alpha, \qquad x_3^2 = \sin^2\theta.$$

This gives

$$\|x\|^2 = (2 + \cos\theta)^2 + \sin^2\theta = 4 + 4\cos\theta + \cos^2\theta + \sin^2\theta = 5 + 4\cos\theta.$$

Therefore $(\|x\|^2 - 5)^2 = 16\cos^2\theta$ and $16x_3^2 = 16\sin^2\theta$. And so

$$(\|x\|^2 - 5)^2 + 16x_3^2 = 16. \tag{4.4}$$

There is a problem, however, in that we have only proved that every $x \in V$ satisfies equation (4.4). Conversely, it can be shown that the toroidal surface T appears as the zero-set N of the function $g : \mathbf{R}^3 \to \mathbf{R}$, defined by

$$g(x) = (\|x\|^2 - 5)^2 + 16(x_3^2 - 1).$$

(The proof, which is not very difficult, is not given here.) Next, we examine whether g is submersive at the points $x \in N$. For this to be so, $Dg(x) \in \mathrm{Lin}(\mathbf{R}^3, \mathbf{R})$ must be of rank 1, which is in fact the case except when

$$Dg(x) = 4(x_1(\|x\|^2 - 5),\ x_2(\|x\|^2 - 5),\ x_3(\|x\|^2 + 3)) = 0.$$

This immediately implies $x_3 = 0$. Also, it follows from the first two coefficients that $\|x\|^2 - 5 = 0$, because $x = 0$ does not satisfy equation (4.4). But these conditions on $x \in N$ violate equation (4.4); consequently, g is submersive in all of N. By the Submersion Theorem it now follows that N is a C^∞ submanifold in $\mathbf{R}^3$ of dimension 2. ☆

4.7 Equivalent definitions of manifold

The preceding yields the useful result that the local descriptions of a manifold as a graph, as a parametrized set, as a zero-set, and as a set which locally looks like $\mathbf{R}^d$ and which is "flat" in $\mathbf{R}^n$, are all entirely equivalent.

Theorem 4.7.1. *Let V be a subset of $\mathbf{R}^n$ and let $x \in V$, and suppose $0 \le d \le n$ and $k \in \mathbf{N}_\infty$. Then the following four assertions are equivalent.*

(i) *There exist an open neighborhood U in $\mathbf{R}^n$ of x, an open subset W of $\mathbf{R}^d$ and a C^k mapping $f : W \to \mathbf{R}^{n-d}$ such that*

$$V \cap U = \mathrm{graph}(f) = \{\, (w, f(w)) \in \mathbf{R}^n \mid w \in W \,\}.$$

(ii) *There exist an open neighborhood U in $\mathbf{R}^n$ of x, an open subset D of $\mathbf{R}^d$ and a C^k embedding $\phi : D \to \mathbf{R}^n$ such that*

$$V \cap U = \mathrm{im}(\phi) = \{\, \phi(y) \in \mathbf{R}^n \mid y \in D \,\}.$$

(iii) *There exist an open neighborhood U in $\mathbf{R}^n$ of x and a C^k submersion $g : U \to \mathbf{R}^{n-d}$ such that*

$$V \cap U = N(g, 0) = \{ u \in U \mid g(u) = 0 \}.$$

(iv) *There exist an open neighborhood U in $\mathbf{R}^n$ of x, a C^k diffeomorphism $\Phi : U \to \mathbf{R}^n$ and an open subset Y of $\mathbf{R}^d$ such that*

$$\Phi(V \cap U) = Y \times \{0_{\mathbf{R}^{n-d}}\}.$$

Proof. **(i) $\Rightarrow$ (ii)** is Remark 1 in Section 4.2. **(ii) $\Rightarrow$ (i)** is Corollary 4.3.2. **(i) $\Rightarrow$ (iii)** is Remark 2 in Section 4.2. **(iii) $\Rightarrow$ (i)** and **(iii) $\Rightarrow$ (iv)** follow by the Submersion Theorem 4.5.2, and **(iv) $\Rightarrow$ (ii)** is trivial. ❑

Remark. Which particular description of a manifold V to choose will depend on the circumstances; the way in which V is initially given is obviously important. But there is a rule of thumb:

- when the "internal" structure of V is important, as in the integration over V in Chapters 7 or 8, a description as an image under an embedding or as a graph is useful;
- when one wishes to study the "external" structure of V, for example, what is the location of V with respect to the ambient space $\mathbf{R}^n$, the description as a zero-set often is to be preferred.

Corollary 4.7.2. *Suppose V is a C^k submanifold in $\mathbf{R}^n$ of dimension d and $\Phi : \mathbf{R}^n \supset\!\to \mathbf{R}^n$ is a C^k diffeomorphism which is defined on an open neighborhood of V, then $\Phi(V)$ is a C^k submanifold in $\mathbf{R}^n$ of dimension d too.*

Proof. We use the local description $V \cap U = \operatorname{im}(\phi)$ according to Theorem 4.7.1.(ii). Then

$$\Phi(V) \cap \Phi(U) = \Phi(V \cap U) = \operatorname{im}(\Phi \circ \phi),$$

where $\Phi \circ \phi : D \to \mathbf{R}^n$ is a C^k embedding, since ϕ is so. ❑

In fact, we can introduce the notion of a C^k mapping of manifolds without recourse to ambient spaces.

Definition 4.7.3. Suppose V is a submanifold in $\mathbf{R}^n$ of dimension d, and V' a submanifold in $\mathbf{R}^{n'}$ of dimension d', and let $k \in \mathbf{N}_\infty$. A *mapping of manifolds* $\Phi : V \to V'$ is said to be a C^k *mapping* if for every $x \in V$ there exist neighborhoods U of x in $\mathbf{R}^n$, and U' of $\Phi(x)$ in $\mathbf{R}^{n'}$, open sets $D \subset \mathbf{R}^d$, and $D' \subset \mathbf{R}^{d'}$, and C^k embeddings $\phi : D \to V \cap U$, and $\phi' : D' \to V' \cap U'$, respectively, such that

$$\phi'^{-1} \circ \Phi \circ \phi : D \supset\!\!\to \mathbf{R}^{d'}$$

is a C^k mapping. It is a consequence of Lemma 4.3.3.(iii) that the particular choices of the embeddings ϕ and ϕ' in this definition are irrelevant. ❍

This definition hints at a theory of manifolds that does not require a manifold a priori to be a subset of some ambient space $\mathbf{R}^n$. In such a theory, all properties of manifolds are formulated in terms of the embeddings ϕ or the charts $\kappa = \phi^{-1}$.

Remark. To say that a subset V of $\mathbf{R}^n$ is a smooth d-dimensional submanifold is to make a statement about the local behavior of V. A global, algebraic, variant of the characterization as a zero-set is the following definition: $V \subset \mathbf{R}^n$ is said to be an *algebraic submanifold* of $\mathbf{R}^n$ if there are polynomials $g_1, \dots, g_{n-d}$ in n variables for which

$$V = \{\, x \in \mathbf{R}^n \mid g_i(x) = 0,\ 1 \le i \le n-d \,\}.$$

Here, on account of the algebraic character of the definition, $\mathbf{R}$ may be replaced by an arbitrary number system (field) k. Because k^n is the standard model of the n-dimensional affine space (over the field k), the terminology *affine algebraic manifold* V is also used; further, if $k = \mathbf{R}$, this manifold is said to be *real*. When doing so, one can also formulate the dimension of V, as well as its being smooth (where applicable), in purely algebraic terms; this will not be pursued here (but see Exercise 5.76). Still, it will be clear that if $k = \mathbf{R}$ and if $Dg_1(x), \dots, Dg_{n-d}(x)$, for every $x \in V$, are linearly independent, the real affine algebraic manifold V is also smooth, and of dimension d. Without the assumption about the gradients, V can have interesting singularities, one example being the quadratic cone $x_1^2 + x_2^2 - x_3^2 = 0$ in $\mathbf{R}^3$. Note that V is always a closed subset of $\mathbf{R}^n$ because polynomials are continuous functions.

4.8 Morse's Lemma

Let $U \subset \mathbf{R}^n$ be open and $f \in C^k(U)$ with $k \ge 1$. In Theorem 2.6.3.(iv) we have seen that $x^0 \in U$ is a critical point of f if f attains a local extremum at x^0. Generally, x^0 is said to be a *singular* (= nonregular, because nonsubmersive, (see the Remark at the end of Section 4.5) or *critical point* of f if $\operatorname{grad} f(x^0) = 0$. From the Submersion Theorem 4.5.2 we know that near regular points the level surfaces

$$N(c) = \{\, x \in U \mid f(x) = c \,\}$$

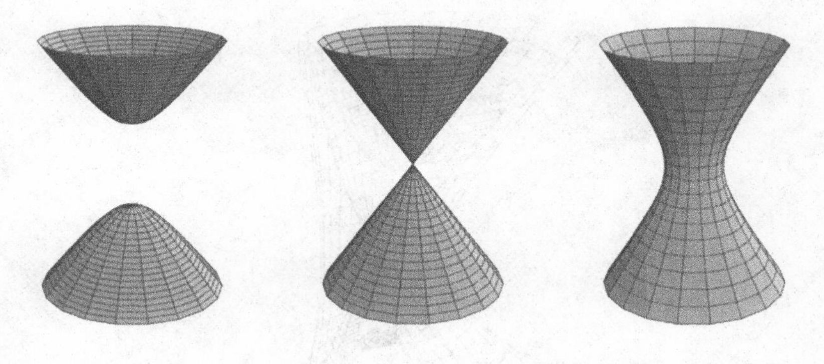

Illustration for Morse's Lemma

appear as graphs of C^k functions: one of the coordinates is a C^k function of the other $n-1$ coordinates. Near a critical point the level surfaces will in general behave in a much more complicated way; the illustration shows the level surfaces for $f : \mathbf{R}^3 \to \mathbf{R}$ with $f(x) = x_1^2 + x_2^2 - x_3^2 = c$ for $c < 0$, $c = 0$, and $c > 0$, respectively.

Now suppose that $U \subset \mathbf{R}^n$ is also convex and that $f \in C^k(U)$, for $k \geq 2$, has a nondegenerate critical point at x^0 (see Definition 2.9.8), that is, the self-adjoint operator $Hf(x^0) \in \operatorname{End}^+(\mathbf{R}^n)$, which is associated with the Hessian $D^2 f(x^0) \in \operatorname{Lin}^2(\mathbf{R}^n, \mathbf{R})$, actually belongs to $\operatorname{Aut}(\mathbf{R}^n)$. Using the Spectral Theorem 2.9.3 we then can find $A \in \operatorname{Aut}(\mathbf{R}^n)$ such that the linear change of variables $x = Ay$ in $\mathbf{R}^n$ puts $Hf(x^0)$ into the normal form

$$y \mapsto y_1^2 + \cdots + y_p^2 - y_{p+1}^2 - \cdots - y_n^2.$$

Note that nondegeneracy forbids having 0 as an eigenvalue.

In Section 2.9 we remarked already that, in this case, the second-derivative test from Theorem 2.9.7 says that $Hf(x^0)$ determines the main features of the behavior of f near x^0: whether f has a relative maximum or minimum or a saddle point at x^0. However, a stronger result is valid. The following result, called *Morse's Lemma*, asserts that the function f can, in a neighborhood of a nondegenerate critical point x^0, be made equal to a quadratic polynomial, the coefficients of which constitute the Hessian matrix of f at x^0 in normal form. Phrased differently, f can be brought into the normal form

$$g(y) = f(x^0) + y_1^2 + \cdots + y_p^2 - y_{p+1}^2 - \cdots - y_n^2$$

by means of a regular substitution of variables $x = \Psi(y)$, such that the level surfaces of f appear as the image of the level surfaces of g under the diffeomorphism Ψ.

The principal importance of Morse's Lemma lies in the detailed description in saddle point situations, when $Hf(x^0)$ has both positive and negative eigenvalues.

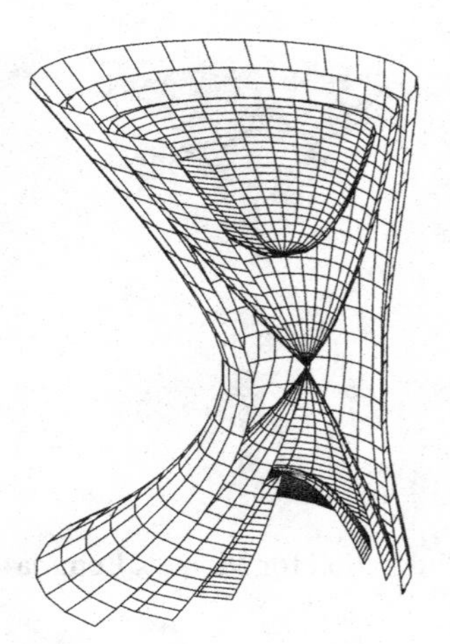

Illustration for Morse's Lemma: Pointed caps

For $n = 3$ and $p = 2$ the level surface $\{ x \in U \mid f(x) = f(x^0) \}$ for the level $f(x^0)$ will thus appear as the image under Ψ of two pointed caps directed at each other and with their apices exactly at the origin. This image under Ψ looks similar, with the difference that the apices of the cones now coincide at x^0 and that the cones may have been extended in some directions and compressed in others; in addition they may have tilt and warp. But we still have two pointed caps directed at each other intersecting at a point. The level surfaces of f for neighboring values c can also be accurately described: if $c > c^0 = f(x^0)$, with c near c^0, they are connected near x^0, whereas this is no longer the case if $c < c^0$, with c near c^0.

If $p = n$, things look altogether different: the level surface near x^0 for the level $c > c^0 = f(x^0)$, with c near c^0, is the Ψ-image of an $(n-1)$-dimensional sphere centered at the origin and with radius $\sqrt{c - c^0}$. If $c = c^0$ it becomes a point, and if $c < c^0$ the level set near x^0 is empty. f attains a local minimum at x^0.

Theorem 4.8.1. *Let $U_0 \subset \mathbf{R}^n$ be open and $f \in C^k(U_0)$ for $k \geq 3$. Let $x^0 \in U_0$ be a nondegenerate critical point of f. Then there exist open neighborhoods $U \subset U_0$ of x^0 and V of 0 in $\mathbf{R}^n$, respectively, and a C^{k-2} diffeomorphism Ψ of V onto U such that $\Psi(0) = x^0$, $D\Psi(0) = I$ and*

$$f \circ \Psi(y) = f(x^0) + \frac{1}{2} \langle Hf(x^0)y, y \rangle \qquad (y \in V).$$

Proof. By means of the substitution of variables $x = x^0 + h$ we reduce this to the case $x^0 = 0$. Taylor's formula for f at 0 with the integral formula for the first remainder from Theorem 2.8.3 yields

$$f(x) = f(0) + \langle Q(x)x, x \rangle \qquad (\|x\| < \delta).$$

Here $\delta > 0$ has been chosen sufficiently small; and

$$Q(x) = \int_0^1 (1-t)\, Hf(tx)\, dt \in \mathrm{End}(\mathbf{R}^n)$$

has C^{k-2} dependence on x, while $Q(0) = \frac{1}{2}Hf(0)$. Next it follows from Theorem 2.7.9 that $Hf(0)$, and hence also $Q(0) \in \mathrm{End}^+(\mathbf{R}^n)$, in other words, these operators are self-adjoint. We now wish to find out whether we may write

$$\langle Q(x)x, x\rangle = \langle Q(0)y, y\rangle$$

with $y = A(x)x$, where $A(x) \in \mathrm{End}(\mathbf{R}^n)$ is suitably chosen. Because

$$\langle Q(0) \circ A(x)x,\ A(x)x\rangle = \langle A(x)^t \circ Q(0) \circ A(x)x,\ x\rangle,$$

we see that this can be arranged by ensuring that $A = A(x)$ satisfies the equation

$$F(A, x) := A^t \circ Q(0) \circ A - Q(x) = 0.$$

Now $F(I, 0) = 0$, and it proves convenient to try $A = I + \frac{1}{2}Q(0)^{-1}B$ with $B \in \mathrm{End}^+(\mathbf{R}^n)$, near 0. In other words, we switch to the equation

$$\begin{aligned} 0 &= G(B, x) := (I + \tfrac{1}{2}Q(0)^{-1}B)^t \circ Q(0) \circ (I + \tfrac{1}{2}Q(0)^{-1}B) - Q(x) \\ &= B + \tfrac{1}{4}B \circ Q(0)^{-1} \circ B + Q(0) - Q(x). \end{aligned}$$

Now $G(0, 0) = 0$, and $D_B(0, 0)$ equals the identity mapping of the linear space $\mathrm{End}^+(\mathbf{R}^n)$ into itself. Application of the Implicit Function Theorem 3.5.1 gives an open neighborhood U_1 of 0 in $\mathbf{R}^n$ and a C^{k-2} mapping B from U_1 into the linear space $\mathrm{End}^+(\mathbf{R}^n)$ with $B(0) = 0$ and $G(B(x), x) = 0$, for all $x \in U_1$. Writing $\Phi(x) = A(x)x = (I + \frac{1}{2}Q(0)^{-1}B(x))x$, we then have $\Phi \in C^{k-2}(U_1, \mathbf{R}^n)$, $\Phi(0) = 0$ and $D\Phi(0) = I$. Furthermore,

$$\langle Q(x)x, x\rangle = \langle Q(0) \circ \Phi(x),\ \Phi(x)\rangle \qquad (x \in U_1).$$

By the Local Inverse Function Theorem 3.2.4 there is an open $U \subset U_1$ around 0 such that $\Phi|_U$ is a C^{k-2} diffeomorphism onto an open neighborhood V of 0 in $\mathbf{R}^n$. Now $\Psi = (\Phi|_U)^{-1}$ meets all requirements. ❑

Lemma 4.8.2 (Morse). *Let the notation be as in Theorem 4.8.1. We make the further assumption that among the eigenvalues of $Hf(x^0)$ there are p that are positive and $n - p$ that are negative. Then there exist open neighborhoods U of x^0 and W of 0, respectively, in $\mathbf{R}^n$, and a C^{k-2} diffeomorphism Ξ from W onto U such that $\Xi(0) = x^0$ and*

$$f \circ \Xi(w) = f(x^0) + w_1^2 + \cdots + w_p^2 - w_{p+1}^2 - \cdots - w_n^2 \qquad (w \in W).$$

Proof. Let $\lambda_1, \dots, \lambda_n$ be the (real) eigenvalues of $Hf(x^0) \in \mathrm{End}^+(\mathbf{R}^n)$, each eigenvalue being repeated a number of times equal to its multiplicity, with $\lambda_1, \dots, \lambda_p$ positive and $\lambda_{p+1}, \dots, \lambda_n$ negative. From Formula (2.29) it is known that there is an orthogonal operator $O \in \mathbf{O}(\mathbf{R}^n)$ such that

$$\frac{1}{2}\langle Hf(x^0) \circ Oz, Oz\rangle = \frac{1}{2}\sum_{1 \le j \le n} \lambda_j z_j^2.$$

Under the substitution $z_j = (2/|\lambda_j|)^{\frac{1}{2}} w_j$ this takes the form

$$w_1^2 + \cdots + w_p^2 - w_{p+1}^2 - \cdots - w_n^2.$$

If $P \in \mathrm{End}(\mathbf{R}^n)$ has a diagonal matrix with $P_{jj} = (2/|\lambda_j|)^{\frac{1}{2}}$, Morse's Lemma immediately follows from the preceding theorem by writing $\Xi = \Psi \circ O \circ P$. ❑

Note that $D\Xi(0) = OP$; this information could have been added to Morse's Lemma. Here P is responsible for the "extension" or the "compression", respectively, in the various directions, and O for the "tilt". The terms of higher order in the Taylor expansion of Ξ at 0 are responsible for the "warp" of the level surfaces discussed in the introduction.

Remark. At this stage, Theorem 2.9.7 in the case of a nondegenerate critical point is a consequence of Morse's Lemma.

Chapter 5

Tangent Spaces

Differentiable mappings can be approximated by affine mappings; similarly, manifolds can locally be approximated by affine spaces, which are called (geometric) tangent spaces. These spaces provide enhanced insight into the structure of the manifold. In Volume II, in Chapter 7 the concept of tangent space plays an essential role in the integration on a manifold; the same applies to the study of the boundary of an open set, an important subject in the extension of the Fundamental Theorem of Calculus to higher-dimensional spaces. In this chapter we consider various other applications of tangent spaces as well: cusps, normal vectors, extrema of the restriction of a function to a submanifold, and the curvature of curves and surfaces. We close this chapter with introductory remarks on one-parameter groups of diffeomorphisms, and on linear Lie groups and their Lie algebras, which are important tools in describing continuous symmetry.

5.1 Definition of tangent space

Let $k \in \mathbf{N}_\infty$, let V be a C^k submanifold in $\mathbf{R}^n$ of dimension d and let $x \in V$. We want to define the (geometric) tangent space of V at the point x; this is, locally at x, the "best" approximation of V by a d-dimensional affine manifold (= translated linear subspace). In Section 2.2 we encountered already the description of the tangent space of $\operatorname{graph}(f)$ at the point $(w, f(w))$ as the graph of the affine mapping $h \mapsto f(w)+Df(w)h$. In the following definition of the tangent space of a manifold at a point, however, we wish to avoid the assumption of a graph as the only local description of the manifold. This is why we shall base our definition of tangent space on the concept of *tangent vector of a curve* in $\mathbf{R}^n$, which we introduce in the following way.

Let $I \subset \mathbf{R}$ be an open interval and let $\gamma : I \to \mathbf{R}^n$ be a differentiable mapping.

The image $\gamma(I)$, and, for that matter, γ itself, is said to be a *differentiable (space) curve* in $\mathbf{R}^n$. The vector

$$\gamma'(t) = \begin{pmatrix} \gamma_1'(t) \\ \vdots \\ \gamma_n'(t) \end{pmatrix} \in \mathbf{R}^n,$$

is said to be a *tangent vector of $\gamma(I)$ at the point $\gamma(t)$*.

Definition 5.1.1. Let V be a C^1 submanifold in $\mathbf{R}^n$ of dimension d, and let $x \in V$. A vector $v \in \mathbf{R}^n$ is said to be a *tangent vector of V at the point x* if there exist a differentiable curve $\gamma : I \to \mathbf{R}^n$ and a $t_0 \in I$ such that

$$\gamma(t) \in V \quad \text{for all} \quad t \in I; \qquad \gamma(t_0) = x; \qquad \gamma'(t_0) = v.$$

The set of the tangent vectors of V at the point x is said to be the *tangent space $T_x V$ of V at the point x*. ❍

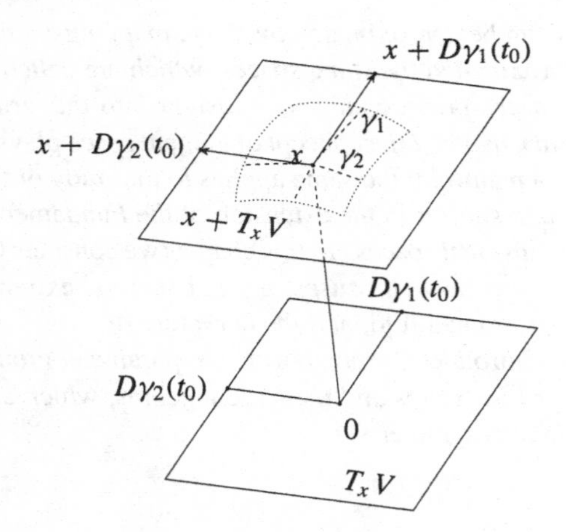

Illustration for Definition 5.1.1

An immediate consequence of this definition is that $\lambda v \in T_x V$, if $\lambda \in \mathbf{R}$ and $v \in T_x V$. Indeed, we may assume $\gamma(t) \in V$ for $t \in I$, with $\gamma(0) = x$ and $\gamma'(0) = v$. Considering the differentiable curve $\gamma_\lambda(t) = \gamma(\lambda t)$ one has, for t sufficiently small, $\gamma_\lambda(t) \in V$, while $\gamma_\lambda(0) = x$ and $\gamma_\lambda'(0) = \lambda v$. It is less obvious that $v + w \in T_x V$, if $v,\ w \in T_x V$. But this can be seen from the following result.

Theorem 5.1.2. *Let $k \in \mathbf{N}_\infty$, let V be a d-dimensional C^k manifold in $\mathbf{R}^n$ and let $x \in V$. Assume that, locally at x, the manifold V is described as in Theorem 4.7.1, in particular, that $x = (w, f(w)) = \phi(y)$ and $g(x) = 0$. Then*

$$T_x V = \operatorname{graph}(Df(w)) = \operatorname{im}(D\phi(y)) = \ker(Dg(x)).$$

In particular, $T_x V$ is a d-dimensional linear subspace of $\mathbf{R}^n$.

Proof. We use the notations from Theorem 4.7.1. Let $h \in \mathbf{R}^d$. Then there exists an $\epsilon > 0$ such that $w + th \in W$, for all $|t| < \epsilon$. Consequently, $\gamma : t \mapsto (w + th,\ f(w + th))$ is a differentiable curve in $\mathbf{R}^n$ such that

$$\gamma(t) \in V \quad (|t| < \epsilon); \qquad \gamma(0) = (w,\ f(w)) = x; \qquad \gamma'(0) = (h,\ Df(w)h).$$

But this implies graph $(Df(w)) \subset T_x V$. And it is equally true that im $(D\phi(y)) \subset T_x V$. Now let $v \in T_x V$, and assume $v = \gamma'(t_0)$. It then follows that $(g \circ \gamma)(t) = 0$, for $t \in I$ (this may require I to be chosen smaller). Differentiation with respect to t at t_0 gives

$$0 = D(g \circ \gamma)(t_0) = Dg(x) \circ \gamma'(t_0) = Dg(x)v.$$

Therefore we now have

$$\operatorname{graph}(Df(w)) \cup \operatorname{im}(D\phi(y)) \subset T_x V \subset \ker(Dg(x)).$$

Since $h \mapsto (h,\ Df(w)h)$ is an injective linear mapping $\mathbf{R}^d \to \mathbf{R}^n$, while $Dg(x) \in \operatorname{Lin}(\mathbf{R}^n, \mathbf{R}^{n-d})$ is surjective, it follows that

$$\dim \operatorname{graph}(Df(w)) = \dim \operatorname{im}(D\phi(y)) = \dim \ker(Dg(x)) = d.$$

This proves the theorem. ❑

Remark. In classical textbooks, and in drawings, it is more common to refer to the linear manifold $x + T_x V$ as the tangent space of V at the point x: the linear manifold which has a contact of order 1 with V at x. What one does is to represent $v \in T_x V$ as an arrow, originating at x and with its head at $x + v$. In case of ambiguity we shall refer to $x + T_x V$ as the *geometric tangent space of* V *at* x. Obviously, x plays a special role in this tangent space. Choosing the point x as the origin, one reobtains the identification of the geometric tangent space with the linear space $T_x V$. Indeed, the origin 0 has a special meaning in $T_x V$: it is the velocity vector of the constant curve $\gamma(t) \equiv x$. Experience shows that it is convenient to regard $T_x V$ as a linear space.

Remark. Consider the curves $\gamma, \delta : \mathbf{R} \to \mathbf{R}^2$ defined by $\gamma(t) = (\cos t,\ \sin t)$ and $\delta(t) = \gamma(t^2)$. The image of both curves is the unit circle in $\mathbf{R}^2$. On the other hand, one has

$$\|\gamma'(t)\| = \|(-\sin t,\ \cos t)\| = 1; \qquad \|\delta'(t)\| = 2t\|(-\sin t^2,\ \cos t^2)\| = 2t.$$

In the study of manifolds as geometric objects, those properties which are independent of the particular description employed to analyze the manifold are of special interest. The theory of tangent spaces of manifolds therefore emphasizes the linear subspace spanned by all tangent vectors, rather than the individual tangent vectors.

For some purposes a local parametrization of a manifold by an open subset of one of its tangent spaces is useful.

Proposition 5.1.3. *Let $k \in \mathbf{N}_\infty$, let V be a C^k submanifold in $\mathbf{R}^n$ of dimension d, and let $x^0 \in V$. Then there exist an open neighborhood D of 0 in $T_{x^0}V$ and a C^{k-1} mapping $\phi : D \to \mathbf{R}^n$ such that*

$$\phi(D) \subset V, \qquad \phi(0) = x^0, \qquad D\phi(0) = \iota \in \mathrm{Lin}(T_{x^0}V, \mathbf{R}^n),$$

where ι is the standard inclusion.

Proof. We use the description $V \cap U = \mathrm{im}(\widetilde{\phi})$ with $\widetilde{\phi} : \widetilde{D} \to \mathbf{R}^n$ and $\widetilde{\phi}(y^0) = x^0$ according to Theorem 4.7.1.(ii). Using a translation in $\mathbf{R}^d$ we may assume that $y^0 = 0$; and we have $T_{x^0}V = \mathrm{im}\,(D\widetilde{\phi}(0)) = \{\, D\widetilde{\phi}(0)h \mid h \in \mathbf{R}^d \,\}$, while $D\widetilde{\phi}(0)$ is injective. Now define, for v in the open neighborhood $D = D\widetilde{\phi}(0)\widetilde{D}$ of 0 in $T_{x^0}V$

$$\phi(v) = \widetilde{\phi}(D\widetilde{\phi}(0)^{-1}v).$$

Using the chain rule we verify at once that ϕ satisfies the conditions. ❑

The following corollary makes explicit that the tangent space approximates the submanifold in a neighborhood of the point under consideration, with the analogous result for mappings.

Corollary 5.1.4. *Let $k \in \mathbf{N}_\infty$, let V be a d-dimensional C^k manifold in $\mathbf{R}^n$ and let $x^0 \in V$.*

(i) *For every $\epsilon > 0$ there exists a neighborhood V_0 of x^0 in V such that for every $x \in V_0$ there exists $v \in T_{x^0}V$ with*

$$\|x - x^0 - v\| < \epsilon\|x - x^0\|.$$

(ii) *Let $f : \mathbf{R}^n \to \mathbf{R}^p$ be a C^k mapping. Then we can select V_0 such that in addition to (i) we have*

$$\|f(x) - f(x^0) - Df(x^0)v\| < \epsilon\|x - x^0\|.$$

Proof. **(i).** Select ϕ as in Proposition 5.1.3. By means of first-order Taylor approximation of ϕ in the open neighborhood D of 0 in $T_{x^0}V$ we then obtain $x = \phi(v) = x^0 + v + o(\|v\|), \quad v \to 0$, where $v \in T_{x^0}V$. This implies at once $x = x^0 + h + o(\|x - x^0\|), \quad x \to x^0$.
(ii). Apply assertion (i) with V given by $\{\,(\phi(v), (f \circ \phi)(v)) \mid v \in D\,\}$, which is a C^k submanifold of $\mathbf{R}^n \times \mathbf{R}^p$. This then, in conjunction with the Mean Value Theorem 2.5.3, gives assertion (ii). ❑

5.2 Tangent mapping

Let V be a C^1 submanifold in $\mathbf{R}^n$ of dimension d, and let $x \in V$; further let W be a C^1 submanifold in $\mathbf{R}^p$ of dimension f. Let Φ be a C^1 mapping $\mathbf{R}^n \to \mathbf{R}^p$ such that $\Phi(V) \subset W$, or weaker, such that $\Phi(V \cap U) \subset W$, for a suitably chosen neighborhood U of x in $\mathbf{R}^n$. According to Definition 5.1.1, every $v \in T_x V$ can be written as

$$v = \gamma'(t_0), \qquad \text{with} \quad \gamma : I \to V \text{ a } C^1 \text{ curve}, \qquad \gamma(t_0) = x.$$

For such γ,

$$\Phi \circ \gamma : I \to W \text{ is a } C^1 \text{ curve}, \qquad \Phi \circ \gamma(t_0) = \Phi(x), \qquad (\Phi \circ \gamma)'(t_0) \in T_{\Phi(x)} W.$$

By virtue of the chain rule,

$$(\Phi \circ \gamma)'(t_0) = D\Phi(x) \circ \gamma'(t_0) = D\Phi(x)v, \tag{5.1}$$

where $D\Phi(x) \in \operatorname{Lin}(\mathbf{R}^n, \mathbf{R}^p)$ is the derivative of Φ at x.

Definition 5.2.1. We now consider Φ as a mapping of V into W, and we define

$$D\Phi(x) \in \operatorname{Lin}(T_x V,\ T_{\Phi(x)} W),$$

the *tangent mapping to* $\Phi : V \to W$ at x, by

$$D\Phi(x)v = (\Phi \circ \gamma)'(t_0) \qquad (v \in T_x V). \qquad ❍$$

On account of Formula (5.1) this definition is independent of the choice of the curve γ used to represent v. See Exercise 5.74 for a proof that the definition of the tangent mapping to $\Phi : V \to W$ at a point $x \in V$ is independent of the behavior of Φ outside V. Identifying the d- and f-dimensional vector spaces $T_x V$ and $T_{\Phi(x)} W$, respectively, with $\mathbf{R}^d$ and $\mathbf{R}^f$, respectively, we sometimes also write

$$D\Phi(x) : \mathbf{R}^d \to \mathbf{R}^f.$$

5.3 Examples of tangent spaces

Example 5.3.1. Let $f : \mathbf{R} \supset\!\!\to \mathbf{R}^2$ be a C^1 mapping. The submanifold V of $\mathbf{R}^3$ is the curve given as the graph of f, that is

$$V = \{\, (t,\ f_1(t),\ f_2(t)) \in \mathbf{R}^3 \mid t \in \operatorname{dom}(f) \,\}.$$

Then

$$\operatorname{graph}(Df(t)) = \mathbf{R}\,(1,\ f_1'(t),\ f_2'(t)).$$

A parametric representation of the geometric tangent line of V at $(t,\ f(t))$ (with $t \in \mathbf{R}$ fixed) is

$$(t,\ f_1(t),\ f_2(t)) + \lambda\,(1,\ f_1'(t),\ f_2'(t)) \qquad (\lambda \in \mathbf{R}). \qquad ☆$$

Example 5.3.2 (Helix). (ἡ ἕλιξ = curl of hair.) Let $V \subset \mathbf{R}^3$ be the *helix* or bi-infinite regularly wound *spiral* such that

$$V \subset \{ x \in \mathbf{R}^3 \mid x_1^2 + x_2^2 = 1 \},$$
$$V \cap \{ x \in \mathbf{R}^3 \mid x_3 = 2k\pi \} = \{ (1, 0, 2k\pi) \} \qquad (k \in \mathbf{Z}).$$

Then V is the graph of the C^∞ mapping $f : \mathbf{R} \to \mathbf{R}^2$ with

$$f(t) = (\cos t, \sin t), \qquad \text{that is,} \qquad V = \{ (\cos t, \sin t, t) \mid t \in \mathbf{R} \}.$$

It follows that V is a C^∞ manifold of dimension 1. Moreover, V is a zero-set. One has, for example

$$x \in V \qquad \Longleftrightarrow \qquad g(x) = \begin{pmatrix} x_1 - \cos x_3 \\ x_2 - \sin x_3 \end{pmatrix} = 0.$$

For $x = (f(t), t)$ we obtain

$$T_x V = \text{graph}\, Df(t) = \mathbf{R}\,(-\sin t, \cos t, 1) = \mathbf{R}\,(-x_2, x_1, 1),$$
$$Dg(x) = \begin{pmatrix} 1 & 0 & \sin x_3 \\ 0 & 1 & -\cos x_3 \end{pmatrix}.$$

Indeed, we have

$$\langle (1, 0, \sin x_3), (-x_2, x_1, 1) \rangle = 0, \qquad \langle (0, 1, -\cos x_3), (-x_2, x_1, 1) \rangle = 0.$$

The parametric representation of the geometric tangent line of V at $x = (f(t), t)$ is

$$(\cos t, \sin t, t) + \lambda(-\sin t, \cos t, 1) \qquad (\lambda \in \mathbf{R});$$

and this line intersects the plane $x_3 = 0$, for $\lambda = -t$. The cosine of the angle at the point of intersection between this plane and the direction vector of the tangent line has the constant value

$$\left\langle \frac{(-\sin t, \cos t, 1)}{\|(-\sin t, \cos t, 1)\|}, (-\sin t, \cos t, 0) \right\rangle = \frac{1}{2}\sqrt{2}. \qquad \text{☆}$$

Example 5.3.3. The submanifold V of $\mathbf{R}^3$ is a surface given by a C^1 embedding $\phi : \mathbf{R}^2 \supset\!\!\to \mathbf{R}^3$, that is

$$V = \{ \phi(y) \in \mathbf{R}^3 \mid y \in \text{dom}(\phi) \}.$$

Then

$$D\phi(y) = \begin{pmatrix} D_1\phi(y) & D_2\phi(y) \end{pmatrix} = \begin{pmatrix} D_1\phi_1(y) & D_2\phi_1(y) \\ \vdots & \vdots \\ D_1\phi_3(y) & D_2\phi_3(y) \end{pmatrix} \in \text{Lin}(\mathbf{R}^2, \mathbf{R}^3).$$

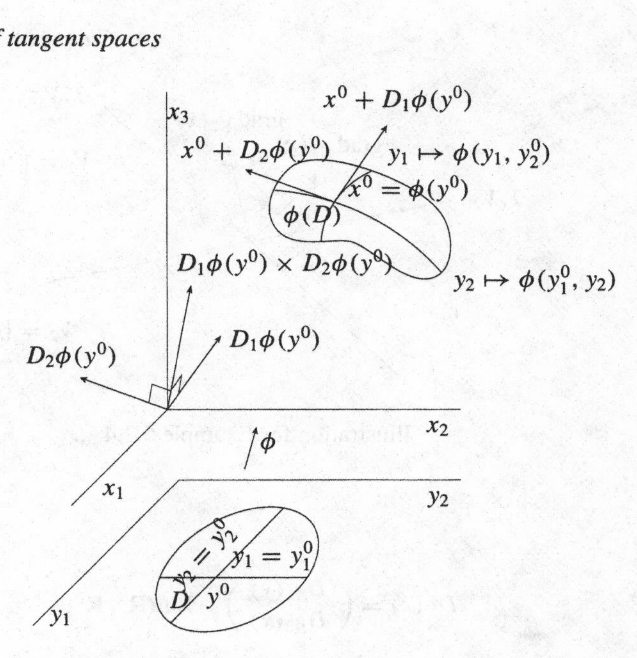

Illustration for Example 5.3.3

The tangent space $T_x V$, with $x = \phi(y)$, is spanned by the vectors $D_1\phi(y)$ and $D_2\phi(y)$ in $\mathbf{R}^3$. Therefore, a parametric representation of the geometric tangent plane of V at $\phi(y)$ is

$$\begin{array}{ccccccc} u_1 & = & \phi_1(y) & + & \lambda_1 D_1\phi_1(y) & + & \lambda_2 D_2\phi_1(y) \\ \vdots & & \vdots & & \vdots & & \vdots \\ u_3 & = & \phi_3(y) & + & \lambda_1 D_1\phi_3(y) & + & \lambda_2 D_2\phi_3(y) \end{array} \qquad (\lambda \in \mathbf{R}^2).$$

From linear algebra it is known that the linear subspace in $\mathbf{R}^3$ orthogonal to $T_x V$, the *normal space to V at x*, is spanned by the cross product (see also Example 5.3.11)

$$\mathbf{R}^3 \ni D_1\phi(y) \times D_2\phi(y) = \begin{pmatrix} D_1\phi_2(y)\, D_2\phi_3(y) - D_1\phi_3(y)\, D_2\phi_2(y) \\ D_1\phi_3(y)\, D_2\phi_1(y) - D_1\phi_1(y)\, D_2\phi_3(y) \\ D_1\phi_1(y)\, D_2\phi_2(y) - D_1\phi_2(y)\, D_2\phi_1(y) \end{pmatrix}.$$

Conversely, therefore, $T_x V = \{\, h \in \mathbf{R}^3 \mid \langle h,\, D_1\phi(y) \times D_2\phi(y) \rangle = 0 \,\}$. ☆

Example 5.3.4. The submanifold V of $\mathbf{R}^3$ is the curve in $\mathbf{R}^3$ given as the zero-set of a C^1 mapping $g : \mathbf{R}^3 \supset\!\!\to \mathbf{R}^2$, in other words, as the intersection of two surfaces, that is

$$x \in V \quad\Longleftrightarrow\quad g(x) = \begin{pmatrix} g_1(x) \\ g_2(x) \end{pmatrix} = 0.$$

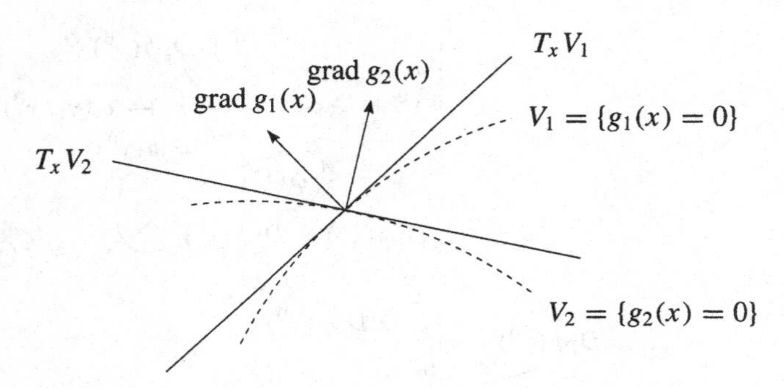

Illustration for Example 5.3.4

Then

$$Dg(x) = \begin{pmatrix} Dg_1(x) \\ Dg_2(x) \end{pmatrix} \in \mathrm{Lin}(\mathbf{R}^3, \mathbf{R}^2),$$

$$\ker (Dg(x)) = \{\, h \in \mathbf{R}^3 \mid \langle\, \mathrm{grad}\, g_1(x), h \rangle = \langle\, \mathrm{grad}\, g_2(x), h \rangle = 0 \,\}.$$

Thus the tangent space $T_x V$ is seen to be the line in $\mathbf{R}^3$ through 0, formed by intersection of the two planes $\{\, h \in \mathbf{R}^3 \mid \langle\, \mathrm{grad}\, g_1(x), h \rangle = 0 \,\}$ and $\{\, h \in \mathbf{R}^3 \mid \langle\, \mathrm{grad}\, g_2(x), h \rangle = 0 \,\}$. ☆

Example 5.3.5. The submanifold V of $\mathbf{R}^n$ is given as the zero-set of a C^1 mapping $g : \mathbf{R}^n \supset\!\!\rightarrow \mathbf{R}^{n-d}$, that is (see Theorem 4.7.1.(iii))

$$V = \{\, x \in \mathbf{R}^n \mid x \in \mathrm{dom}(g),\ g(x) = 0 \,\}.$$

Then $h \in T_x V$, for $x \in V$, if and only if

$$D(g)(x)h = 0 \quad \Longleftrightarrow \quad \langle\, \mathrm{grad}\, g_1(x), h \rangle = \cdots = \langle\, \mathrm{grad}\, g_{n-d}(x), h \rangle = 0.$$

This once again proves the property from Theorem 2.6.3.(iii), namely that the vectors $\mathrm{grad}\, g_i(x) \in \mathbf{R}^n$ are orthogonal to the level set $g^{-1}(\{0\})$ through x, in the sense that $\mathrm{grad}\, g_i(x)$ is orthogonal to $T_x V$, for $1 \le i \le n-d$. Phrased differently,

$$\mathrm{grad}\, g_i(x) \in (T_x V)^{\perp} \qquad (1 \le i \le n-d),$$

the orthocomplement of $T_x V$ in $\mathbf{R}^n$. Furthermore, $\dim (T_x V)^{\perp} = n - \dim T_x V = n - d$, while the $n-d$ vectors $\mathrm{grad}\, g_1(x)$, …, $\mathrm{grad}_{n-d}(x)$ are linearly independent by virtue of the fact that g is a submersion at x. As a consequence, $(T_x V)^{\perp}$ is spanned by these gradient vectors. ☆

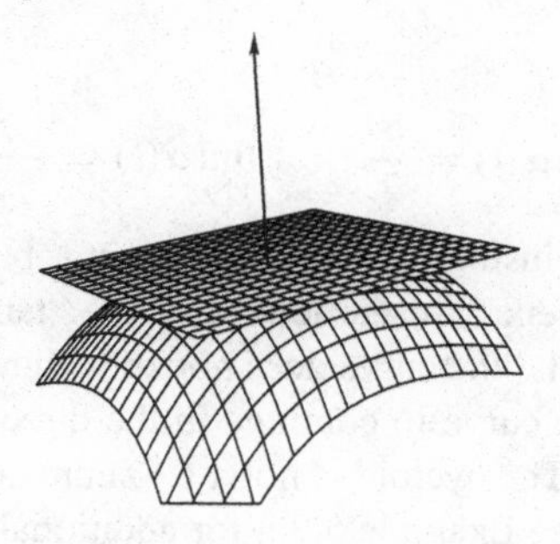

Illustration for Example 5.3.5

Example 5.3.6 (Cycloid). (ὁ κύκλος = wheel.) Consider the circle $x_1^2+(x_2-1)^2 = 1$ in $\mathbf{R}^2$. The *cycloid* C in $\mathbf{R}^2$ is defined as the curve in $\mathbf{R}^2$ traced out by the point $(0, 0)$ on the circle when the latter, from time $t = 0$ onward, rolls with constant velocity 1 to the right along the x_1-axis. The location of the center of the circle at time t is $(t, 1)$. The point $(0, 0)$ then has rotated clockwise with respect to the center by an angle of t radians; in other words, in a moving Cartesian coordinate system with origin at $(t, 1)$, the point $(0, 0)$ has been mapped into $(-\sin t, -\cos t)$. By superposition of these two motions one obtains that the cycloid C is the curve given by

$$\phi : \mathbf{R} \to \mathbf{R}^2 \qquad \text{with} \qquad \phi(t) = \begin{pmatrix} t - \sin t \\ 1 - \cos t \end{pmatrix}.$$

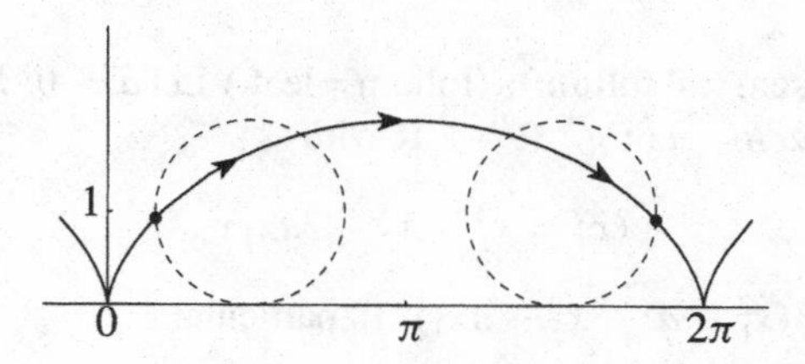

Illustration for Example 5.3.6: Cycloid

One has

$$D\phi(t) = \begin{pmatrix} 1 - \cos t \\ \sin t \end{pmatrix} \in \mathrm{Lin}(\mathbf{R}, \mathbf{R}^2).$$

If $\alpha(t)$ denotes the angle between this tangent vector and the positive direction of the x_1-axis, one may therefore write

$$\tan\alpha(t) = \frac{\sin t}{1 - \cos t} = \frac{2 + \mathcal{O}(t^2)}{t + \mathcal{O}(t^3)}, \qquad t \to 0.$$

Consequently

$$\lim_{t\downarrow 0}\alpha(t) = \frac{\pi}{2}, \qquad \lim_{t\uparrow 0}\alpha(t) = -\frac{\pi}{2}.$$

Hence the following conclusion: ϕ is a C^1 mapping, but ϕ is not an immersion for $t \in 2\pi\mathbf{Z}$. In the present case the **length** of the "tangent vector" vanishes at those points, and as a result "the curve does not know how to go from there". This makes it possible for "the curve to continue in the direction exactly opposite the one in which it arrived". The cycloid is not a C^1 submanifold in $\mathbf{R}^2$ of dimension 1, because it has cusps (see Example 5.3.8 for additional information) orthogonal to the x_1-axis for $t \in 2\pi\mathbf{Z}$. But the cycloid is a C^1 submanifold in $\mathbf{R}^2$ of dimension 1 at all points $\phi(t)$ with $t \notin 2\pi\mathbf{Z}$. This once again demonstrates the importance of immersivity in Corollary 4.3.2. ☆

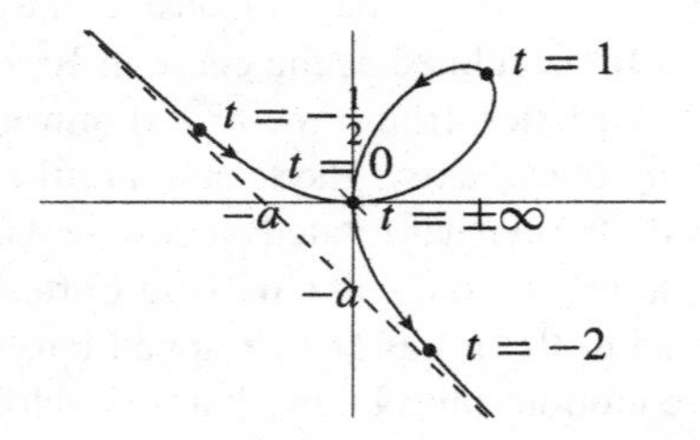

Illustration for Example 5.3.7: Descartes' folium

Example 5.3.7 (Descartes' folium). (folium = leaf.) Let $a > 0$. *Descartes' folium* F is defined as the zero-set of $g : \mathbf{R}^2 \to \mathbf{R}$ with

$$g(x) = x_1^3 + x_2^3 - 3ax_1x_2.$$

One has $Dg(x) = 3(x_1^2 - ax_2,\ x_2^2 - ax_1)$. In particular

$$Dg(x) = 0 \quad\Longrightarrow\quad x_1^4 = a^2x_2^2 = a^3x_1 \quad\Longrightarrow\quad (x_1 = 0 \quad\text{or}\quad x_1 = a).$$

Now $0 = (0, 0) \in F$, while $(a, a) \notin F$. Using the Submersion Theorem 4.5.2 we see that F is a C^∞ submanifold in $\mathbf{R}^2$ of dimension 1 at every point $x \in F \setminus \{0\}$. To study the point $0 \in F$ more closely, we intersect F with the lines $x_2 = tx_1$, for $t \in \mathbf{R}$, all of which run through 0. The points of intersection x are found from

$$x_1^3 + t^3x_1^3 - 3atx_1^2 = 0, \qquad \text{hence} \qquad x_1 = 0 \quad\text{or}\quad x_1 = \frac{3at}{1+t^3} \qquad (t \neq -1).$$

That is, the points of intersection x are

$$x = x(t) = \frac{3at}{1+t^3}\begin{pmatrix} 1 \\ t \end{pmatrix} \qquad (t \in \mathbf{R} \setminus \{-1\}).$$

Therefore $\operatorname{im}(\phi) \subset F$, with $\phi : \mathbf{R} \setminus \{-1\} \to \mathbf{R}^2$ given by

$$\phi(t) = \frac{3a}{1+t^3}\begin{pmatrix} t \\ t^2 \end{pmatrix}.$$

We have $\phi(0) = 0$, and furthermore

$$D\phi(t) = \frac{3a}{(1+t^3)^2}\begin{pmatrix} 1+t^3-3t^3 \\ 2t(1+t^3)-3t^4 \end{pmatrix} = \frac{3a}{(1+t^3)^2}\begin{pmatrix} 1-2t^3 \\ 2t-t^4 \end{pmatrix}.$$

In particular

$$D\phi(0) = 3a\begin{pmatrix} 1 \\ 0 \end{pmatrix}.$$

Consequently, ϕ is an immersion at the point 0 in $\mathbf{R}$. By the Immersion Theorem 4.3.1 it follows that there exists an open neighborhood D of 0 in $\mathbf{R}$ such that $\phi(D)$ is a C^∞ submanifold of $\mathbf{R}^2$ with $0 \in \phi(D) \subset F$. Moreover, the tangent line of $\phi(D)$ at 0 is horizontal.

This does not complete the analysis of the structure of F in neighborhoods of 0, however. Indeed,

$$\lim_{t\to\pm\infty} \phi(t) = 0 = \phi(0),$$

which suggests that F intersects with itself at 0. Furthermore, the one-parameter family of lines $\{\, x_2 = t x_1 \mid t \in \mathbf{R} \,\}$ does not comprise all lines through the origin: it excludes the x_2-axis ($t = \pm\infty$). Therefore we now intersect F with the lines $x_1 = u x_2$, for $u \in \mathbf{R}$. We then find the points of intersection

$$x = x(u) = \frac{3au}{1+u^3}\begin{pmatrix} u \\ 1 \end{pmatrix} \qquad (u \in \mathbf{R} \setminus \{-1\}).$$

Therefore $\operatorname{im}(\widetilde{\phi}) \subset F$, with $\widetilde{\phi} : \mathbf{R} \setminus \{-1\} \to \mathbf{R}^2$ given by

$$\widetilde{\phi}(u) = \frac{3a}{1+u^3}\begin{pmatrix} u^2 \\ u \end{pmatrix} = \phi\Big(\frac{1}{u}\Big).$$

Then

$$D\widetilde{\phi}(u) = \frac{3a}{(1+u^3)^2}\begin{pmatrix} 2u-u^4 \\ 1-2u^3 \end{pmatrix}, \qquad \text{in particular} \qquad D\widetilde{\phi}(0) = 3a\begin{pmatrix} 0 \\ 1 \end{pmatrix}.$$

For the same reasons as before there exists an open neighborhood $\widetilde{D}$ of 0 in $\mathbf{R}$ such that $\widetilde{\phi}(\widetilde{D})$ is a C^∞ submanifold of $\mathbf{R}^2$ with $0 \in \widetilde{\phi}(\widetilde{D}) \subset F$. However, the tangent line of $\widetilde{\phi}(\widetilde{D})$ at 0 is vertical.

In summary: F is a C^∞ curve at all its points, except at 0; F intersects with itself at 0, in such a way that one part of F at 0 is orthogonal to the other. There are two lines that run through 0 in linearly independent directions and that are both "tangent to" F at 0. This makes it plausible that $\operatorname{grad} g(0) = 0$, because $\operatorname{grad} g(0)$ must be orthogonal to both "tangent lines" at the same time. ☆

Example 5.3.8 (Cusp of a plane curve). Consider the curve $\gamma : \mathbf{R} \to \mathbf{R}^2$ given by $\gamma(t) = (t^2, t^3)$. The image $\mathrm{im}(\gamma) = \{x \in \mathbf{R}^2 \mid x_1^3 = x_2^2\}$ is said to be the *semicubic parabola* (the behavior observed in the parts above and below the x_1-axis is like that of a parabola). Note that $\gamma'(t) = (2t, 3t^2)^t \in \mathrm{Lin}(\mathbf{R}, \mathbf{R}^2)$ is injective, unless $t = 0$. Furthermore, $\|\gamma'(t)\| = 2|t|\sqrt{1 + (\frac{3}{2}t)^2}$. For $t \neq 0$ we therefore have the normalized tangent vector

$$T(t) = \|\gamma'(t)\|^{-1}\gamma'(t) = \frac{\operatorname{sgn} t}{\sqrt{1 + (\frac{3}{2}t)^2}} \begin{pmatrix} 1 \\ \frac{3}{2}t \end{pmatrix}.$$

Obviously, then

$$\lim_{t \downarrow 0} T(t) = (1, 0) = -\lim_{t \uparrow 0} T(t).$$

This equality tells us that the unit tangent vector field $t \mapsto T(t)$ along the curve abruptly reverses its direction at the point 0. We further note that the second derivative $\gamma''(0) = (2, 0)$ and the third derivative $\gamma'''(0) = (0, 6)$ are linearly independent vectors in $\mathbf{R}^2$. Finally, $C := \mathrm{im}(\gamma)$ is not a C^1 submanifold in $\mathbf{R}^2$ of dimension 1 at $(0, 0)$. Indeed, suppose there exist a neighborhood U of $(0, 0)$ in $\mathbf{R}^2$ and a C^1 function $f : W \to \mathbf{R}$ with $0 \in W$, $f(0) = 0$ and $C \cap U = \{(f(w), w) \mid w \in W\}$. Necessarily, then $f(w) = f(-w)$, for all $w \in W$. But this implies that the tangent space at $(0, 0)$ of C is the x_2-axis, which forms a contradiction.

The reversal of the direction of the unit tangent vector field is characteristic of a cusp, and the semicubic parabola is considered the **prototype** of the simplest type of cusp (there are also cusps of the form $t \mapsto (t^2, t^5)$, etc). If $\phi : I \to \mathbf{R}^2$ is a C^3 curve, a possible first definition of an *ordinary cusp* at a point x^0 of ϕ is that, after application of a local C^3 diffeomorphism Φ of $\mathbf{R}^2$ defined in a neighborhood of x^0 with $\Phi(x^0) = (0, 0)$, the image curve $\Phi \circ \phi : I \to \mathbf{R}^2$ in a neighborhood of $(0, 0)$ coincides with the curve γ above. Such a definition, however, has the drawback of not being formulated in terms of the curve ϕ itself. This is overcome by the following second definition, which can be shown to be a consequence of the first one. ☆

Definition 5.3.9. Let $\phi : I \to \mathbf{R}^2$ be a C^3 curve, $0 \in I$ and $\phi(0) = x^0$. Then ϕ is said to have an *ordinary cusp* at x^0 if

$$\phi'(0) = 0, \qquad \text{and if} \qquad \phi''(0) \quad \text{and} \quad \phi'''(0)$$

are linearly independent vectors in $\mathbf{R}^2$. ❍

Note that for the parabola $t \mapsto (t^2, t^4)$ the second and third derivative at 0 are not linearly independent vectors in $\mathbf{R}^2$. It can be proved that Definition 5.3.9 is in fact independent of the choice of the parametrization ϕ. In this book we omit proof of the fact that the second definition is implied by the first one. Instead, we add the following remark. By Taylor expansion we find that for a C^4 curve $\phi : I \to \mathbf{R}^2$ with $0 \in I$, $\phi(0) = x^0$ and $\phi'(0) = 0$, the following equality of vectors holds in $\mathbf{R}^2$, for $t \in I$:

$$\phi(t) = x^0 + \frac{t^2}{2!}\phi''(0) + \frac{t^3}{3!}\phi'''(0) + \mathcal{O}(t^4), \qquad t \to 0.$$

If ϕ satisfies Definition 5.3.9, it follows that

$$\widetilde{\phi}(t) = x^0 + (t^2, t^3) + \mathcal{O}(t^4), \qquad t \to 0,$$

which becomes apparent upon application of the linear coordinate transformation in $\mathbf{R}^2$ which maps $\frac{1}{2!}\phi''(0)$ into $(1, 0)$ and $\frac{1}{3!}\phi'''(0)$ into $(0, 1)$. To prove that the first definition is a consequence of Definition 5.3.9, one evidently has to show that the terms in t of order higher than 3 can be removed by means of a suitable (nonlinear) coordinate transformation in $\mathbf{R}^2$. Also note that the terms of higher order in no way affect the reversal of the direction of the unit tangent vector field. Incidentally, the error is of the order of $\frac{1}{10.000}$ for $t = \frac{1}{10}$, which will go unnoticed in most illustrations.

For practical applications we formulate the following:

Lemma 5.3.10. *A curve in $\mathbf{R}^2$ has an ordinary cusp at x^0 if it possesses a C^4 parametrization $\phi : I \to \mathbf{R}^2$ with $0 \in I$ and $x^0 = \phi(0)$, and if there are numbers $a, b \in \mathbf{R} \setminus \{0\}$ with*

$$\phi(t) = x^0 + \begin{pmatrix} a\,t^2 + \mathcal{O}(t^3) \\ b\,t^3 + \mathcal{O}(t^4) \end{pmatrix}, \qquad t \to 0.$$

Furthermore, $\mathrm{im}(\phi)$ *at x^0 is not a C^1 submanifold in $\mathbf{R}^2$ of dimension* 1.

Proof. $\frac{1}{2!}\phi''(0) = (a, 0)$ and $\frac{1}{3!}\phi'''(0) = (*, b)$ are linearly independent vectors in $\mathbf{R}^2$. The last assertion follows by similar arguments as for the semicubic parabola.❑

By means of this lemma we can once again verify that the cycloid $\phi : \mathbf{R} \to \mathbf{R}^2$ from Example 5.3.6 has an ordinary cusp at $(0, 0) \in \mathbf{R}^2$, and, on account of the periodicity, at all points of the form $(2\pi k, 0) \in \mathbf{R}^2$ for $k \in \mathbf{Z}$; indeed, Taylor expansion gives

$$\phi(t) = \begin{pmatrix} t - \sin t \\ 1 - \cos t \end{pmatrix} = \begin{pmatrix} \frac{1}{6}t^3 + \mathcal{O}(t^4) \\ \frac{1}{2}t^2 + \mathcal{O}(t^3) \end{pmatrix}, \qquad t \to 0.$$

Example 5.3.11 (Hypersurface in $\mathbf{R}^n$). This example will especially be used in Chapter 7. Recall that a C^k *hypersurface* V in $\mathbf{R}^n$ is a C^k submanifold of dimension

$n - 1$. For such a hypersurface $\dim T_x V = n - 1$, for all $x \in V$; and therefore $\dim(T_x V)^\perp = 1$. A *normal to V at x* is a vector $n(x) \in \mathbf{R}^n$ with

$$n(x) \perp T_x V \qquad \text{and} \qquad \|n(x)\| = 1.$$

A normal is thus uniquely determined, to within its sign. We now write

$$\mathbf{R}^n \ni x = (x',\, x_n), \qquad \text{with} \qquad x' = (x_1, \dots, x_{n-1}) \in \mathbf{R}^{n-1}.$$

On account of Theorem 4.7.1 there exists for every $x^0 \in V$ a neighborhood U of x^0 in $\mathbf{R}^n$ such that the following assertions are equivalent (this may require permutation of the coordinates of $\mathbf{R}^n$).

(i) There exist an open neighborhood U' of $(x^0)'$ in $\mathbf{R}^{n-1}$ and a C^k function $h : U' \to \mathbf{R}$ with

$$V \cap U = \text{graph}(h) = \{\, (x',\, h(x')) \mid x' \in U' \,\}.$$

(ii) There exist an open subset $D \subset \mathbf{R}^{n-1}$ and a C^k embedding $\phi : D \to \mathbf{R}^n$ with

$$V \cap U = \text{im}(\phi) = \{\, \phi(y) \mid y \in D \,\}.$$

(iii) There exists a C^k function $g : U \to \mathbf{R}$ with $Dg(x) \neq 0$ for $x \in U$, such that

$$V \cap U = N(g, 0) = \{\, x \in U \mid g(x) = 0 \,\}.$$

In Description (i) of V one has, if $x = (x',\, h(x'))$,

$$T_x V = \text{graph}\,(Dh(x')) = \{\, (v,\; Dh(x')v) \mid v \in \mathbf{R}^{n-1} \,\},$$

where

$$Dh(x') = (D_1 h(x'), \dots,\, D_{n-1} h(x')).$$

Therefore, if $e_1, \dots, e_{n-1}$ are the standard basis vectors of $\mathbf{R}^{n-1}$, it follows that $T_x V$ is spanned by the vectors $u_j := (e_j,\; D_j h(x'))$, for $1 \le j \le n-1$. Hence $(T_x V)^\perp$ is spanned by the vector $w = (w_1, \dots, w_{n-1}, 1) \in \mathbf{R}^n$ satisfying $0 = \langle w, u_j \rangle = w_j + D_j h(x')$, for $1 \le j < n$. Therefore $w = (-Dh(x'),\, 1)$, and consequently

$$n(x) = \pm(1 + \|Dh(x')\|^2)^{-1/2}\,(-Dh(x'),\, 1) \qquad (x = (x',\, h(x'))).$$

In Description (iii) one has

$$T_x V = \ker\,(Dg(x)).$$

In particular, on the basis of (i) we can choose the function g in (iii) as follows:

$$g(x) = x_n - h(x'); \qquad \text{then} \qquad Dg(x) = (-Dh(x'),\, 1) \neq 0.$$

This once again confirms the formula for $n(x)$ given above.

In Description (ii) $T_x V$ is spanned by the vectors $v_j := D_j \phi(y)$, for $1 \le j < n$. Hence, a vector $w \in (T_x V)^\perp$ is determined, to within a scalar, by the equations

$$\langle v_j, w \rangle = 0 \qquad (1 \le j < n).$$

Remark on linear algebra. In linear algebra the method of solving w from the system of equations $\langle v_j, w\rangle = 0$, for $1 \le j < n$, is formalized as follows. Consider $n-1$ linearly independent vectors $v_1, \dots, v_{n-1} \in \mathbf{R}^n$. Then the mapping in $\mathrm{Lin}(\mathbf{R}^n, \mathbf{R})$ with

$$v \mapsto \det(v \; v_1 \; \cdots \; v_{n-1})$$

is nontrivial (the vectors $v, v_1, \dots, v_{n-1} \in \mathbf{R}^n$ occur as columns of the matrix on the right–hand side). According to Lemma 2.6.1 there exists a unique w in $\mathbf{R}^n \setminus \{0\}$ such that

$$\det(v \; v_1 \; \cdots \; v_{n-1}) = \langle v, w\rangle \qquad (v \in \mathbf{R}^n).$$

Our choice is here to include the test vector v in the determinant in front position, not at the back. This has all kinds of consequences for signs, but in this way one ensures that the theory is compatible with that of differential forms (see Example 8.6.5).

The vector w thus found is said to be the *cross product* of $v_1, \dots, v_{n-1}$, notation

$$v_1 \times \cdots \times v_{n-1} \in \mathbf{R}^n, \qquad \text{with} \qquad \det(v \, v_1 \; \cdots \; v_{n-1}) = \langle v, \; v_1 \times \cdots \times v_{n-1}\rangle. \tag{5.2}$$

Since

$$\langle v_j, w\rangle = \det(v_j \, v_1 \; \cdots \; v_{n-1}) = 0 \qquad (1 \le j < n);$$
$$\det(w \, v_1 \; \cdots \; v_{n-1}) = \langle w, w\rangle = \|w\|^2 > 0,$$

the vector w is perpendicular to the linear subspace in $\mathbf{R}^n$ spanned by $v_1, \dots, v_{n-1}$; also, the n-tuple of vectors $(w, v_1, \dots, v_{n-1})$ is positively oriented. The vector $w = (w_1, \dots, w_n)$ is calculated using

$$w_j = \langle e_j, w\rangle = \det(e_j \, v_1 \; \cdots \; v_{n-1}) \qquad (1 \le j \le n),$$

where $(e_1, \dots, e_n)$ is the standard basis for $\mathbf{R}^n$. The length of w is given by

$$\|v_1 \times \cdots \times v_{n-1}\|^2 = \begin{vmatrix} \langle v_1, v_1\rangle & \cdots & \langle v_1, v_{n-1}\rangle \\ \vdots & & \vdots \\ \langle v_{n-1}, v_1\rangle & \cdots & \langle v_{n-1}, v_{n-1}\rangle \end{vmatrix} = \det(V^t \, V), \tag{5.3}$$

where

$$V := (v_1 \cdots v_{n-1}) \in \mathrm{Mat}\,(n \times (n-1), \mathbf{R}) \text{ has } v_1, \dots, v_{n-1} \in \mathbf{R}^n \text{ as columns,}$$

while $V^t \in \mathrm{Mat}\,((n-1) \times n, \mathbf{R})$ is the transpose matrix.

The proof of (5.3) starts with the following remark. Let $A \in \mathrm{Lin}(\mathbf{R}^p, \mathbf{R}^n)$. Then, as in the intermezzo on linear algebra in Section 2.1, the composition $A^t A \in \mathrm{End}(\mathbf{R}^p)$ has the symmetric matrix

$$(\langle a_i, a_j\rangle)_{1 \le i,\, j \le p} \qquad \text{with} \qquad a_j \qquad \text{the } j\text{-th column of } A. \tag{5.4}$$

Recall that this is Gram's matrix associated with the vectors $a_1, \dots, a_p \in \mathbf{R}^n$. Applying this remark first with $A = (w \, v_1 \; \cdots \; v_{n-1}) \in \mathrm{Mat}(n, \mathbf{R})$, and then with

$A = V \in \mathrm{Mat}\,(n \times (n-1), \mathbf{R})$, we now have, using the fact that $\langle v_j, w\rangle = 0$,

$$\|w\|^4 = (\det(w\; v_1\; \cdots\; v_{n-1}))^2 = (\det A)^2 = \det A^t\, \det A = \det(A^t A)$$

$$= \begin{vmatrix} \langle w, w\rangle & 0 & \cdots & 0 \\ 0 & \langle v_1, v_1\rangle & \cdots & \langle v_1, v_{n-1}\rangle \\ \vdots & \vdots & & \vdots \\ 0 & \langle v_{n-1}, v_1\rangle & \cdots & \langle v_{n-1}, v_{n-1}\rangle \end{vmatrix} = \|w\|^2 \det(V^t\, V).$$

This proves Formula (5.3).

In particular, we may consider these results for $\mathbf{R}^3$: if $x, y \in \mathbf{R}^3$, then $x \times y \in \mathbf{R}^3$ equals

$$x \times y = (x_2 y_3 - x_3 y_2,\; x_3 y_1 - x_1 y_3,\; x_1 y_2 - x_2 y_1).$$

Furthermore,

$$\langle x, y\rangle^2 + \|x \times y\|^2 = \langle x, y\rangle^2 + \begin{vmatrix} \|x\|^2 & \langle x, y\rangle \\ \langle y, x\rangle & \|y\|^2 \end{vmatrix} = \|x\|^2\,\|y\|^2.$$

Therefore $-1 \le \frac{\langle x, y\rangle}{\|x\|\,\|y\|} \le 1$ (which is the Cauchy–Schwarz inequality), and thus there exists a unique number $0 \le \alpha \le \pi$, called the *angle* $\angle(x, y)$ between x and y, satisfying (see Example 7.4.1)

$$\frac{\langle x, y\rangle}{\|x\|\,\|y\|} = \cos\alpha, \qquad \text{and then} \qquad \frac{\|x \times y\|}{\|x\|\,\|y\|} = \sin\alpha.$$

Sequel to Example 5.3.11. In Description (ii) it follows from the foregoing that if $x = \phi(y)$,

$$D_1\phi(y) \times \cdots \times D_{n-1}\phi(y) \in (T_x V)^{\perp}. \tag{5.5}$$

In the particular case where the description in (ii) coincides with that in (i), that is, if $\phi(y) = (y, h(y))$, we have already seen that $D_1\phi(y) \times \cdots \times D_{n-1}\phi(y)$ and $(-Dh(y),\ 1)$ are equal to within a scalar factor. This factor is $(-1)^{n-1}$. To see this, we note that

$$D_j\phi(y) = (0, \dots, 0, 1, 0, \dots, D_j h(y))^t \qquad (1 \le j < n).$$

And so $\det\,(e_n\; D_1\phi(y)\; \cdots\; D_{n-1}\phi(y))$ equals

$$\begin{vmatrix} 0 & 1 & \cdots & 0 & \cdots & 0 \\ \vdots & 0 & \ddots & \ddots & & \\ \vdots & \vdots & \ddots & 1 & \ddots & \\ \vdots & \vdots & & \ddots & \ddots & 0 \\ 0 & 0 & \cdots & 0 & & 1 \\ 1 & D_1 h(y) & \cdots & D_j h(y) & \cdots & D_{n-1} h(y) \end{vmatrix} = (-1)^{n-1}.$$

Using this we find

$$D_1\phi(y) \times \cdots \times D_{n-1}\phi(y) = (-1)^{n-1}(-Dh(y), 1); \tag{5.6}$$

and therefore

$$\begin{aligned} \sqrt{\det(D\phi(y)^t \circ D\phi(y))} &= \|D_1\phi(y) \times \cdots \times D_{n-1}\phi(y)\| \\ &= \sqrt{1 + \|Dh(y)\|^2}. \end{aligned} \tag{5.7}$$

☆

5.4 Method of Lagrange multipliers

Consider the following problem. Determine the minimal distance in $\mathbf{R}^2$ of a point x on the circle $\{ x \in \mathbf{R}^2 \mid \|x\| = 1 \}$ to a point y on the line $\{ y \in \mathbf{R}^2 \mid y_1 + 2y_2 = 5 \}$. To do this, we have to find minima of the function

$$f(x, y) = \|x - y\|^2 = (x_1 - y_1)^2 + (x_2 - y_2)^2$$

under the *constraints*

$$g_1(x, y) = \|x\|^2 - 1 = x_1^2 + x_2^2 - 1 = 0 \qquad \text{and} \qquad g_2(x, y) = y_1 + 2y_2 - 5 = 0.$$

Phrased in greater generality, we want to determine the extrema of the restriction $f|_V$ of a function f to a subset V that is defined by the vanishing of some vector-valued function g. And this V is a submanifold of $\mathbf{R}^n$ under the condition that g be a submersion. The following theorem contains a necessary condition.

Definition 5.4.1. Assume $U \subset \mathbf{R}^n$ open and $k \in \mathbf{N}_\infty$, $f \in C^k(U)$; let $d \leq n$ and $g \in C^k(U, \mathbf{R}^{n-d})$. We define the *Lagrange function* L of f and g by

$$L : U \times \mathbf{R}^{n-d} \to \mathbf{R} \qquad \text{with} \qquad L(x, \lambda) = f(x) - \langle \lambda, g(x) \rangle.$$

❍

Theorem 5.4.2. *Let the notation be that of Definition 5.4.1. Let $V = \{ x \in U \mid g(x) = 0 \}$, let $x^0 \in V$ and suppose g is a submersion at x^0. If $f|_V$ is extremal at x^0 (in other words, $f(x) \leq f(x^0)$ or $f(x) \geq f(x^0)$, respectively, if $x \in U$ satisfies the constraint $x \in V$), then there exists $\lambda \in \mathbf{R}^{n-d}$ such that Lagrange's function L of f and g as a function of x has a critical point at (x^0, λ), that is*

$$D_x L(x^0, \lambda) = 0 \in \operatorname{Lin}(\mathbf{R}^n, \mathbf{R}).$$

Expressed in more explicit form, $x^0 \in U$ and $\lambda \in \mathbf{R}^{n-d}$ then satisfy

$$Df(x^0) = \sum_{1 \leq i \leq n-d} \lambda_i \, Dg_i(x^0), \qquad g_i(x^0) = 0 \qquad (1 \leq i \leq n - d).$$

Proof. Because g is a C^1 submersion at x^0, it follows by Theorem 4.7.1 that V is, locally at x^0, a C^1 manifold in $\mathbf{R}^n$ of dimension d. Then, again by the same theorem, there exist an open subset $D \subset \mathbf{R}^d$ and a C^1 embedding $\phi : D \to \mathbf{R}^n$ such that, locally at x^0, the manifold V can be described as

$$V = \{\, \phi(y) \in \mathbf{R}^n \mid y \in D \,\}, \qquad \phi(y^0) = x^0.$$

Because of the assumption that $f|_V$ is extremal at x^0, it follows that $f \circ \phi : D \to \mathbf{R}$ is extremal at y^0, where D is now **open** in $\mathbf{R}^d$. According to Theorem 2.6.3.(iv), therefore,

$$0 = D(f \circ \phi)(y^0) = Df(x^0) \circ D\phi(y^0).$$

That is, for every $v \in \operatorname{im} D\phi(y^0) = T_{x^0}V$ (see Theorem 5.1.2) we get $0 = Df(x^0)v = \langle \operatorname{grad} f(x^0), v \rangle$. And this means

$$\operatorname{grad} f(x^0) \in (T_{x^0}V)^{\perp}.$$

The theorem now follows from Example 5.3.5. ❑

Remark. Observe that the proof above comes down to the assertion that the restriction $f|_V$ has a stationary point at $x^0 \in V$ if and only if the restriction $Df(x^0)|_{T_{x^0}V}$ vanishes.

Remark. The numbers $\lambda_1, \dots, \lambda_{n-d}$ are known as the *Lagrange multipliers*. The system of $2n - d$ equations

$$D_j f(x^0) = \sum_{1 \le i \le n-d} \lambda_i \, D_j g_i(x^0) \quad (1 \le j \le n); \qquad g_i(x^0) = 0 \quad (1 \le i \le n-d) \tag{5.8}$$

in the $2n - d$ unknowns $x_1^0, \dots, x_n^0, \lambda_1, \dots, \lambda_{n-d}$ forms a necessary condition for f to have an extremum at x^0 under the constraints $g_1 = \cdots = g_{n-d} = 0$, if in addition $Dg_1(x^0), \dots, Dg_{n-d}(x^0)$ are known to be linearly independent. In many cases a few isolated points only are found as possibilities for x^0. One should also be aware of the possibility that points x^0 satisfying Equations (5.8) are saddle points for $f|_V$. In general, therefore, additional arguments are needed, to show that f is in fact extremal at a point x^0 satisfying Equations (5.8) (see Section 5.6).

A further question that may arise is whether one should not rather look for the solutions y of $D(f \circ \phi)(y) = 0$ instead. That problem would, in fact, seem far simpler to solve, involving as it does d equations in d unknowns only. Indeed, such an approach is preferable in cases where one can find an explicit embedding ϕ; generally speaking, however, there are equations to be solved before ϕ is obtained, and one then prefers to start directly from the method of multipliers.

5.5 Applications of the method of multipliers

Example 5.5.1. We return to the problem from the beginning of Section 5.4. We discuss it in a more general context and consider the question: what is the minimal distance between the unit sphere $S = \{ x \in \mathbf{R}^n \mid \|x\| = 1 \}$ and the $(n-1)$-dimensional linear submanifold $L = \{ y \in \mathbf{R}^n \mid \langle y, a \rangle = c \}$, here $a \in S$ and $c \in \mathbf{R}$ are nonnegative. (Verify that every linear submanifold of dimension $n-1$ can be represented in this form.) Define $f : \mathbf{R}^{2n} \to \mathbf{R}$ and $g : \mathbf{R}^{2n} \to \mathbf{R}^2$ by

$$f(x, y) = \|x - y\|^2, \qquad g(x, y) = \begin{pmatrix} \|x\|^2 - 1 \\ \langle y, a \rangle - c \end{pmatrix}.$$

Application of Theorem 5.4.2 shows that an extremal point $(x^0, y^0) \in \mathbf{R}^{2n}$ for the restriction of f to $g^{-1}(\{0\})$ satisfies the following conditions: there exists $\lambda \in \mathbf{R}^2$ such that

$$\begin{pmatrix} 2(x^0 - y^0) \\ 2(y^0 - x^0) \end{pmatrix} = \lambda_1 \begin{pmatrix} 2x^0 \\ 0 \end{pmatrix} + \lambda_2 \begin{pmatrix} 0 \\ a \end{pmatrix}, \qquad \text{and} \qquad g(x^0, y^0) = 0.$$

If $\lambda_1 = 0$ or $\lambda_2 = 0$, then $x^0 = y^0$, which gives $x^0 \in S \cap L$; in that case the minimum equals 0. Further, the Cauchy–Schwarz inequality then implies $|c| = |\langle x^0, a \rangle| \leq \|x^0\| \, \|a\| = 1$. From now on we assume $c > 1$, which implies $S \cap L = \emptyset$. Then $\lambda_1 \neq 0$ and $\lambda_2 \neq 0$, and we conclude

$$x^0 = \mu \, a := -\frac{\lambda_2}{2\lambda_1} a \qquad \text{and} \qquad y^0 = (1 - \lambda_1)\mu \, a.$$

But $g_1(x^0, y^0) = 0$ gives $\mu = \pm 1$ and $g_2(x^0, y^0) = 0$ gives $(1 - \lambda_1)\mu = c$. We obtain $\|x^0 - y^0\| = \| \pm a - c\, a\| = c \pm 1$ and so the minimal value equals $c - 1$. In the problem preceding Definition 5.4.1 we have $a = \frac{1}{\sqrt{5}}(1, 2)$ and $c = \sqrt{5}$, the minimal distance therefore is $\sqrt{5} - 1$, if we use a geometrical argument to conclude that we actually have a minimum.

Strictly speaking, the reasoning above only shows that $c - 1$ is a critical value, see the next section for a systematic discussion of this problem. Now we give an argument ad hoc that we have a minimum. We decompose arbitrary elements $x \in S$ and $y \in L$ in components along a and perpendicular to a, that is, we write $x = \alpha\, a + v$ and $y = \beta\, a + w$, where $\langle v, a \rangle = \langle w, a \rangle = 0$. From $g(x, y) = 0$ we obtain $|\alpha| \leq 1$ and $\beta = c$, and accordingly the Pythagorean property from Lemma 1.1.5.(iv) gives

$$\|x - y\|^2 = (\alpha - \beta)^2 + \|v - w\|^2 = (c - \alpha)^2 + \|v - w\|^2 \geq (c - 1)^2. \qquad ☆$$

Example 5.5.2. Here we generalize the arguments from Example 5.5.1. Let V_1 and V_2 be C^1 submanifolds in $\mathbf{R}^2$ of dimension 1. Assume there exist $x_i^0 \in V_i$ with

$$\|x_1^0 - x_2^0\| = \inf / \sup\{\, \|x_1 - x_2\| \mid x_i \in V_i \ (i = 1, 2)\,\},$$

and further assume that, locally at x_i^0, the V_i are described by $g_i(x) = 0$, for $i = 1, 2$. Then $f(x_1, x_2) = \|x_1 - x_2\|^2$ has an extremum at (x_1^0, x_2^0) under the constraints $g_1(x_1) = g_2(x_2) = 0$. Consequently,

$$Df(x_1^0, x_2^0) = 2(x_1^0 - x_2^0, -(x_1^0 - x_2^0)) = (\lambda_1\, Dg_1(x_1^0), \lambda_2\, Dg_2(x_2^0)).$$

That is, $x_1^0 - x_2^0 \in \mathbf{R}\,\mathrm{grad}\, g_1(x_1^0) = \mathbf{R}\,\mathrm{grad}\, g_2(x_2^0)$. In other words (see Theorem 5.1.2),

$$x_1^0 - x_2^0 \text{ is orthogonal to } T_{x_1^0} V_1 \text{ and orthogonal to } T_{x_2^0} V_2. \qquad ☆$$

Example 5.5.3 (Hadamard's inequality). A parallelepiped has maximal volume (see Example 6.6.3) when it is rectangular. That is, for $a_j \in \mathbf{R}^n$, with $1 \le j \le n$,

$$|\det(a_1 \cdots a_n)| \le \prod_{1 \le j \le n} \|a_j\|;$$

and equality obtains if and only if the vectors a_j are mutually orthogonal.

In fact, consider $A := (a_1 \cdots a_n) \in \mathrm{Mat}(n, \mathbf{R})$, the matrix with the a_j as its column vectors; then $a_j = Ae_j$ where $e_j \in \mathbf{R}^n$ is the j-th standard basis vector. Note that a proof is needed only in the case where $\|a_j\| \ne 0$; moreover, because the mapping $A \mapsto \det A$ is n-linear in the column vectors, we may assume $\|a_j\| = 1$, for all $1 \le j \le n$. Now define f_j and $f : \mathrm{Mat}(n, \mathbf{R}) \to \mathbf{R}$ by

$$f_j(A) = \frac{1}{2}(\|a_j\|^2 - 1), \qquad \text{and} \qquad f(A) = \det A.$$

The problem therefore is to prove that $|f(A)| \le 1$ if $f_1(A) = \ldots = f_n(A) = 0$. We have $Df_j(A)H = \langle a_j, h_j \rangle$, and therefore

$$\mathrm{grad}\, f_j(A) = (0 \cdots 0\, a_j\, 0 \cdots 0) \in \mathrm{Mat}(n, \mathbf{R}).$$

Note that the matrices $\mathrm{grad}\, f_j(A)$, for $1 \le j \le n$, are linearly independent in $\mathrm{Mat}(n, \mathbf{R})$. Set $A^\sharp = (a_{ij}^\sharp)$ with

$$a_{ij}^\sharp = \det(a_1 \cdots a_{i-1}\, e_j\, a_{i+1} \cdots a_n), \qquad \text{and} \qquad a_j^* = (A^\sharp)^t\, e_j \in \mathbf{R}^n.$$

The identity $a_j = Ae_j$ and Cramer's rule $\det A \cdot I = A\, A^\sharp = A^\sharp A$ from (2.6) now yield

$$f(A) = \det A = \langle a_j^*, a_j \rangle \qquad (1 \le j \le n). \tag{5.9}$$

Because the coefficients of A occurring in a_j do not occur in a_j^*, we obtain (see also Exercise 2.44.(i))

$$\operatorname{grad} f(A) = (a_1^* \; a_2^* \; \cdots \; a_n^*) = (A^\sharp)^t \in \operatorname{Mat}(n, \mathbf{R}).$$

According to Lagrange there exists a $\lambda \in \mathbf{R}^n$ such that for the $A \in \operatorname{Mat}(n, \mathbf{R})$ where $f(A)$ possibly is extremal, $\operatorname{grad} f(A) = \sum_{1\leq j\leq n} \lambda_j \operatorname{grad} f_j(A)$. We therefore examine the following identity of matrices:

$$(a_1^* \; a_2^* \; \cdots \; a_n^*) = (\lambda_1 a_1 \; \lambda_2 a_2 \; \cdots \; \lambda_n a_n), \qquad \text{in particular} \qquad a_j^* = \lambda_j a_j.$$

Accordingly, using Formula (5.9) we find that $\lambda_j = \det A$, but applying A^t to both sides of the latter identity above and using Cramer's rule once more then gives

$$\det A \, e_j = \det A \, (A^t A) e_j \qquad (1 \leq j \leq n).$$

Therefore there are two possibilities: either $\det A = 0$, which implies $f(A) = 0$; or $A^t A = I$, which implies $A \in \mathbf{O}(n, \mathbf{R})$ and also $f(A) = \det A = \pm 1$. Therefore the absolute maximum of f on $\{ A \in \operatorname{Mat}(n, \mathbf{R}) \mid f_1(A) = \cdots = f_n(A) = 0 \}$ is either 0 or ± 1, and so we certainly have $f(A) \leq 1$ on that set. Now $f(I) = 1$, therefore $\max f = 1$; and from the preceding it is evident that A is orthogonal if $f(A) = 1$. ✩

Remark. In applications one often encounters the following, seemingly more general, problem. Let U be an open subset of $\mathbf{R}^n$ and let g_i, for $1 \leq i \leq p$, and h_k, for $1 \leq k \leq q$, be functions in $C^1(U)$. Define

$$F = \{ x \in U \mid g_i(x) = 0, \; 1 \leq i \leq p, \; h_k(x) \leq 0, \; 1 \leq k \leq q \}.$$

In other words, F is a subset of an open set defined by p equalities and q inequalities. Again the problem is to determine the extremal values of the restriction of $f \in C^1(U)$ to F, and if necessary, to locate the extremal points in F. This can be reduced to the problem of finding the extremal values of f under constraints on an open set as follows. For every subset $Q \subset Q_0 = \{ 1, \dots, q \}$ define

$$F_Q = \{ x \in U \mid g_i(x) = 0, \; 1 \leq i \leq p, \; h_k(x) = 0, \; k \in Q, \; h_l(x) < 0, \; l \notin Q \}.$$

Then F is the union of the mutually disjoint sets F_Q, for Q running through all the subsets of Q_0. With the notation

$$U_Q = \{ x \in U \mid h_l(x) < 0, \; l \notin Q \},$$

we see that K_Q takes the form, familiar from the method of Lagrange multipliers,

$$K_Q = \{ x \in U_Q \mid g_i(x) = 0, \; 1 \leq i \leq p, \; h_k(x) = 0, \; k \in Q \}.$$

When $f|_F$ attains an extremum at $x \in F_Q$, then $f|_{F_Q}$ certainly attains an extremum at x; and if applicable, the method of multipliers can now be used to find those points $x \in F_Q$, for all $Q \subset Q_0$. The set of points where the method of multipliers does not apply because the gradients are linearly dependent is closed and therefore small generically. These cases should be treated separately.

5.6 Closer investigation of critical points

Let the notation be that of Theorem 5.4.2. In the proof of this theorem it is shown that the restriction $f|_V$ has a critical point at $x^0 \in V$ if and only if the restriction $Df(x^0)|_{T_{x^0}V}$ vanishes. Our next goal is to define the Hessian of $f|_V$ at a critical punt x^0, see Section 2.9, and to use it for the investigation of critical points. We expect it to be a bilinear form on $T_{x^0}V$, in other words, to be an element of $\mathrm{Lin}^2(T_{x^0}V, \mathbf{R})$, see Definition 2.7.3.

As preparation we choose a local C^2 parametrization $V \cap U = \mathrm{im}\,\phi$ with $\phi : D \to \mathbf{R}^n$ according to Theorem 4.7.1, and write $x = \phi(y)$. For $f \in C^2(U, \mathbf{R})$, and ϕ, the Hessian $D^2 f(x) \in \mathrm{Lin}^2(\mathbf{R}^n, \mathbf{R})$, and $D^2\phi(y) \in \mathrm{Lin}^2(\mathbf{R}^d, \mathbf{R}^n)$, respectively, is well-defined. Differentiating the identity

$$D(f \circ \phi)(y)w = Df(\phi(y))D\phi(y)w \qquad (y \in D,\ w \in \mathbf{R}^d)$$

with respect to y we find, for $v, w \in \mathbf{R}^d$,

$$D^2(f \circ \phi)(y)(v, w) = D^2 f(x)(D\phi(y)v, D\phi(y)w) + Df(x)D^2\phi(y)(v, w). \tag{5.10}$$

Note that $D^2\phi(y)(v, w) \in \mathbf{R}^n$, and that $Df(x)$ does map this vector into $\mathbf{R}$. Since $D^2\phi(y^0)(v, w)$ does not necessarily belong to $T_{x^0}V$, the second term at the right–hand side does not vanish automatically at x^0, even though $Df(x^0)|_{T_{x^0}V}$ vanishes. Yet there exists an expression for the Hessian of $f \circ \phi$ at y^0 as a bilinear form on $T_{x^0}V$.

Definition 5.6.1. From Definition 5.4.1 we recall the Lagrange function $L : U \times \mathbf{R}^{n-d} \to \mathbf{R}$ of f and g, which satisfies $L(x, \lambda) = f(x) - \langle \lambda, g(x) \rangle$. Then $D_x^2 L(x^0, \lambda) \in \mathrm{Lin}^2(\mathbf{R}^n, \mathbf{R})$. The bilinear form

$$H(f|_V)(x^0) := D_x^2 L(x^0, \lambda)|_{T_{x^0}V} \in \mathrm{Lin}^2(T_{x^0}V, \mathbf{R}).$$

is said to be the Hessian of $f|_V$ at x^0. ❍

Theorem 5.6.2. *Let the notation be that of Theorem 5.4.2, but with $k \geq 2$, assume $f|_V$ has a critical point at x^0, and let $\lambda \in \mathbf{R}^{n-d}$ be the associated Lagrange multiplier. The definition of $H(f|_V)(x^0)$ is independent of the choice of the submersion g such that $V = g^{-1}(\{0\})$. Further, we have, for every C^2 embedding $\phi : D \to \mathbf{R}^n$ with D open in $\mathbf{R}^d$ and $x^0 = \phi(y^0) \in V$ and every $v, w \in T_{x^0}V$*

$$H(f|_V)(x^0)(v, w) = D^2(f \circ \phi)(y^0)(D\phi(y^0)^{-1}v, D\phi(y^0)^{-1}w).$$

Proof. The vanishing of g on V implies that $f = L$ on V, whence $f \circ \phi = L \circ \phi$ on the open set D in $\mathbf{R}^d$. Accordingly we have $D^2(f \circ \phi)(y^0) = D^2(L \circ$

$\phi)(y^0)$. Furthermore, from Theorem 5.4.2 we know $D_x L(x^0, \lambda) = 0 \in \mathrm{Lin}(\mathbf{R}^n, \mathbf{R})$. Also, $D\phi(y^0) \in \mathrm{Lin}(\mathbf{R}^d, T_{x^0} V)$ is bijective, because ϕ is an immersion at y^0. Formula (5.10), with f replaced by L, therefore gives, for v and $w \in T_{x^0} V$,

$$\begin{aligned} & D^2(f \circ \phi)(y^0)(D\phi(y^0)^{-1}v, D\phi(y^0)^{-1}w) \\ &= D^2(L \circ \phi)(y^0)(D\phi(y^0)^{-1}v, D\phi(y^0)^{-1}w) \\ &= D_x^2 L(x^0, \lambda)(v, w) = H(f|_V)(x^0)(v, w). \end{aligned}$$

The definition of $H(f|_V)(x^0)$ is independent of the choice of g, because the left–hand side of the formula above does not depend on g. On the other hand, the left–hand side is independent of the choice of ϕ since the right–hand side is so. ❑

Example 5.6.3. Let the notation be that of Theorem 5.4.2, let $g(x) = -1 + \sum_{1 \le i \le 3} x_i^{-1}$ and $f(x) = \sum_{1 \le i \le 3} x_i^3$. One has

$$Dg(x) = -(x_1^{-2}, x_2^{-2}, x_3^{-2}), \qquad Df(x) = 3(x_1^2, x_2^2, x_3^2).$$

Using Theorem 5.4.2 one finds that $f|_V$ has critical points x^0 for $x^0 = 3(1, 1, 1)$ corresponding to $\lambda = -3^5$, or $x^0 = (-1, 1, 1)$, $(1, -1, 1)$ or $(1, 1, -1)$, all corresponding to $\lambda = -3$. Furthermore,

$$D^2 g(x) = 2 \begin{pmatrix} x_1^{-3} & 0 & 0 \\ 0 & x_2^{-3} & 0 \\ 0 & 0 & x_3^{-3} \end{pmatrix}, \qquad D^2 f(x) = 6 \begin{pmatrix} x_1 & 0 & 0 \\ 0 & x_2 & 0 \\ 0 & 0 & x_3 \end{pmatrix}.$$

For $\lambda = -3^5$ and $x^0 = 3(1, 1, 1)$ this yields

$$D^2(f - \lambda g)(x^0) = 36 \begin{pmatrix} 1 & 0 & 0 \\ 0 & 1 & 0 \\ 0 & 0 & 1 \end{pmatrix}, \qquad T_{x^0} V = \{\, v \in \mathbf{R}^3 \mid \sum_{1 \le i \le 3} v_i = 0 \,\}.$$

With respect to the basis $(1, -1, 0)$ and $(1, 0, -1)$ for $T_{x^0} V$ we find

$$D^2(f - \lambda g)(x^0)|_{T_{x^0} V} = 36 \begin{pmatrix} 2 & 1 \\ 1 & 2 \end{pmatrix}.$$

The matrix on the right–hand side is positive definite because its eigenvalues are 1 and 3. Consequently, $f|_V$ has a nondegenerate minimum at $3(1, 1, 1)$ with the value 3^4. For $\lambda = -3$ and $x^0 = (-1, 1, 1)$ we find

$$D^2(f - \lambda g)(x^0) = 12 \begin{pmatrix} -1 & 0 & 0 \\ 0 & 1 & 0 \\ 0 & 0 & 1 \end{pmatrix}, \qquad T_{x^0} V = \{\, v \in \mathbf{R}^3 \mid \sum_{1 \le i \le 3} v_i = 0 \,\}.$$

With respect to the basis $(1, -1, 0)$ and $(1, 0, -1)$ for $T_{x^0} V$ we obtain

$$D^2(f - \lambda g)(x^0)|_{T_{x^0} V} = 12 \begin{pmatrix} 0 & -1 \\ -1 & 0 \end{pmatrix}.$$

The matrix on the right–hand side is indefinite because its eigenvalues are -12 and 12. Consequently, $f|_V$ has a saddle point at $(-1, 1, 1)$ with the value 1. The same conclusions apply for $x^0 = (1, -1, 1)$ and $x^0 = (1, 1, -1)$. ☆

5.7 Gaussian curvature of surface

Let V be a surface in $\mathbf{R}^3$, or, more accurately, a C^2 submanifold in $\mathbf{R}^3$ of dimension 2. Assume $x \in V$ and let $\phi : D \to V$ be a C^2 parametrization of a neighborhood of x in V; in particular, let $x = \phi(y)$ with $y \in D \subset \mathbf{R}^2$. The tangent space $T_x V$ is spanned by the vectors $D_1\phi(y)$ and $D_2\phi(y) \in \mathbf{R}^3$.

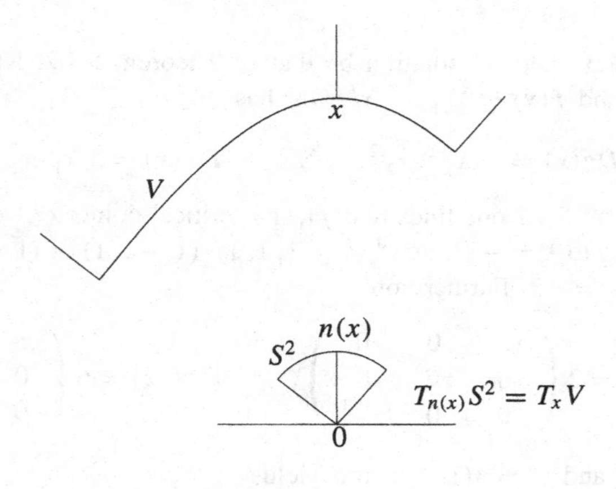

Gauss mapping

Define the *Gauss mapping* $n : V \to S^2$, with S^2 the unit sphere in $\mathbf{R}^3$, by

$$n(x) = \|D_1\phi(y) \times D_2\phi(y)\|^{-1}\, D_1\phi(y) \times D_2\phi(y).$$

Except for multiplication of the vector $n(x)$ by the scalar -1, the definition of $n(x)$ is independent of the choice of the parametrization ϕ, because $T_x V = (\mathbf{R}\, n(x))^{\perp}$. In particular

$$\langle\, n \circ \phi(y),\ D_j\phi(y)\,\rangle = \langle\, n(x),\ D_j\phi(y)\,\rangle = 0 \qquad (1 \le j \le 2).$$

For $j = 2$ and 1 differentiation with respect to y_1 and y_2, respectively, yields

$$\langle\, D_1(n \circ \phi)(y),\ D_2\phi(y)\,\rangle + \langle\, n \circ \phi(y),\ D_1 D_2\phi(y)\,\rangle = 0,$$

$$\langle\, D_2(n \circ \phi)(y),\ D_1\phi(y)\,\rangle + \langle\, n \circ \phi(y),\ D_2 D_1\phi(y)\,\rangle = 0.$$

Because ϕ has continuous second-order partial derivatives, Theorem 2.7.2 implies

$$\langle\, D_1(n \circ \phi)(y),\ D_2\phi(y)\,\rangle = \langle\, D_2(n \circ \phi)(y),\ D_1\phi(y)\,\rangle. \tag{5.11}$$

Furthermore,

$$D_j(n \circ \phi)(y) = Dn(x)D_j\phi(y) \qquad (1 \le j \le 2), \tag{5.12}$$

where $Dn(x) : T_x V \to T_{n(x)} S^2$ is the tangent mapping to the Gauss mapping at x. And so, by Formula (5.11)

$$\langle\, Dn(x)D_1\phi(y),\ D_2\phi(y)\,\rangle = \langle\, D_1\phi(y),\ Dn(x)D_2\phi(y)\,\rangle. \tag{5.13}$$

Also, of course

$$\langle\, Dn(x)D_j\phi(y),\ D_j\phi(y)\,\rangle = \langle\, D_j\phi(y),\ Dn(x)D_j\phi(y)\,\rangle \qquad (1 \le j \le 2). \tag{5.14}$$

One has $Dn(x)v \in T_{n(x)} S^2$, for every $v \in T_x V$. Now S^2 possesses the well-known property that the tangent space to S^2 at a point z is orthogonal to the vector z. Therefore $Dn(x)v$ is orthogonal to $n(x)$. In addition, $T_x V$ is the orthogonal complement of $n(x)$ in $\mathbf{R}^3$. Therefore we now identify $Dn(x)v$ with an element from $T_x V$, that is, we interpret $Dn(x)$ as $Dn(x) \in \mathrm{End}(T_x V)$. In this context $Dn(x)$ is called the *Weingarten mapping*. Formulae (5.13) and (5.14) then tell us that in addition $Dn(x)$ is self-adjoint. According to the Spectral Theorem 2.9.3 the operator

$$Dn(x) \in \mathrm{End}^+(T_x V)$$

has two real eigenvalues $k_1(x)$ and $k_2(x)$; these are known as the *principal curvatures* of V at x. We now define the *Gaussian curvature* $K(x)$ of V at x by

$$K(x) = k_1(x)\, k_2(x) = \det Dn(x).$$

Note that the Gaussian curvature is independent of the choice of the sign of the normal. Further, it follows from the Spectral Theorem that the tangent vectors along which the principal curvatures are attained are mutually orthogonal.

Example 5.7.1. For the sphere S^2 in $\mathbf{R}^3$ the Gauss mapping n is the identical mapping $S^2 \to S^2$ (verify); therefore the Gaussian curvature at every point of S^2 equals 1. For the surface of a cylinder in $\mathbf{R}^3$, the image of n is a circle on S^2. Therefore the tangent mapping $Dn(x)$ is not surjective, and as a consequence the Gaussian curvature of a cylindrical surface vanishes at every point. For similar reasons the Gaussian curvature of a conical surface equals 0 identically. ☆

Example 5.7.2 (Gaussian curvature of torus). Let V be the toroidal surface as in Example 4.4.3, given by, for $-\pi \le \alpha \le \pi$ and $-\pi \le \theta \le \pi$,

$$x = \phi(\alpha, \theta) = ((2 + \cos\theta)\cos\alpha,\ (2 + \cos\theta)\sin\alpha,\ \sin\theta).$$

The tangent space $T_x V$ to V at $x = \phi(\alpha, \theta)$ is spanned by

$$\frac{\partial \phi}{\partial \alpha}(\alpha, \theta) = \begin{pmatrix} -(2+\cos\theta)\sin\alpha \\ (2+\cos\theta)\cos\alpha \\ 0 \end{pmatrix}, \qquad \frac{\partial \phi}{\partial \theta}(\alpha, \theta) = \begin{pmatrix} -\sin\theta\cos\alpha \\ -\sin\theta\sin\alpha \\ \cos\theta \end{pmatrix}.$$

Now

$$\frac{\partial \phi}{\partial \alpha}(\alpha, \theta) \times \frac{\partial \phi}{\partial \theta}(\alpha, \theta) = (2+\cos\theta) \begin{pmatrix} \cos\alpha\cos\theta \\ \sin\alpha\cos\theta \\ \sin\theta \end{pmatrix},$$

and so

$$n(x) = n \circ \phi(\alpha, \theta) = \begin{pmatrix} \cos\alpha\cos\theta \\ \sin\alpha\cos\theta \\ \sin\theta \end{pmatrix}.$$

Because in spherical coordinates $(\alpha,\ \theta)$ the sphere S^2 is parametrized with α running from $-\pi$ to π and θ running from $-\frac{\pi}{2}$ to $\frac{\pi}{2}$, we see that the Gauss mapping always maps two different points on V onto a single point on S^2. In addition (compare with Formula (5.12))

$$\begin{aligned} Dn(x)\frac{\partial \phi}{\partial \alpha}(y) &= \frac{\partial (n \circ \phi)}{\partial \alpha}(\alpha, \theta) = \begin{pmatrix} -\sin\alpha\cos\theta \\ \cos\alpha\cos\theta \\ 0 \end{pmatrix} \\ &= \frac{\cos\theta}{2+\cos\theta} \begin{pmatrix} -(2+\cos\theta)\sin\alpha \\ (2+\cos\theta)\cos\alpha \\ 0 \end{pmatrix} = \frac{\cos\theta}{2+\cos\theta}\frac{\partial \phi}{\partial \alpha}(y), \end{aligned}$$

while

$$Dn(x)\frac{\partial \phi}{\partial \theta}(y) = \frac{\partial (n \circ \phi)}{\partial \theta}(\alpha, \theta) = \begin{pmatrix} -\cos\alpha\sin\theta \\ -\sin\alpha\sin\theta \\ \cos\theta \end{pmatrix} = \frac{\partial \phi}{\partial \theta}(y).$$

As a result we now have the matrix representation and the Gaussian curvature:

$$Dn(\phi(\alpha, \theta)) = \begin{pmatrix} \dfrac{\cos\theta}{2+\cos\theta} & 0 \\ 0 & 1 \end{pmatrix}, \qquad K(x) = K(\phi(\alpha, \theta)) = \frac{\cos\theta}{2+\cos\theta},$$

respectively. Thus we see that $K(x) = 0$ for x on the parallel circles $\theta = -\frac{\pi}{2}$ or $\theta = \frac{\pi}{2}$, while the Gaussian curvature $K(x)$ is positive, or negative, for x on the "outer" part $(-\frac{\pi}{2} < \theta < \frac{\pi}{2})$, or on the "inner" part $(-\pi \le \theta < -\frac{\pi}{2},\ \frac{\pi}{2} < \theta \le \pi)$, respectively, of the toroidal surface. ☆

5.8 Curvature and torsion of curve in $\mathbf{R}^3$

For a curve in $\mathbf{R}^3$ we shall define its curvature and torsion and prove that these functions essentially determine the curve. In this section we need the results on parametrization by arc length from Example 7.4.4.

Definition 5.8.1. Suppose $\gamma : J \to \mathbf{R}^n$ is a C^3 parametrization by arc length of the curve $\mathrm{im}(\gamma)$; in particular, $T(s) := \gamma'(s) \in \mathbf{R}^n$ is a tangent vector of unit length for all $s \in J$. Then the *acceleration* $\gamma''(s) \in \mathbf{R}^n$ is perpendicular to $\mathrm{im}(\gamma)$ at $\gamma(s)$, as follows by differentiating the identity $\langle \gamma'(s), \gamma'(s) \rangle = 1$. Accordingly, the unit vector $N(s) \in \mathbf{R}^n$ in the direction of $\gamma''(s)$ (assuming that $\gamma''(s) \neq 0$) is called the *principal normal* to $\mathrm{im}(\gamma)$ at $\gamma(s)$, and $\kappa(s) := \|\gamma''(s)\| \geq 0$ is called the *curvature* of $\mathrm{im}(\gamma)$ at $\gamma(s)$. It follows that $\gamma''(s) = \kappa(s)N(s)$.

Now suppose $n = 3$. Then define the *binormal* $B(s) \in \mathbf{R}^3$ by $B(s) := T(s) \times N(s)$, and note that $\|B(s)\| = 1$. Hence $(T(s)\, N(s)\, B(s))$ is a positively oriented triple of mutually orthogonal unit vectors in $\mathbf{R}^3$, in other words, the matrix

$$O(s) = (T(s)\, N(s)\, B(s)) \in \mathbf{SO}(3, \mathbf{R}) \qquad (s \in J),$$

in the notation of Example 4.6.2 and Exercise 2.5.

In this context, the following terminology is usual. The linear subspace in $\mathbf{R}^3$ spanned by $T(s)$ and $N(s)$ is called the *osculating plane* (osculum = kiss) of $\mathrm{im}(\gamma)$ at $\gamma(s)$, that by $N(s)$ and $B(s)$ the *normal plane*, and that by $B(s)$ and $T(s)$ the *rectifying plane*. ○

If we differentiate the identity $O(s)^t\, O(s) = I$ and use $(O^t)' = (O')^t$ we find, with $O'(s) = \frac{dO}{ds}(s) \in \mathrm{Mat}(3, \mathbf{R})$

$$(O(s)^t\, O'(s))^t + O(s)^t\, O'(s) = 0 \qquad (s \in J).$$

Therefore there exists a mapping $J \to \mathrm{A}(3, \mathbf{R})$, the linear subspace in $\mathrm{Mat}(3, \mathbf{R})$ consisting of antisymmetric matrices, with $s \mapsto A(s)$ such that

$$O(s)^t\, O'(s) = A(s), \qquad \text{hence} \qquad O'(s) = O(s)A(s) \qquad (s \in J). \tag{5.15}$$

In view of Lemma 8.1.8 we can find $a : J \to \mathbf{R}^3$ so that we have the following equality of matrix-valued functions on J:

$$(T'\, N'\, B') = (T\, N\, B) \begin{pmatrix} 0 & -a_3 & a_2 \\ a_3 & 0 & -a_1 \\ -a_2 & a_1 & 0 \end{pmatrix}.$$

In particular, $\gamma'' = T' = a_3 N - a_2 B$. On the other hand, $\gamma'' = \kappa N$, and this implies $\kappa = a_3$ and $a_2 = 0$. We write $\tau(s) := a_1(s)$, the *torsion* of $\mathrm{im}(\gamma)$ at $\gamma(s)$.

It follows that $a = \tau T + \kappa B$. We now have obtained the following *formulae of Frenet–Serret*, with X equal to T, N and $B : J \to \mathbf{R}^3$, respectively:

$$(T'\,N'\,B') = (T\,N\,B)\begin{pmatrix} 0 & -\kappa & 0 \\ \kappa & 0 & -\tau \\ 0 & \tau & 0 \end{pmatrix},$$

$$\text{thus} \qquad X' = a \times X, \qquad \begin{aligned} T' &= \kappa N \\ N' &= -\kappa T + \tau B \\ B' &= -\tau N \end{aligned}$$

In particular, if $\operatorname{im}(\gamma)$ is a *planar curve* (that is, lies in some plane), then B is constant, thus $B' = 0$, and this implies $\tau = 0$.

Next we drop the assumption that $\gamma : I \to \mathbf{R}^3$ is a parametrization by arc length, that is, we do not necessarily suppose $\|\gamma'\| = 1$. Instead, we assume γ to be a *biregular* C^3 parametrization, meaning that $\gamma'(t)$ and $\gamma''(t) \in \mathbf{R}^3$ are linearly independent for all $t \in I$. We shall express T, N, B, κ and τ all in terms of the velocity γ', the speed $v := \|\gamma'\|$, the acceleration γ'', and its derivative γ'''. From Example 7.4.4 we get $\frac{ds}{dt}(t) = v(t)$ if $s = \lambda(t)$ denotes the arc-length function, and therefore we obtain for the derivatives with respect to $t \in I$, using the formulae of Frenet–Serret,

$$\begin{aligned} \gamma' &= vT, \\ \gamma'' &= v'T + vT' = v'T + v\frac{dT}{ds}\frac{ds}{dt} = v'T + v^2\kappa N, \\ \gamma' \times \gamma'' &= v^3\kappa T \times N = v^3\kappa B. \end{aligned} \tag{5.16}$$

This implies

$$\begin{gathered} T = \frac{1}{\|\gamma'\|}\gamma', \qquad B = \frac{1}{\|\gamma' \times \gamma''\|}\gamma' \times \gamma'', \qquad N = B \times T, \\ \kappa = \frac{\|\gamma' \times \gamma''\|}{\|\gamma'\|^3}, \qquad \tau = \frac{\det(\gamma'\,\gamma''\,\gamma''')}{\|\gamma' \times \gamma''\|^2}. \end{gathered} \tag{5.17}$$

Only the formula for the torsion τ still needs a proof. Observe that $\gamma''' = (v'T + v^2\kappa N)'$. The contribution to γ''' that involves B comes from evaluating

$$v^2\kappa N' = v^3\kappa\frac{dN}{ds} = v^3\kappa(\tau B - \kappa T).$$

It follows that $(\gamma'\,\gamma''\,\gamma''')$ is an upper triangular matrix with respect to the basis (T, N, B); and this implies that its determinant equals the product of the coefficients on the main diagonal. Using (5.16) we therefore derive the desired formula:

$$\det(\gamma'\,\gamma''\,\gamma''') = v^6\kappa^2\tau = \tau\|\gamma' \times \gamma''\|^2.$$

Example 5.8.2. For the helix $\mathrm{im}(\gamma)$ from Example 5.3.2 with $\gamma : \mathbf{R} \to \mathbf{R}^3$ given by $\gamma(t) = (a\cos t,\ a\sin t,\ bt)$ where $a > 0$ and $b \in \mathbf{R}$, we obtain the constant values

$$\kappa = \frac{a}{a^2+b^2}, \qquad \tau = \frac{b}{a^2+b^2}.$$

If $b = 0$, the helix reduces to a circle of radius a and its curvature reduces to $\frac{1}{a}$. It is a consequence of the following theorem that the only curves with constant, nonzero curvature and constant, arbitrary torsion are the helices.

Now suppose that γ is a biregular C^3 parametrization by arc length s for which the ratio of torsion to curvature is constant. Then there exists $0 < \alpha < \pi$ with $(\kappa \cos\alpha - \tau \sin\alpha)N = 0$. Integrating this equality with respect to the variable s and using the Frenet–Serret formulae we obtain the existence of a fixed unit vector $c \in \mathbf{R}^3$ with

$$\cos\alpha\, T + \sin\alpha\, B = c, \qquad \text{whence} \qquad \langle N, c\rangle = 0.$$

Integrating this once again we find $\langle T, c\rangle = \cos\alpha$. That is, the tangent line to $\mathrm{im}(\gamma)$ makes a constant angle with a fixed vector in $\mathbf{R}^3$. In particular, helices have this property (compare with Example 5.3.2). ☆

Theorem 5.8.3. *Consider two curves with biregular C^3 parametrizations γ_1 and γ_2 by arc length, both of which have the same curvature and torsion. Then one of the curves can be rotated and translated so as to coincide exactly with the other.*

Proof. We may assume that the $\gamma_i : J \to \mathbf{R}^3$ for $1 \le i \le 2$ are parametrizations by arc length s both starting at s_0. We have $O_i' = O_i A_i$, but the $A_i : J \to \mathrm{A}(3, \mathbf{R})$ associated with both curves coincide, being in terms of the curvature and torsion. Define $C = O_2 O_1^{-1} : J \to \mathbf{SO}(3, \mathbf{R})$, thus $O_2 = C O_1$. Differentiation with respect to the arc length s gives

$$O_2 A = O_2' = C' O_1 + C O_1 A = C' O_1 + O_2 A, \qquad \text{so} \qquad C' O_1 = 0.$$

Hence $C' = 0$, and therefore C is a constant mapping. From $O_2 = C O_1$ we obtain by considering the first column vectors

$$\gamma_2' = T_2 = C T_1 = C\gamma_1', \qquad \text{therefore} \qquad \gamma_2(s) = C\,\gamma_1(s) + d \qquad (s \in J),$$

with the rotation $C \in \mathbf{SO}(3, \mathbf{R})$ (see Exercise 2.5) and the vector $d \in \mathbf{R}^3$ both constant. ❑

Remark. Let V be a C^2 submanifold in $\mathbf{R}^3$ of dimension 2 and let $n : V \to S^2$ be the corresponding Gauss mapping. Suppose $0 \in J$ and $\gamma : J \to \mathbf{R}^3$ is a C^2 parametrization by arc length of the curve $\mathrm{im}(\gamma) \subset V$. We now make a few remarks on the relation between the properties of the Gauss mapping n and the curvature κ of γ.

Suppose $x = \gamma(0) \in V$, then $n(x)$ is a choice of the normal to V at x. If $\gamma''(0) \neq 0$, let $N(0)$ be the principal normal to $\mathrm{im}(\gamma)$ at x. Then we define the *normal curvature* $k_n(x) \in \mathbf{R}$ of $\mathrm{im}(\gamma)$ in V at x as

$$k_n(x) = \kappa(0)\, \langle n(x), N(0)\rangle.$$

From $\langle (n \circ \gamma)(s), T(s)\rangle = 0$, for s near 0, we find

$$\langle (n \circ \gamma)'(s), T(s)\rangle + \langle (n \circ \gamma)(s), \kappa(s) N(s)\rangle = 0,$$

and this implies, if $v = T(0) = \gamma'(0) \in T_x V$,

$$-\langle Dn(x)v, v\rangle = -\langle (n \circ \gamma)'(0), T(0)\rangle = \kappa(0)\, \langle n(x), N(0)\rangle = k_n(x).$$

It follows that all curves lying on the surface V and having the same unit tangent vector v at x have the same normal curvature at x. This allows us to speak of the *normal curvature* of V at x along a tangent vector at x. Given a unit vector $v \in T_x V$, the intersection of V with the plane through x which is spanned by $n(x)$ and v is called the *normal section* of V at x along v. In a neighborhood of x, a normal section of V at x is an embedded plane curve on V that can be parametrized by arc length and whose principal normal $N(0)$ at x equals $\pm n(x)$ or 0. With this terminology, we can say that the absolute value of the normal curvature of V at x along $v \in T_x V$ is equal to the curvature of the normal section of V at x along v. We now see that the two principal curvatures $k_1(x)$ and $k_2(x)$ of V at x are the extreme values of the normal curvature of V at x.

5.9 One-parameter groups and infinitesimal generators

In this section we deal with some theory, which in addition may serve as background to Exercises 2.41, 2.50, 4.22, 5.58 and 5.60. Example 2.4.10 is a typical example of this theory. A *one-parameter group* or *group action of C^1 diffeomorphisms* $(\Phi^t)_{t\in\mathbf{R}}$ of $\mathbf{R}^n$ is defined as a C^1 mapping

$$\Phi : \mathbf{R} \times \mathbf{R}^n \to \mathbf{R}^n \qquad \text{with} \qquad \Phi^t(x) := \Phi(t, x) \qquad ((t, x) \in \mathbf{R} \times \mathbf{R}^n), \quad (5.18)$$

with the property

$$\Phi^0 = I \qquad \text{and} \qquad \Phi^t \circ \Phi^{t'} = \Phi^{t+t'} \qquad (t,\, t' \in \mathbf{R}). \qquad (5.19)$$

It then follows that, for every $t \in \mathbf{R}$, the mapping $\Phi^t : \mathbf{R}^n \to \mathbf{R}^n$ is a C^1 diffeomorphism, with inverse Φ^{-t}; that is

$$(\Phi^t)^{-1} = \Phi^{-t} \qquad (t \in \mathbf{R}). \qquad (5.20)$$

Note that, for every $x \in \mathbf{R}^n$, the curve $t \mapsto \Phi^t(x) = \Phi(t, x) : \mathbf{R} \to \mathbf{R}^n$ is differentiable on $\mathbf{R}$, and $\Phi^0(x) = x$. The set $\{ \Phi^t(x) \mid t \in \mathbf{R} \}$ is said to be the *orbit* of x under the action of $(\Phi^t)_{t \in \mathbf{R}}$.

Next, we define the *infinitesimal generator* or *tangent vector field* of the one-parameter group of diffeomorphisms $(\Phi^t)_{t \in \mathbf{R}}$ on $\mathbf{R}^n$ as the mapping $\phi : \mathbf{R}^n \to \mathbf{R}^n$ with

$$\phi(x) = \frac{\partial \Phi}{\partial t}(0, x) \in \mathrm{Lin}(\mathbf{R}, \mathbf{R}^n) \simeq \mathbf{R}^n \qquad (x \in \mathbf{R}^n). \tag{5.21}$$

By means of Formula (5.19) we find, for $(t, x) \in \mathbf{R} \times \mathbf{R}^n$,

$$\frac{d}{dt}\Phi^t(x) = \left.\frac{d}{ds}\right|_{s=0} \Phi^{s+t}(x) = \left.\frac{d}{ds}\right|_{s=0} \Phi^s(\Phi^t(x)) = \phi(\Phi^t(x)). \tag{5.22}$$

In other words, for every $x \in \mathbf{R}^n$ the curve $t \mapsto \Phi^t(x)$ is an integral curve γ of the *ordinary differential equation* on $\mathbf{R}^n$, with *initial condition* respectively

$$\gamma'(t) = \phi(\gamma(t)) \quad (t \in \mathbf{R}), \qquad \gamma(0) = x. \tag{5.23}$$

Conversely, the mapping Φ^t is known as the *flow over time t of the vector field* ϕ. In the theory of differential equations one studies the problem to what extent it is possible, under reasonable conditions on the vector field ϕ, to construct the one-parameter group $(\Phi^t)_{t \in \mathbf{R}}$ of flows, starting from ϕ, and to what extent $(\Phi^t)_{t \in \mathbf{R}}$ is uniquely determined by ϕ. This is complicated by the fact that Φ^t cannot always be defined for all $t \in \mathbf{R}$. In addition, one often deals with the more general problem

$$\gamma'(t) = \phi(t, \gamma(t)),$$

where the vector field ϕ itself now also depends on the time variable t. To distinguish between these situations, the present case of a time-independent vector field is called *autonomous*.

We introduce the *induced action* of $(\Phi^t)_{t \in \mathbf{R}}$ on $C^\infty(\mathbf{R}^n)$ by assigning $\Phi^t f \in C^\infty(\mathbf{R}^n)$ to Φ^t and $f \in C^\infty(\mathbf{R}^n)$, where

$$(\Phi^t f)(x) = f(\Phi^{-t}(x)) \qquad (x \in \mathbf{R}^n). \tag{5.24}$$

Here we apply the rule: "the value of the transformed function at the transformed point equals the value of the original function at the original point". We then have a *group action*, that is

$$(\Phi^t \Phi^{t'}) f = \Phi^t(\Phi^{t'} f) \qquad (t, t' \in \mathbf{R},\ f \in C^\infty(\mathbf{R}^n)).$$

At a more fundamental level even, one identifies a function $f : X \to \mathbf{R}$ with $\mathrm{graph}(f) \subset X \times \mathbf{R}$. Accordingly, the natural definition of the function $g = \Phi f : Y \to \mathbf{R}$, for a bijection $\Phi : X \to Y$, is via $\mathrm{graph}(g) := (\Phi \times I)\,\mathrm{graph}(f)$. This gives

$$(y, g(y)) = (\Phi(x), f(x)), \qquad \text{hence} \qquad \Phi f(y) = g(y) = f(x) = f(\Phi^{-1}(y)).$$

Note that in general the action of Φ^t on $\mathbf{R}^n$ is nonlinear, that is, one does not necessarily have $\Phi^t(x+x') = \Phi^t(x) + \Phi^t(x')$, for $x, x' \in \mathbf{R}^n$. In contrast, the induced action of Φ^t on $C^\infty(\mathbf{R}^n)$ is linear; indeed, $\Phi^t(f+\lambda f') = \Phi^t(f)+\lambda\Phi^t(f')$, because $(f+\lambda f')(\Phi^{-t}(x)) = f(\Phi^{-t}(x)) + \lambda f'(\Phi^{-t}(x))$, for $f, f' \in C^\infty(\mathbf{R}^n)$, $\lambda \in \mathbf{R}$, $x \in \mathbf{R}^n$. On the other hand, $\mathbf{R}^n$ is finite-dimensional, while $C^\infty(\mathbf{R}^n)$ is infinite-dimensional.

In its turn the definition in (5.24) leads to the induced action ∂_ϕ on $C^\infty(\mathbf{R}^n)$ of the infinitesimal generator ϕ of $(\Phi^t)_{t\in\mathbf{R}}$, given by

$$\partial_\phi \in \mathrm{End}\,(C^\infty(\mathbf{R}^n)) \qquad \text{with} \qquad (\partial_\phi f)(x) = \frac{d}{dt}\bigg|_{t=0} (\Phi^t f)(x). \tag{5.25}$$

This said, from Formulae (5.24) and (5.21) readily follows

$$(\partial_\phi f)(x) = \frac{d}{dt}\bigg|_{t=0} f(\Phi^{-t}(x)) = -Df(x)\phi(x) = -\sum_{1\le j\le n} \phi_j(x)\, D_j f(x).$$

We see that ∂_ϕ is a *partial differential operator with variable coefficients* on $C^\infty(\mathbf{R}^n)$, with the notation

$$\partial_\phi(x) = -\sum_{1\le j\le n} \phi_j(x)\, D_j. \tag{5.26}$$

Note that, but for the $-$sign, $\partial_\phi(x)$ is the directional derivative in the direction $\phi(x)$. Warning: most authors omit the $-$sign in this identification of tangent vector field and partial differential operator.

Example 5.9.1. Let $\Phi : \mathbf{R} \times \mathbf{R}^2 \to \mathbf{R}^2$ be given by

$$\Phi(t, x) = (x_1 \cos t - x_2 \sin t,\ x_1 \sin t + x_2 \cos t),$$

$$\text{that is,} \qquad \Phi^t = \begin{pmatrix} \cos t & -\sin t \\ \sin t & \cos t \end{pmatrix} \in \mathrm{Mat}(2, \mathbf{R}).$$

It readily follows that we have a one-parameter group of diffeomorphisms of $\mathbf{R}^2$. The orbits are the circles in $\mathbf{R}^2$ of center 0 and radius ≥ 0. Moreover, $\phi(x) = (-x_2, x_1)$. The said circles do in fact satisfy the differential equation

$$\begin{pmatrix} \gamma_1'(t) \\ \gamma_2'(t) \end{pmatrix} = \begin{pmatrix} -\gamma_2(t) \\ \gamma_1(t) \end{pmatrix} \qquad (t \in \mathbf{R}).$$

By means of Formula (5.26) we find $\partial_\phi(x) = x_2 D_1 - x_1 D_2$, for $x \in \mathbf{R}^2$. ☆

Example 5.9.2. See Exercise 2.50. Let $a \in \mathbf{R}^n$ be fixed, and define $\Phi : \mathbf{R} \times \mathbf{R}^n \to \mathbf{R}^n$ by $\Phi(t, x) = x + ta$, hence

$$\phi(x) = a, \qquad \text{and so} \qquad \partial_\phi(x) = -\sum_{1\le j\le n} a_j D_j \qquad (x \in \mathbf{R}^n).$$

In this case the orbits are straight lines. Further note that ∂_ϕ is a partial differential operator with constant coefficients, which we encountered in Exercise 2.50.(ii) under the name t_a. ☆

Example 5.9.3. See Example 2.4.10 and Exercises 2.41, 4.22, 5.58 and 5.60. Consider $a \in \mathbf{R}^3$ with $\|a\| = 1$, and let $\Phi_a : \mathbf{R}\times\mathbf{R}^3 \to \mathbf{R}^3$ be given by $\Phi_a(t, x) = R_{t,a}x$, as in Exercise 4.22. It is evident that we then find a one-parameter group in $\mathbf{SO}(3, \mathbf{R})$. The orbit of $x \in \mathbf{R}^3$ is the circle in $\mathbf{R}^3$ of center $\langle x, a\rangle\, a$ and radius $\|a \times x\|$, lying in the plane through x and orthogonal to a. Using Euler's formula from Exercise 4.22.(iii) we obtain

$$\phi_a(x) = a \times x =: r_a(x).$$

Therefore the orbit of x is the integral curve through x of the following differential equation for the rotations $R_{t,a}$:

$$\frac{dR_{t,a}}{dt}(x) = r_a \circ R_{t,a}(x) \quad (t \in \mathbf{R}), \qquad R_{0,a}(x) = I. \tag{5.27}$$

According to Formula (5.26) we have

$$\partial_{\phi_a}(x) = \langle a, (\mathrm{grad}) \times x\, \rangle = \langle a, L(x)\, \rangle \qquad (x \in \mathbf{R}^3).$$

Here L is the angular momentum operator from Exercises 2.41 and 5.60. In Exercise 5.58 we verify again that the solution of the differential equation (5.27) is given by $R_{t,\,a} = e^{t\, r_a}$, where the exponential mapping is defined as in Example 2.4.10. ☆

Finally, we derive Formula (5.31) below, which we shall use in Example 6.6.9. We assume that $(\Phi^t)_{t\in\mathbf{R}}$ is a one-parameter group of C^2 diffeomorphisms. We write $D\Phi^t(x) \in \mathrm{End}(\mathbf{R}^n)$ for the derivative of Φ^t at x with respect to the variable in $\mathbf{R}^n$. Using Formula (5.19) for the first equality below, the chain rule for the second, and the product rule for determinants for the third, we obtain

$$\begin{aligned} \frac{d}{dt}\det D\Phi^t(x) &= \frac{d}{ds}\Big|_{s=0} \det D(\Phi^s \circ \Phi^t)(x) \\ &= \frac{d}{ds}\Big|_{s=0} \det\big((D\Phi^s)(\Phi^t(x)) \circ (D\Phi^t)(x)\big) \\ &= \det D\Phi^t(x) \frac{d}{ds}\Big|_{s=0} (\det \circ D\Phi^s)(\Phi^t(x)). \end{aligned} \tag{5.28}$$

Apply the chain rule once again, note that $D\Phi^0 = DI = I$, change the order of differentiation, then use Formula (5.21), and conclude from Exercise 2.44.(i) that $(D\det)(I) = \mathrm{tr}$. This gives

$$\begin{aligned} \frac{d}{ds}\Big|_{s=0} (\det \circ D\Phi^s)(\Phi^t(x)) &= (D\det)(I) \circ D\Big(\frac{d}{ds}\Big|_{s=0} \Phi^s\Big)(\Phi^t(x)) \\ &= \mathrm{tr}(D\phi)(\Phi^t(x)). \end{aligned} \tag{5.29}$$

We now define the *divergence* of the vector field ϕ, notation: $\operatorname{div}\phi$, as the function $\mathbf{R}^n \to \mathbf{R}$, with

$$\operatorname{div}\phi = \operatorname{tr} D\phi = \sum_{1\le j\le n} D_j\phi_j. \tag{5.30}$$

Combining Formulae (5.28) – (5.30) we find

$$\frac{d}{dt}\det D\Phi^t(x) = \operatorname{div}\phi(\Phi^t(x))\,\det D\Phi^t(x) \qquad ((t,x)\in\mathbf{R}\times\mathbf{R}^n). \tag{5.31}$$

Because $\det D\Phi^0(x) = 1$, solving the differential equation in (5.31) we obtain

$$\det D\Phi^t(x) = e^{\int_0^t \operatorname{div}\phi(\Phi^\tau(x))\,d\tau} \qquad ((t,x)\in\mathbf{R}\times\mathbf{R}^n). \tag{5.32}$$

Example 5.9.4. Let $A\in\operatorname{End}(\mathbf{R}^n)$ be fixed, and define $\Phi : \mathbf{R}\times\mathbf{R}^n \to \mathbf{R}^n$ by $\Phi(t,x) = e^{tA}x$. According to Example 2.4.10 we thus obtain a one-parameter group in $\operatorname{Aut}(\mathbf{R}^n)$, and furthermore we have

$$\phi(x) = Ax, \qquad \text{and so} \qquad \partial_\phi(x) = -\sum_{1\le i\le n}\Big(\sum_{1\le j\le n} a_{ij}x_j\Big)D_i \qquad (x\in\mathbf{R}^n).$$

Because $\Phi^t\in\operatorname{End}(\mathbf{R}^n)$, we have $D\Phi^t(x) = \Phi^t = e^{tA}$. Moreover, $D\phi(x) = A$, and so $\operatorname{div}\phi(x) = \operatorname{tr}A$, for $x\in\mathbf{R}^n$. It follows that $\operatorname{div}\phi(\Phi^\tau(x)) = \operatorname{tr}A$; consequently, Formula (5.32) gives (compare with Exercise 2.44.(ii))

$$\det(e^{tA}) = e^{t\operatorname{tr}A} \qquad (t\in\mathbf{R},\ A\in\operatorname{End}(\mathbf{R}^n)). \tag{5.33}$$

☆

5.10 Linear Lie groups and their Lie algebras

We recall that $\operatorname{Mat}(n,\mathbf{R})$ is a linear space, which is linearly isomorphic to $\mathbf{R}^{n^2}$, and further that $\mathbf{GL}(n,\mathbf{R}) = \{A\in\operatorname{Mat}(n,\mathbf{R}) \mid \det A\ne 0\}$ is an open subset of $\operatorname{Mat}(n,\mathbf{R})$.

Definition 5.10.1. A *linear Lie group* is a subgroup G of $\mathbf{GL}(n,\mathbf{R})$ which is also a C^2 submanifold of $\operatorname{Mat}(n,\mathbf{R})$. If this is the case, write $\mathfrak{g} = T_I G \subset \operatorname{Mat}(n,\mathbf{R})$ for the tangent space of G at $I\in G$. ❍

It is an important result in the theory of Lie groups that the definition above can be weakened substantially while the same conclusions remain valid, viz., a linear Lie group is a subgroup of $\mathbf{GL}(n,\mathbf{R})$ that is also a closed subset of $\mathbf{GL}(n,\mathbf{R})$, see Exercise 5.64. Further, there is a more general concept of Lie group, but to keep the exposition concise we restrict ourselves to the subclass of linear Lie groups.

According to Example 2.4.10 we have

$$\exp : \mathrm{Mat}(n, \mathbf{R}) \to \mathbf{GL}(n, \mathbf{R}), \qquad \exp X = e^X = \sum_{k \in \mathbf{N}_0} \frac{1}{k!} X^k \in \mathbf{GL}(n, \mathbf{R}).$$

Then $e^{X_1} e^{X_2} = e^{X_1 + X_2}$ if X_1 and $X_2 \in \mathrm{Mat}(n, \mathbf{R})$ commute. We note that $\gamma(t) = e^{tX}$ is a differentiable curve in $\mathbf{GL}(n, \mathbf{R})$ with tangent vector at I equal to

$$\gamma'(0) = \frac{d}{dt}\Big|_{t=0} e^{tX} = X \qquad (X \in \mathrm{Mat}(n, \mathbf{R})). \tag{5.34}$$

Because $D\exp(0) = I \in \mathrm{End}(\mathrm{Mat}(n, \mathbf{R}))$, it follows from the Local Inverse Function Theorem 3.2.4 that there exists an open neighborhood of 0 in $\mathrm{Mat}(n, \mathbf{R})$ such that the restriction of exp to U is a diffeomorphism onto an open neighborhood of I in $\mathbf{GL}(n, \mathbf{R})$. As another consequence of (5.34) we see, in view of the definition of tangent space,

$$\widetilde{\mathfrak{g}} := \{ X \in \mathrm{Mat}(n, \mathbf{R}) \mid e^{tX} \in G, \text{ for all } t \in \mathbf{R} \text{ with } |t| \text{ sufficiently small} \} \subset \mathfrak{g}. \tag{5.35}$$

Theorem 5.10.2. *Let G be a linear Lie group, with corresponding tangent space $\mathfrak{g}$ at I. Then we have the following assertions.*

(i) *G is a closed subset of* $\mathbf{GL}(n, \mathbf{R})$.

(ii) *The restriction of the exponential mapping to $\mathfrak{g}$ maps $\mathfrak{g}$ into G, hence* $\exp : \mathfrak{g} \to G$. *In particular,*

$$\mathfrak{g} = \{ X \in \mathrm{Mat}(n, \mathbf{R}) \mid e^{tX} \in G, \textit{ for all } t \in \mathbf{R} \}.$$

Proof. (i). Because G is a submanifold of $\mathbf{GL}(n, \mathbf{R})$ there exists an open neighborhood U of I in $\mathbf{GL}(n, \mathbf{R})$ such that $G \cap U$ is a closed subset of U. As $x \mapsto x^{-1}$ is a homeomorphism of $\mathbf{GL}(n, \mathbf{R})$, U^{-1} is also an open neighborhood of I in $\mathbf{GL}(n, \mathbf{R})$. It follows that xU^{-1} is an open neighborhood of x in $\mathbf{GL}(n, \mathbf{R})$. Given x in the closure $\overline{G}$ of G in $\mathbf{GL}(n, \mathbf{R})$, choose $y \in xU^{-1} \cap G$; then $y^{-1}x \in \overline{G} \cap U = G \cap U$, whence $x \in G$. This implies assertion (i).
(ii). G is a submanifold at I, hence, on account of Proposition 5.1.3, there exist an open neighborhood D of 0 in $\mathfrak{g}$ and a C^1 mapping $\phi : D \to G$, such that $\phi(0) = I$ and $D\phi(0)$ equals the inclusion mapping $\mathfrak{g} \to \mathrm{Mat}(n, \mathbf{R})$. Since $\exp : \mathrm{Mat}(n, \mathbf{R}) \to \mathbf{GL}(n, \mathbf{R})$ is a local diffeomorphism at 0, we may adapt D to arrange that there exists a C^1 mapping $\psi : D \to \mathrm{Mat}(n, \mathbf{R})$ with $\psi(0) = 0$ and

$$\phi(X) = e^{\psi(X)} \qquad (X \in D).$$

By application of the chain rule we see that $D\phi(0) = D\exp(0) D\psi(0) = D\psi(0)$, hence $D\psi(0)$ equals the inclusion map $\mathfrak{g} \to \mathrm{Mat}(n, \mathbf{R})$ too. Now let $X \in \mathfrak{g}$ be arbitrary. For $k \in \mathbf{N}$ sufficiently large we have $\frac{1}{k}X \in D$, which implies $\phi(\frac{1}{k}X)^k \in G$. Furthermore,

$$\phi(\frac{1}{k}X)^k = \big(e^{\psi(\frac{1}{k}X)}\big)^k = e^{k\psi(\frac{1}{k}X)} \to e^X \qquad \text{as} \qquad k \to \infty,$$

since $\lim_{k\to\infty} k\psi(\frac{1}{k}X) = D\psi(0)X = X$ by the definition of derivative. Because G is closed, we obtain $e^X \in G$, which implies $\mathfrak{g} \subset \widetilde{\mathfrak{g}}$, see Formula (5.35). This yields the equality $\mathfrak{g} = \widetilde{\mathfrak{g}} = \{\, X \in \mathrm{Mat}(n, \mathbf{R}) \mid e^{tX} \in G, \text{ for all } t \in \mathbf{R} \,\}$. ❑

Example 5.10.3. In terms of endomorphisms of $\mathbf{R}^n$ instead of matrices, $\mathfrak{g} = \mathrm{End}(\mathbf{R}^n)$ if $G = \mathrm{Aut}(\mathbf{R}^n)$. Further, Formula (5.33) implies $\mathfrak{g} = \mathfrak{sl}(n, \mathbf{R})$ if $G = \mathbf{SL}(n, \mathbf{R})$, where

$$\mathfrak{sl}(n, \mathbf{R}) = \{\, X \in \mathrm{Mat}(n, \mathbf{R}) \mid \mathrm{tr}\, X = 0 \,\},$$
$$\mathbf{SL}(n, \mathbf{R}) = \{\, A \in \mathbf{GL}(n, \mathbf{R}) \mid \det A = 1 \,\}.$$ ☆

We need some more definitions. Given $g \in G$, we define the mapping

$$\mathbf{Ad}\, g : G \to G \qquad \text{by} \qquad (\mathbf{Ad}\, g)(x) = gxg^{-1} \qquad (x \in G).$$

Obviously $\mathbf{Ad}\, g$ is a C^2 diffeomorphism of G leaving I fixed. Next we define, see Definition 5.2.1,

$$\mathrm{Ad}\, g = D(\mathbf{Ad}\, g)(I) \in \mathrm{End}(\mathfrak{g}).$$

$\mathrm{Ad}\, g \in \mathrm{End}(\mathfrak{g})$ is called the *adjoint mapping* of $g \in G$. Because $gY^kg^{-1} = (gYg^{-1})^k$ for all $g \in G$, $Y \in \mathfrak{g}$ and $k \in \mathbf{N}$, we have $\mathbf{Ad}\, g(e^{tY}) = g\, e^{tY} g^{-1} = e^{t\, gYg^{-1}}$, and therefore, on the strength of Formula (5.34), the chain rule, and Theorem 5.10.2.(ii)

$$(\mathrm{Ad}\, g)Y = \frac{d}{dt}\Big|_{t=0} (\mathbf{Ad}\, g)(e^{tY}) = \frac{d}{dt}\Big|_{t=0} e^{t\, gYg^{-1}} = gYg^{-1} \in \mathfrak{g}.$$

Hence, Ad acts by conjugation, as does **Ad**. Further, we find, for $g, h \in G$ and $Y \in \mathfrak{g}$,

$$\mathrm{Ad}\, gh = \mathrm{Ad}\, g \circ \mathrm{Ad}\, h, \qquad \mathrm{Ad} : G \to \mathrm{Aut}(\mathfrak{g}). \tag{5.36}$$

$\mathrm{Ad} : G \to \mathrm{Aut}(\mathfrak{g})$ is said to be the *adjoint representation* of G in the linear space of automorphisms of $\mathfrak{g}$. Furthermore, since Ad is a C^1 diffeomorphism we can define

$$\mathrm{ad} = D(\mathrm{Ad})(I) : \mathfrak{g} \to \mathrm{End}(\mathfrak{g}).$$

For the same reasons as above we obtain, for all $X, Y \in \mathfrak{g}$,

$$(\mathrm{ad}\, X)Y = \frac{d}{dt}\Big|_{t=0} (\mathrm{Ad}\, e^{tX})Y = \frac{d}{dt}\Big|_{t=0} e^{tX}\, Y\, e^{-tX} = XY - YX =: [\, X, Y\,].$$

Thus we see that $\mathfrak{g}$ in addition to being a vector space carries the structure of a Lie algebra, which is defined in the following:

Definition 5.10.4. A vector space $\mathfrak{g}$ is said to be a *Lie algebra* if it is provided with a bilinear mapping $\mathfrak{g} \times \mathfrak{g} \to \mathfrak{g}$, called the *Lie brackets* or *commutator*, which is *anticommutative*, and satisfies *Jacobi's identity*, for X_1, X_2 and $X_3 \in \mathfrak{g}$,

$$[X_1, X_2] = -[X_2, X_1],$$
$$[X_1, [X_2, X_3]] + [X_2, [X_3, X_1]] + [X_3, [X_1, X_2]] = 0. \qquad ❍$$

Therefore we are justified in calling $\mathfrak{g}$ the Lie algebra of G. Note that Jacobi's identity can be reformulated as the assertion that ad X_1 satisfies Leibniz' rule, that is

$$(\mathrm{ad}\, X_1)[X_2, X_3] = [(\mathrm{ad}\, X_1)X_2, X_3] + [X_2, (\mathrm{ad}\, X_1)X_3].$$

This property (partially) explains why ad $X \in \mathrm{End}(\mathfrak{g})$ is called the *inner derivation* determined by $X \in \mathfrak{g}$. (As above, ad is coming from adjoint.) Jacobi's identity can also be phrased as

$$\mathrm{ad}[X_1, X_2] = [\mathrm{ad}\, X_1, \mathrm{ad}\, X_2] \in \mathrm{End}(\mathfrak{g}).$$

Remark. If the group G is *Abelian*, or in other words, commutative, then Ad $g = I$, for all $g \in G$. In turn, this implies ad $X = 0$, for all $X \in \mathfrak{g}$, and therefore $[X_1, X_2] = 0$ for all X_1 and $X_2 \in \mathfrak{g}$ in this case. On the other hand, if G is connected and not Abelian, the Lie algebra structure of $\mathfrak{g}$ is nontrivial.

At this stage, the Lie algebra $\mathfrak{g}$ arises as tangent space of the group G, but Lie algebras do occur in mathematics in their own right, without accompanying group. It is a difficult theorem that every finite-dimensional Lie algebra is the Lie algebra of a linear Lie group.

Lemma 5.10.5. *Every one-parameter subgroup* $(\Phi^t)_{t\in\mathbf{R}}$ *that acts in* $\mathbf{R}^n$ *and consists of operators in* $\mathbf{GL}(n, \mathbf{R})$ *is of the form* $\Phi^t = e^{tX}$, *for* $t \in \mathbf{R}$, *where* $X = \left.\frac{d}{dt}\right|_{t=0} \Phi^t \in \mathrm{Mat}(n, \mathbf{R})$, *the infinitesimal generator of* $(\Phi^t)_{t\in\mathbf{R}}$.

Proof. From Formula (5.22) we obtain that Φ^t is an operator-valued solution to the initial-value problem

$$\frac{d\Phi^t}{dt} = X\,\Phi^t, \qquad \Phi^0 = I.$$

According to Example 2.4.10 the unique and globally defined solution to this first-order linear system of ordinary differential equations with constant coefficients is given by $\Phi^t = e^{tX}$. ❑

With the definitions above we now derive the following results on preservation of structures by suitable mappings. We will see concrete examples of this in, for instance, Exercises 5.59, 5.60, 5.67.(x) and 5.70.

Theorem 5.10.6. *Let G and G' be linear Lie groups with Lie algebras $\mathfrak{g}$ and $\mathfrak{g}'$, respectively, and let $\Phi : G \to G'$ be a homomorphism of groups that is of class C^1 at $I \in G$. Write $\phi = D\Phi(I) : \mathfrak{g} \to \mathfrak{g}'$. Then we have the following properties.*

(i) $\phi \circ \operatorname{Ad} g = \operatorname{Ad} \Phi(g) \circ \phi : \mathfrak{g} \to \mathfrak{g}'$, *for every* $g \in G$.

(ii) $\phi : \mathfrak{g} \to \mathfrak{g}'$ *is a* homomorphism of Lie algebras, *that is, it is a linear mapping satisfying in addition*

$$\phi([X_1, X_2]) = [\phi(X_1), \phi(X_2)] \qquad (X_1, X_2 \in \mathfrak{g}).$$

(iii) $\Phi \circ \exp = \exp \circ \phi : \mathfrak{g} \to G'$. *In particular, with* $\Phi = \operatorname{Ad} : G \to \operatorname{Aut}(\mathfrak{g})$ *we obtain* $\operatorname{Ad} \circ \exp = \exp \circ \operatorname{ad} : \mathfrak{g} \to \operatorname{Aut}(\mathfrak{g})$.

Accordingly, we have the following commutative diagrams:

$$\begin{array}{ccc} \mathfrak{g}' & \xrightarrow{\operatorname{Ad}\Phi g} & \mathfrak{g}' \\ \phi\uparrow & & \uparrow\phi \\ \mathfrak{g} & \xrightarrow{\operatorname{Ad} g} & \mathfrak{g} \end{array} \qquad \begin{array}{ccc} G & \xrightarrow{\Phi} & G' \\ \exp\uparrow & & \uparrow\exp \\ \mathfrak{g} & \xrightarrow{\phi} & \mathfrak{g}' \end{array} \qquad \begin{array}{ccc} G & \xrightarrow{\operatorname{Ad}} & \operatorname{Aut}(\mathfrak{g}) \\ \exp\uparrow & & \uparrow\exp \\ \mathfrak{g} & \xrightarrow{\operatorname{ad}} & \operatorname{End}(\mathfrak{g}) \end{array}$$

Proof. (i). Differentiating

$$\Phi\big((\mathbf{Ad}\, g)(x)\big) = \Phi(gxg^{-1}) = \Phi(g)\Phi(x)\Phi(g)^{-1} = \big(\mathbf{Ad}\, \Phi(g)\big)(\Phi(x))$$

with respect to x at $x = I$ in the direction of $X_2 \in \mathfrak{g}$ and using the chain rule, we get

$$(\phi \circ \operatorname{Ad} g)X_2 = (\operatorname{Ad} \Phi(g) \circ \phi)X_2.$$

(ii). Differentiating this equality with respect to g at $g = I$ in the direction of $X_1 \in \mathfrak{g}$, we obtain the equality in (ii).
(iii). The result follows from Theorem 5.10.2.(ii) and Lemma 5.10.5. Indeed, for given $X \in \mathfrak{g}$, both one-parameter groups $t \mapsto \Phi(\exp tX)$ and $t \mapsto \exp t\phi(X)$ in G' have the same infinitesimal generator $\phi(X) \in \mathfrak{g}'$. ❑

Finally, we give a geometric meaning for the addition and the commutator in $\mathfrak{g}$. The sum in $\mathfrak{g}$ of two vectors each of which is tangent to a curve in G is tangent to the product curve in G, and the commutator in $\mathfrak{g}$ of tangent vectors is tangent to the commutator of the curves in G having a modified parametrization.

Proposition 5.10.7. *Let G be a linear Lie group with Lie algebra $\mathfrak{g}$, and let X and Y be in $\mathfrak{g}$.*

(i) *$X + Y$ is the tangent vector at I to the curve* $\mathbf{R} \ni t \mapsto e^{tX}e^{tY} \in G$.

(ii) $[X, Y] \in \mathfrak{g}$ *is the tangent vector at* I *to the curve* $\mathbf{R}_+ \to G$ *given by* $t \mapsto e^{\sqrt{t}X} e^{\sqrt{t}Y} e^{-\sqrt{t}X} e^{-\sqrt{t}Y}$.

(iii) *We have* Lie's product formula $e^{X+Y} = \lim_{k\to\infty} (e^{\frac{1}{k}X} e^{\frac{1}{k}Y})^k$.

Proof. In this proof we write $\|\cdot\|$ for the Euclidean norm on $\mathrm{Mat}(n, \mathbf{R})$.
(i). From the estimate $\| \sum_{k\geq 2} \frac{1}{k!} X^k \| \leq \sum_{k\geq 2} \frac{1}{k!} \|X\|^k$ we obtain $e^X = I + X + \mathcal{O}(\|X\|^2)$, $\quad X \to 0$. Multiplying the expressions for e^X and e^Y we see

$$e^X e^Y = e^{X+Y} + \mathcal{O}((\|X\| + \|Y\|)^2), \qquad (X, Y) \to (0, 0). \tag{5.37}$$

This implies assertion (i).
(ii). Modulo terms in $\mathrm{Mat}(n, \mathbf{R})$ in X and $Y \in \mathfrak{g}$ of order ≥ 2 we have

$$\begin{aligned}
e^{\pm X} e^{\pm Y} &\equiv (I \pm X + \tfrac{1}{2}X^2 + \cdots)(I \pm Y + \tfrac{1}{2}Y^2 + \cdots) \\
&\equiv I \pm (X + Y) + XY + \tfrac{1}{2}X^2 + \tfrac{1}{2}Y^2 + \cdots \\
&= I \pm (X + Y) + \tfrac{1}{2}(XY - YX) + \tfrac{1}{2}(X^2 + Y^2 + XY + YX + \cdots) \\
&\equiv I + (\pm (X + Y) + \tfrac{1}{2}[X, Y]) + \tfrac{1}{2}(\pm (X + Y) + \tfrac{1}{2}[X, Y])^2 + \cdots \\
&\equiv e^{\pm(X+Y) + \frac{1}{2}[X, Y] + \cdots}.
\end{aligned} \tag{5.38}$$

Therefore assertion (ii) follows from $e^{\sqrt{t}X} e^{\sqrt{t}Y} e^{-\sqrt{t}X} e^{-\sqrt{t}Y} = e^{t[X, Y] + o(t)}$, $\quad t \downarrow 0$.
(iii). We begin the proof with the equality

$$A^k - B^k = \sum_{0\leq l<k} A^{k-1-l}(A - B)B^l \qquad (A, B \in \mathfrak{g}, k \in \mathbf{N}).$$

Setting $m = \max\{\, \|A\|, \|B\| \,\}$ we obtain

$$\|A^k - B^k\| \leq k m^{k-1} \|A - B\|. \tag{5.39}$$

Next let

$$A_k = e^{\frac{1}{k}(X+Y)}, \qquad B_k = e^{\frac{1}{k}X} e^{\frac{1}{k}Y} \qquad (k \in \mathbf{N}).$$

Then $\|A_k\|$ and $\|B_k\|$ both are bounded above by $e^{\frac{\|X\|+\|Y\|}{k}}$. From Formula (5.37) we get $A_k - B_k = \mathcal{O}(\frac{1}{k^2})$, $\quad k \to \infty$. Hence (5.39) implies

$$\|A_k^k - B_k^k\| \leq k e^{\|X\|+\|Y\|} \, \mathcal{O}\Big(\frac{1}{k^2}\Big) = \mathcal{O}\Big(\frac{1}{k}\Big), \qquad k \to \infty.$$

Assertion (iii) now follows as $A_k^k = e^{X+Y}$, for all $k \in \mathbf{N}$. ❑

Remark. Formula (5.38) shows that in first-order approximation multiplication of exponentials is commutative, but that at the second-order level already commutators occur, which cause noncommutative behavior.

5.11 Transversality

Let $m \geq n$ and let $f \in C^k(U, \mathbf{R}^n)$ with $k \in \mathbf{N}$ and $U \subset \mathbf{R}^m$ open. By the Submersion Theorem 4.5.2 the solutions $x \in U$ of the equation $f(x) = y$, for $y \in \mathbf{R}^n$ fixed, form a manifold in U, provided that f is regular at x. Now let $Y \subset \mathbf{R}^n$ be a C^k submanifold of codimension $n - d$, and assume

$$X = f^{-1}(Y) = \{ x \in U \mid f(x) \in Y \} \neq \emptyset.$$

We investigate under what conditions X is a manifold in $\mathbf{R}^m$. Because this is a local problem, we start with $x \in X$ fixed; let $y = f(x)$. By Theorem 4.7.1 there exist functions $g_1, \dots, g_{n-d}$, defined on a neighborhood of y in $\mathbf{R}^n$, such that

(i) grad $g_1(y), \dots,$ grad $g_{n-d}(y)$ are linearly independent vectors in $\mathbf{R}^n$,

(ii) near y, the submanifold Y is the zero-set of $g_1, \dots, g_{n-d}$.

Near x, therefore, X is the zero-set of

$$g \circ f := (g_1 \circ f, \dots, g_{n-d} \circ f) : U \to \mathbf{R}^{n-d}.$$

According to the Submersion Theorem 4.5.2, we have that X is a manifold near x, if

$$D(g \circ f)(x) = Dg(y) \circ Df(x) \in \mathrm{Lin}(\mathbf{R}^m, \mathbf{R}^{n-d})$$

is surjective. From Theorem 5.1.2 we get that $Dg(y) \in \mathrm{Lin}(\mathbf{R}^n, \mathbf{R}^{n-d})$ is a surjection with kernel exactly equal to $T_y Y$. Hence it follows that $D(g \circ f)(x)$ is surjective if

$$\mathrm{im}\, Df(x) + T_y Y = \mathbf{R}^n. \tag{5.40}$$

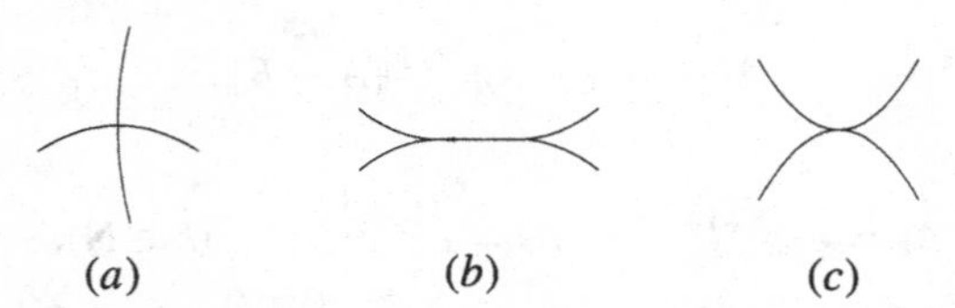

Illustration for Section 5.11: Curves in $\mathbf{R}^2$
(a) transversal, (b) and (c) nontransversal

We say that f is *transversal* to Y at y if condition (5.40) is met for every $x \in f^{-1}(\{y\})$. If this is the case, it follows that X is a manifold in $\mathbf{R}^m$ near every $x \in f^{-1}(\{y\})$, while

$$\mathrm{codim}_{\mathbf{R}^m} X = n - d = \mathrm{codim}_{\mathbf{R}^n} Y.$$

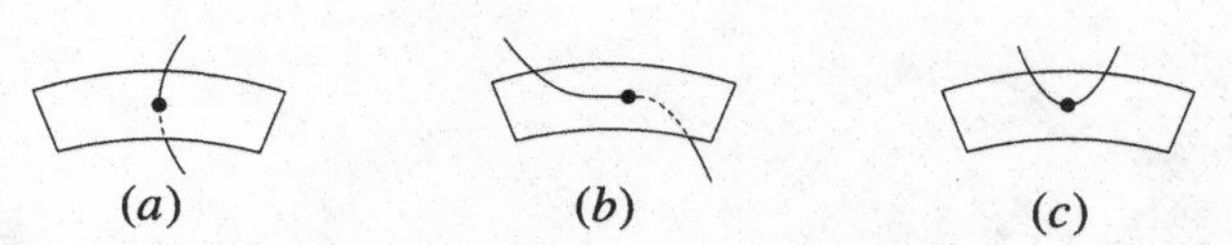

Illustration for Section 5.11: Curves and surfaces in $\mathbf{R}^3$
(a) transversal, (b) and (c) nontransversal

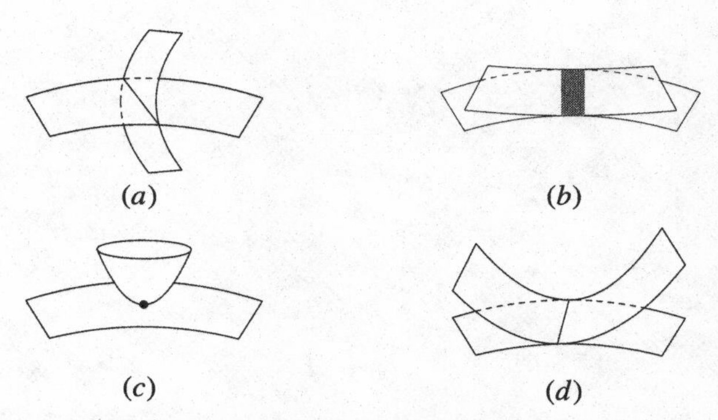

Illustration for Section 5.11: Surfaces in $\mathbf{R}^3$
(a) transversal, (b), (c) and (d) nontransversal

A variant of the preceding argumentation may be applied to the embedding mapping $i : Z \to \mathbf{R}^n$ of a second submanifold in $\mathbf{R}^n$. Then

$$i^{-1}(Y) = Y \cap Z,$$

while $Di(x) \in \mathrm{Lin}(T_x Z, \mathbf{R}^n)$ is the embedding mapping. Condition (5.40) now means

$$T_x Y + T_x Z = \mathbf{R}^n,$$

and here again the conclusion is: $Y \cap Z$ is a manifold in Z, and therefore in $\mathbf{R}^n$; in addition one has $\mathrm{codim}_Z(Y \cap Z) = \mathrm{codim}_{\mathbf{R}^n} Y$. That is

$$\dim Z - \dim(Y \cap Z) = n - \dim Y.$$

Hence, for the codimension in $\mathbf{R}^n$ one has

$$\mathrm{codim}(Y \cap Z) = n - \dim(Y \cap Z) = (n - \dim Y) + (n - \dim Z) = \mathrm{codim}\, Y + \mathrm{codim}\, Z.$$

Exercises

Review Exercises

Exercise 0.1 (Rational parametrization of circle – needed for Exercises 2.71, 2.72, 2.79 and 2.80). For purposes of integration it is useful to parametrize points of the circle $\{ (\cos\alpha, \sin\alpha) \mid \alpha \in \mathbf{R} \}$ by means of rational functions of a variable in $\mathbf{R}$.

(i) Verify for all $\alpha \in \,]-\pi, \pi\,[\setminus \{0\}$ we have $\frac{\sin\alpha}{1+\cos\alpha} = \frac{1-\cos\alpha}{\sin\alpha}$. Deduce that both quotients equal a number $t \in \mathbf{R}$. Next prove

$$\cos\alpha = \frac{1-t^2}{1+t^2}, \qquad \sin\alpha = \frac{2t}{1+t^2}.$$

Now use the identity $\tan\alpha = \frac{2\tan\alpha/2}{1-\tan^2\alpha/2}$ to conclude $t = \tan\frac{\alpha}{2}$, that is $\alpha = 2\arctan t$.
Background. Note that $\frac{\sin\alpha}{1+\cos\alpha} = t = \tan\frac{\alpha}{2}$ is the slope of the straight line in $\mathbf{R}^2$ through the points $(-1, 0)$ and $(\cos\alpha, \sin\alpha)$. For another derivation of this parametrization see Exercise 5.36.

(ii) Show for $0 \le \alpha < \frac{\pi}{2}$ that $t = \tan\alpha$, thus $\alpha = \arctan t$, implies

$$\cos^2\alpha = \frac{1}{1+t^2}, \qquad \sin^2\alpha = \frac{t^2}{1+t^2}.$$

Exercise 0.2 (Characterization of function by functional equation).

(i) Suppose $f : \mathbf{R} \to \mathbf{R}$ is differentiable and satisfies the *functional equation* $f(x+y) = f(x) + f(y)$ for all x and $y \in \mathbf{R}$. Prove that $f(x) = f(1)\,x$ for all $x \in \mathbf{R}$.
Hint: Differentiate with respect to y and set $y = 0$.

(ii) Suppose $g : \mathbf{R} \to \mathbf{R}_+$ is differentiable and satisfies $g(\frac{x+y}{2}) = \frac{1}{2}(g(x)+g(y))$ for all x and $y \in \mathbf{R}$. Show that $g(x) = g(1)x + g(0)(1-x)$ for all $x \in \mathbf{R}$.
Hint: Replace x by $x + y$ and y by 0 in the functional equation, and reduce to part (i).

(iii) Suppose $g : \mathbf{R} \to \mathbf{R}_+$ is differentiable and satisfies $g(x+y) = g(x)g(y)$ for all x and $y \in \mathbf{R}$. Show that $g(x) = g(1)^x$ for all $x \in \mathbf{R}$ (see Exercise 6.81 for a related result).
Hint: Consider $f = \log_{g(1)} \circ g$, where $\log_{g(1)}$ is the base $g(1)$ logarithmic function.

(iv) Suppose $h : \mathbf{R}_+ \to \mathbf{R}_+$ is differentiable and satisfies $h(xy) = h(x)h(y)$ for all x and $y \in \mathbf{R}_+$. Verify that $h(x) = x^{\log h(e)} = x^{h'(1)}$ for all $x \in \mathbf{R}_+$.
Hint: Consider $g = h \circ \exp$, that is, $g(y) = h(e^y)$.

(v) Suppose $k : \mathbf{R}_+ \to \mathbf{R}$ is differentiable and satisfies $k(xy) = k(x) + k(y)$ for all x and $y \in \mathbf{R}_+$. Show that $k(x) = k(e) \log x = \log_b x$ for all $x \in \mathbf{R}_+$, where $b = e^{k(e)^{-1}}$.
Hint: Consider $f = k \circ \exp$ or $h = \exp \circ k$.

(vi) Let $f : \mathbf{R} \to \mathbf{R}$ be Riemann integrable over every closed interval in $\mathbf{R}$ and suppose $f(x+y) = f(x) + f(y)$ for all x and $y \in \mathbf{R}$. Verify that the conclusion of (i) still holds.
Hint: Integration of $f(x+t) = f(x) + f(t)$ with respect to t over $[0, y]$ gives

$$\int_0^y f(x+t)\,dt = yf(x) + \int_0^y f(t)\,dt.$$

Substitution of variables now implies

$$\int_0^{x+y} f(t)\,dt = \int_0^x f(t)\,dt + \int_x^{x+y} f(t)\,dt = \int_0^x f(t)\,dt + \int_0^y f(x+t)\,dt.$$

Hence

$$yf(x) = \int_0^{x+y} f(t)\,dt - \int_0^x f(t)\,dt - \int_0^y f(t)\,dt.$$

The technique from part (i) allows us to handle more complicated functional equations too.

(vii) Suppose $f : \mathbf{R} \to \mathbf{R} \cup \{\pm\infty\}$ is differentiable where real-valued and satisfies

$$f(x+y) = \frac{f(x) + f(y)}{1 - f(x)f(y)} \qquad (x, y \in \mathbf{R})$$

where well-defined. Check that $f(x) = \tan(f'(0)\,x)$ for $x \in \mathbf{R}$ with $f'(0)x \notin \frac{\pi}{2} + \pi\mathbf{Z}$. (Note that the constant function $f(x) = i$ satisfies the equation if we allow complex-valued solutions.)

(viii) Suppose $f : \mathbf{R} \to \mathbf{R}$ is differentiable and satisfies $|f(x)| \le 1$ for all $x \in \mathbf{R}$ and

$$f(x+y) = f(x)\sqrt{1 - f(y)^2} + f(y)\sqrt{1 - f(x)^2} \qquad (x, y \in \mathbf{R}).$$

Prove that $f(x) = \sin(f'(0)x)$ for $x \in \mathbf{R}$.

Exercise 0.3 (Needed for Exercise 3.20). We have

$$\arctan x + \arctan \frac{1}{x} = \begin{cases} \dfrac{\pi}{2}, & x > 0; \\ -\dfrac{\pi}{2}, & x < 0. \end{cases}$$

Prove this by means of the following four methods.

(i) Set $\arctan x = \alpha$ and express $\frac{1}{x}$ in terms of α.

(ii) Use differentiation.

(iii) Recall that $\arctan x = \int_0^x \frac{1}{1+t^2}\, dt$, and make a substitution of variables.

(iv) Use the formula $\arctan x + \arctan y = \arctan\left(\frac{x+y}{1-xy}\right)$, for x and $y \in \mathbf{R}$ with $xy \neq 1$, which can be deduced from Exercise 0.2.(vii).

Deduce $\lim_{x\to\infty} x\left(\frac{\pi}{2} - \arctan x\right) = 1$.

Exercise 0.4 (Legendre polynomials and associated Legendre functions – needed for Exercises 3.17 and 6.63). Let $l \in \mathbf{N}_0$. The polynomial $f(x) = (x^2 - 1)^l$ satisfies the following *ordinary differential equation*, which gives a relation among various derivatives of f:

$$(\star) \qquad (x^2 - 1)\, f'(x) - 2lx\, f(x) = 0.$$

Define the *Legendre polynomial* P_l on $\mathbf{R}$ by *Rodrigues' formula*

$$P_l(x) = \frac{1}{2^l\, l!}\left(\frac{d}{dx}\right)^l (x^2 - 1)^l.$$

(i) Prove by $(l+1)$-fold differentiation of $(\star)$ that P_l satisfies the following, known as *Legendre's equation* for a twice differentiable function $u : \mathbf{R} \to \mathbf{R}$:

$$(1 - x^2)\, u''(x) - 2x\, u'(x) + l(l+1)\, u(x) = 0.$$

Let $m \in \mathbf{N}_0$ with $m \leq l$.

(ii) Prove by m-fold differentiation of Legendre's equation that $\frac{d^m P_l}{dx^m}$ satisfies the differential equation

$$(1 - x^2)\, v''(x) - 2(m+1)x\, v'(x) + (l(l+1) - m(m+1))\, v(x) = 0.$$

Now define the *associated Legendre function* $P_l^m : [-1, 1] \to \mathbf{R}$ by

$$P_l^m(x) = (1 - x^2)^{\frac{m}{2}} \frac{d^m P_l}{dx^m}(x).$$

(iii) Verify that P_l^m satisfies

$$\frac{d}{dx}\Big((1-x^2)\frac{dw}{dx}(x)\Big)+\Big(l(l+1)-\frac{m^2}{1-x^2}\Big)\,w(x)=0.$$

Note that in the case $m=0$ this reduces to Legendre's equation.

(iv) Demonstrate that

$$\frac{dP_l^m}{dx}(x)=\frac{1}{\sqrt{1-x^2}}P_l^{m+1}(x)-\frac{mx}{1-x^2}P_l^m(x).$$

Verify

$$P_l^m(x)=\frac{(-1)^m}{2^l\,l!}\frac{(l+m)!}{(l-m)!}(1-x^2)^{-\frac{m}{2}}\Big(\frac{d}{dx}\Big)^{l-m}(x^2-1)^l,$$

and use this to derive

$$\frac{dP_l^m}{dx}(x)=-\frac{(l+m)(l-m+1)}{\sqrt{1-x^2}}P_l^{m-1}(x)+\frac{mx}{1-x^2}P_l^m(x).$$

Exercise 0.5 (Parametrization of circle and hyperbola by area). Let $e_1=(1,0)\in\mathbf{R}^2$.

(i) Prove that $\frac{1}{2}t$ is the area of the bounded part of $\mathbf{R}^2$ bounded by the half-line R_+e_1, the unit circle $\{x\in\mathbf{R}^2\mid x_1^2+x_2^2=1\}$, and the half-line $\mathbf{R}_+(\cos t,\sin t)$, if $0\le t\le 2\pi$.

(ii) Prove that $\frac{1}{2}t$ is the area of the bounded part of $\mathbf{R}^2$ bounded by R_+e_1, the branch $\{x\in\mathbf{R}^2\mid x_1^2-x_2^2=1,\ x_1>0\}$ of the unit hyperbola, and $\mathbf{R}_+(\cosh t,\sinh t)$, if $t\in\mathbf{R}$.

Exercise 0.6 (Needed for Exercise 6.108). Finding antiderivatives for a function by means of different methods may lead to seemingly distinct expressions for these antiderivatives. For a striking example of this phenomenon, consider

$$I=\int\frac{dx}{\sqrt{(x-a)(b-x)}}\qquad(a<x<b).$$

(i) Compute I by completing the square in the function $x\mapsto(x-a)(b-x)$.

(ii) Compute I by means of the substitution $x=a\cos^2\theta+b\sin^2\theta$, for $0<\theta<\frac{\pi}{2}$.

(iii) Compute I by means of the substitution $\frac{b-x}{x-a}=t^2$, for $0<t<\infty$.

(iv) Show that for a suitable choice of the constants c_1, c_2 and $c_3 \in \mathbf{R}$ we have, for all x with $a < x < b$,

$$\arcsin\left(\frac{2x-b-a}{b-a}\right)+c_1 = \arccos\left(\frac{b+a-2x}{b-a}\right)+c_2$$

$$= -2\arctan\sqrt{\frac{b-x}{x-a}}+c_3.$$

Express two of the constants in the third, for instance by substituting $x = b$.

(v) Show that the following improper integral is convergent, and prove

$$\int_a^b \frac{dx}{\sqrt{(x-a)(b-x)}} = \pi.$$

Exercise 0.7 (Needed for Exercises 3.10, 5.30 and 5.31). Show

$$\int \frac{1}{\cos\alpha}\,d\alpha = \int \frac{\cos\alpha}{1-\sin^2\alpha}\,d\alpha = \frac{1}{2}\log\left|\frac{1+\sin\alpha}{1-\sin\alpha}\right| + c.$$

Use $\cos 2\alpha = 2\cos^2\alpha - 1 = 1 - 2\sin^2\alpha$ to prove

$$\int \frac{1}{\cos\alpha}\,d\alpha = \log\left|\tan\left(\frac{\alpha}{2}+\frac{\pi}{4}\right)\right| + c.$$

The number $\tan(\frac{\alpha}{2}+\frac{\pi}{4})$ arises in trigonometry also as follows. The triangle in $\mathbf{R}^2$ with vertices $(0,0)$, $(0,1)$ and $(\cos\alpha, \sin\alpha)$ is isosceles. This implies that the line connecting $(0,1)$ and $(\cos\alpha,\sin\alpha)$ has x_1-intercept equal to $\tan(\frac{\alpha}{2}+\frac{\pi}{4})$.

Exercise 0.8 (Needed for Exercises 6.60 and 7.30). For $p \in \mathbf{R}_+$ and $q \in \mathbf{R}$, deduce from

$$\int_{\mathbf{R}_+} e^{(-p+iq)x}\,dx = \frac{p+iq}{p^2+q^2}$$

that

$$\int_{\mathbf{R}_+} e^{-px}\cos qx\,dx = \frac{p}{p^2+q^2}, \qquad \int_{\mathbf{R}_+} e^{-px}\sin qx\,dx = \frac{q}{p^2+q^2}.$$

Exercise 0.9 (Needed for Exercise 8.21). Let a and $b \in \mathbf{R}$ with $a > |b|$ and $n \in \mathbf{N}$, and prove

$$\int_0^\pi (a+b\cos x)^{-n}\,dx = (a^2-b^2)^{\frac{1}{2}-n}\int_0^\pi (a-b\cos y)^{n-1}\,dy.$$

Hint: Introduce the new variable y by means of

$$(a+b\cos x)(a-b\cos y) = a^2-b^2; \qquad \text{use} \qquad \sin x = (a^2-b^2)^{\frac{1}{2}}\frac{\sin y}{a-b\cos y}.$$

Exercise 0.10 (Duplication formula for (lemniscatic) sine). Set $I = [\,0, 1\,]$.

(i) Let $\alpha \in [\,0, \frac{\pi}{4}\,]$, write $x = \sin\alpha \in [\,0, \frac{1}{2}\sqrt{2}\,]$ and $f(x) = \sin 2\alpha$. Deduce for $x \in [\,0, \frac{1}{2}\sqrt{2}\,]$ that $f(x) = 2x\sqrt{1-x^2} \in I$ and

$$\int_0^{f(x)} \frac{1}{\sqrt{1-t^2}}\,dt = 2\int_0^x \frac{1}{\sqrt{1-t^2}}\,dt.$$

(ii) Prove that the mapping $\psi_1 : I \to I$ is an order-preserving bijection if $\psi_1(u) = \frac{2u^2}{1+u^4}$. Verify that the substitution of variables $t^2 = \psi_1(u)$ yields

$$\int_0^z \frac{1}{\sqrt{1-t^4}}\,dt = \sqrt{2}\int_0^y \frac{1}{\sqrt{1+u^4}}\,du,$$

where for all $y \in I$ we define $z \in I$ by $z^2 = \psi_1(y)$.

(iii) Prove that the mapping $\psi_2 : [\,0, \sqrt{\sqrt{2}-1}\,] \to I$ is an order-preserving bijection if $\psi_2(t) = \frac{2t^2}{1-t^4}$. Verify that the substitution of variables $u^2 = \psi_2(t)$ yields

$$\int_0^y \frac{1}{\sqrt{1+u^4}}\,du = \sqrt{2}\int_0^x \frac{1}{\sqrt{1-t^4}}\,dt,$$

where for all $x \in [\,0, \sqrt{\sqrt{2}-1}\,]$ we define $y \in I$ by $y^2 = \psi_2(x)$.

(iv) Combine (ii) and (iii) to obtain for all $x \in [\,0, \sqrt{\sqrt{2}-1}\,]$ (compare with part (i))

$$\int_0^{g(x)} \frac{1}{\sqrt{1-t^4}}\,dt = 2\int_0^x \frac{1}{\sqrt{1-t^4}}\,dt, \qquad g(x) = \frac{2x\sqrt{1-x^4}}{1+x^4} \in I.$$

Background. The mapping $x \mapsto \int_0^x \frac{1}{\sqrt{1-t^4}}\,dt$ arises in the computation of the length of the lemniscate (see Exercise 7.5.(i)); its inverse function, the *lemniscatic sine*, plays a role similar to that of the sine function. The *duplication formula* in part (iv) is a special case of the addition formula from Exercises 3.44 and 4.34.(iii).

Exercise 0.11 (Binomial series – needed for Exercise 6.69). Let $\alpha \in \mathbf{R}$ be fixed and define $f : \,]-\infty, 1\,[$ by $f(x) = (1-x)^{-\alpha}$.

(i) For $k \in \mathbf{N}_0$ show $f^{(k)}(x) = (\alpha)_k(1-x)^{-\alpha-k}$ with

$$(\alpha)_0 = 1; \qquad (\alpha)_k = \alpha(\alpha+1)\cdots(\alpha+k-1) \qquad (k \in \mathbf{N}).$$

The numbers $(\alpha)_k$ are called *shifted factorials* or *Pochhammer symbols*. Next introduce the MacLaurin series $F(x)$ of f by

$$F(x) = \sum_{k \in \mathbf{N}_0} \frac{(\alpha)_k}{k!}x^k.$$

In the following two parts we shall prove that $F(x) = f(x)$ if $|x| < 1$.

(ii) Using the ratio test show that the series $F(x)$ has radius of convergence equal to 1.

(iii) For $|x| < 1$, prove by termwise differentiation that $(1-x)F'(x) = \alpha F(x)$ and deduce that $f(x) = F(x)$ from

$$\frac{d}{dx}((1-x)^{\alpha} F(x)) = 0.$$

(iv) Conclude from (iii) that

$$(1-x)^{-(n+1)} = \sum_{k \in \mathbf{N}_0} \binom{n+k}{n} x^k \qquad (|x| < 1,\ n \in \mathbf{N}_0).$$

Show that this identity also follows by n-fold differentiation of the geometric series $(1-x)^{-1} = \sum_{k \in \mathbf{N}_0} x^k$.

(v) For $|x| < 1$, prove

$$(1+x)^{\alpha} = \sum_{k \in \mathbf{N}_0} \binom{\alpha}{k} x^k, \qquad \text{where} \qquad \binom{\alpha}{k} = \frac{\alpha(\alpha-1)\cdots(\alpha-k+1)}{k!}.$$

In particular, show for $|x| < 1$

$$(1-4x)^{-\frac{1}{2}} = \sum_{k \in \mathbf{N}_0} \binom{2k}{k} x^k.$$

For $|x| < |y|$ deduce the following identity, which generalizes Newton's Binomial Theorem:

$$(x+y)^{\alpha} = \sum_{k \in \mathbf{N}_0} \binom{\alpha}{k} x^k\, y^{\alpha-k}.$$

The power series for $(1+x)^{\alpha}$ is called a *binomial series*, and its coefficients *generalized binomial coefficients*.

Exercise 0.12 (Power series expansion of tangent and $\zeta(2n)$ – needed for Exercises 0.13, 0.14 and 0.20). Define $f : \mathbf{R} \setminus \mathbf{Z} \to \mathbf{R}$ by

$$f(x) = \frac{\pi^2}{\sin^2(\pi x)} - \sum_{k \in \mathbf{Z}} \frac{1}{(x-k)^2}.$$

(i) Check that f is well-defined. Verify that the series converges uniformly on bounded and closed subsets of $\mathbf{R}\setminus\mathbf{Z}$, and conclude that $f:\mathbf{R}\setminus\mathbf{Z}\to\mathbf{R}$ is continuous. Prove, by Taylor expansion of the function sin, that f can be continued to a function, also denoted by f, that is continuous at 0. Conclude that $f:\mathbf{R}\to\mathbf{R}$ thus defined is a continuous periodic function, and that consequently f is bounded on $\mathbf{R}$.

(ii) Show that

$$f\Big(\frac{x}{2}\Big)+f\Big(\frac{x+1}{2}\Big)=4f(x)\qquad (x\in\mathbf{R}).$$

Use this, and the boundedness of f, to prove that $f=0$ on $\mathbf{R}$, that is, for $x\in\mathbf{R}\setminus\mathbf{Z}$, and $x-\frac{1}{2}\in\mathbf{R}\setminus\mathbf{Z}$, respectively,

$$\begin{aligned}\frac{\pi^2}{\sin^2(\pi x)} &= \sum_{k\in\mathbf{Z}}\frac{1}{(x-k)^2},\\ \pi^2\tan^{(1)}(\pi x) &= \frac{\pi^2}{\cos^2(\pi x)}=2^2\sum_{k\in\mathbf{Z}}\frac{1}{(2x-2k-1)^2}.\end{aligned}$$

(iii) Prove $\sum_{k\in\mathbf{N}}\frac{1}{k^2}=\frac{\pi^2}{6}$ by setting $x=0$ in the equality in (ii) for $\pi^2\tan^{(1)}(\pi x)$ and using

$$\sum_{k\in\mathbf{N}}\frac{1}{(2k-1)^2}=\sum_{k\in\mathbf{N}}\frac{1}{k^2}-\sum_{k\in\mathbf{N}}\frac{1}{(2k)^2}=\frac{3}{4}\sum_{k\in\mathbf{N}}\frac{1}{k^2}.$$

(iv) Prove by $(2n-2)$-fold differentiation

$$\pi^{2n}\frac{\tan^{(2n-1)}(\pi x)}{(2n-1)!}=2^{2n}\sum_{k\in\mathbf{Z}}\frac{1}{(2x-2k-1)^{2n}}\qquad \Big(n\in\mathbf{N},\ x-\frac{1}{2}\in\mathbf{R}\setminus\mathbf{Z}\Big).$$

In particular, for $n\in\mathbf{N}$,

$$\begin{aligned}\pi^{2n}\frac{\tan^{(2n-1)}(0)}{(2n-1)!} &= 2^{2n}\sum_{k\in\mathbf{Z}}\frac{1}{(2k-1)^{2n}}=2^{2n+1}\sum_{k\in\mathbf{N}}\frac{1}{(2k-1)^{2n}}\\ &= 2^{2n+1}(1-2^{-2n})\sum_{k\in\mathbf{N}}\frac{1}{k^{2n}}=2(2^{2n}-1)\,\zeta(2n).\end{aligned}$$

Here we have defined $\zeta(2n)=\sum_{k\in\mathbf{N}}\frac{1}{k^{2n}}$. Conclude that $\zeta(2)=\frac{\pi^2}{6}$ and $\zeta(4)=\frac{\pi^4}{90}$.

(v) Now deduce that

$$\tan x=\sum_{n\in\mathbf{N}}\frac{2(2^{2n}-1)\,\zeta(2n)}{\pi^{2n}}\,x^{2n-1}\qquad \Big(|x|<\frac{\pi}{2}\Big).$$

The values of the $\zeta(2n) \in \mathbf{R}$, for $n > 1$, can be obtained from that of $\zeta(2)$ as follows.

(vi) Define $g : \mathbf{N} \times \mathbf{N} \to \mathbf{R}$ by

$$g(k, l) = \frac{1}{kl^3} + \frac{1}{2k^2l^2} + \frac{1}{k^3l}.$$

Verify that

$$(\star) \qquad g(k, l) - g(k + l, l) - g(k, k + l) = \frac{1}{k^2l^2},$$

and that summation over all $k, l \in \mathbf{N}$ gives

$$\zeta(2)^2 = \Big(\sum_{k,l\in\mathbf{N}} - \sum_{k,l\in\mathbf{N},\, k>l} - \sum_{k,l\in\mathbf{N},\, l>k}\Big) g(k, l) = \sum_{k\in\mathbf{N}} g(k, k) = \frac{5}{2}\zeta(4).$$

Conclude that $\zeta(4) = \frac{\pi^4}{90}$. Similarly introduce, for $n \in \mathbf{N} \setminus \{1\}$,

$$g(k, l) = \frac{1}{kl^{2n-1}} + \frac{1}{2} \sum_{2\le i\le 2n-2} \frac{1}{k^i l^{2n-i}} + \frac{1}{k^{2n-1}l} \qquad (k,\ l \in \mathbf{N}).$$

In this case the left–hand side of $(\star)$ takes the form $\sum_{2\le i\le 2n-2,\ i \text{ even}} \frac{1}{k^i l^{2n-i}}$, hence

$$\zeta(2n) = \frac{2}{2n+1} \sum_{1\le i\le n-1} \zeta(2i)\,\zeta(2n - 2i) \qquad (n \in \mathbf{N} \setminus \{1\}).$$

See Exercise 0.21.(iv) for a different proof.

Background. More methods for the computation of the $\zeta(2n) \in \mathbf{R}$, for $n \in \mathbf{N}$, can be found in Exercises 0.20 (two different ones), 0.21, 6.40 and 6.89.

Exercise 0.13 (Partial-fraction decomposition of trigonometric functions, and Wallis' product – sequel to Exercise 0.12 – needed for Exercises 0.15 and 0.21). We write $\sum'_{k\in\mathbf{Z}}$ for $\lim_{n\to\infty} \sum_{k\in\mathbf{Z},\, -n\le k\le n}$.

(i) Verify that antidifferentiation of the identity from Exercise 0.12.(ii) leads to

$$\pi \tan(\pi x) = -\sum_{k\in\mathbf{Z}}{}' \frac{1}{x - k - \frac{1}{2}} = 8x \sum_{k\in\mathbf{N}} \frac{1}{(2k-1)^2 - 4x^2},$$

for $x - \frac{1}{2} \in \mathbf{R} \setminus \mathbf{Z}$. Using $\cot(\pi x) = -\tan \pi(x + \frac{1}{2})$, conclude that one has the following *partial-fraction decomposition* of the cotangent (see Exercises 0.21.(ii) and 6.96.(vi) for other proofs):

$$\pi \cot(\pi x) = \sum_{k\in\mathbf{Z}}{}' \frac{1}{x - k} = \frac{1}{x} + 2x \sum_{k\in\mathbf{N}} \frac{1}{x^2 - k^2} \qquad (x \in \mathbf{R} \setminus \mathbf{Z}).$$

Use $2\frac{\pi}{\sin(\pi x)} = \pi \cot(\pi \frac{x}{2}) - \pi \cot(\pi \frac{x+1}{2})$ to verify that

$$\frac{\pi}{\sin(\pi x)} = {\sum_{k \in \mathbf{Z}}}' \frac{(-1)^k}{x-k} = \frac{1}{x} + 2x \sum_{k \in \mathbf{N}} \frac{(-1)^k}{x^2 - k^2} \qquad (x \in \mathbf{R} \setminus \mathbf{Z}).$$

Replace x by $\frac{1}{2} - x$, then factor each term, and show

$$\frac{\pi}{2\cos(\frac{\pi}{2}x)} = 2 \sum_{k \in \mathbf{N}} (-1)^k \frac{2k-1}{x^2 - (2k-1)^2} \qquad (\frac{x-1}{2} \in \mathbf{R} \setminus \mathbf{Z}).$$

(ii) Demonstrate that $\log(\frac{\sin(\pi x)}{\pi x})$ is the antiderivative of $\pi \cot(\pi x) - \frac{1}{x}$ which has limit 0 at 0. Now, using part (i), prove the following:

$$\sin(\pi x) = \pi x \prod_{k \in \mathbf{N}} \Big(1 - \frac{x^2}{k^2}\Big).$$

At the outset, this result is obtained for $0 < x < 1$, but on account of the antisymmetry and the periodicity of the expressions on the left and on the right, the identity now holds for all $x \in \mathbf{R}$.

(iii) Prove *Wallis' product* (see also Exercise 6.56.(iv))

$$\frac{2}{\pi} = \prod_{n \in \mathbf{N}} \Big(1 - \frac{1}{4n^2}\Big), \qquad \text{equivalently} \qquad \lim_{n \to \infty} \frac{(2^n n!)^2}{(2n)!\,\sqrt{2n+1}} = \sqrt{\frac{\pi}{2}}.$$

Background. Obviously, the cotangent is essentially determined by the prescription that it is the function on $\mathbf{R}$ which has a singularity of the form $x \mapsto \frac{1}{x - k\pi}$ at all points $k\pi$, for $k \in \mathbf{Z}$, and similar results apply to the tangent, the sine and the cosine. Furthermore, the sine is the bounded function on $\mathbf{R}$ which has a simple zero at all points $k\pi$, for $k \in \mathbf{Z}$.

Let $0 \neq \omega_0 \in \mathbf{R}$, and let $\Omega = \mathbf{Z} \cdot \omega_0 \subset \mathbf{R}$ be the *period lattice* determined by ω_0. Define $f : \mathbf{R} \setminus \Omega \to \mathbf{R}$ by

$$\begin{aligned} f(x) &= \frac{\pi}{\omega_0} \cot\Big(\frac{\pi}{\omega_0} x\Big) = {\sum_{k \in \mathbf{Z}}}' \frac{1}{x - k\omega_0} \\ &= {\sum_{\omega \in \Omega}}' \frac{1}{x - \omega} = \frac{1}{x} + \sum_{\omega \in \Omega,\, \omega \neq 0} \Big(\frac{1}{x - \omega} + \frac{1}{\omega}\Big). \end{aligned}$$

(iv) Use Exercise 0.12.(iv) or 0.20 to verify that $]\,0, \omega_0\,[\, \ni x \mapsto (f(x),\ f'(x)) \in \mathbf{R}^2$ is a parametrization of the parabola

$$\{\,(x, y) \in \mathbf{R}^2 \mid y = -x^2 - g\,\} \qquad \text{with} \qquad g = 3 \sum_{\omega \in \Omega,\, \omega \neq 0} \frac{1}{\omega^2}.$$

Background. For the generalization to $\mathbf{C}$ of the technique described above one introduces *periods* ω_1 and $\omega_2 \in \mathbf{C}$ that are linearly independent over $\mathbf{R}$, and in addition the *period lattice* $\Omega = \mathbf{Z}\cdot\omega_1 + \mathbf{Z}\cdot\omega_2 \subset \mathbf{C}$. The *Weierstrass* $\wp$ *function*: $\mathbf{C}\setminus\Omega \to \mathbf{C}$ associated with Ω, which is a *doubly-periodic function* on $\mathbf{C}\setminus\Omega$, is defined by

$$\wp(z) = \frac{1}{z^2} + \sum_{\omega\in\Omega,\,\omega\neq 0}\Big(\frac{1}{(z-\omega)^2} - \frac{1}{\omega^2}\Big).$$

It gives a parametrization $\{\, t_1\omega_1 + t_2\omega_2 \in \mathbf{C} \mid 0 < t_i < 1,\ i = 1, 2\,\} \ni z \mapsto (\wp(z), \wp'(z)) \in \mathbf{C}^2$ of the *elliptic curve*

$$\{\,(x, y) \in \mathbf{C}^2 \mid y^2 = 4x^3 - g_2x - g_3\,\},$$

$$g_2 = 60\sum_{\omega\in\Omega,\,\omega\neq 0}\frac{1}{\omega^4}, \qquad g_3 = 140\sum_{\omega\in\Omega,\,\omega\neq 0}\frac{1}{\omega^6}.$$

Exercise 0.14 ($\int_{\mathbf{R}_+}\frac{\sin x}{x}\,dx = \frac{\pi}{2}$ – sequel to Exercise 0.12). Deduce from Exercise 0.12.(ii)

$$1 = \sum_{k\in\mathbf{Z}}\frac{\sin^2(x+k\pi)}{(x+k\pi)^2} \qquad (x\in\mathbf{R}).$$

Integrate termwise over $[\,0, \pi\,]$ to obtain

$$\pi = \sum_{k\in\mathbf{Z}}\int_{k\pi}^{(k+1)\pi}\frac{\sin^2 x}{x^2}\,dx = \int_{\mathbf{R}}\frac{\sin^2 x}{x^2}\,dx.$$

Integration by parts yields, for any $a > 0$,

$$\int_0^a\frac{\sin^2 x}{x^2}\,dx = -\frac{\sin^2 a}{a} + \int_0^{2a}\frac{\sin x}{x}\,dx.$$

Deduce the convergence of the improper integral $\int_{\mathbf{R}_+}\frac{\sin x}{x}\,dx$ as well as the evaluation $\int_{\mathbf{R}_+}\frac{\sin x}{x}\,dx = \frac{\pi}{2}$ (see Example 2.10.14 and Exercises 6.60 and 8.19 for other proofs).

Exercise 0.15 (Special values of Beta function – sequel to Exercise 0.13 – needed for Exercises 2.83 and 6.59). Let $0 < p < 1$.

(i) Verify that the following series is uniformly convergent for $x \in [\,\epsilon, 1-\epsilon'\,]$, where ϵ and $\epsilon' > 0$ are arbitrary:

$$\frac{x^{p-1}}{1+x} = \sum_{k\in\mathbf{N}_0}(-1)^k x^{p+k-1}.$$

(ii) Deduce

$$\int_0^1 \frac{x^{p-1}}{1+x}\,dx = \sum_{k\in\mathbf{N}_0} \frac{(-1)^k}{p+k}.$$

(iii) Using the substitution $x = \frac{1}{y}$, prove

$$\int_1^\infty \frac{x^{p-1}}{1+x}\,dx = \int_0^1 \frac{y^{(1-p)-1}}{1+y}\,dy = \sum_{k\in\mathbf{N}} \frac{(-1)^k}{p-k}.$$

(iv) Apply Exercise 0.13.(i) to show (see Exercise 6.58.(v) for another proof and for the relation to the Beta function)

$$\int_{\mathbf{R}_+} \frac{x^{p-1}}{x+1}\,dx = \frac{\pi}{\sin(\pi p)} \qquad (0 < p < 1).$$

(v) Imitate the steps (i) – (iv) and use Exercise 0.12.(ii) to show

$$\int_{\mathbf{R}_+} \frac{x^{p-1}\log x}{x-1}\,dx = \sum_{k\in\mathbf{Z}} \frac{1}{(p-k)^2} = \frac{\pi^2}{\sin^2(\pi p)} \qquad (0 < p < 1).$$

Exercise 0.16 (Bernoulli polynomials, Bernoulli numbers and power series expansion of tangent – needed for Exercises 0.18, 0.20, 0.21, 0.22, 0.23, 0.25, 6.29 and 6.62).

(i) Prove the sequence of polynomials $(b_n)_{n\geq 0}$ on $\mathbf{R}$ to be uniquely determined by the following conditions:

$$b_0 = 1; \qquad b_n' = nb_{n-1}; \qquad \int_0^1 b_n(x)\,dx = 0 \qquad (n \in \mathbf{N}).$$

The numbers $B_n = b_n(0) \in \mathbf{R}$, for $n \in \mathbf{N}_0$, are known as the *Bernoulli numbers*. Show that these conditions imply $b_n(0) = b_n(1) = B_n$, for $n \geq 2$, and moreover $b_n(1-x) = (-1)^n b_n(x)$, for $n \in \mathbf{N}_0$ and $x \in \mathbf{R}$. Deduce that $B_{2n+1} = 0$, for $n \in \mathbf{N}$.

(ii) Prove that the polynomial b_n is of degree n, for all $n \in \mathbf{N}$.

(iii) Verify

$$b_0(x) = 1, \qquad b_1(x) = x - \frac{1}{2}, \qquad b_2(x) = x^2 - x + \frac{1}{6},$$

$$b_3(x) = x^3 - \frac{3}{2}x^2 + \frac{1}{2}x = x(x-\frac{1}{2})(x-1),$$

$$b_4(x) = x^4 - 2x^3 + x^2 - \frac{1}{30},$$

$$b_5(x) = x^5 - \frac{5}{2}x^4 + \frac{5}{3}x^3 - \frac{1}{6}x = x(x-\frac{1}{2})(x-1)(x^2-x-\frac{1}{3}).$$

The b_n are known as the *Bernoulli polynomials*. We now introduce these polynomials in a different way, thereby illuminating some new aspects; for convenience they are again denoted as b_n from the start. For every $x \in \mathbf{R}$ we define $f_x : \mathbf{R} \to \mathbf{R}$ by

$$f_x(t) = \begin{cases} \dfrac{te^{xt}}{e^t - 1}, & t \in \mathbf{R} \setminus \{0\}; \\ 1, & t = 0. \end{cases}$$

(iv) Prove that there exists a number $\delta > 0$ such that $f_x(t)$, for $t \in \,]-\delta,\, \delta[$, is given by a convergent power series in t.

Now define the functions b_n on $\mathbf{R}$, for $n \in \mathbf{N}_0$, by

$$f_x(t) = \sum_{n \in \mathbf{N}_0} \frac{b_n(x)}{n!}\, t^n \qquad (|t| < \delta).$$

(v) Prove that $b_0(x) = 1$, $b_1(x) = x - \frac{1}{2}$, $B_0 = 1$ and $B_1 = -\frac{1}{2}$, and furthermore

$$1 = \Big(\sum_{n \in \mathbf{N}} \frac{1}{n!} t^{n-1} \Big) \Big(\sum_{n \in \mathbf{N}_0} \frac{B_n}{n!} t^n \Big) \qquad (|t| < \delta).$$

(vi) Use the identity $\sum_{n \in \mathbf{N}_0} b_n(x) \frac{t^n}{n!} = e^{xt} \sum_{n \in \mathbf{N}_0} B_n \frac{t^n}{n!}$, for $|t| < \delta$, to show that

$$b_n(x) = \sum_{0 \le k \le n} \binom{n}{k} B_k\, x^{n-k}.$$

Conclude that every function b_n is a polynomial on $\mathbf{R}$ of degree n.

(vii) Using complex function theory it can be shown that the series for $f_x(t)$ may be differentiated termwise with respect to x. Prove by termwise differentiation, and integration, respectively, that the polynomials b_n just defined coincide with those from part (i).

(viii) Demonstrate that

$$\frac{t}{e^t - 1} + \frac{1}{2} t = 1 + \sum_{n \in \mathbf{N} \setminus \{1\}} \frac{B_n}{n!}\, t^n \qquad (|t| < \delta).$$

Check that the expression on the left is an even function in t, and conclude (compare with part (i)) that $B_{2n+1} = 0$, for $n \in \mathbf{N}$.

(ix) Prove, by means of the identity $\frac{te^{(x+1)t}}{e^t-1} - \frac{te^{xt}}{e^t-1} = te^{xt}$,

$$(\star) \qquad b_n(x+1) - b_n(x) = nx^{n-1} \qquad (n \in \mathbf{N},\ x \in \mathbf{R}).$$

Conclude that one has, for $n \geq 2$,

$$b_n(1) = B_n, \qquad \sum_{0 \leq k < n} \binom{n}{k} B_k = 0.$$

Note that the latter relation allows the Bernoulli numbers to be calculated by recursion. In particular

$$B_2 = \frac{1}{6}, \qquad B_4 = -\frac{1}{30}, \qquad B_6 = \frac{1}{42}, \qquad B_8 = -\frac{1}{30},$$

$$B_{10} = \frac{5}{66}, \qquad B_{12} = -\frac{691}{2730}.$$

(x) Prove by means of part (ix)

$$\frac{te^t}{e^t-1} = 1 + \frac{1}{2}t + \sum_{n \in \mathbf{N}} \frac{B_{2n}}{(2n)!} t^{2n} \qquad (|t| < \delta).$$

(xi) Prove by means of the identity $(\star)$, for $p, k \in \mathbf{N}$,

$$\sum_{1 \leq k < n} k^p = \frac{1}{p+1}(b_{p+1}(n) - b_{p+1}(1)).$$

Use part (vi) to derive the following, known as *Bernoulli's summation formula*:

$$\sum_{1 \leq k \leq n} k^p = \frac{n^{p+1}}{p+1} + \frac{n^p}{2} + \sum_{2 \leq k \leq p} \frac{B_k}{k} \binom{p}{k-1} n^{p+1-k}.$$

Or, equivalently,

$$(p+1) \sum_{1 \leq k \leq n} k^p = n^p + n^{p+1} \sum_{0 \leq k \leq p} \binom{p+1}{k} \frac{B_k}{n^k}.$$

(xii) Verify, for $|t| < \pi$,

$$\frac{t}{2}\frac{e^t-1}{e^t+1} = t\frac{e^t}{e^t+1} - \frac{t}{2} = t\frac{e^{2t}-e^t}{e^{2t}-1} - \frac{t}{2} = \frac{2te^{2t}}{e^{2t}-1} - \frac{te^t}{e^t-1} - \frac{t}{2}.$$

Then use part (x) to derive

$$\frac{t}{2}\frac{e^{t/2}-e^{-t/2}}{e^{t/2}+e^{-t/2}} = \sum_{n \in \mathbf{N}} (2^{2n}-1) \frac{B_{2n}}{(2n)!} t^{2n} \qquad (|t| < \pi).$$

(xiii) Now apply the preceding part to $\tan x = \frac{1}{i}\frac{e^{ix}-e^{-ix}}{e^{ix}+e^{-ix}}$, and confirm that the result is the following power series expansion of tan about 0:

$$\tan x = \sum_{n\in\mathbf{N}}(-1)^{n-1}2^{2n}(2^{2n}-1)\frac{B_{2n}}{(2n)!}\,x^{2n-1} \qquad \Big(|x| < \frac{\pi}{2}\Big).$$

In particular,

$$\tan x = x + \frac{1}{3}x^3 + \frac{2}{15}x^5 + \frac{17}{315}x^7 + \cdots.$$

Exercise 0.17 (Dirichlet's test for uniform convergence – needed for Exercise 0.18). Let $A \subset \mathbf{R}^n$ and let $(f_j)_{j\in\mathbf{N}}$ be a sequence of mappings $A \to \mathbf{R}^p$. Assume that the partial sums are uniformly bounded on A, in the sense that there exists a constant $m > 0$ such that $\left|\sum_{1\le j\le k} f_j(x)\right| \le m$, for every $k \in \mathbf{N}$ and $x \in A$. Further, suppose that $(a_j)_{j\in\mathbf{N}}$ is a sequence in $\mathbf{R}$ which decreases monotonically to 0 as $j \to \infty$. Then there exists $f : A \to \mathbf{R}^p$ such that

$$\sum_{j\in\mathbf{N}} a_j f_j = f \qquad \text{uniformly on } A.$$

As a consequence, the mapping f is continuous if each of the mappings f_j is continuous.

Indeed, write $A_k = \sum_{1\le j\le k} a_j f_j$ and $F_k = \sum_{1\le j\le k} f_j$, for $k \in \mathbf{N}$. Prove, for any $1 \le k \le l$, *Abel's partial summation formula*, which is the analog for series of the formula for integration by parts:

$$A_l - A_k = \sum_{k<j\le l}(a_j - a_{j+1})F_j - a_{k+1}F_k + a_{l+1}F_l.$$

Using $a_j - a_{j+1} \ge 0$ and $a_j \ge 0$, conclude that $|A_l - A_k| \le 2a_{k+1}m$, and deduce that $(A_k)_{k\in\mathbf{N}}$ is a uniform Cauchy sequence.

Exercise 0.18 (Fourier series of Bernoulli functions – sequel to Exercises 0.16 and 0.17 – needed for Exercises 0.19, 0.20, 0.21, 0.23, 6.62 and 6.87). The notation is that of Exercise 0.16.

(i) Antidifferentiation of the geometric series $(1-z)^{-1} = \sum_{k\in\mathbf{N}_0} z^k$ yields the power series $-\log(1-z) = \sum_{k\in\mathbf{N}} \frac{z^k}{k}$ which converges for $z \in \mathbf{C}$, $|z| \le 1$, $z \ne 1$. Substitute $z = e^{i\alpha}$ with $0 < \alpha < 2\pi$, use De Moivre's formula, and decompose into real and imaginary parts; this results in, for $0 < \alpha < 2\pi$,

$$\sum_{k\in\mathbf{N}}\frac{\cos k\alpha}{k} = -\log\Big(2\sin\frac{1}{2}\alpha\Big), \qquad \sum_{k\in\mathbf{N}}\frac{\sin k\alpha}{k} = \frac{1}{2}(\pi - \alpha).$$

Now substitute $\alpha = 2\pi x$, with $0 < x < 1$ and use Exercise 0.16.(iii) to find

$$(\star) \qquad b_1(x - [x]) = -\sum_{k \in \mathbf{Z}\setminus\{0\}} \frac{e^{2k\pi i x}}{2k\pi i} = -\sum_{k \in \mathbf{N}} \frac{\sin 2k\pi x}{k\pi} \qquad (x \in \mathbf{R} \setminus \mathbf{Z}).$$

(ii) Use Dirichlet's test for uniform convergence from Exercise 0.17 to show that the series in $(\star)$ converges uniformly on any closed interval that does not contain $\mathbf{Z}$. Next apply successive antidifferentiation and Exercise 0.16.(i) to verify the following Fourier series for the n-th *Bernoulli function*

$$\frac{b_n(x - [x])}{n!} = -\sum_{k \in \mathbf{Z}\setminus\{0\}} \frac{e^{2k\pi i x}}{(2k\pi i)^n} \qquad (n \in \mathbf{N} \setminus \{1\},\ x \in \mathbf{R}).$$

In particular,

$$\frac{b_{2n}(x - [x])}{(2n)!} = 2(-1)^{n-1} \sum_{k \in \mathbf{N}} \frac{\cos 2k\pi x}{(2k\pi)^{2n}} \qquad (n \in \mathbf{N},\ x \in \mathbf{R}).$$

Prove that the n-th Bernoulli function belongs to $C^{n-2}(\mathbf{R})$, for $n \geq 2$.

(iii) In $(\star)$ in part (i) replace x by $2x$ and divide by 2, then subtract the resulting identity from $(\star)$ in order to obtain

$$\sum_{k \in \mathbf{N}} \frac{\sin(2k-1)\pi x}{2k-1} = \begin{cases} \dfrac{\pi}{4}, & x \in \,]\,0, 1\,[\, + 2\mathbf{Z}; \\ 0, & x \in \mathbf{Z}; \\ -\dfrac{\pi}{4}, & x \in \,]\,-1, 0\,[\, + 2\mathbf{Z}. \end{cases}$$

Take $x = \frac{1}{2}$ and deduce Leibniz' series $\frac{\pi}{4} = \sum_{k \in \mathbf{N}_0} \frac{(-1)^k}{2k+1}$.

Background. We will use the series for b_2 for proving the Fourier inversion theorem, both in the periodic (Exercise 0.19) and the nonperiodic case (Exercise 6.90), as well as Poisson's summation formula (Exercise 6.87) to be valid for suitable classes of functions.

Exercise 0.19 (Fourier series – sequel to Exercises 0.16 and 0.18 – needed for Exercises 6.67 and 6.88). Suppose $f \in C^2(\mathbf{R})$ is *periodic* with period 1, that is $f(x+1) = f(x)$, for all $x \in \mathbf{R}$. Then we have the following *Fourier series* representation for f, valid for $x \in \mathbf{R}$:

$$(\star) \qquad f(x) = \sum_{k \in \mathbf{Z}} \widehat{f}(k) e^{2\pi i k x}, \qquad \text{with} \qquad \widehat{f}(k) = \int_0^1 f(x) e^{-2\pi i k x}\, dx.$$

Here $\widehat{f}(k) \in \mathbf{C}$ is said to be the k-th *Fourier coefficient* of f, for $k \in \mathbf{Z}$.

(i) Use integration by parts to obtain the absolute and uniform convergence of the series on the right–hand side; indeed, verify

$$|\widehat{f}(k)| \leq \frac{1}{4\pi^2 k^2}\int_0^1 |f''(x)|\,dx \qquad (k \in \mathbf{Z}\setminus\{0\}).$$

(ii) First prove the equality $(\star)$ for $x = 0$. In order to do so, deduce from Exercise 0.18.(ii) that

$$\frac{b_2(x-[x])}{2}f''(x) = -\sum_{k\in\mathbf{Z}\setminus\{0\}} \frac{e^{2\pi i k x}}{(2\pi i k)^2} f''(x).$$

Note that this series converges uniformly on the interval $[\,0,1\,]$. Therefore, integrate the series termwise over $[\,0,1\,]$ and apply integration by parts and Exercise 0.16.(i) and (iii) to obtain

$$\int_0^1 b_1(x-[x])\,f'(x)\,dx = -\sum_{k\in\mathbf{Z}\setminus\{0\}}\int_0^1 e_k(x)\,f'(x)\,dx,$$

where $e_k'(x) = e^{2\pi i k x}$. Integrate by parts once more to get

$$f(0) - \int_0^1 f(x)\,dx = \sum_{k\in\mathbf{Z}\setminus\{0\}} \widehat{f}(k).$$

(iii) Finally, in order to find $(\star)$ at an arbitrary point $x^0 \in\ ]\,0,1\,[$, apply the preceding result with f replaced by $x \mapsto f(x+x^0)$, and verify

$$\int_0^1 f(x+x^0)e^{-2\pi i k x}\,dx = e^{2\pi i k x^0}\int_0^1 f(x)e^{-2\pi i k x}\,dx = \widehat{f}(k)e^{2\pi i k x^0}.$$

(iv) Verify that $(\star)$ in Exercise 0.18.(i) is the Fourier series of $x \mapsto b_1(x-[x])$.

Background. In *Fourier analysis* one studies the relation between a function and its Fourier series. The result above says there is equality if the function is of class C^2.

Exercise 0.20 ($\zeta(2n)$ – sequel to Exercises 0.12, 0.16 and 0.18 – needed for Exercises 0.21, 0.24 and 6.96). The notation is that of Exercise 0.16. In parts (i) and (ii) we shall give two different proofs of the formula

$$\zeta(2n) =: \sum_{k\in\mathbf{N}} \frac{1}{k^{2n}} = (-1)^{n-1}\frac{1}{2}(2\pi)^{2n}\frac{B_{2n}}{(2n)!}.$$

(i) Use Exercise 0.12.(iv) and Exercise 0.16.(xiii).

(ii) Substitute $x = 0$ in the formula for $b_{2n}(x - [x])$ in Exercise 0.18.(ii)

(iii) From Exercise 0.18.(ii) deduce $B_{2n+1} = b_{2n+1}(0) = 0$ for $n \in \mathbf{N}$, and

$$|b_{2n}(x - [x])| \leq |B_{2n}| \qquad (n \in \mathbf{N},\ x \in \mathbf{R}).$$

Furthermore, derive the following estimates from the formula for $\zeta(2n)$:

$$2\,\frac{1}{(2\pi)^{2n}} < \frac{|B_{2n}|}{(2n)!} \leq \frac{\pi^2}{3}\,\frac{1}{(2\pi)^{2n}} \qquad (n \in \mathbf{N}).$$

Exercise 0.21 (Partial-fraction decomposition of cotangent and $\zeta(2n)$ – sequel to Exercises 0.16 and 0.18).

(i) Using Exercise 0.16.(v) and (viii) prove, for $x \in \mathbf{R}$ with $|x|$ sufficiently small,

$$\begin{aligned} \pi x \cot(\pi x) &= \pi i x + \frac{2\pi i x}{e^{2\pi i x} - 1} = \pi i x + \sum_{n \in \mathbf{N}_0} \frac{B_n}{n!}(2\pi i x)^n \\ &= \sum_{n \in \mathbf{N}_0} (-1)^n (2\pi)^{2n} \frac{B_{2n}}{(2n)!} x^{2n}. \end{aligned}$$

Note that we have extended the power series to an open neighborhood of 0 in $\mathbf{C}$.

(ii) Show that the partial-fraction decomposition

$$\pi \cot(\pi\xi) = \frac{1}{\xi} + \xi \sum_{n \in \mathbf{Z}\setminus\{0\}} \frac{1}{n(\xi - n)}$$

from Exercise 0.13.(i) can be obtained as follows. Let $\delta > 0$ and note that the series in ($\star$) from Exercise 0.18 converges uniformly on $[\,\delta, 1-\delta\,]$. Therefore, evaluation of

$$\frac{2\pi i}{e^{2\pi i \xi} - 1} \int_\delta^{1-\delta} f(x) e^{2\pi i \xi x}\, dx,$$

where $f(x)$ equals the left and right hand side using termwise integration, respectively, in ($\star$) is admissible. Next use the uniform convergence of the resulting series in δ to take the limit for $\delta \downarrow 0$ again term-by-term.

(iii) By expansion of a geometric series and by interchange of the order of summation find the following series expansion (compare with Exercise 0.13.(i)):

$$\pi \cot(\pi x) = \frac{1}{x} - 2x \sum_{n \in \mathbf{N}} \frac{1}{n^2(1 - \frac{x^2}{n^2})} = \frac{1}{x} - 2 \sum_{n \in \mathbf{N}} \zeta(2n) x^{2n-1} \qquad (|x| < 1).$$

Deduce $\zeta(2n) = (-1)^{n-1} \frac{1}{2} (2\pi)^{2n} \frac{B_{2n}}{(2n)!}$ by equating the coefficients of x^{2n-1}.

(iv) Let $f : \mathbf{R} \to \mathbf{R}$ be a twice differentiable function. Verify

$$\Big(\frac{f'}{f}\Big)^2 = \frac{f''}{f} - \Big(\frac{f'}{f}\Big)'.$$

Apply this identity with $f(x) = \sin(\pi x)$, and derive by multiplying the series expansion for $\pi \cot(\pi x)$ by itself (compare with Exercise 0.12.(vi))

$$\zeta(2n) = \frac{2}{2n+1} \sum_{1 \le i \le n-1} \zeta(2i)\,\zeta(2n-2i) \qquad (n \in \mathbf{N} \setminus \{1\}).$$

Deduce

$$-(2n+1)B_{2n} = \sum_{1 \le i \le n-1} \binom{2n}{2i} B_{2i}\, B_{2n-2i}.$$

Note that the summands are invariant under the symmetry $i \leftrightarrow n-i$; using this one can halve the number of terms in the summation.

Exercise 0.22 (Euler–MacLaurin summation formula – sequel to Exercise 0.16 – needed for Exercises 0.24 and 0.25). We employ the notations from part (i) in Exercise 0.16 and from its part (iii) the fact that $b_1(0) = -\frac{1}{2} = -b_1(1)$. Let $k \in \mathbf{N}$ and $f \in C^k(\mathbf{R})$.

(i) Prove using repeated integration by parts

$$\begin{aligned}\int_0^1 f(x)\,dx &= \Big[b_1(x) f(x)\Big]_0^1 - \int_0^1 b_1(x) f'(x)\,dx \\ &= \sum_{1 \le i \le k} (-1)^{i-1} \Big[\frac{b_i(x)}{i!} f^{(i-1)}(x)\Big]_0^1 + (-1)^k \int_0^1 \frac{b_k(x)}{k!} f^{(k)}(x)\,dx.\end{aligned}$$

(ii) Demonstrate

$$\begin{aligned}f(1) = &\int_0^1 f(x)\,dx + \sum_{1 \le i \le k} (-1)^i \frac{B_i}{i!} (f^{(i-1)}(1) - f^{(i-1)}(0)) \\ &+ (-1)^{k-1} \int_0^1 \frac{b_k(x)}{k!} f^{(k)}(x)\,dx.\end{aligned}$$

Now replace f by $x \mapsto f(j-1+x)$, with $j \in \mathbf{Z}$, and sum j from $m+1$ to n, for $m, n \in \mathbf{Z}$ with $m < n$.

(iii) Prove

$$\sum_{m < j \le n} f(j) = \int_m^n f(x)\,dx + \sum_{1 \le i \le k} (-1)^i \frac{B_i}{i!} (f^{(i-1)}(n) - f^{(i-1)}(m)) + R_k;$$

here we have, with $[x]$ the greatest integer in x,

$$R_k = (-1)^{k-1} \int_m^n \frac{b_k(x-[x])}{k!} f^{(k)}(x)\,dx.$$

Verify that the *Bernoulli summation formula* from Exercise 0.16.(xi) is a special case.

(iv) Now use Exercise 0.16.(viii), and set $k = 2l+1$, assuming k to be odd. Verify that this leads to the following *Euler–MacLaurin summation formula*:

$$\begin{aligned}\frac{1}{2}f(m) + \sum_{m<j<n} f(j) + \frac{1}{2}f(n) &= \int_m^n f(x)\,dx \\ &+ \sum_{1\le i\le l} \frac{B_{2i}}{(2i)!}(f^{(2i-1)}(n) - f^{(2i-1)}(m)) \\ &+ \int_m^n \frac{b_{2l+1}(x-[x])}{(2l+1)!} f^{(2l+1)}(x)\,dx.\end{aligned}$$

Here

$$\frac{B_2}{2!} = \frac{1}{12}, \qquad \frac{B_4}{4!} = -\frac{1}{720}, \qquad \frac{B_6}{6!} = \frac{1}{30240}, \qquad \frac{B_8}{8!} = -\frac{1}{1209600}.$$

Exercise 0.23 (Another proof of the Euler–MacLaurin summation formula – sequel to Exercise 0.16). Let $k \in \mathbf{N}$ and $f \in C^k(\mathbf{R})$. Multiply the function $x \mapsto b_1(x-[x]) = x - [x] - \frac{1}{2}$ in Exercise 0.16 by the derivative f' and integrate the product over the interval $[\,m, n\,]$. Next use integration by parts to obtain

$$\sum_{m\le i\le n} f(i) = \int_m^n f(x)\,dx + \frac{1}{2}(f(m)+f(n)) + \int_m^n b_1(x-[x])\,f'(x)\,dx.$$

Now repeat integration by parts in the last integral and use the parts (i) and (viii) from Exercise 0.16 in order to find the Euler–MacLaurin summation formula from Exercise 0.22.

Exercise 0.24 (Stirling's asymptotic expansion – sequel to Exercises 0.20 and 0.22). We study the growth properties of the factorials $n!$, for $n \to \infty$.

(i) Prove that $\log^{(i)} x = (-1)^{i-1}(i-1)!\,x^{-i}$, for $i \in \mathbf{N}$ and $x > 0$, and conclude by Exercise 0.22.(iv), for every $l \in \mathbf{N}$,

$$\begin{aligned}\log n! = \sum_{2\le i\le n} \log i &= \int_1^n \log x\,dx + \frac{1}{2}\log n \\ &+ \sum_{1\le i\le l} \frac{B_{2i}}{2i(2i-1)}\Big(\frac{1}{n^{2i-1}} - 1\Big) + \frac{1}{2l}\int_1^n b_{2l}(x-[x])x^{-2l}\,dx.\end{aligned}$$

(ii) Conclude that

$$(\star) \qquad \log n! = \Big(n+\frac{1}{2}\Big)\log n - n + C(l) + \sum_{1\le i\le l} \frac{B_{2i}}{2i(2i-1)}\frac{1}{n^{2i-1}} - E(n,l),$$

where

$$C(l) = \frac{1}{2l}\int_1^\infty b_{2l}(x-[x])x^{-2l}\,dx + 1 - \sum_{1\le i\le l}\frac{B_{2i}}{2i(2i-1)},$$

$$E(n,l) = \frac{1}{2l}\int_n^\infty b_{2l}(x-[x])x^{-2l}\,dx.$$

(iii) Demonstrate that

$$\lim_{n\to\infty} \log n! - \Big(n+\frac{1}{2}\Big)\log n + n = C(l),$$

and conclude that $C(l) = C$, independent of l.

We now write

$$(\star\star) \qquad \log n! = \Big(n+\frac{1}{2}\Big)\log n - n + R(n).$$

(iv) Prove that $\lim_{n\to\infty} R(n) = C$.

(v) Finally we calculate C, making use of Wallis' product from Exercise 0.13.(iii) or 6.56.(iv). Conclude that

$$\lim_{n\to\infty} 2(n\log 2 + \log n!) - \log(2n)! - \frac{1}{2}\log(2n+1) = \frac{1}{2}\log\frac{\pi}{2}.$$

Then use the identity $(\star\star)$, and prove by means of part (iv) that $C = \log\sqrt{2\pi}$.

Despite appearances the absolute value of the general term in

$$\frac{1}{12n} - \frac{1}{360n^3} + \frac{1}{1260n^5} - \frac{1}{1680n^7} + \cdots + \frac{B_{2l}}{2l(2l-1)}\frac{1}{n^{2l-1}} + \cdots$$

diverges to ∞ for $l \to \infty$. Indeed, Exercise 0.20.(iii) implies

$$\frac{1}{2\pi^2 n}\frac{1\cdot 2\cdots(2l-2)}{(2\pi n)(2\pi n)\cdots(2\pi n)} < \frac{|B_{2l}|}{2l(2l-1)}\frac{1}{n^{2l-1}}.$$

In particular, we do not obtain a convergent series from $(\star)$ by sending $l \to \infty$. We now study the properties of the remainder term $E(n,l)$ in $(\star)$.

(vi) Use Exercise 0.20.(iii) to show

$$|E(n,l)| \le \frac{|B_{2l}|}{2l}\int_n^\infty x^{-2l}\,dx = \frac{|B_{2l}|}{2l(2l-1)}\frac{1}{n^{2l-1}}.$$

This estimate cannot be essentially improved. For verifying this write

$$R(n,l) = \frac{B_{2l}}{2l(2l-1)}\frac{1}{n^{2l-1}} - E(n,l).$$

(vii) Conclude from part (vi) that $R(n,l)$ has the same sign as B_{2l}, and that there exists $\theta_l > 0$ satisfying

$$R(n,l) = \theta_l \frac{B_{2l}}{2l(2l-1)}\frac{1}{n^{2l-1}}.$$

Using ($\star$) and replacing l by $l+1$ in the formula above, verify

$$\begin{aligned} R(n,l) &= \frac{B_{2l}}{2l(2l-1)}\frac{1}{n^{2l-1}} + R(n,l+1) \\ &= \frac{B_{2l}}{2l(2l-1)}\frac{1}{n^{2l-1}} + \theta_{l+1}\frac{B_{2l+2}}{(2l+2)(2l+1)}\frac{1}{n^{2l+1}}. \end{aligned}$$

Since B_{2l} and B_{2l+2} have opposite signs one can write this as

$$R(n,l) = \frac{B_{2l}}{2l(2l-1)}\frac{1}{n^{2l-1}}\left(1 - \theta_{l+1}\left|\frac{B_{2l+2}}{B_{2l}}\right|\frac{2l(2l-1)}{(2l+2)(2l+1)}\frac{1}{n^2}\right).$$

Comparing this expression for $R(n,l)$ with the one containing the definition of θ_l, deduce that θ_l is given by the expression in $\left(\cdots\right)$, so that $\theta_l < 1$.

Background. Given n and $l \in \mathbf{N}$ we have obtained $0 < \theta_{l+1} < 1$, depending on n, such that

$$\begin{aligned} \log n! &= \left(n + \tfrac{1}{2}\right)\log n - n + \tfrac{1}{2}\log 2\pi + \sum_{1\le i\le l}\frac{B_{2i}}{2i(2i-1)}\frac{1}{n^{2i-1}} \\ &\quad + \theta_{l+1}\frac{B_{2l+2}}{(2l+2)(2l+1)}\frac{1}{n^{2l+1}}. \end{aligned}$$

The remainder term is a positive fraction of the first neglected term. Therefore the value of $\log n!$ can be calculated from this formula with great accuracy for large values of n by neglecting the remainder term, when l is suitably chosen.

The preceding argument can be formalized as follows. Let $f : \mathbf{R}_+ \to \mathbf{R}$ be a given function. A formal power series in $\frac{1}{x}$, for $x \in \mathbf{R}_+$, with coefficients $a_i \in \mathbf{R}$,

$$\sum_{i\in\mathbf{N}_0} a_i \frac{1}{x^i},$$

is said to be an *asymptotic expansion* for f if the following condition is met. Write $s_k(x)$ for the sum of the first $k+1$ terms of the series, and $r_k(x) = x^k(f(x) - s_k(x))$. Then one must have, for each $k \in \mathbf{N}$,

$$f(x) - s_k(x) = o(x^{-k}), \quad x \to \infty, \qquad \text{that is} \qquad \lim_{x\to\infty} r_k(x) = 0.$$

Note that no condition is imposed on $\lim_{k\to\infty} r_k(x)$, for x fixed. When the foregoing definition is satisfied, we write

$$f(x) \sim \sum_{i\in\mathbf{N}_0} a_i \frac{1}{x^i}, \quad x \to \infty.$$

Thus

$$\log n! \sim \Big(n + \frac{1}{2}\Big)\log n - n + \frac{1}{2}\log 2\pi + \sum_{i\in\mathbf{N}} \frac{B_{2i}}{2i(2i-1)}\frac{1}{n^{2i-1}}, \qquad n \to \infty.$$

We note the final result, which is known as *Stirling's asymptotic expansion* (see Exercise 6.55 for a different approach)

$$n! \sim n^n e^{-n}\sqrt{2\pi n}\exp\Big(\sum_{i\in\mathbf{N}} \frac{B_{2i}}{2i(2i-1)}\frac{1}{n^{2i-1}}\Big), \qquad n \to \infty.$$

(viii) Among the binomial coefficients $\binom{2n}{k}$ the middle coefficient with $k = n$ is the largest. Prove

$$\binom{2n}{n} \sim \frac{4^n}{\sqrt{\pi n}}, \qquad n \to \infty.$$

Exercise 0.25 (Continuation of zeta function – sequel to Exercises 0.16 and 0.22 – needed for Exercises 6.62 and 6.89). Apply the Euler–MacLaurin summation formula from Exercise 0.22.(iii) with $f(x) = x^{-s}$ for $s \in \mathbf{C}$. We have, in the notation from Exercise 0.11,

$$f^{(i)}(x) = (-1)^i (s)_i\, x^{-s-i} \qquad (i \in \mathbf{N}).$$

We find, for every $N \in \mathbf{N}$, $s \neq 1$, and $k \in \mathbf{N}$,

$$\sum_{1\le j\le N} j^{-s} = \frac{1}{s-1}(1 - N^{-s+1}) + \frac{1}{2}(1 + N^{-s})$$
$$- \sum_{2\le i\le k} \frac{B_i}{i!}(s)_{i-1}(N^{-s-i+1} - 1) - \frac{(s)_k}{k!}\int_1^N b_k(x - [x])x^{-s-k}\,dx.$$

Under the condition $\operatorname{Re} s > 1$ we may take the limit for $N \to \infty$; this yields, for every $k \in \mathbf{N}$,

$$(\star) \qquad \zeta(s) := \sum_{j \in \mathbf{N}} j^{-s} = \frac{1}{s-1} + \frac{1}{2} + \sum_{2 \le i \le k} \frac{B_i}{i!} (s)_{i-1} - \frac{(s)_k}{k!} \int_1^\infty b_k(x - [x]) x^{-s-k}\, dx.$$

We note that the integral on the right converges for $s \in \mathbf{C}$ with $\operatorname{Re} s > 1 - k$, in fact, uniformly on compact domains in this half-plane. Accordingly, for $s \neq 1$ the function $\zeta(s)$ may be **defined** in that half-plane by means of $(\star)$, and it is then a complex-differentiable function of s, except at $s = 1$. Since $k \in \mathbf{N}$ is arbitrary, ζ can be continued to a complex-differentiable function on $\mathbf{C} \setminus \{1\}$. To calculate $\zeta(-n)$, for $n \in \mathbf{N}_0$, we set $k = n + 1$. The factor in front of the integral then vanishes, and applying Exercise 0.16.(i) and (ix) we obtain

$$-(n+1)\zeta(-n) = \sum_{0 \le i \le n+1} B_i \binom{n+1}{i} = \begin{cases} \dfrac{1}{2}, & n = 0; \\ B_{n+1}, & n \in \mathbf{N}. \end{cases}$$

That is

$$\zeta(0) = -\frac{1}{2}, \qquad \zeta(-2n) = 0, \qquad \zeta(1 - 2n) = -\frac{B_{2n}}{2n} \qquad (n \in \mathbf{N}).$$

Exercise 0.26 (Regularization of integral). Assume $f \in C^\infty(\mathbf{R})$ vanishes outside a bounded set.

(i) Verify that $s \mapsto \int_{\mathbf{R}_+} x^s f(x)\, dx$ is well-defined for $s \in \mathbf{C}$ with $-1 < \operatorname{Re} s$, and gives a complex-differentiable function on that domain.

(ii) For $-1 < \operatorname{Re} s$ and $n \in \mathbf{N}$, prove that

$$(\star) \qquad \int_{\mathbf{R}_+} x^s f(x)\, dx = \int_0^1 x^s \Big(f(x) - \sum_{0 \le i < n} \frac{f^{(i)}(0)}{i!} x^i \Big) dx + \sum_{0 \le i < n} \frac{f^{(i)}(0)}{i!(s+i+1)} + \int_1^\infty x^s f(x)\, dx.$$

(iii) Verify that the expression on the right in $(\star)$ is well-defined for $-n-1 < \operatorname{Re} s$, $s \neq -1, \dots, -n$.

(iv) Next, assume $-n - 1 < s < -n$. Check that $\int_1^\infty x^{s+i}\, dx = -\frac{1}{s+i+1}$, for $0 \le i < n$. Use this to prove that the right–hand side in $(\star)$ equals

$$\int_{\mathbf{R}_+} x^s \Big(f(x) - \sum_{0 \le i < n} \frac{f^{(i)}(0)}{i!} x^i \Big) dx.$$

Thus $s \mapsto \int_{\mathbf{R}_+} x^s f(x)\, dx$ has been extended as a complex-differentiable function to a function on $\{ s \in \mathbf{C} \mid s \neq -n,\ n \in \mathbf{N} \}$.

Background. One notes that in this case *regularization*, that is, assigning a meaning to a divergent integral, is achieved by omitting some part of the integrand, while keeping the part that does give a finite result. Hence the term *taking the finite part* for this construction.

Exercises for Chapter 1

Exercise 1.1 (Orthogonal decomposition – needed for Exercise 2.65). Let L be a linear subspace of $\mathbf{R}^n$. Then we define $L^\perp$, the *orthocomplement* of L, by $L^\perp = \{ x \in \mathbf{R}^n \mid \langle x, y \rangle = 0 \text{ for all } y \in L \}$.

(i) Verify that $L^\perp$ is a linear subspace of $\mathbf{R}^n$ and that $L \cap L^\perp = (0)$.

Suppose that $(v_1, \dots, v_l)$ is an orthonormal basis for L. Given $x \in \mathbf{R}^n$, define y and $z \in \mathbf{R}^n$ by

$$y = \sum_{1 \le j \le l} \langle x, v_j \rangle v_j, \qquad z = x - \sum_{1 \le j \le l} \langle x, v_j \rangle v_j.$$

(ii) Prove that $x = y + z$, that $y \in L$ and $z \in L^\perp$. Using (i) show that y and z are uniquely determined by these properties. We say that y is the *orthogonal projection* of x onto L. Deduce that we have the orthogonal direct sum decomposition $\mathbf{R}^n = L \oplus L^\perp$, that is, $\mathbf{R}^n = L + L^\perp$ and $L \cap L^\perp = (0)$.

(iii) Verify

$$\|y\|^2 = \sum_{1 \le j \le l} \langle x, v_j \rangle^2, \qquad \|z\|^2 = \|x\|^2 - \sum_{1 \le j \le l} \langle x, v_j \rangle^2.$$

For any $v \in L$, prove $\|x - v\|^2 = \|x - y\|^2 + \|y - v\|^2$. Deduce that y is the unique point in L at the shortest distance to x.

(iv) By means of (ii) show that $(L^\perp)^\perp = L$.

(v) If M is also a linear subspace of $\mathbf{R}^n$, prove that $(L + M)^\perp = L^\perp \cap M^\perp$, and using (iv) deduce $(L \cap M)^\perp = L^\perp + M^\perp$.

Exercise 1.2 (Parallelogram identity). Verify, for x and $y \in \mathbf{R}^n$,

$$\|x + y\|^2 + \|x - y\|^2 = 2(\|x\|^2 + \|y\|^2).$$

That is, the sum of the squares of the diagonals of a parallelogram equals the sum of the squares of the sides.

Exercise 1.3 (Symmetry identity – needed for Exercise 7.70). Show, for x and $y \in \mathbf{R}^n \setminus \{0\}$,

$$\left\| \frac{1}{\|x\|} x - \|x\| y \right\| = \left\| \frac{1}{\|y\|} y - \|y\| x \right\|.$$

Give a geometric interpretation of this identity.

Exercise 1.4. Prove, for x and $y \in \mathbf{R}^n$

$$\langle x, y\rangle(\|x\| + \|y\|) \le \|x\|\,\|y\|\,\|x + y\|.$$

Verify that the inequality does not hold if $\langle x, y\rangle$ is replaced by $|\langle x, y\rangle|$.
Hint: The inequality needs a proof only if $\langle x, y\rangle \ge 0$; in that case, square both sides.

Exercise 1.5. Construct a countable family of open subsets of $\mathbf{R}$ whose intersection is not open, and also a countable family of closed subsets of $\mathbf{R}$ whose union is not closed.

Exercise 1.6. Let A and B be any two subsets of $\mathbf{R}^n$.

(i) Prove $\mathrm{int}(A) \subset \mathrm{int}(B)$ if $A \subset B$.

(ii) Show $\mathrm{int}(A \cap B) = \mathrm{int}(A) \cap \mathrm{int}(B)$.

(iii) From (ii) deduce $\overline{A \cup B} = \overline{A} \cup \overline{B}$.

Exercise 1.7 (Boundary of closed set – needed for Exercise 3.48). Let $n \in \mathbf{N}\setminus\{1\}$. Let F be a closed subset of $\mathbf{R}^n$ with $\mathrm{int}(F) \ne \emptyset$. Prove that the boundary ∂F of F contains infinitely many points, unless $F = \mathbf{R}^n$ (in which case we have $\partial F = \emptyset$).
Hint: Assume $x \in \mathrm{int}(F)$ and $y \in F^{\mathrm{c}}$. Since both these sets are open in $\mathbf{R}^n$, there exist neighborhoods U of x and V of y with $U \subset \mathrm{int}(F)$ and $V \subset F^{\mathrm{c}}$. Check, for all $z \in V$, that the line segment from x to z contains a point z' with $z' \ne x$ and $z' \in \partial F$.
Background. It is possible for an open subset of $\mathbf{R}^n$ to have a boundary consisting of finitely many points, for example, a set consisting of $\mathbf{R}^n$ minus a finite number of points. The condition $n > 1$ is essential, as is obvious from the fact that a closed interval in $\mathbf{R}$ has zero, one or two boundary points in $\mathbf{R}$.

Exercise 1.8. Set $V = \,]-2, 0\,[\, \cup \,]\,0, 2\,] \subset \mathbf{R}$.

(i) Show that both $]-2, 0\,[$ and $]\,0, 2\,]$ are open in V. Prove that both are also closed in V.

(ii) Let $A = \,]-1, 1\,[\, \cap V$. Prove that A is open and not closed in V.

Exercise 1.9 (Inverse image of union and intersection). Let $f : A \to B$ be a mapping between sets. Let I be an index set and suppose for every $i \in I$ we have $B_i \subset B$. Show

$$f^{-1}\Big(\bigcup_{i\in I} B_i\Big) = \bigcup_{i\in I} f^{-1}(B_i), \qquad f^{-1}\Big(\bigcap_{i\in I} B_i\Big) = \bigcap_{i\in I} f^{-1}(B_i).$$

Exercise 1.10. We define $f : \mathbf{R}^2 \to \mathbf{R}$ by

$$f(x) = \begin{cases} \dfrac{x_1 x_2^2}{x_1^2 + x_2^6}, & x \neq 0; \\ 0, & x = 0. \end{cases}$$

Show that $\text{im}(f) = \mathbf{R}$ and prove that f is not continuous.
Hint: Consider $x_1 = x_2^l$, for a suitable $l \in \mathbf{N}$.

Exercise 1.11 (Homogeneous function). A function $f : \mathbf{R}^n \setminus \{0\} \to \mathbf{R}$ is said to be *positively homogeneous* of degree $d \in \mathbf{R}$, if $f(tx) = t^d f(x)$, for all $x \neq 0$ and $t > 0$. Assume f to be continuous. Prove that f has an extension as a continuous function to $\mathbf{R}^n$ precisely in the following cases: (i) if $d < 0$, then $f \equiv 0$; (ii) if $d = 0$, then f is a constant function; (iii) if $d > 0$, then there is no further condition on f. In each case, indicate which value f has to assume at 0.

Exercise 1.12. Suppose $A \subset \mathbf{R}^n$ and let f and $g : A \to \mathbf{R}^p$ be continuous.

(i) Show that $\{ x \in A \mid f(x) = g(x) \}$ is a closed set in A.

(ii) Let $p = 1$. Prove that $\{ x \in A \mid f(x) > g(x) \}$ is open in A.

Exercise 1.13 (Needed for Exercise 1.33). Let $A \subset V \subset \mathbf{R}^n$ and suppose $\overline{A}^V = V$. In this case, A is said to be *dense* in V. Let f and $g : V \to \mathbf{R}^p$ be continuous mappings. Show that $f = g$ if f and g coincide on A.

Exercise 1.14 (Graph of continuous mapping – needed for Exercise 1.24). Suppose $F \subset \mathbf{R}^n$ is closed, let $f : F \to \mathbf{R}^p$ be continuous, and define its *graph* by

$$\text{graph}(f) = \{ (x, f(x)) \mid x \in F \} \subset \mathbf{R}^n \times \mathbf{R}^p \simeq \mathbf{R}^{n+p}.$$

(i) Show that graph(f) is closed in $\mathbf{R}^{n+p}$.
Hint: Use the Lemmata 1.2.12 and 1.3.6, or the fact that graph(f) is the inverse image of the closed set $\{ (y, y) \mid y \in \mathbf{R}^p \} \subset \mathbf{R}^{2p}$ under the continuous mapping: $F \times \mathbf{R}^p \to \mathbf{R}^{2p}$ given by $(x, y) \mapsto (f(x), y)$.

Now define $f : \mathbf{R} \to \mathbf{R}$ by

$$f(t) = \begin{cases} \sin \dfrac{1}{t}, & t \neq 0; \\ 0, & t = 0. \end{cases}$$

(ii) Prove that graph(f) is not closed in $\mathbf{R}^2$, whereas $F = \text{graph}(f) \cup \{ (0, y) \in \mathbf{R}^2 \mid -1 \le y \le 1 \}$ is closed in $\mathbf{R}^2$. Deduce that f is not continuous at 0.

Exercise 1.15 (Homeomorphism between open ball and $\mathbf{R}^n$ – needed for Exercise 6.23). Let $r > 0$ be arbitrary and let $B = \{ x \in \mathbf{R}^n \mid \|x\| < r \}$, and define $f : B \to \mathbf{R}^n$ by

$$f(x) = (r^2 - \|x\|^2)^{-1/2}\, x.$$

Prove that $f : B \to \mathbf{R}^n$ is a homeomorphism, with inverse $f^{-1} : \mathbf{R}^n \to B$ given by

$$f^{-1}(y) = (1 + \|y\|^2)^{-1/2}\, ry.$$

Exercise 1.16. Let $f : \mathbf{R}^n \to \mathbf{R}^p$ be a continuous mapping.

(i) Using Lemma 1.3.6 prove that $f(\overline{A}) \subset \overline{f(A)}$, for every subset A of $\mathbf{R}^n$.

(ii) Show that continuity of f does not imply any inclusion relations between $f(\text{int}(A))$ and $\text{int}\,(f(A))$.

(iii) Show that a mapping $f : \mathbf{R}^n \to \mathbf{R}^p$ is closed and continuous if $f(\overline{A}) = \overline{f(A)}$ for every subset A of $\mathbf{R}^n$.

Exercise 1.17 (Completeness). A subset A of $\mathbf{R}^n$ is said to be *complete* if every Cauchy sequence in A is convergent in A. Prove that A is complete if and only if A is closed in $\mathbf{R}^n$.

Exercise 1.18 (Addition to Theorem 1.6.2). Show that $a \in \mathbf{R}$ as in the proof of the Theorem of Bolzano–Weierstrass 1.6.2 is also equal to

$$a = \inf_{N \in \mathbf{N}} \Big(\sup_{k \ge N} x_k \Big) = \limsup_{k \to \infty} x_k.$$

Prove that $b \le a$ if b is the limit of any subsequence of $(x_k)_{k \in \mathbf{N}}$.

Exercise 1.19. Set $B = \{ x \in \mathbf{R}^n \mid \|x\| < 1 \}$ and suppose $f : B \to B$ is continuous and satisfies $\|f(x)\| < \|x\|$, for all $x \in B \setminus \{0\}$. Select $0 \ne x_1 \in B$ and define the sequence $(x_k)_{k \in \mathbf{N}}$ by setting $x_{k+1} = f(x_k)$, for $k \in \mathbf{N}$. Prove that $\lim_{k \to \infty} x_k = 0$.

Hint: Continuity implies $f(0) = 0$; hence, if any $x_k = 0$, then so are all subsequent terms. Assume therefore that all $x_k \ne 0$. As $(\|x_k\|)_{k \in \mathbf{N}}$ is decreasing, it has a limit $r \ge 0$. Now $(x_k)_{k \in \mathbf{N}}$ has a convergent subsequence $(x_{k_l})_{l \in \mathbf{N}}$, with limit $a \in B$. Then $r = 0$, since

$$\|a\| = \| \lim_{l \to \infty} x_{k_l} \| = \lim_{l \to \infty} \|x_{k_l}\| = r \le \lim_{l \to \infty} \|x_{k_l + 1}\| = \lim_{l \to \infty} \|f(x_{k_l})\| = \|f(a)\|.$$

Exercise 1.20. Suppose $(x_k)_{k\in\mathbf{N}}$ is a convergent sequence in $\mathbf{R}^n$ with limit $a \in \mathbf{R}^n$. Show that $\{a\} \cup \{\, x_k \mid k \in \mathbf{N}\,\}$ is a compact subset of $\mathbf{R}^n$.

Exercise 1.21. Let $F \subset \mathbf{R}^n$ be closed and $\delta > 0$. Show that A is closed in $\mathbf{R}^n$ if

$$A = \{\, a \in \mathbf{R}^n \mid \text{there exists } x \in F \text{ with } \|x - a\| = \delta \,\}.$$

Exercise 1.22. Show that $K \subset \mathbf{R}^n$ is compact if and only if every continuous function: $K \to \mathbf{R}$ is bounded.

Exercise 1.23 (Projection along compact factor is proper – needed for Exercise 1.24). Let $A \subset \mathbf{R}^n$ be arbitrary, let $K \subset \mathbf{R}^p$ be compact, and let $p : A \times K \to A$ be the projection with $p(x, y) = x$. Show that p is proper and deduce that it is a closed mapping.

Exercise 1.24 (Closed graph – sequel to Exercises 1.14 and 1.23). Let $A \subset \mathbf{R}^n$ be closed and $K \subset \mathbf{R}^p$ be compact, and let $f : A \to K$ be a mapping.

(i) Show that f is continuous if and only if graph(f) (see Exercise 1.14) is closed in $\mathbf{R}^{n+p}$.
Hint: Use Exercise 1.14. Next, suppose graph(f) is closed. Write $p_1 : \mathbf{R}^{n+p} \to \mathbf{R}^n$, and $p_2 : \mathbf{R}^{n+p} \to \mathbf{R}^p$, for the projection $p_1(x, y) = x$, and $p_2(x, y) = y$, for $x \in \mathbf{R}^n$ and $y \in \mathbf{R}^p$. Let $F \subset \mathbf{R}^p$ be closed and show $p_1(p_2^{-1}(F) \cap \text{graph}(f)) = f^{-1}(F)$. Now apply Exercise 1.23.

(ii) Verify that compactness of K is necessary for the validity of (i) by considering $f : \mathbf{R} \to \mathbf{R}$ with

$$f(x) = \begin{cases} \dfrac{1}{x}, & x \neq 0; \\ 0, & x = 0. \end{cases}$$

Exercise 1.25 (Polynomial functions are proper – needed for Exercises 1.28 and 3.48). A nonconstant polynomial function: $\mathbf{C} \to \mathbf{C}$ is proper. Deduce that im(p) is closed in $\mathbf{C}$.
Hint: Suppose $p(z) = \sum_{0\le i\le n} a_i z^i$ with $n \in \mathbf{N}$ and $a_n \neq 0$. Then $|p(z)| \to \infty$ as $|z| \to \infty$, because

$$|p(z)| \geq |z|^n \Big(|a_n| - \sum_{0\le i<n} |a_i|\,|z|^{i-n} \Big).$$

Exercise 1.26 (Extension of definition of proper mapping). Let $A \subset \mathbf{R}^n$ and $B \subset \mathbf{R}^p$, and let $f : A \to B$ be a mapping. Extending Definition 1.8.5 we say that f is *proper* if the inverse image under f of every compact set in B is compact in A. In contrast to the case of continuity, this property of f depends on the target space B of f. In order to see this, consider $A = \,]-1, 1[$ and $f(x) = x$.

(i) Prove that $f : A \to A$ is proper.

(ii) Prove that $f : A \to \mathbf{R}$ is not proper, by considering $f^{-1}([-2, 2])$.

Now assume that $f : A \to B$ is a continuous injection, and denote by $g : f(A) \to A$ the inverse mapping of f. Prove that the following assertions are equivalent.

(iii) f is a proper mapping.

(iv) f is a closed mapping.

(v) $f(A)$ is a closed subset in B and g is continuous.

Exercise 1.27 (Lipschitz continuity of inverse – needed for Exercises 1.49 and 3.24). Let $f : \mathbf{R}^n \to \mathbf{R}^n$ be continuous and suppose there exists $k > 0$ such that $\|f(x) - f(x')\| \geq k\|x - x'\|$, for all x and $x' \in \mathbf{R}^n$.

(i) Show that f is injective and proper. Deduce that $f(\mathbf{R}^n)$ is closed in $\mathbf{R}^n$.

(ii) Show that $f^{-1} : \operatorname{im}(f) \to \mathbf{R}^n$ is Lipschitz continuous and conclude that $f : \mathbf{R}^n \to \operatorname{im}(f)$ is a homeomorphism.

Exercise 1.28 (Cauchy's Minimum Theorem – sequel to Exercise 1.25). For every polynomial function $p : \mathbf{C} \to \mathbf{C}$ there exists $w \in \mathbf{C}$ with $|p(w)| = \inf\{\,|p(z)| \mid z \in \mathbf{C}\,\}$. Prove this using Exercise 1.25.

Exercise 1.29 (Another proof of Contraction Lemma 1.7.2). The notation is as in that Lemma. Define $g : F \to \mathbf{R}$ by $g(x) = \|x - f(x)\|$.

(i) Verify $|g(x) - g(x')| \leq (1 + \epsilon)\|x - x'\|$ for all $x, x' \in F$, and deduce that g is continuous on F.

If F is bounded the continuous function g assumes its minimum at a point p belonging to the compact set F.

(ii) Show $g(p) \leq g(f(p)) \leq \epsilon g(p)$, and conclude that $g(p) = 0$. That is, $p \in F$ is a fixed point of f.

If F is not bounded, set $F_0 = \{\,x \in F \mid g(x) \leq g(x_0)\,\}$. Then $F_0 \subset F$ is nonempty and closed.

(iii) Prove, for $x \in F_0$,

$$\begin{aligned}\|x - x_0\| &\leq \|x - f(x)\| + \|f(x) - f(x_0)\| + \|f(x_0) - x_0\| \\ &\leq 2g(x_0) + \epsilon\|x - x_0\|.\end{aligned}$$

Hence $\|x - x_0\| \leq \frac{2g(x_0)}{1-\epsilon}$ for $x \in F_0$, and therefore F_0 is bounded.

(iv) Show that f is a mapping of F_0 in F_0 by noting $g(f(x)) \leq \epsilon g(x) \leq g(x_0)$ for $x \in F_0$; and proceed as above to find a fixed point $p \in F_0$.

Exercise 1.30. Let $A \subset \mathbf{R}^n$ be bounded and $f : A \to \mathbf{R}^p$ uniformly continuous. Show that f is bounded on A.

Exercise 1.31. Let f and $g : \mathbf{R}^n \to \mathbf{R}$ be uniformly continuous.

(i) Show that $f + g$ is uniformly continuous.

(ii) Show that fg is uniformly continuous if f and g are bounded. Give a counterexample to show that the condition of boundedness cannot be dropped in general.

(iii) Assume $n = 1$. Is $g \circ f$ uniformly continuous?

Exercise 1.32. Let $g : \mathbf{R} \to \mathbf{R}$ be continuous and define $f : \mathbf{R}^2 \to \mathbf{R}$ by $f(x) = g(x_1 x_2)$. Show that f is uniformly continuous only if g is a constant function.

Exercise 1.33 (Uniform continuity and Cauchy sequences – sequel to Exercise 1.13 – needed for Exercise 1.34). Let $A \subset \mathbf{R}^n$ and let $f : A \to \mathbf{R}^p$ be uniformly continuous.

(i) Show that $(f(x_k)_{k\in\mathbf{N}})$ is a Cauchy sequence in $\mathbf{R}^p$ whenever $(x_k)_{k\in\mathbf{N}}$ is a Cauchy sequence in A.

(ii) Using (i) prove that f can be extended in a unique fashion as a continuous function to $\overline{A}$.
Hint: Uniqueness follows from Exercise 1.13.

(iii) Give an example of a set $A \subset \mathbf{R}$ and a continuous function: $A \to \mathbf{R}$ that does not take Cauchy sequences in A to Cauchy sequences in $\mathbf{R}$.

Suppose that $g : \mathbf{R}^n \to \mathbf{R}^p$ is continuous.

(iv) Prove that g takes Cauchy sequences in $\mathbf{R}^n$ to Cauchy sequences in $\mathbf{R}^p$.

Exercise 1.34 (Sequel to Exercise 1.33). Let $f : \mathbf{R}^2 \to \mathbf{R}$ be the function from Example 1.3.10.

(i) Using Exercise 1.33.(iii) show that f is not uniformly continuous on $\mathbf{R}^2 \setminus \{0\}$.

(ii) Verify that f is uniformly continuous on $\{ x \in \mathbf{R}^2 \mid \|x\| \geq r \}$, for every $r > 0$.

Exercise 1.35 (Determining extrema by reduction to dimension one). Assume that $K \subset \mathbf{R}^n$ and $L \subset \mathbf{R}^p$ are compact sets, and that $f : K \times L \to \mathbf{R}$ is continuous.

(i) Show that for every $\epsilon > 0$ there exists $\delta > 0$ such that

$$a, b \in K \quad \text{with} \quad \|a - b\| < \delta, \quad y \in L \quad \Longrightarrow \quad |f(a, y) - f(b, y)| < \epsilon.$$

Next, define $m : K \to \mathbf{R}$ by $m(x) = \max\{ f(x, y) \mid y \in L \}$.

(ii) Prove that for every $\epsilon > 0$ there exists $\delta > 0$ such that $m(a) > m(b) - \epsilon$, for all $a, b \in K$ with $\|a - b\| < \delta$.
Hint: Use that $m(b) = f(b, y_b)$ for a suitable $y_b \in L$ which depends on b.

(iii) Deduce from (ii) that $m : K \to \mathbf{R}$ is uniformly continuous.

(iv) Show that $\max\{ m(x) \mid x \in K \} = \max\{ f(x, y) \mid x \in K,\ y \in L \}$.

Consider a continuous function f on $\mathbf{R}^n$ that is defined on a product of n intervals in $\mathbf{R}$. It is a consequence of (iv) that the problem of finding maxima for f can be reduced to the problem of successively finding maxima for functions associated with f that depend on one variable only. Under suitable conditions the latter problem might be solved by using the differential calculus in one variable.

(v) Apply this method for proving that the function

$$f : [-\frac{1}{2}, 1] \times [0, 2] \to \mathbf{R} \qquad \text{with} \qquad f(x) = \|x\|^2 e^{-x_1 - x_2^2}$$

attains its maximum value $e^{-\frac{1}{4}}$ at $\frac{1}{2}(-1, \sqrt{3})$.
Hint: $m(x_1) = f(x_1, \sqrt{1 - x_1^2}) = e^{-1 - x_1 + x_1^2}$.

Exercise 1.36. For disjoint nonempty subsets A and B of $\mathbf{R}^2$ define $d(A, B) = \inf\{ \|a - b\| \mid a \in A, b \in B \}$. Prove there exist such A and B which are closed and satisfy $d(A, B) = 0$.

Exercise 1.37 (Distance between sets – needed for Exercises 1.38 and 1.39). For $x \in \mathbf{R}^n$ and $\emptyset \neq A \subset \mathbf{R}^n$ define the distance from x to A as $d(x, A) = \inf\{ \|x - a\| \mid a \in A \}$.

(i) Prove that $\overline{A} = \{ x \in \mathbf{R}^n \mid d(x, A) = 0 \}$. Deduce that $d(x, A) > 0$ if $x \notin A$ and A is closed.

(ii) Show that the function $x \mapsto d(x, A)$ is uniformly continuous on $\mathbf{R}^n$.
Hint: For all x and $x' \in \mathbf{R}^n$ and all $a \in A$, we have $d(x, A) \le \|x - a\| \le \|x - x'\| + \|x' - a\|$, so that $d(x, A) \le \|x - x'\| + d(x', A)$.

(iii) Suppose A is a closed set in $\mathbf{R}^n$. Deduce from (ii) and (i) that there exist open sets $O_k \subset \mathbf{R}^n$, for $k \in \mathbf{N}$, such that $A = \cap_{k \in \mathbf{N}} O_k$. In other words, every closed set is the intersection of countably many open sets. Deduce that every open set in $\mathbf{R}^n$ is the union of countably many closed sets in $\mathbf{R}^n$.

(iv) Assume K and A are disjoint nonempty subsets of $\mathbf{R}^n$ with K compact and A closed. Using (ii) prove that there exists $\delta > 0$ such that $\|x - a\| \ge \delta$, for all $x \in K$ and $a \in A$.

(v) Consider disjoint closed nonempty subsets A and $B \subset \mathbf{R}^n$. Verify that $f : \mathbf{R}^n \to \mathbf{R}$ defined by

$$f(x) = \frac{d(x, A)}{d(x, A) + d(x, B)}$$

is continuous, with $0 \le f \le 1$, and $A = f^{-1}(\{0\})$ and $B = f^{-1}(\{1\})$. Deduce that there exist disjoint open sets U and V in $\mathbf{R}^n$ with $A \subset U$ and $B \subset V$.

Exercise 1.38 (Sequel to Exercise 1.37). Let $K \subset \mathbf{R}^n$ be compact and let $f : K \to K$ be a distance-preserving mapping, that is, $\|f(x) - f(x')\| = \|x - x'\|$, for all $x, x' \in K$. Show that f is surjective and hence a homeomorphism.
Hint: If $K \setminus f(K) \neq \emptyset$, select $x_0 \in K \setminus f(K)$. Set $\delta = d(x_0, f(K))$; then $\delta > 0$ on account of Exercise 1.37.(i). Define $(x_k)_{k \in \mathbf{N}}$ inductively by $x_k = f(x_{k-1})$, for $k \in \mathbf{N}$. Show that $(x_k)_{k \in \mathbf{N}}$ is a sequence in $f(K)$ satisfying $\|x_k - x_l\| \ge \delta$, for all k and $l \in \mathbf{N}$.

Exercise 1.39 (Sum of sets – sequel to Exercise 1.37). For A and $B \subset \mathbf{R}^n$ we define $A + B = \{ a + b \mid a \in A, \, b \in B \}$. Assuming that A is closed and B is compact, show that $A + B$ is closed.
Hint: Consider $x \notin A + B$, and set $C = \{ x - a \mid a \in A \}$. Then C is closed, and B and C are disjoint. Now apply Exercise 1.37.(iv).

Exercise 1.40 (Total boundedness). A set $A \subset \mathbf{R}^n$ is said to be *totally bounded* if for every $\delta > 0$ there exist finitely many points $x_k \in A$, for $1 \le k \le l$, with $A \subset \cup_{1 \le k \le l} B(x_k; \delta)$.

(i) Show that A is totally bounded if and only if its closure $\overline{A}$ is totally bounded.

(ii) Show that a closed set is totally bounded if and only if it is compact.
Hint: If A is not totally bounded, then there exists $\delta > 0$ such that A cannot be covered by finitely many open balls of radius δ centered at points of A. Select $x_1 \in A$ arbitrary. Then there exists $x_2 \in A$ with $\|x_2 - x_1\| \geq \delta$. Continuing in this fashion construct a sequence $(x_k)_{k\in\mathbf{N}}$ with $x_k \in A$ and $\|x_k - x_l\| \geq \delta$, for $k \neq l$.

Exercise 1.41 (Lebesgue number of covering – needed for Exercise 1.42). Let K be a compact subset in $\mathbf{R}^n$ and let $\mathcal{O} = \{ O_i \mid i \in I \}$ be an open covering of K. Prove that there exists a number $\delta > 0$, called a *Lebesgue number* of $\mathcal{O}$, with the following property. If $x, y \in K$ and $\|x - y\| < \delta$, then there exists $i \in I$ with x, $y \in O_i$.
Hint: Proof by reductio ad absurdum: choose $\delta_k = \frac{1}{k}$, then use the Bolzano–Weierstrass Theorem 1.6.3.

Exercise 1.42 (Continuous mapping on compact set is uniformly continuous – sequel to Exercise 1.41). Let $K \subset \mathbf{R}^n$ be compact and let $f : K \to \mathbf{R}^p$ be a continuous mapping. Once again prove the result from Theorem 1.8.15 that f is uniformly continuous, in the following two ways.

(i) On the basis of Definition 1.8.16 of compactness.
Hint: $\forall x \in K\ \forall \epsilon > 0\ \exists \delta(x) > 0 : x' \in B(x;\ \delta(x)) \Longrightarrow \|f(x') - f(x)\| < \frac{1}{2}\epsilon$. Then $K \subset \bigcup_{x\in K} B(x;\ \frac{1}{2}\delta(x))$.

(ii) By means of Exercise 1.41.

Exercise 1.43 (Product of compact sets is compact). Let $K \subset \mathbf{R}^n$ and $L \subset \mathbf{R}^p$ be compact subsets in the sense of Definition 1.8.16. Prove that the Cartesian product $K \times L$ is a compact subset in $\mathbf{R}^{n+p}$ in the following two ways.

(i) Use the Heine–Borel Theorem 1.8.17.

(ii) On the basis of Definition 1.8.16 of compactness.
Hint: Let $\{ O_i \mid i \in I \}$ be an open covering of $K \times L$. For every $z = (x, y)$ in $K \times L$ there exists an index $i(z) \in I$ such that $z \in O_{i(z)}$. Then there exist open neighborhoods U_z in $\mathbf{R}^n$ of x and V_z in $\mathbf{R}^p$ of y with $U_z \times V_z \subset O_{i(z)}$. Now choose $x \in K$, fixed for the moment. Then $\{ V_z \mid z \in \{x\} \times L \}$ is an open covering of the compact set L, hence there exist points $z_1, \dots, z_k \in K \times L$ (which depend on x) such that $L \subset \bigcup_{1\leq j\leq k} V_{z_j}$. Now $B(x) := \bigcap_{1\leq j\leq k} U_{z_j}$ is an open set in $\mathbf{R}^n$ containing x. Verify that $B(x) \times L$ is covered by a finite number of sets O_i. Observe that $\{ B(x) \mid x \in K \}$ is an open covering of K, then use the compactness of K.

Exercise 1.44 (Cantor's Theorem – needed for Exercise 1.45). Let $(F_k)_{k\in\mathbf{N}}$ be a sequence of closed sets in $\mathbf{R}^n$ such that $F_k \supset F_{k+1}$, for all $k \in \mathbf{N}$, and $\lim_{k\to\infty} \text{diameter}(F_k) = 0$.

(i) Prove *Cantor's Theorem*, which asserts that $\bigcap_{k\in\mathbf{N}} F_k$ contains a single point. **Hint:** See the proof of the Theorem of Heine–Borel 1.8.17 or use Proposition 1.8.21.

Let $I = [\,0, 1\,] \subset \mathbf{R}$, then the n-fold Cartesian product $I^n \subset \mathbf{R}^n$ is a unit hypercube. For every $k \in \mathbf{N}$, let $\mathcal{B}_k$ be the collection of the 2^{nk} congruent hypercubes contained in I^n of the form, for $1 \leq j \leq n$, $i_j \in \mathbf{N}_0$ and $0 \leq i_j < 2^k$,

$$\{\, x \in I^n \mid \frac{i_j}{2^k} \leq x_j \leq \frac{i_j+1}{2^k} \quad (1 \leq j \leq n)\,\}.$$

(ii) Using part (i) prove that for every $x \in I^n$ there exists a sequence $(B_k)_{k\in\mathbf{N}}$ satisfying $B_k \in \mathcal{B}_k$; $B_k \supset B_{k+1}$, for $k \in \mathbf{N}$; and $\bigcap_{k\in\mathbf{N}} B_k = \{x\}$.

Exercise 1.45 (Space-filling curve – sequel to Exercise 1.44). Let $I = [\,0, 1\,] \subset \mathbf{R}$ and $S = I \times I \subset \mathbf{R}^2$. In this exercise we describe Hilbert's construction of a mapping $\Gamma : I \to S$ which is continuous but nowhere differentiable and satisfies $\Gamma(I) = S$.

Let $i \in F = \{0, 1, 2, 3\}$ and partition I into four congruent subintervals I_i with left endpoints $t_i = \frac{i}{4}$. Define the affine bijections

$$\alpha_i : I_i \to I \qquad \text{by} \qquad \alpha_i(t) = 4(t - t_i) = 4t - i \qquad (i \in F).$$

Let (e_1, e_2) be the standard basis for $\mathbf{R}^2$ and partition S into four congruent subsquares S_i with lower left vertices $b_0 = 0$, $b_1 = \frac{1}{2}e_2$, $b_2 = \frac{1}{2}(e_1 + e_2)$, and $b_3 = \frac{1}{2}e_1$, respectively. Let $T_i : \mathbf{R}^2 \to \mathbf{R}^2$ be the translation of $\mathbf{R}^2$ sending 0 to b_i. Let R_0 be the reflection of $\mathbf{R}^2$ in the line $\{\, x \in \mathbf{R}^2 \mid x_2 = x_1 \,\}$, let $R_1 = R_2$ be the identity in $\mathbf{R}^2$, and R_3 the reflection of $\mathbf{R}^2$ in the line $\{\, x \in \mathbf{R}^2 \mid x_2 = 1 - x_1 \,\}$. Now introduce the affine bijections

$$\beta_i : S \to S_i \qquad \text{by} \qquad \beta_i = T_i \circ R_i \circ \frac{1}{2} \qquad (i \in F).$$

Next, suppose

$$(\star) \qquad \gamma : I \to S \quad \text{is a continuous mapping with} \quad \gamma(0) = 0 \quad \text{and} \quad \gamma(1) = e_1.$$

Then define $\Phi\gamma : I \to S$ by

$$\Phi\gamma|_{I_i} = \beta_i \circ \gamma \circ \alpha_i : I_i \to S_i \qquad (i \in F).$$

(i) Verify (most readers probably prefer to draw a picture at this stage)

$$\Phi\gamma(t) = \begin{cases} \frac{1}{2}(\gamma_2(4t), \gamma_1(4t)), & 0 \le t \le \frac{1}{4}; \\ \frac{1}{2}(\gamma_1(4t-1), 1+\gamma_2(4t-1)), & \frac{1}{4} \le t \le \frac{2}{4}; \\ \frac{1}{2}(1+\gamma_1(4t-2), 1+\gamma_2(4t-2)), & \frac{2}{4} \le t \le \frac{3}{4}; \\ \frac{1}{2}(2-\gamma_2(4t-3), 1-\gamma_1(4t-3)), & \frac{3}{4} \le t \le 1. \end{cases}$$

Prove that $\Phi\gamma : I \to S$ satisfies ($\star$).

Furthermore, define $\Phi^k\gamma : I \to S$ by mathematical induction over $k \in \mathbf{N}$ as $\Phi(\Phi^{k-1}\gamma)$, where $\Phi^0\gamma = \gamma$.

(ii) Deduce from (i) that $\Phi^k\gamma : I \to S$ satisfies ($\star$), for every $k \in \mathbf{N}$.

Given $(i_1, \dots, i_k) \in F^k$, for $k \in \mathbf{N}$, define the finite 4-adic expansion $t_{i_1 \cdots i_k} \in I$ and the corresponding subinterval $I_{i_1 \cdots i_k} \subset I$ by, respectively,

$$t_{i_1 \cdots i_k} = \sum_{1 \le j \le k} i_j 4^{-j}, \qquad I_{i_1 \cdots i_k} = [\, t_{i_1 \cdots i_k},\; t_{i_1 \cdots i_k} + 4^{-k}\,].$$

(iii) By mathematical induction over $k \in \mathbf{N}$ prove that we have, for $t \in I_{i_1 \cdots i_k}$,

$$t \in I_{i_1}, \quad \alpha_{i_1}(t) \in I_{i_2}, \quad \dots \quad , \alpha_{i_{j-1}} \circ \cdots \circ \alpha_{i_1}(t) \in I_{i_j} \quad (1 \le j \le k+1),$$

where $I_{i_{k+1}} = I$. Furthermore, show that $\alpha_{i_k} \circ \cdots \circ \alpha_{i_1} : I_{i_1 \cdots i_k} \to I$ is an affine bijection.

Similarly, define the corresponding subsquare $S_{i_1 \cdots i_k} \subset S$ by $S_{i_1 \cdots i_k} = \beta_{i_1} \circ \cdots \circ \beta_{i_k}(S)$.

(iv) Show that $S_{i_1 \cdots i_k}$ is a square with sides of length equal to 2^{-k}. Using (iii) verify

$$\Phi^k\gamma|_{I_{i_1 \cdots i_k}} = \beta_{i_1} \circ \cdots \circ \beta_{i_k} \circ \gamma \circ \alpha_{i_k} \circ \cdots \circ \alpha_{i_1}.$$

Deduce $\Phi^k\gamma(I_{i_1 \cdots i_k}) \subset S_{i_1 \cdots i_k}$.

(v) Let $t \in I$. Prove that there exists a sequence $(i_j)_{j \in \mathbf{N}}$ in F such that we have the infinite 4-adic expansion

$$t = \sum_{j \in \mathbf{N}} i_j 4^{-j}.$$

(Note that, in general, the i_j are not uniquely determined by t.) Write $S^{(k)} = S_{i_1 \cdots i_k}$. Show that $S^{(k+1)} \subset S^{(k)}$. Use part (iv) and Exercise 1.44.(i) to verify that there exists a unique $x \in S$ satisfying $x \in S^{(k)}$, for every $k \in \mathbf{N}$.

Now define $\Gamma : I \to S$ by $\Gamma(t) = x$.

(vi) Conclude from (iv) and (v) that

$$\|\Phi^k(\gamma)(t) - \Gamma(t)\| \leq \sqrt{2}\, 2^{-k} \qquad (k \in \mathbf{N}).$$

Prove that $(\Phi^k(\gamma)(t))_{k\in\mathbf{N}}$ converges to $\Gamma(t)$, for all $t \in I$. Show that the curves $\Phi^k(\gamma)$ converge to Γ uniformly on I, for $k \to \infty$. Conclude that $\Gamma : I \to S$ is continuous.

Note that the construction in part (v) makes the definition of $\Gamma(t)$ independent of the choice of the initial curve γ, whereas part (vi) shows that this definition does not depend on the particular representation of t as a 4-adic expansion.

(vii) Prove by induction on k that $S_{i_1\cdots i_k}$ and $S_{i'_1\cdots i'_k}$ do not overlap, if $(i_1, \dots, i_k) \neq (i'_1, \dots, i'_k)$ (first show this assuming $i_1 \neq i'_1$, and use the induction hypothesis when $i_1 = i'_1$). Verify that the $S_{i_1\cdots i_k}$, for all admissible $(i_1, \cdots, i_k)$, yield the subdivision of S into the subsquares belonging to $\mathcal{B}_k$ as in Exercise 1.44. Now show that for every $x \in S$ there exists $t \in I$ such that $x = \Gamma(t)$, i.e. Γ is surjective.

(viii) Show that $\Gamma^{-1}(\{b_2\})$ consists at least of two elements (compare with Exercise 1.53).

Finally we derive rather precise information on the variation of Γ, which enables us to demonstrate that Γ is nowhere differentiable.

(ix) Given distinct t and $t' \in I$ we can find $k \in \mathbf{N}_0$ satisfying $4^{-k-1} < |t - t'| \leq 4^{-k}$. Deduce from the second inequality that t and t' both belong to the union of at most two consecutive intervals of the type $I_{i_1\cdots i_k}$. The two corresponding squares $S_{i_1\cdots i_k}$ are connected by means of the continuous curve $\Phi^k(\gamma)$. These subsquares are actually adjacent because the mappings β_i, up to translations, leave the diagonals of the subsquares invariant as sets, while the entrance and exit points of $\Phi^k(\gamma)$ in a subsquare never belong to one diagonal. Derive

$$\|\Gamma(t) - \Gamma(t')\| \leq \sqrt{5}\, 2^{-k} < 2\sqrt{5}\, |t - t'|^{\frac{1}{2}} \qquad (t,\ t' \in I).$$

One says that Γ is *uniformly Hölder continuous* of exponent $\frac{1}{2}$.

(x) For each $(i_1, \dots, i_k)$ we have that $\Gamma(I_{i_1\cdots i_k}) = S_{i_1\cdots i_k}$. In particular, there are $u_\pm \in I_{i_1\cdots i_k}$ such that $\Gamma(u_-)$ and $\Gamma(u_+)$ are diametrically opposite points in $S_{i_1\cdots i_k}$, which implies that the coordinate functions of Γ satisfy $|\Gamma_j(u_-) - \Gamma_j(u_+)| = 2^{-k}$, for $1 \leq j \leq 2$. Let $t \in I_{i_1\cdots i_k}$ be arbitrary. Verify for one of the $u_\pm$, say u_+, that

$$|\Gamma_j(t) - \Gamma_j(u_+)| \geq \frac{1}{2} 2^{-k} \geq \frac{1}{2} |t - u_+|^{\frac{1}{2}}.$$

Note that the first inequality implies that $t \neq u_+$. Deduce that for every $t \in I$ and every $\delta > 0$ there exists $t' \in I$ such that $0 < |t - t'| < \delta$ and $|\Gamma_j(t) - \Gamma_j(t')| \geq \frac{1}{2}|t - t'|^{\frac{1}{2}}$. This shows that Γ_j is no better than Hölder continuous of exponent $\frac{1}{2}$ at every point of I. In particular, prove that both coordinate functions of Γ are nowhere differentiable.

(xi) Show that the Hölder exponent $\frac{1}{2}$ in (ix) is optimal in the following sense. Suppose that there exist a mapping $\gamma : I \to S$ and positive constants c and α such that

$$\|\gamma(t) - \gamma(t')\| \leq c|t - t'|^\alpha \qquad (t,\, t' \in I).$$

By subdividing I into n consecutive intervals of length $\frac{1}{n}$, verify that $\gamma(I)$ is contained in the union F_n of n disks of radius $c(\frac{2}{n})^\alpha$. If $\alpha > \frac{1}{2}$, then the total area of F_n converges to zero as $n \to \infty$, which implies that $\gamma(I)$ has no interior points.

Background. Peano's discovery of continuous *space-filling curves* like Γ came as a great shock in 1890.

Exercise 1.46 (Polynomial approximation of absolute value – needed for Exercise 1.55). Suppose we want to approximate the absolute value function from below by nonnegative polynomial functions p_k on the interval $I = [\,-1, 1\,] \subset \mathbf{R}$. An algorithm that successively reduces the error is

$$|t| - p_{k+1}(t) = (|t| - p_k(t))(1 - \frac{1}{2}(|t| + p_k(t))).$$

Accordingly we define the sequence $(p_k(t))_{k\in\mathbf{N}}$ inductively by

$$p_1(t) = 0, \qquad p_{k+1}(t) = \frac{1}{2}(t^2 + 2p_k(t) - p_k(t)^2) \qquad (t \in I).$$

(i) By mathematical induction on $k \in \mathbf{N}$ prove that $p_k : t \mapsto p_k(t)$ is a polynomial function in t^2 with zero constant term.

(ii) Verify, for $t \in I$ and $k \in \mathbf{N}$,

$$2(|t| - p_{k+1}(t)) = |t|(2 - |t|) - p_k(t)(2 - p_k(t)),$$
$$2(p_{k+1}(t) - p_k(t)) = t^2 - p_k(t)^2.$$

(iii) Show that $x \mapsto x(2 - x)$ is monotonically increasing on $[\,0, 1\,]$. Deduce that $0 \leq p_k(t) \leq |t|$, for $k \in \mathbf{N}$ and $t \in I$, and that $(p_k(t))_{k\in\mathbf{N}}$ is monotonically increasing, with $\lim p_k(t) = |t|$, for $t \in I$.

(iv) Use Dini's Theorem to prove that the sequence of polynomial functions $(p_k)_{k\in\mathbf{N}}$ converges uniformly on I to the absolute value function $t \mapsto |t|$.

Let $K \subset \mathbf{R}^n$ be compact and $f : K \to \mathbf{R}$ be a nonzero continuous function. Then $0 < \|f\| := \sup\{\,|f(x)| \mid x \in K\,\} < \infty$.

(v) Show that the function $|f|$ with $x \mapsto |f(x)|$ is the limit of a sequence of polynomials in f that converges uniformly on K.
Hint: Note that $\frac{f(x)}{\|f\|} \in I$, for all $x \in K$. Next substitute $\frac{f(x)}{\|f\|}$ for the variable t in the polynomials $p_k(t)$.

Background. The uniform convergence of $(p_k)_{k\in\mathbf{N}}$ on I can be proved without appeal to Dini's Theorem as in (iv). In fact, it can be shown that

$$0 \le |t| - p_k(t) \le |t|(1 - \frac{1}{2}|t|)^{k-1} \le \frac{2}{k} \qquad (t \in I).$$

Exercise 1.47 (Polynomial approximation of absolute value – needed for Exercise 1.55). Another idea for approximating the absolute value function by polynomial functions (see Exercise 1.46) is to note that $|t| = \sqrt{t^2}$ and to use Taylor expansion. Unfortunately, $\sqrt{\cdot}$ is not differentiable at 0. The following argument bypasses this difficulty.

Let $I = [0, 1]$ and $J = [-1, 1]$ and let $\epsilon > 0$ be arbitrary but fixed. Consider $f : I \to \mathbf{R}$ given by $f(t) = \sqrt{\epsilon^2 + t}$. Show that the Taylor series for f about $\frac{1}{2}$ converges uniformly on I. Deduce the existence of a polynomial function p such that $|p(t^2) - \sqrt{\epsilon^2 + t^2}| < \epsilon$, for $t \in J$. Prove $\sqrt{\epsilon^2 + t^2} - |t| \le \epsilon$ and deduce $|p(t^2) - |t|| < 2\epsilon$, for all $t \in J$.

Exercise 1.48 (Connectedness of graph). Consider a mapping $f : \mathbf{R}^n \to \mathbf{R}^p$.

(i) Prove that graph(f) is connected in $\mathbf{R}^{n+p}$ if f is continuous.

(ii) Show that $F \subset \mathbf{R}^2$ is connected if F is defined as in Exercise 1.14.(ii).

(iii) Is f continuous if graph(f) is connected in $\mathbf{R}^{n+p}$?

Exercise 1.49 (Addition to Exercise 1.27). In the notation of that exercise, show that $f : \mathbf{R}^n \to \mathbf{R}^n$ is a homeomorphism if $\mathrm{im}(f)$ is open in $\mathbf{R}^n$.

Exercise 1.50. Let $I = [\,0, 1\,]$ and let $f : I \to I$ be continuous. Prove there exists $x \in I$ with $f(x) = x$.

Exercise 1.51. Prove the following assertions. In $\mathbf{R}$ each closed interval is homeomorphic to $[\,-1, 1\,]$, each open interval to $]\,-1, 1\,[$, and each half-open interval to $]\,-1, 1\,]$. Furthermore, no two of these three intervals are homeomorphic.
Hint: We can remove 2, 0, and 1 points, respectively, without destroying the connectedness.

Exercise 1.52. Show that $S^1 = \{ x \in \mathbf{R}^2 \mid \|x\| = 1 \| \}$ is not homeomorphic to any interval in $\mathbf{R}$.
Hint: Use that S^1 with an arbitrary point omitted is still connected.

Exercise 1.53 (Injective space-filling curves do not exist). Let $I = [\,0, 1\,]$ and let $f : I \to I^2$ be a continuous surjection (like Γ constructed in Exercise 1.45). Suppose that f is injective. Using the compactness of I deduce that f is a homeomorphism, and by omitting one point from I and I^2, respectively, show that we have arrived at a contradiction.

Exercise 1.54 (Approximation by piecewise affine function – needed for Exercise 1.55). Let $I = [\,a, b\,] \subset \mathbf{R}$ and $g : I \to \mathbf{R}$. We say that g is an *affine function* on I if there exist c and $\lambda \in \mathbf{R}$ with $g(t) = c + \lambda\, t$, for all $t \in I$. And g is said to be *piecewise affine* if there exists a finite sequence $(t_k)_{0 \le k \le l}$ with $a = t_0 < \cdots < t_l = b$ such that the restriction of g to $\left]\, t_{k-1}, t_k \,\right[$ is affine, for $1 \le k \le l$. If g is continuous and piecewise affine, then the restriction of g to each $[\, t_{k-1}, t_k\,]$ is affine.

Let $f : I \to \mathbf{R}$ be continuous. Show that for every $\epsilon > 0$ there exists a continuous piecewise affine function g such that $|\, f(t) - g(t)| < \epsilon$, for all $t \in I$.
Hint: In view of the uniform continuity of f we find $\delta > 0$ with $|\, f(t) - f(t')| < \epsilon$ if $|t - t'| < \delta$. Choose $l > \frac{b-a}{\delta}$, and set $t_k = a + k\frac{b-a}{l}$. Then define g to be the continuous piecewise affine function satisfying $g(t_k) = f(t_k)$, for $0 \le k \le l$. Next, for any $t \in I$ there exists k with $t \in [\, t_{k-1}, t_k\,]$, hence $g(t)$ lies between $f(t_{k-1})$ and $f(t_k)$. Finally, use the Intermediate Value Theorem 1.9.5.

Exercise 1.55 (Weierstrass' Approximation Theorem on R – sequel to Exercises 1.46 and 1.54). Let $I = [\,a, b\,] \subset \mathbf{R}$ and let $f : I \to \mathbf{R}$ be a continuous function. Then there exists a sequence $(p_k)_{k \in \mathbf{N}}$ of polynomial functions that converges to f uniformly on I, that is, for every $\epsilon > 0$ there exists $N \in \mathbf{N}$ such that $|\, f(x) - p_k(x)| < \epsilon$, for every $x \in I$ and $k \ge N$. (See Exercise 6.103 for a generalization to $\mathbf{R}^n$.) We shall prove this result in the following three steps.

(i) Apply Exercise 1.54 to approximate f uniformly on I by a continuous and piecewise affine function g.

Suppose that $a = t_0 < \cdots < t_l = b$ and that $g(t) = g(t_{k-1}) + \lambda_k(t - t_{k-1})$, for $1 \le k \le l$ and $t \in [\, t_{k-1}, t_k\,]$.

(ii) Set $\lambda_0 = 0$ and $x^+ = \frac{1}{2}(x + |x|)$, for all $x \in \mathbf{R}$. Using mathematical induction over the index k show that

$$g(t) = g(a) + \sum_{1 \le k \le l} (\lambda_k - \lambda_{k-1})(t - t_{k-1})^+ \qquad (t \in I).$$

(iii) Deduce Weierstrass' Approximation Theorem from parts (i) and (ii) and Exercise 1.46.(v).

Exercises for Chapter 2

Exercise 2.1 (Trace and inner product on $\mathrm{Mat}(n, \mathbf{R})$ **– needed for Exercise 2.44).** We study some properties of the operation of taking the trace, see Section 2.1; this will provide some background for the Euclidean norm. All the results obtained below apply, mutatis mutandis, to $\mathrm{End}(\mathbf{R}^n)$ as well.

(i) Verify that $\mathrm{tr} : \mathrm{Mat}(n, \mathbf{R}) \to \mathbf{R}$ is a linear mapping, and that, for $A, B \in \mathrm{Mat}(n, \mathbf{R})$,

$$\mathrm{tr}\, A = \mathrm{tr}\, A^t, \qquad \mathrm{tr}\, AB = \mathrm{tr}\, BA; \qquad \mathrm{tr}\, BAB^{-1} = \mathrm{tr}\, A$$

if $B \in \mathbf{GL}(n, \mathbf{R})$. Deduce that the mapping tr vanishes on *commutators* in $\mathrm{Mat}(n, \mathbf{R})$, that is (compare with Exercise 2.41), for A and $B \in \mathrm{Mat}(n, \mathbf{R})$,

$$\mathrm{tr}([\,A, B\,]) = 0, \qquad \text{where} \qquad [\,A, B\,] = AB - BA.$$

Define a bilinear functional $\langle \cdot, \cdot \rangle : \mathrm{Mat}(n, \mathbf{R}) \times \mathrm{Mat}(n, \mathbf{R}) \to \mathbf{R}$ by $\langle A, B \rangle = \mathrm{tr}(A^t B)$ (see Definition 2.7.3).

(ii) Show that $\langle A, B \rangle = \mathrm{tr}(B^t A) = \mathrm{tr}(AB^t) = \mathrm{tr}(BA^t)$.

(iii) Let a_j be the j-th column vector of $A \in \mathrm{Mat}(n, \mathbf{R})$, thus $a_j = Ae_j$. Verify, for $A, B \in \mathrm{Mat}(n, \mathbf{R})$

$$A^t B = (\langle a_i, b_j \rangle)_{1 \le i,j \le n}, \qquad \langle A, B \rangle = \sum_{1 \le j \le n} \langle a_j, b_j \rangle.$$

(iv) Prove that $\langle \cdot, \cdot \rangle$ defines an inner product on $\mathrm{Mat}(n, \mathbf{R})$, and that the corresponding norm equals the Euclidean norm on $\mathrm{Mat}(n, \mathbf{R})$.

(v) Using part (ii), show that the decomposition from Lemma 2.1.4 is an orthogonal direct sum, in other words, $\langle A, B \rangle = 0$ if $A \in \mathrm{End}^+(\mathbf{R}^n)$ and $B \in \mathrm{End}^-(\mathbf{R}^n)$. Verify that $\dim \mathrm{End}^\pm(\mathbf{R}^n) = \frac{1}{2}n(n \pm 1)$.

Define the bilinear functional $B : \mathrm{Mat}(n, \mathbf{R}) \times \mathrm{Mat}(n, \mathbf{R}) \to \mathbf{R}$ by $B(A, A') = \mathrm{tr}(AA')$.

(vi) Prove that B satisfies $B(A_\pm, A_\pm) \gtrless 0$, for $0 \ne A_\pm \in \mathrm{End}^\pm(\mathbf{R}^n)$.

Finally, we show that the trace is completely determined by some of the properties listed in part (i).

(vii) Let $\tau : \mathrm{Mat}(n, \mathbf{R}) \to \mathbf{R}$ be a linear mapping satisfying $\tau(AB) = \tau(BA)$, for all $A, B \in \mathrm{Mat}(n, \mathbf{R})$. Prove

$$\tau = \frac{1}{n}\tau(I)\, \mathrm{tr}\,.$$

Hint: Note that τ vanishes on commutators. Let $E_{ij} \in \mathrm{Mat}(n, \mathbf{R})$ be given by having a single 1 at position (i, j) and zeros elsewhere. Then the statement is immediate from $[\,E_{ij}, E_{jj}\,] = E_{ij}$ for $i \ne j$, and $[\,E_{ij}, E_{ji}\,] = E_{ii} - E_{jj}$.

Exercise 2.2 (Another proof of Lemma 2.1.2). Let $A \in \mathrm{Aut}(\mathbf{R}^n)$ and $B \in \mathrm{End}(\mathbf{R}^n)$. Let $\|\cdot\|$ denote the Euclidean or operator norm on $\mathrm{End}(\mathbf{R}^n)$.

(i) Prove $AB \in \mathrm{End}(\mathbf{R}^n)$ and $\|AB\| \le \|A\|\,\|B\|$, for $A, B \in \mathrm{End}(\mathbf{R}^n)$.

(ii) Deduce from (i), for all $x \in \mathbf{R}^n$,

$$\begin{aligned}\|Bx\| &\ge \|Ax\| - \|(A-B)x\| \ge \|Ax\| - \|A-B\|\,\|A^{-1}\|\,\|Ax\| \\ &= (1 - \|A-B\|\,\|A^{-1}\|)\|Ax\|.\end{aligned}$$

(iii) Conclude from (ii) that $B \in \mathrm{Aut}(\mathbf{R}^n)$ if $\|A - B\| < \frac{1}{\|A^{-1}\|}$, and prove the assertion of Lemma 2.1.2.

Exercise 2.3 (Another proof of Lemma 2.1.2 – needed for Exercise 2.46). Let $\|\cdot\|$ be the Euclidean norm on $\mathrm{End}(\mathbf{R}^n)$.

(i) Prove $AB \in \mathrm{End}(\mathbf{R}^n)$ and $\|AB\| \le \|A\|\,\|B\|$, for $A, B \in \mathrm{End}(\mathbf{R}^n)$.

Next, let $A \in \mathrm{End}(\mathbf{R}^n)$ with $\|A\| < 1$.

(ii) Show that $\left(\sum_{0\le k\le l} A^k\right)_{l\in\mathbf{N}}$ is a Cauchy sequence in $\mathrm{End}(\mathbf{R}^n)$, and conclude that the following element is well-defined in the complete space $\mathrm{End}(\mathbf{R}^n)$, see Theorem 1.6.5:

$$\sum_{k\in\mathbf{N}_0} A^k := \lim_{l\to\infty} \sum_{0\le k\le l} A^k \in \mathrm{End}(\mathbf{R}^n).$$

(iii) Prove that $(I-A)\sum_{0\le k\le l} A^k = I - A^{l+1}$, that $I - A$ is invertible in $\mathrm{End}(\mathbf{R}^n)$, and that

$$(I-A)^{-1} = \sum_{k\in\mathbf{N}_0} A^k.$$

(iv) Let $L_0 \in \mathrm{Aut}(\mathbf{R}^n)$, and assume $L \in \mathrm{End}(\mathbf{R}^n)$ satisfies $\epsilon := \|I - L_0^{-1}L\| < 1$. Deduce $L \in \mathrm{Aut}(\mathbf{R}^n)$. Set $A = I - L_0^{-1}L$ and show

$$L^{-1} - L_0^{-1} = ((I-A)^{-1} - I)L_0^{-1} = \Big(\sum_{k\in\mathbf{N}} A^k\Big)L_0^{-1},$$

$$\text{so} \qquad \|L^{-1} - L_0^{-1}\| \le \frac{\epsilon}{1-\epsilon}\|L_0^{-1}\|.$$

Exercise 2.4 (Orthogonal transformation and $\mathbf{O}(n, \mathbf{R})$ – needed for Exercises 2.5 and 2.39). Suppose that $A \in \mathrm{End}(\mathbf{R}^n)$ is an *orthogonal* transformation, that is, $\|Ax\| = \|x\|$, for all $x \in \mathbf{R}^n$.

(i) Show that $A \in \mathrm{Aut}(\mathbf{R}^n)$ and that A^{-1} is orthogonal.

(ii) Deduce from the polarization identity in Lemma 1.1.5.(iii)

$$\langle A^t A x, y \rangle = \langle Ax, Ay \rangle = \langle x, y \rangle \qquad (x, y \in \mathbf{R}^n).$$

(iii) Prove $A^t A = I$ and deduce that $\det A = \pm 1$. Furthermore, using (i) show that A^t is orthogonal, thus $AA^t = I$; and also obtain $\langle Ae_i, Ae_j \rangle = \langle A^t e_i, A^t e_j \rangle = \langle e_i, e_j \rangle = \delta_{ij}$, where $(e_1, \dots, e_n)$ is the standard basis for $\mathbf{R}^n$. Conclude that the column and row vectors, respectively, in the corresponding matrix of A form an orthonormal basis for $\mathbf{R}^n$.

(iv) Deduce from (iii) that the corresponding matrix $A = (a_{ij}) \in \mathbf{GL}(n, \mathbf{R})$ is *orthogonal* with coefficients in $\mathbf{R}$ and belongs to the *orthogonal group* $\mathbf{O}(n, \mathbf{R})$, which means that it satisfies $A^t A = I$; in addition, deduce

$$\sum_{1 \le k \le n} a_{ki} a_{kj} = \sum_{1 \le k \le n} a_{ik} a_{jk} = \delta_{ij} \qquad (1 \le i, j \le n).$$

Exercise 2.5 (Euler's Theorem – sequel to Exercise 2.4 – needed for Exercises 4.22 and 5.65). Let $\mathbf{SO}(3, \mathbf{R})$ be the *special orthogonal group* in $\mathbf{R}^3$, consisting of the orthogonal matrices in $\mathrm{Mat}(3, \mathbf{R})$ with determinant equal to 1. One then has, for every $R \in \mathbf{SO}(3, \mathbf{R})$, see Exercise 2.4,

$$R \in \mathbf{GL}(3, \mathbf{R}), \qquad R^t R = I, \qquad \det R = 1.$$

Prove that for every $R \in \mathbf{SO}(3, \mathbf{R})$ there exist $\alpha \in \mathbf{R}$ with $0 \le \alpha \le \pi$ and $a \in \mathbf{R}^3$ with $\|a\| = 1$ with the following properties: R fixes a and maps the linear subspace N_a orthogonal to a into itself; in N_a the action of R is that of rotation by the angle α such that, for $0 < \alpha < \pi$ and for all $y \in N_a$, one has $\det(a\ y\ Ry) > 0$.

We write $R = R_{\alpha,a}$, which is referred to as the counterclockwise *rotation* in $\mathbf{R}^3$ by the *angle* α about the *axis of rotation* $\mathbf{R}a$. (For the exact definition of the concept of angle, see Example 7.4.1.)

Hint: From $R^t R = I$ one derives $(R^t - I)R = I - R$. Hence

$$\det(R - I) = \det(R^t - I) = \det(I - R) = -\det(R - I).$$

It follows that R has an eigenvalue 1; let $a \in \mathbf{R}^3$ be a corresponding eigenvector with $\|a\| = 1$. Now choose $b \in N_a$ with $\|b\| = 1$. Further define $c = a \times b \in \mathbf{R}^3$ (for the definition of the cross product $a \times b$ of a and b, see the Remark on linear algebra in Section 5.3). One then has $c \in N_a$, $c \perp b$ and $\|c\| = 1$, while (b, c) is a basis for the linear subspace N_a. Replace a by $-a$ if $\langle a \times b, Rb \rangle < 0$; thus we may assume $\langle c, Rb \rangle \ge 0$. Check that $Rb \in N_a$, and use $\|Rb\| = 1$ to conclude that $0 \le \alpha \le \pi$ exists with $Rb = (\cos\alpha)\, b + (\sin\alpha)\, c$. Now use $Rc \in N_a$, $\|Rc\| = 1$,

$Rc \perp Rb$ and $\det R = 1$, to arrive at $Rc = -(\sin\alpha)\, b + (\cos\alpha)\, c$. With respect to the basis (a, b, c) for $\mathbf{R}^3$ one has

$$R = \begin{pmatrix} 1 & 0 & 0 \\ 0 & \cos\alpha & -\sin\alpha \\ 0 & \sin\alpha & \cos\alpha \end{pmatrix}.$$

Exercise 2.6 (Approximating a zero – needed for Exercises 2.47 and 3.28). Assume a differentiable function $f : \mathbf{R} \to \mathbf{R}$ has a zero, so $f(x) = 0$ for some $x \in \mathbf{R}$. Suppose $x_0 \in \mathbf{R}$ is a first approximation to this zero and consider the tangent line to the graph $\{\, (x, f(x)) \mid x \in \operatorname{dom}(f) \,\}$ of f at $(x_0, f(x_0))$; this is the set

$$\{\, (x, y) \in \mathbf{R}^2 \mid y - f(x_0) = f'(x_0)(x - x_0) \,\}.$$

Then determine the intercept x_1 of that line with the x-axis, in other words $x_1 \in \mathbf{R}$ for which $-f(x_0) = f'(x_0)(x_1 - x_0)$, or, under the assumption that $A^{-1} := f'(x_0) \neq 0$,

$$x_1 = F(x_0) \qquad \text{where} \qquad F(x) = x - Af(x).$$

It seems plausible that x_1 is nearer the required zero x than x_0, and that iteration of this procedure will get us nearer still. In other words, we hope that the sequence $(x_k)_{k \in \mathbf{N}_0}$ with $x_{k+1} := F(x_k)$ converges to the required zero x, for which then $F(x) = x$. Now we formalize this heuristic argument.

Let $x_0 \in \mathbf{R}^n$, $\delta > 0$, and let $V = V(x_0; \delta)$ be the closed ball in $\mathbf{R}^n$ of center x_0 and radius δ; furthermore, let $f : V \to \mathbf{R}^n$. Assume $A \in \operatorname{Aut}(\mathbf{R}^n)$ and a number ϵ with $0 \leq \epsilon < 1$ to exist such that:

(i) the mapping $F : V \to \mathbf{R}^n$ with $F(x) = x - A(f(x))$ is a contraction with contraction factor $\leq \epsilon$;

(ii) $\|Af(x_0)\| \leq (1 - \epsilon)\, \delta$.

Prove that there exists a unique $x \in V$ with

$$f(x) = 0; \qquad \text{and also} \qquad \|x - x_0\| \leq \frac{1}{1 - \epsilon} \|A\, f(x_0)\|.$$

Exercise 2.7. Calculate (without writing out into coordinates) the total derivatives of the mappings f defined below on $\mathbf{R}^n$; that is, describe, for every point $x \in \mathbf{R}^n$, the linear mapping $\mathbf{R}^n \to \mathbf{R}$, or $\mathbf{R}^n \to \mathbf{R}^n$, respectively, with $h \mapsto Df(x)h$. Assume one has $a, b \in \mathbf{R}^n$; successively define $f(x)$, for $x \in \mathbf{R}^n$, by

$$\langle a, x \rangle; \qquad \langle a, x \rangle a; \qquad \langle x, x \rangle; \qquad \langle a, x \rangle \langle b, x \rangle; \qquad \langle x, x \rangle x.$$

Exercise 2.8. Let $f : \mathbf{R}^n \to \mathbf{R}$ be differentiable.

(i) Prove that f is constant if $Df = 0$.
Hint: Recall that the result is known if $n = 1$, and use directional derivatives.

Let $f : \mathbf{R}^n \to \mathbf{R}^p$ be differentiable.

(ii) Prove that f is constant if $Df = 0$.

(iii) Let $L \in \mathrm{Lin}(\mathbf{R}^n, \mathbf{R}^p)$, and suppose $Df(x) = L$, for every $x \in \mathbf{R}^n$. Prove the existence of $c \in \mathbf{R}^p$ with $f(x) = Lx + c$, for every $x \in \mathbf{R}^n$.

Exercise 2.9. Let $f : \mathbf{R}^n \to \mathbf{R}$ be homogeneous of degree 1, in the sense that $f(tx) = t\,f(x)$ for all $x \in \mathbf{R}^n$ and $t \in \mathbf{R}$. Show that f has directional derivatives at 0 in all directions. Prove that f is differentiable at 0 if and only if f is linear, which is the case if and only if f is additive.

Exercise 2.10. Let g be as in Example 1.3.11. Then we define $f : \mathbf{R}^2 \to \mathbf{R}$ by

$$f(x) = x_2 g(x) = \begin{cases} \dfrac{x_1 x_2^3}{x_1^2 + x_2^4}, & x \neq 0; \\ 0, & x = 0. \end{cases}$$

Show that $D_v f(0) = 0$, for all $v \in \mathbf{R}^2$; and deduce that $Df(0) = 0$, if f were differentiable at 0. Prove that $D_v f(x)$ is well-defined, for all v and $x \in \mathbf{R}^2$, and that $v \mapsto D_v f(x)$ belongs to $\mathrm{Lin}(\mathbf{R}^2, \mathbf{R})$, for all $x \in \mathbf{R}^2$. Nevertheless, verify that f is not differentiable at 0 by showing

$$\lim_{x_2 \to 0} \frac{|f(x_2^2, x_2)|}{\|(x_2^2, x_2)\|} = \frac{1}{2}.$$

Exercise 2.11. Define $f : \mathbf{R}^2 \to \mathbf{R}$ by

$$f(x) = \begin{cases} \dfrac{x_1^3}{\|x\|^2}, & x \neq 0; \\ 0, & x = 0. \end{cases}$$

Prove that f is continuous on $\mathbf{R}^2$. Show that directional derivatives of f at 0 in all directions do exist. Verify, however, that f is not differentiable at 0. Compute

$$D_1 f(x) = 1 + x_2^2 \frac{x_1^2 - x_2^2}{\|x\|^4}, \qquad D_2 f(x) = -\frac{2x_1^3 x_2}{\|x\|^4} \qquad (x \in \mathbf{R}^2 \setminus \{0\}).$$

In particular, both partial derivatives of f are well-defined on all of $\mathbf{R}^2$. Deduce that at least one of these partial derivatives has to be discontinuous at 0. To see this explicitly, note that

$$D_1 f(tx) = D_1 f(x), \qquad D_2 f(tx) = D_2 f(x) \qquad (t \in \mathbf{R} \setminus \{0\},\ x \in \mathbf{R}^2).$$

This means that the partial derivatives assume constant values along every line through the origin under omission of the origin. More precisely, in every neighborhood of 0 the function $D_1 f$ assumes every value in $[\,0, \frac{9}{8}\,]$ while $D_2 f$ assumes every value in $[\,-\frac{3}{8}\sqrt{3}, \frac{3}{8}\sqrt{3}\,]$. Accordingly, in this case both partial derivatives are discontinuous at 0.

Exercise 2.12 (Stronger version of Theorem 2.3.4). Let the notation be as in that theorem, but with $n \geq 2$. Show that the conclusion of the theorem remains valid under the weaker assumption that $n-1$ partial derivatives of f exist in a neighborhood of a and are continuous at a while the remaining partial derivative merely exists at a.
Hint: Write

$$f(a+h) - f(a) = \sum_{n \geq j \geq 2} (f(a+h^{(j)}) - f(a+h^{(j-1)})) + f(a+h^{(1)}) - f(a).$$

Apply the method of the theorem to the sum over j and the definition of derivative to the remaining difference.
Background. As a consequence of this result we see that at least two different partial derivatives of f must be discontinuous at a if f fails to be differentiable at a, compare with Example 2.3.5.

Exercise 2.13. Let $f_j : \mathbf{R} \to \mathbf{R}$ be differentiable, for $1 \leq j \leq n$, and define $f : \mathbf{R}^n \to \mathbf{R}$ by $f(x) = \sum_{1 \leq j \leq n} f_j(x_j)$. Show that f is differentiable with derivative

$$Df(a) = (f_1'(a_1) \ \cdots \ f_n'(a_n)) \in \mathrm{Lin}(\mathbf{R}^n, \mathbf{R}) \qquad (a \in \mathbf{R}^n).$$

Background. The f_i might have discontinuous derivatives, which implies that the partial derivatives of f are not necessarily continuous. Accordingly, continuity of the partial derivatives is not a necessary condition for the differentiability of f, compare with Theorem 2.3.4.

Exercise 2.14. Show that the mappings ν_1 and $\nu_2 : \mathbf{R}^n \to \mathbf{R}$ given by

$$\nu_1(x) = \sum_{1 \leq j \leq n} |x_j|, \qquad \nu_2(x) = \sup_{1 \leq j \leq n} |x_j|,$$

respectively, are norms on $\mathbf{R}^n$. Determine the set of points where ν_i is differentiable, for $1 \leq i \leq 2$.

Exercise 2.15. Let $f : \mathbf{R}^n \to \mathbf{R}$ be a C^1 function and $k > 0$, and assume that $|D_j f(x)| \leq k$, for $1 \leq j \leq n$ and $x \in \mathbf{R}^n$.

(i) Prove that f is Lipschitz continuous with $\sqrt{n}\,k$ as a Lipschitz constant.

Next, let $g : \mathbf{R}^n \setminus \{0\} \to \mathbf{R}$ be a C^1 function and $k > 0$, and assume that $|D_j g(x)| \le k$, for $1 \le j \le n$ and $x \in \mathbf{R}^n \setminus \{0\}$.

(ii) Show that if $n \ge 2$, then g can be extended to a continuous function defined on all of $\mathbf{R}^n$. Show that this is false if $n = 1$ by giving a counterexample.

Exercise 2.16. Define $f : \mathbf{R}^2 \to \mathbf{R}$ by $f(x) = \log(1 + \|x\|^2)$. Compute the partial derivatives $D_1 f$, $D_2 f$, and all second-order partial derivatives $D_1^2 f$, $D_1 D_2 f$, $D_2 D_1 f$ and $D_2^2 f$.

Exercise 2.17. Let g and $h : \mathbf{R} \to \mathbf{R}$ be twice differentiable functions. Define $U = \{ x \in \mathbf{R}^2 \mid x_1 \ne 0 \}$ and $f : U \to \mathbf{R}$ by $f(x) = x_1 g(\frac{x_2}{x_1}) + h(\frac{x_2}{x_1})$. Prove

$$x_1^2 D_1^2 f(x) + 2x_1 x_2 D_1 D_2 f(x) + x_2^2 D_2^2 f(x) = 0 \qquad (x \in U).$$

Exercise 2.18 (Independence of variable). Let $U = \mathbf{R}^2 \setminus \{ (0, x_2) \mid x_2 \ge 0 \}$ and define $f : U \to \mathbf{R}$ by

$$f(x) = \begin{cases} x_2^2, & x_1 > 0 \quad \text{and} \quad x_2 \ge 0; \\ 0, & x_1 < 0 \quad \text{or} \quad x_2 < 0. \end{cases}$$

(i) Show that $D_1 f = 0$ on all of U but that f is not independent of x_1.

Now suppose that $U \subset \mathbf{R}^2$ is an open set having the property that for each $x_2 \in \mathbf{R}$ the set $\{ x_1 \in \mathbf{R} \mid (x_1, x_2) \in U \}$ is an interval.

(ii) Prove that $D_1 f = 0$ on all of U implies that f is independent of x_1.

Exercise 2.19. Let $h(t) = (\cos t)^{\sin t}$ with $\mathrm{dom}(h) = \{ t \in \mathbf{R} \mid \cos t > 0 \}$. Prove $h'(t) = (\cos t)^{-1+\sin t}(\cos^2 t \log\cos t - \sin^2 t)$ in two different ways: by using the chain rule for $\mathbf{R}$; and by writing

$$h = g \circ f \qquad \text{with} \qquad f(t) = (\cos t, \sin t) \qquad \text{and} \qquad g(x) = x_1^{x_2}.$$

Exercise 2.20. Define $f : \mathbf{R}^2 \to \mathbf{R}^2$ and $g : \mathbf{R}^2 \to \mathbf{R}^2$ by

$$f(x) = (x_1^2 + x_2^2, \sin x_1 x_2), \qquad g(x) = (x_1 x_2, e^{x_2}).$$

Then $D(g \circ f)(x) : \mathbf{R}^2 \to \mathbf{R}^2$ has the matrix

$$\begin{pmatrix} 2x_1 \sin x_1 x_2 + x_2(x_1^2 + x_2^2)\cos x_1 x_2 & 2x_2 \sin x_1 x_2 + x_1(x_1^2 + x_2^2)\cos x_1 x_2 \\ x_2 e^{\sin x_1 x_2} \cos x_1 x_2 & x_1 e^{\sin x_1 x_2} \cos x_1 x_2 \end{pmatrix}.$$

Prove this in two different ways: by using the chain rule; and by explicitly computing $g \circ f$.

Exercise 2.21 (Eigenvalues and eigenvectors – needed for Exercise 3.41). Let $A^0 \in \mathrm{End}(\mathbf{R}^n)$ be fixed, and assume that $x^0 \in \mathbf{R}^n$ and $\lambda^0 \in \mathbf{R}$ satisfy

$$A^0x^0 = \lambda^0 x^0, \qquad \langle x^0, x^0\rangle = 1,$$

that is, x^0 is a normalized *eigenvector* of A^0 corresponding to the *eigenvalue* λ^0. Define the mapping $f : \mathbf{R}^n \times \mathbf{R} \to \mathbf{R}^n \times \mathbf{R}$ by

$$f(x, \lambda) = (A^0x - \lambda x,\ \langle x, x\rangle - 1).$$

The total derivative $Df(x, \lambda)$ of f at (x, λ) belongs to $\mathrm{End}(\mathbf{R}^n \times \mathbf{R})$. Choose a basis $(v_1, \dots, v_n)$ for $\mathbf{R}^n$ consisting of pairwise orthogonal vectors, with $v_1 = x^0$. The vectors $(v_1, 0), \dots, (v_n, 0),\ (0, 1)$ then form a basis for $\mathbf{R}^n \times \mathbf{R}$.

(i) Prove

$$\begin{aligned} Df(x, \lambda)(v_j, 0) &= ((A^0 - \lambda I)v_j,\ 2\langle x, v_j\rangle) \quad (1 \le j \le n),\\ Df(x, \lambda)(0, 1) &= (-x, 0). \end{aligned}$$

(ii) Show that the matrix of $Df(x^0, \lambda)$ with respect to the basis above, for all $\lambda \in \mathbf{R}$, is of the following form:

$$\left(\begin{array}{c|c|c|c|c} \lambda^0 - \lambda & & & & -1 \\ 0 & & & & 0 \\ \vdots & (A^0 - \lambda I)v_2 & \cdots & (A^0 - \lambda I)v_n & \vdots \\ 0 & & & & 0 \\ \hline 2 & 0 & \cdots & 0 & 0 \end{array}\right).$$

(iii) Demonstrate by expansion according to the last column, and then according to the bottom row, that $\det Df(x^0, \lambda) = 2 \det B$, with

$$A^0 - \lambda I = \left(\begin{array}{c|ccc} \lambda^0 - \lambda & \star & \cdots & \star \\ \hline 0 & & & \\ \vdots & & B & \\ 0 & & & \end{array}\right).$$

Conclude that, for all $\lambda \neq \lambda^0$, we have $\det Df(x^0, \lambda) = 2\,\dfrac{\det(A^0 - \lambda I)}{\lambda^0 - \lambda}$.

(iv) Prove

$$\det Df(x^0, \lambda^0) = 2 \lim_{\lambda \to \lambda^0} \frac{\det(A^0 - \lambda I)}{\lambda^0 - \lambda}.$$

Exercise 2.22. Let $U \subset \mathbf{R}^n$ and $V \subset \mathbf{R}^p$ be open and let $f : U \to V$ be a C^1 mapping. Define $Tf : U \times \mathbf{R}^n \to \mathbf{R}^p \times \mathbf{R}^p$, the *tangent mapping* of f, to be the mapping given by (see for more details Section 5.2)

$$Tf(x, h) = (f(x), Df(x)h).$$

Let $V \subset \mathbf{R}^p$ be open and let $g : V \to \mathbf{R}^q$ be a C^1 mapping. Show that the chain rule takes the natural form

$$T(g \circ f) = Tg \circ Tf : U \times \mathbf{R}^n \to \mathbf{R}^q \times \mathbf{R}^q.$$

Exercise 2.23 (Another proof of Mean Value Theorem 2.5.3). Let the notation be as in that theorem. Let x' and $x \in U$ be arbitrary and consider x_t and $x_s \in L(x', x)$ as in Formula (2.15). Suppose $m > k$.

(i) By means of the definition of differentiability verify, for $t > s$ and $t - s$ sufficiently small,

$$\|f(x_t) - f(x_s) - (t - s)Df(x_s)(x - x')\| \leq (m - k)(t - s)\|x - x'\|.$$

Applying the reverse triangle inequality deduce

$$\|f(x_t) - f(x_s)\| \leq m(t - s)\|x - x'\|.$$

Set $I = \{\, t \in \mathbf{R} \mid 0 \leq t \leq 1,\ \|f(x_t) - f(x')\| \leq mt\|x - x'\|\,\}$.

(ii) Use the continuity of f to obtain that I is closed. Note that $0 \in I$ and deduce that I has a largest element $0 \leq s \leq 1$.

(iii) Suppose that $s < 1$. Then prove by means of (i) and (ii), for $s < t \leq 1$ with $t - s$ sufficiently small,

$$\|f(x_t) - f(x')\| \leq \|f(x_t) - f(x_s)\| + \|f(x_s) - f(x')\| \leq mt\|x - x'\|.$$

Conclude that $s = 1$, and hence that $\|f(x) - f(x')\| \leq m\|x - x'\|$. Now use the validity of this estimate for every $m > k$ to obtain the Mean Value Theorem.

Exercise 2.24 (Lipschitz continuity – needed for Exercise 3.26). Let U be a convex open subset of $\mathbf{R}^n$ and let $f : U \to \mathbf{R}^p$ be a differentiable mapping. Prove that the following assertions are equivalent.

(i) The mapping f is Lipschitz continuous on U with Lipschitz constant k.

(ii) $\|Df(x)\| \leq k$, for all $x \in U$, where $\|\cdot\|$ denotes the operator norm.

Exercise 2.25. Let $U \subset \mathbf{R}^n$ be open and $a \in U$, and let $f : U \setminus \{a\} \to \mathbf{R}^p$ be differentiable. Suppose there exists $L \in \mathrm{Lin}(\mathbf{R}^n, \mathbf{R}^p)$ with $\lim_{x \to a} Df(x) = L$. Prove that f is differentiable at a, with $Df(a) = L$.
Hint: Apply the Mean Value Theorem to $x \mapsto f(x) - Lx$.

Exercise 2.26 (Criterion for injectivity). Let $U \subset \mathbf{R}^n$ be convex and open and let $f : U \to \mathbf{R}^n$ be differentiable. Assume that the self-adjoint part of $Df(x)$ is positive-definite for each $x \in U$, that is, $\langle h, Df(x)h \rangle > 0$ if $0 \neq h \in \mathbf{R}^n$. Prove that f is injective.
Hint: Suppose $f(x) = f(x')$ where $x' = x + h$. Consider $g : [0, 1] \to \mathbf{R}$ with $g(t) = \langle h, f(x + th) \rangle$. Note that $g(0) = g(1)$ and deduce $g'(\tau) = 0$ for some $0 < \tau < 1$.
Background. For $n = 1$, this is the well-known result that a function $f : I \to \mathbf{R}$ on an interval I is strictly increasing and hence injective if $f' > 0$ on I.

Exercise 2.27 (Needed for Exercise 6.36). Let $U \subset \mathbf{R}^n$ be an open set and consider a mapping $f : U \to \mathbf{R}^p$.

(i) Suppose $f : U \to \mathbf{R}^p$ is differentiable on U and $Df : U \to \mathrm{Lin}(\mathbf{R}^n, \mathbf{R}^p)$ is continuous at a. Using Example 2.5.4, deduce, for x and x' sufficiently close to a,

$$f(x) - f(x') = Df(a)(x - x') + \|x - x'\| \psi(x, x'),$$

with $\lim_{(x,x') \to (a,a)} \psi(x, x') = 0$.

(ii) Let $f \in C^1(U, \mathbf{R}^p)$. Prove that for every compact $K \subset U$ there exists a function $\lambda : [\,0, \infty\,[\; \to [\,0, \infty\,[$ with the following properties. First, $\lim_{t \downarrow 0} \lambda(t) = 0$; and further, for every $x \in K$ and $h \in \mathbf{R}^n$ for which $x + h \in K$,

$$\|f(x + h) - f(x) - Df(x)(h)\| \leq \|h\| \, \lambda(\|h\|).$$

(iii) Let I be an open interval in $\mathbf{R}$ and let $f \in C^1(I, \mathbf{R}^p)$. Define $g : I \times I \to \mathbf{R}^p$ by

$$g(x, x') = \begin{cases} \dfrac{1}{x - x'}(f(x) - f(x')), & x \neq x'; \\ Df(x), & x = x'. \end{cases}$$

Using part (i) prove that g is a C^0 function on $I \times I$ and a C^1 function on $I \times I \setminus \cup_{x \in I} \{(x, x)\}$.

(iv) In the notation of part (iii) suppose that $D^2 f(a)$ exists at $a \in I$. Prove that g as in (iii) is differentiable at (a, a) with derivative $Dg(a, a)(h_1, h_2) =$

$\frac{1}{2}(h_1+h_2)D^2 f(a)$, for $(h_1, h_2) \in \mathbf{R}^2$.
Hint: Consider $h : I \to \mathbf{R}^p$ with

$$h(x) = f(x) - xDf(a) - \frac{1}{2}(x-a)^2 D^2 f(a).$$

Verify, for x and $x' \in I$,

$$\frac{1}{x-x'}(h(x)-h(x')) = g(x, x') - g(a, a) - \frac{1}{2}(x-a+x'-a)D^2 f(a).$$

Next, show that for any $\epsilon > 0$ there exists $\delta > 0$ such that $|h(x) - h(x')| < \epsilon|x-x'|(|x-a| + |x'-a|)$, for x and $x' \in B(a;\delta)$. To this end, apply the definition of the differentiability of Df at a to

$$Dh(x) = Df(x) - Df(a) - (x-a)D^2 f(a).$$

Exercise 2.28. Let $f : \mathbf{R}^n \to \mathbf{R}^n$ be a differentiable mapping and assume that

$$\sup\{\, \|Df(x)\|_{\mathrm{Eucl}} \mid x \in \mathbf{R}^n \,\} \le \epsilon < 1.$$

Prove that f is a contraction with contraction factor $\le \epsilon$. Suppose $F \subset \mathbf{R}^n$ is a closed subset satisfying $f(F) \subset F$. Show that f has a unique fixed point $x \in F$.

Exercise 2.29. Define $f : \mathbf{R} \to \mathbf{R}$ by $f(x) = \log(1+e^x)$. Show that $|f'(x)| < 1$ and that f has no fixed point in $\mathbf{R}$.

Exercise 2.30. Let $U = \mathbf{R}^n \setminus \{0\}$, and define $f \in C^\infty(U)$ by

$$f(x) = \begin{cases} \log\|x\|, & \text{if } n = 2; \\ \dfrac{1}{(2-n)}\dfrac{1}{\|x\|^{n-2}}, & \text{if } n \ne 2. \end{cases}$$

Prove $\operatorname{grad} f(x) = \frac{1}{\|x\|^n}x$, for $n \in \mathbf{N}$ and $x \in U$.

Exercise 2.31.

(i) For f_1, f_2 and $f : \mathbf{R}^n \to \mathbf{R}$ differentiable, prove $\operatorname{grad}(f_1 f_2) = f_1 \operatorname{grad} f_2 + f_2 \operatorname{grad} f_1$, and $\operatorname{grad}\frac{1}{f} = -\frac{1}{f^2}\operatorname{grad} f$ on $\mathbf{R}^n \setminus N(0)$.

(ii) Consider differentiable $f : \mathbf{R}^n \to \mathbf{R}$ and $g : \mathbf{R} \to \mathbf{R}$. Verify $\operatorname{grad}(g \circ f) = (g' \circ f)\operatorname{grad} f$.

(iii) Let $f : \mathbf{R}^n \to \mathbf{R}^p$ and $g : \mathbf{R}^p \to \mathbf{R}$ be differentiable. Show $\operatorname{grad}(g \circ f) = (Df)^t(\operatorname{grad} g) \circ f$. Deduce for the j-th component function $(\operatorname{grad}(g \circ f))_j = \langle (\operatorname{grad} g) \circ f, D_j f \rangle$, where $1 \le j \le n$.

Exercise 2.32 (Homogeneous function – needed for Exercises 2.40, 7.46, 7.62 and 7.66). Let $f : \mathbf{R}^n \setminus \{0\} \to \mathbf{R}$ be a differentiable function.

(i) Let $x \in \mathbf{R}^n \setminus \{0\}$ be a fixed vector and define $g : \mathbf{R}_+ \to \mathbf{R}$ by $g(t) = f(t\,x)$. Show that $g'(t) = Df(tx)(x)$.

Assume f to be *positively homogeneous* of degree $d \in \mathbf{R}$, that is, $f(tx) = t^d\, f(x)$, for all $x \neq 0$ and $t \in \mathbf{R}_+$.

(ii) Prove the following, known as *Euler's identity*:

$$Df(x)(x) = \langle\, x,\ \mathrm{grad}\, f(x)\,\rangle = d\, f(x) \qquad (x \in \mathbf{R}^n \setminus \{0\}).$$

(iii) Conversely, prove that f is positively homogeneous of degree d if we have $Df(x)(x) = d\, f(x)$, for every $x \in \mathbf{R}^n \setminus \{0\}$.
Hint: Calculate the derivative of $t \mapsto t^{-d} g(t)$, for $t \in \mathbf{R}_+$.

Exercise 2.33 (Analog of Rolle's Theorem). Let $U \subset \mathbf{R}^n$ be an open set with compact closure K. Suppose $f : K \to \mathbf{R}$ is continuous on K, differentiable on U, and satisfies $f(x) = 0$, for all $x \in K \setminus U$. Show that there exists $a \in U$ with $\mathrm{grad}\, f(a) = 0$.

Exercise 2.34. Consider $f : \mathbf{R}^2 \to \mathbf{R}$ with $f(x) = 4 - (1 - x_1)^2 - (1 - x_2)^2$ on the bounded set $K \subset \mathbf{R}^2$ bounded by the lines in $\mathbf{R}^2$ with the equations $x_1 = 0$, $x_2 = 0$, and $x_1 + x_2 = 9$, respectively. Prove that the absolute extrema of f on D are: 4 assumed at $(1, 1)$, and -61 at $(9, 0)$ and $(0, 9)$.
Hint: The remaining candidates for extrema are: 3 at $(1, 0)$ and $(0, 1)$, 2 at 0, and $-\frac{41}{2}$ at $\frac{1}{2}(9, 9)$.

Exercise 2.35 (Linear regression). Consider the data points $(x_j, y_j) \in \mathbf{R}^2$, for $1 \le j \le n$. The *regression line* for these data is the line in $\mathbf{R}^2$ that minimizes the sum of the squares of the vertical distances from the points to the line. Prove that the equation $y = \lambda x + c$ of this line is given by

$$\lambda = \frac{\overline{xy} - \overline{x}\;\overline{y}}{\overline{x^2} - \overline{x}^2}, \qquad c = \overline{y} - \lambda \overline{x} = \frac{\overline{x^2}\;\overline{y} - \overline{x}\;\overline{xy}}{\overline{x^2} - \overline{x}^2}.$$

Here we have used a bar to indicate the mean value of a variable in $\mathbf{R}^n$, that is

$$\overline{x} = \frac{1}{n} \sum_{1 \le j \le n} x_j, \qquad \overline{x^2} = \frac{1}{n} \sum_{1 \le j \le n} x_j^2, \qquad \overline{xy} = \frac{1}{n} \sum_{1 \le j \le n} x_j y_j, \qquad \text{etc.}$$

Exercise 2.36. Define $f : \mathbf{R}^2 \to \mathbf{R}$ by $f(x) = (x_1^2 - x_2)(3x_1^2 - x_2)$.

(i) Prove that f has 0 as a critical point, but not as a local extremum, by considering $f(0, t)$ and $f(t, 2t^2)$ for t near 0.

(ii) Show that the restriction of f to $\{ x \in \mathbf{R}^2 \mid x_2 = \lambda x_1 \}$ attains a local minimum 0 at 0, and a local minimum and maximum, $-\frac{1}{12}\lambda^4(1 \pm \frac{2}{3}\sqrt{3})$, at $\frac{1}{6}\lambda(3 \pm \sqrt{3})$, respectively.

Exercise 2.37 (Criterion for surjectivity – needed for Exercises 3.23 and 3.24). Let $\Phi : \mathbf{R}^n \to \mathbf{R}^p$ be a differentiable mapping such that $D\Phi(x) \in \mathrm{Lin}(\mathbf{R}^n, \mathbf{R}^p)$ is surjective, for all $x \in \mathbf{R}^n$ (thus $n \geq p$). Let $y \in \mathbf{R}^p$ and define $f : \mathbf{R}^n \to \mathbf{R}$ by setting $f(x) = \|\Phi(x) - y\|$. Furthermore, suppose that $K \subset \mathbf{R}^n$ is compact and that there exists $\delta > 0$ with

$$\exists x \in K \quad \text{with} \quad f(x) < \delta, \qquad \text{and} \qquad f(x) \geq \delta \qquad (\forall x \in \partial K).$$

Now prove that $\mathrm{int}(K) \neq \emptyset$ and that there is $x \in \mathrm{int}(K)$ with $\Phi(x) = y$.
Hint: Note that the restriction to K of the square of f assumes its minimum in $\mathrm{int}(K)$, and differentiate.

Exercise 2.38. Define $f : \mathbf{R}^2 \to \mathbf{R}$ by

$$f(x) = \begin{cases} x_1^2 \arctan \dfrac{x_2}{x_1} - x_2^2 \arctan \dfrac{x_1}{x_2}, & x_1 x_2 \neq 0; \\ 0, & x_1 x_2 = 0. \end{cases}$$

Show $D_2 D_1 f(0) \neq D_1 D_2 f(0)$.

Exercise 2.39 (Linear transformation and Laplacian – sequel to Exercise 2.4 – needed for Exercises 5.61, 7.61 and 8.32). Consider $A \in \mathrm{End}(\mathbf{R}^n)$, with corresponding matrix $(a_{ij})_{1 \leq i,j \leq n} \in \mathrm{Mat}(n, \mathbf{R})$.

(i) Corroborate the known fact that $DA(x) = A$ by proving $D_j A_i(x) = a_{ij}$, for all $1 \leq i, j \leq n$ and $x \in \mathbf{R}^n$.

Next, assume that $A \in \mathrm{Aut}(n, \mathbf{R})$.

(ii) Verify that $A : \mathbf{R}^n \to \mathbf{R}^n$ is a bijective C^∞ mapping, and that this is also true of A^{-1}.

Let $g : \mathbf{R}^n \to \mathbf{R}$ be a C^2 function. The *Laplace operator* or *Laplacian* Δ assigns to g the C^0 function

$$\Delta g = \sum_{1 \leq k \leq n} D_k^2 g : \mathbf{R}^n \to \mathbf{R}.$$

(iii) Prove that $g \circ A : \mathbf{R}^n \to \mathbf{R}$ again is a C^2 function.

(iv) Prove the following identities of C^1 functions and of C^0 functions, respectively, on $\mathbf{R}^n$, for $1 \leq k \leq n$:

$$\begin{aligned} D_k(g \circ A) &= \sum_{1 \leq j \leq n} a_{jk}(D_j g) \circ A, \\ \Delta(g \circ A) &= \sum_{1 \leq i, j \leq n} \Big(\sum_{1 \leq k \leq n} a_{ik} a_{jk} \Big)(D_i D_j g) \circ A. \end{aligned}$$

(v) Using Exercise 2.4.(iv) prove the following *invariance of the Laplacian under orthogonal transformations*:

$$\Delta(g \circ A) = (\Delta g) \circ A \qquad ((A \in \mathbf{O}(n, \mathbf{R})).$$

Exercise 2.40 (Sequel to Exercise 2.32). Let Δ be the Laplacian as in Exercise 2.39 and let f and $g : \mathbf{R}^n \to \mathbf{R}$ be twice differentiable.

(i) Show $\Delta(fg) = f\Delta g + 2\langle \text{grad } f, \text{grad } g \rangle + g\Delta f$.

Consider the norm $\| \cdot \| : \mathbf{R}^n \setminus \{0\} \to \mathbf{R}$ and $p \in \mathbf{R}$.

(ii) Prove that $\text{grad} \| \cdot \|^p(x) = p\|x\|^{p-2}x$, for $x \in \mathbf{R}^n \setminus \{0\}$, and furthermore, show $\Delta(\| \cdot \|^p) = p(p + n - 2)\| \cdot \|^{p-2}$.

(iii) Demonstrate by means of part (i), for $x \in \mathbf{R}^n \setminus \{0\}$,

$$\begin{aligned} \Delta(\| \cdot \|^p f)(x) &= \|x\|^p(\Delta f)(x) + 2p\|x\|^{p-2}\langle x, \text{grad } f(x) \rangle \\ &\quad + p(p + n - 2)\|x\|^{p-2} f(x). \end{aligned}$$

Now suppose f to be positively homogeneous of degree $d \in \mathbf{R}$, see Exercise 2.32.

(iv) On the strength of Euler's identity verify that $\Delta(\|\cdot\|^{2-n-2d} f) = \|\cdot\|^{2-n-2d} \Delta f$ on $\mathbf{R}^n \setminus \{0\}$. In particular, $\Delta(\| \cdot \|^{2-n}) = 0$ and $\Delta(\| \cdot \|^{-n}) = 2n\| \cdot \|^{-n-2}$.

Exercise 2.41 (Angular momentum, Casimir and Euler operators – needed for Exercises 3.9, 5.60 and 5.61). In *quantum physics* one defines the *angular momentum operator*

$$L = (L_1, L_2, L_3) : C^\infty(\mathbf{R}^3) \to C^\infty(\mathbf{R}^3, \mathbf{R}^3)$$

by (modulo a fourth root of unity), for $f \in C^\infty(\mathbf{R}^3)$ and $x \in \mathbf{R}^3$,

$$(Lf)(x) = (\text{grad } f(x)) \times x \in \mathbf{R}^3.$$

(See the Remark on linear algebra in Section 5.3 for the definition of the cross product $y \times x$ of two vectors y and x in $\mathbf{R}^3$.)

(i) Verify $(L_1 f)(x) = x_3 D_2 f(x) - x_2 D_3 f(x)$.

The *commutator* $[A_1, A_2]$ of two linear operators A_1 and $A_2 : C^\infty(\mathbf{R}^3) \to C^\infty(\mathbf{R}^3)$ is defined by

$$[A_1, A_2] = A_1 A_2 - A_2 A_1 : C^\infty(\mathbf{R}^3) \to C^\infty(\mathbf{R}^3).$$

(ii) Using Theorem 2.7.2 prove that the components of L satisfy the following *commutator relations*:

$$[L_j, L_{j+1}] = L_{j+2} \qquad (1 \le j \le 3),$$

where the indices j are taken modulo 3.

Below, the elements of $C^\infty(\mathbf{R}^3)$ will be taken to be complex-valued functions. Set $i = \sqrt{-1}$ and define $H, X, Y : C^\infty(\mathbf{R}^3) \to C^\infty(\mathbf{R}^3)$ by

$$H = 2iL_3, \qquad X = L_1 + iL_2, \qquad Y = -L_1 + iL_2.$$

(iii) Demonstrate that

$$\begin{aligned} \frac{1}{i}X &= (x_1 + i\,x_2)D_3 - x_3(D_1 + i\,D_2) =: \; z\frac{\partial}{\partial x_3} - 2x_3\frac{\partial}{\partial \bar{z}}, \\ \frac{1}{i}Y &= (x_1 - i\,x_2)D_3 - x_3(D_1 - i\,D_2) =: \; \bar{z}\frac{\partial}{\partial x_3} - 2x_3\frac{\partial}{\partial z}, \end{aligned}$$

where the notation is self-explanatory (see Lemma 8.3.10 as well as Exercise 8.17.(i)). Also prove

$$[H, X] = 2X, \qquad [H, Y] = -2Y, \qquad [X, Y] = H.$$

Background. These commutator relations are of great importance in the theory of *Lie algebras* (see Definition 5.10.4 and Exercises 5.26.(iii) and 5.59).

The *Casimir operator* $C : C^\infty(\mathbf{R}^3) \to C^\infty(\mathbf{R}^3)$ is defined by $C = -\sum_{1\le j\le 3} L_j^2$.

(iv) Prove $[C, L_j] = 0$, for $1 \le j \le 3$.

Denote by $\|\cdot\|^2 : C^\infty(\mathbf{R}^3) \to C^\infty(\mathbf{R}^3)$ the multiplication operator satisfying $(\|\cdot\|^2 f)(x) = \|x\|^2 f(x)$, and let $E : C^\infty(\mathbf{R}^3) \to C^\infty(\mathbf{R}^3)$ be the *Euler operator* given by (see Exercise 2.32.(ii))

$$(Ef)(x) = \langle x, \operatorname{grad} f(x)\rangle = \sum_{1\le j\le 3} x_j D_j f(x).$$

(v) Prove $-C + E(E + I) = \| \cdot \|^2 \Delta$, where Δ denotes the Laplace operator from Exercise 2.39.

(vi) Show using (iii)

$$C = \frac{1}{2}(XY + YX + \frac{1}{2}H^2) = YX + \frac{1}{2}H + \frac{1}{4}H^2.$$

Exercise 2.42 (Another proof of Proposition 2.7.6). The notation is as in that proposition. Set $x_i = a_i + h_i \in \mathbf{R}^n$, for $1 \le i \le k$. Then the k-linearity of T gives

$$T(x_1, \dots, x_k) = T(a_1, \dots, a_k) + \sum_{1 \le i \le k} T(a_1, \dots, a_{i-1}, h_i, x_{i+1}, \dots, x_k).$$

Each of the mappings $(x_{i+1}, \dots, x_k) \mapsto T(a_1, \dots, a_{i-1}, h_i, x_{i+1}, \dots, x_k)$ is $(k-i)$-linear. Deduce

$$\begin{aligned} T(a_1, \dots, a_{i-1}, h_i, x_{i+1}, \dots, x_k) &= T(a_1, \dots, a_{i-1}, h_i, a_{i+1}, \dots, a_k) \\ &+ \sum_{1 \le j \le k-i} T(a_1, \dots, a_{i-1}, h_i, a_{i+1}, \dots, a_{i+j-1}, h_{i+j}, x_{i+j+1}, \dots, x_k). \end{aligned}$$

Now complete the proof of the proposition using Corollary 2.7.5.

Exercise 2.43 (Higher-order derivatives of bilinear mapping).

(i) Let $A \in \mathrm{Lin}(\mathbf{R}^n, \mathbf{R}^p)$. Prove that $D^2 A(a) = 0$, for all $a \in \mathbf{R}^n$.

(ii) Let $T \in \mathrm{Lin}^2(\mathbf{R}^n, \mathbf{R}^p)$. Use Lemma 2.7.4 to show that $D^2 T(a_1, a_2) \in \mathrm{Lin}^2(\mathbf{R}^n \times \mathbf{R}^n, \mathbf{R}^p)$ and prove that it is given by, for $a_i, h_i, k_i \in \mathbf{R}^n$ with $1 \le i \le 2$,

$$D^2 T(a_1, a_2)(h_1, h_2, k_1, k_2) = T(h_1, k_2) + T(k_1, h_2).$$

Deduce $D^3 T(a_1, a_2) = 0$, for all $(a_1, a_2) \in \mathbf{R}^n \times \mathbf{R}^n$.

Exercise 2.44 (Derivative of determinant – sequel to Exercise 2.1 – needed for Exercises 2.51, 4.24 and 5.69). Let $A \in \mathrm{Mat}(n, \mathbf{R})$. We write $A = (a_1 \cdots a_n)$, if $a_j \in \mathbf{R}^n$ is the j-th column vector of A, for $1 \le j \le n$. This means that we identify $\mathrm{Mat}(n, \mathbf{R})$ with $\mathbf{R}^n \times \cdots \times \mathbf{R}^n$ (n copies). Furthermore, we denote by $A^\sharp$ the complementary matrix as in Cramer's rule (2.6).

(i) Consider the function $\det : \mathrm{Mat}(n, \mathbf{R}) \to \mathbf{R}$ as an element of $\mathrm{Lin}^n(\mathbf{R}^n, \mathbf{R})$ via the identification above. Now prove that its derivative $(D\det)(A) \in \mathrm{Lin}(\mathrm{Mat}(n, \mathbf{R}), \mathbf{R})$ is given by

$$(D\det)(A) = \mathrm{tr} \circ A^\sharp \qquad (A \in \mathrm{Mat}(n, \mathbf{R})),$$

where tr denotes the trace of a matrix. More explicitly,

$$(D\det)(A)H = \operatorname{tr}(A^\sharp H) = \langle (A^\sharp)^t, H \rangle \qquad (H \in \operatorname{Mat}(n, \mathbf{R})).$$

Here we use the result from Exercise 2.1.(ii). In particular, verify that $\operatorname{grad}(\det)(A) = (A^\sharp)^t$, the *cofactor matrix*, and

$$(\star) \qquad (D\det)(I) = \operatorname{tr}; \qquad (\star\star) \qquad (D\det)(A) = \det A\,(\operatorname{tr}\circ A^{-1}),$$

for $A \in \mathbf{GL}(n, \mathbf{R})$, by means of Cramer's rule.

(ii) Let $X \in \operatorname{Mat}(n, \mathbf{R})$ and define $e^X \in \mathbf{GL}(n, \mathbf{R})$ as in Example 2.4.10. Show, for $t \in \mathbf{R}$,

$$\begin{aligned}\frac{d}{dt}\det(e^{tX}) &= \left.\frac{d}{ds}\right|_{s=0} \det(e^{(s+t)X}) = \left.\frac{d}{ds}\right|_{s=0} \det(e^{sX})\det(e^{tX}) \\ &= \operatorname{tr} X \det(e^{tX}).\end{aligned}$$

Deduce by solving this differential equation that (compare also with Formula (5.33))

$$\det(e^{tX}) = e^{t \operatorname{tr} X} \qquad (t \in \mathbf{R},\ X \in \operatorname{Mat}(n, \mathbf{R})).$$

(iii) Assume that $A \in \mathbf{GL}(n, \mathbf{R})$. Using $\det(A + H) = \det A \det(I + A^{-1}H)$, deduce $(\star\star)$ from $(\star)$ in (i). Now also derive $(\det A)A^{-1} = A^\sharp$ from (i).

(iv) Assume that $A : \mathbf{R} \to \mathbf{GL}(n, \mathbf{R})$ is a differentiable mapping. Derive from part (i) the following formula for the derivative of $\det \circ A : \mathbf{R} \to \mathbf{R}$:

$$\frac{1}{\det \circ A}(\det \circ A)' = \operatorname{tr}(A^{-1} \circ DA),$$

in other words (compare with part (ii))

$$\frac{d}{dt}(\log \det A)(t) = \operatorname{tr}\Big(A^{-1}\frac{dA}{dt}\Big)(t) \qquad (t \in \mathbf{R}).$$

(v) Suppose that the mapping $A : \mathbf{R} \to \operatorname{Mat}(n, \mathbf{R})$ is differentiable and that $F : \mathbf{R} \to \operatorname{Mat}(n, \mathbf{R})$ is continuous, while we have the differential equation

$$\frac{dA}{dt}(t) = F(t)\,A(t) \qquad (t \in \mathbf{R}).$$

In this context, the *Wronskian* $w \in C^1(\mathbf{R})$ of A is defined as $w(t) = \det A(t)$. Use part (i), Exercise 2.1.(i) and Cramer's rule to show, for $t \in \mathbf{R}$,

$$\frac{dw}{dt}(t) = \operatorname{tr} F(t)\,w(t), \qquad \text{and deduce} \qquad w(t) = w(0)\,e^{\int_0^t \operatorname{tr} F(\tau)\,d\tau}.$$

Exercise 2.45 (Derivatives of inverse acting in $\mathrm{Aut}(\mathbf{R}^n)$ **– needed for Exercises 2.46, 2.48, 2.49 and 2.51).** We recall that $\mathrm{Aut}(\mathbf{R}^n)$ is an open subset of the linear space $\mathrm{End}(\mathbf{R}^n)$ and define the mapping $F : \mathrm{Aut}(\mathbf{R}^n) \to \mathrm{Aut}(\mathbf{R}^n)$ by $F(A) = A^{-1}$. We want to prove that F is a C^∞ mapping and to compute its derivatives.

(i) Deduce from Formula (2.7) that F is differentiable. By differentiating the identity $AF(A) = I$ prove that $DF(A)$, for $A \in \mathrm{Aut}(\mathbf{R}^n)$, satisfies

$$DF(A) \in \mathrm{Lin}\,(\mathrm{End}(\mathbf{R}^n), \mathrm{End}(\mathbf{R}^n)) = \mathrm{End}\,(\mathrm{End}(\mathbf{R}^n)),$$
$$DF(A)H = -A^{-1}HA^{-1}.$$

Generally speaking, we may not assume that A^{-1} and H commute, that is, $A^{-1}H \neq HA^{-1}$. Next define

$$G : \mathrm{End}(\mathbf{R}^n) \times \mathrm{End}(\mathbf{R}^n) \to \mathrm{End}\,(\mathrm{End}(\mathbf{R}^n)) \qquad \text{by} \qquad G(A, B)(H) = -AHB.$$

(ii) Prove that G is bilinear and, using Proposition 2.7.6, deduce that G is a C^∞ mapping.

Define $\Delta : \mathrm{End}(\mathbf{R}^n) \to \mathrm{End}(\mathbf{R}^n) \times \mathrm{End}(\mathbf{R}^n)$ by $\Delta(B) = (B, B)$.

(iii) Verify that $D\Delta(B) = \Delta$, for all $B \in \mathrm{End}(\mathbf{R}^n)$, and deduce that Δ is a C^∞ mapping.

(iv) Conclude from (i) that F satisfies the differential equation $DF = G \circ \Delta \circ F$. Using (ii), (iii) and mathematical induction conclude that F is a C^∞ mapping.

(v) By means of the chain rule and (iv) show

$$D^2F(A) = DG(A^{-1}, A^{-1}) \circ \Delta \circ G(A^{-1}, A^{-1}) \in \mathrm{Lin}^2\,(\mathrm{End}(\mathbf{R}^n), \mathrm{End}(\mathbf{R}^n)).$$

Now use Proposition 2.7.6 to obtain, for H_1 and $H_2 \in \mathrm{End}(\mathbf{R}^n)$,

$$\begin{aligned} D^2F(A)(H_1, H_2) &= (DG(A^{-1}, A^{-1}) \circ \Delta \circ G(A^{-1}, A^{-1})H_1)H_2 \\ &= DG(A^{-1}, A^{-1})(-A^{-1}H_1A^{-1}, -A^{-1}H_1A^{-1})(H_2) \\ &= G(-A^{-1}H_1A^{-1}, A^{-1})(H_2) + G(A^{-1}, -A^{-1}H_1A^{-1})(H_2) \\ &= A^{-1}H_1A^{-1}H_2A^{-1} + A^{-1}H_2A^{-1}H_1A^{-1}. \end{aligned}$$

(vi) Using (i) and mathematical induction over $k \in \mathbf{N}$ show that $D^kF(A) \in \mathrm{Lin}^k\,(\mathrm{End}(\mathbf{R}^n), \mathrm{End}(\mathbf{R}^n))$ is given, for $H_1, \ldots, H_k \in \mathrm{End}(\mathbf{R}^n)$, by

$$\begin{aligned} &D^kF(A)(H_1, \ldots, H_k) \\ &\quad = (-1)^k \sum_{\sigma \in S_k} A^{-1}H_{\sigma(1)}A^{-1}H_{\sigma(2)}A^{-1} \cdots A^{-1}H_{\sigma(k)}A^{-1}. \end{aligned}$$

Verify that for $n = 1$ and $A = t \in \mathbf{R}$ we recover the known formula $f^{(k)}(t) = (-1)^k k!\, t^{-(k+1)}$ where $f(t) = t^{-1}$.

Exercise 2.46 (Another proof for derivatives of inverse acting in $\mathrm{Aut}(\mathbf{R}^n)$ – sequel to Exercises 2.3 and 2.45). Let the notation be as in the latter exercise. In particular, we have $F : \mathrm{Aut}(\mathbf{R}^n) \to \mathrm{Aut}(\mathbf{R}^n)$ with $F(A) = A^{-1}$. For $K \in \mathrm{End}(\mathbf{R}^n)$ with $\|K\| < 1$ deduce from Exercise 2.3.(iii)

$$F(I + K) = \sum_{0 \le j \le k} (-K)^j + (-K)^{k+1} F(I + K).$$

Furthermore, for $A \in \mathrm{Aut}(\mathbf{R}^n)$ and $H \in \mathrm{End}(\mathbf{R}^n)$, prove the equality $F(A + H) = F(I + A^{-1}H)F(A)$. Conclude, for $\|H\| < \|A^{-1}\|^{-1}$,

$$(\star) \qquad \begin{aligned} F(A + H) &= \sum_{0 \le j \le k} (-A^{-1}H)^j A^{-1} + (-A^{-1}H)^{k+1} F(I + A^{-1}H)F(A) \\ &=: \sum_{0 \le j \le k} (-1)^j A^{-1}HA^{-1} \cdots A^{-1}HA^{-1} + R_k(A, H). \end{aligned}$$

Demonstrate

$$\|R_k(A, H)\| = \mathcal{O}(\|H\|^{k+1}), \qquad H \to 0.$$

Using this and the fact that the sum in $(\star)$ is a polynomial of degree k in the variable H, conclude that $(\star)$ actually is the Taylor expansion of F at I. But this implies

$$D^k F(A)(H^k) = (-1)^k k!\, A^{-1}HA^{-1} \cdots A^{-1}HA^{-1}.$$

Now set, for $H_1, \dots, H_k \in \mathrm{End}(\mathbf{R}^n)$,

$$T(H_1, \dots, H_k) = (-1)^k \sum_{\sigma \in S_k} A^{-1}H_{\sigma(1)}A^{-1}H_{\sigma(2)}A^{-1} \cdots A^{-1}H_{\sigma(k)}A^{-1}.$$

Then $T \in \mathrm{Lin}^k(\mathrm{End}(\mathbf{R}^n), \mathrm{End}(\mathbf{R}^n))$ is symmetric and $T(H^k) = D^k F(A)(H^k)$. Since a generalization of the polarization identity from Lemma 1.1.5.(iii) implies that a symmetric k-linear mapping T is uniquely determined by $H \mapsto T(H^k)$, we see that $D^k F(A) = T$.

Exercise 2.47 (Newton's iteration method – sequel to Exercise 2.6 – needed for Exercise 2.48). We come back to the approximation construction of Exercise 2.6, and we use its notation. In particular, $V = V(x_0; \delta)$ and we now require $f \in C^2(V, \mathbf{R}^n)$. Near a zero of f to be determined, a much faster variant is *Newton's iteration method*, given by, for $k \in \mathbf{N}_0$,

$$(\star) \qquad x_{k+1} = F(x_k) \qquad \text{where} \qquad F(x) = x - Df(x)^{-1}(f(x)).$$

This assumes that numbers $c_1 > 0$ and $c_2 > 0$ exist such that, with $\|\cdot\|$ equal to the Euclidean or the operator norm,

$$Df(x) \in \mathrm{Aut}(\mathbf{R}^n), \qquad \|Df(x)^{-1}\| \le c_1, \qquad \|D^2 f(x)\| \le c_2 \qquad (x \in V).$$

(i) Prove $f(x) = f(x_0) + Df(x)(x - x_0) - R_1(x, x_0 - x)$ by Taylor expansion of f at x; and based on the estimate for the remainder R_1 from Theorem 2.8.3 deduce that $F : V \to V$, if in addition it is assumed

$$c_1\Big(\|f(x_0)\| + \frac{1}{2}c_2\,\delta^2\Big) \leq \delta.$$

This is the case if δ, and in its turn $\|f(x_0)\|$, is sufficiently small. Deduce that $(x_k)_{k\in\mathbf{N}_0}$ is a sequence in V.

(ii) Apply f to $(\star)$ and use Taylor expansion of f of order 2 at x_k to obtain

$$\|f(x_{k+1})\| \leq \frac{1}{2}c_2\,\|Df(x_k)^{-1}(f(x_k))\|^2 \leq c_3\,\|f(x_k)\|^2,$$

with the notation $c_3 = \frac{1}{2}c_1^2c_2$. Verify by mathematical induction over $k \in \mathbf{N}_0$

$$\|f(x_k)\| \leq \frac{1}{c_3}\,\epsilon^{2^k} \qquad \text{if} \qquad \epsilon := c_3\,\|f(x_0)\|.$$

If $\epsilon < 1$ this proves that $(f(x_k))_{k\in\mathbf{N}_0}$ very rapidly converges to 0. In numerical terms: the number of decimal zeros doubles with each iteration.

(iii) Deduce from part (ii), with $d = \frac{c_1}{c_3} = \frac{2}{c_1c_2}$,

$$\|x_{k+1} - x_k\| \leq d\,\epsilon^{2^k} \qquad (k \in \mathbf{N}_0),$$

from which one easily concludes that $(x_k)_{k\in\mathbf{N}_0}$ constitutes a Cauchy sequence in V. Prove that the limit x is the unique zero of f in V and that

$$\|x_k - x\| \leq d\sum_{l\in\mathbf{N}_0}\epsilon^{2^{k+l}} \leq d\,\epsilon^{2^k}\,\frac{1}{1-\epsilon} \qquad (k \in \mathbf{N}_0).$$

Again, (x_k) converges to the desired solution x at such a rate that each additional iteration doubles the number of correct decimals. Newton's iteration is the method of choice when the calculation of $Df(x)^{-1}$ (dependent on x) does not present too much difficulty, see Example 1.7.3. The following figure gives a geometrical illustration of both the method of Exercise 2.6 and Newton's iteration method.

Exercise 2.48 (Division by means of Newton's iteration method – sequel to Exercises 2.45 and 2.47). Given $0 \neq y \in \mathbf{R}$ we want to approximate $\frac{1}{y} \in \mathbf{R}$ by means of Newton's iteration method. To this end, consider $f(x) = \frac{1}{x} - y$ and choose a suitable initial value x_0 for $\frac{1}{y}$.

(i) Show that the iteration takes the form

$$x_{k+1} = 2x_k - yx_k^2 \qquad (k \in \mathbf{N}_0).$$

Observe that this approximation to division requires only multiplication and addition.

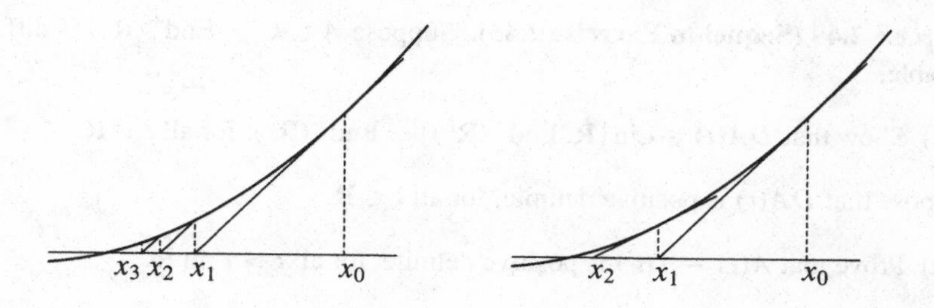

$$x_{k+1} = x_k - A\,f(x_k) \qquad\qquad x_{k+1} = x_k - f'(x_k)^{-1} f(x_k)$$

Illustration for Exercise 2.47

(ii) Introduce the k-th remainder $r_k = 1 - yx_k$. Show

$$r_{k+1} = r_k^2, \qquad \text{and thus} \qquad r_k = r_0^{2^k} \qquad (k \in \mathbf{N}_0).$$

Prove that $|x_k - \frac{1}{y}| = |\frac{r_k}{y}|$, and deduce that $(x_k)_{k\in\mathbf{N}_0}$ converges, with limit $\frac{1}{y}$, if and only if $|r_0| < 1$. If this is the case, the convergence rate is exactly quadratic.

(iii) Verify that

$$x_k = x_0 \frac{1 - r_0^{2^k}}{1 - r_0} = x_0 \sum_{0\le j<2^k} r_0^j \qquad (k \in \mathbf{N}_0),$$

which amounts to summing the first 2^k terms of the geometric series

$$\frac{1}{y} = \frac{x_0}{1 - r_0} = x_0 \sum_{j\in\mathbf{N}_0} r_0^j.$$

Finally, if $g(x) = 1 - xy$, application of Newton's iteration would require that we know $g'(x)^{-1} = -\frac{1}{y}$, which is the quantity we want to approximate. Therefore, consider the iteration (which is next best)

$$x'_{k+1} = x'_k + x_0 g(x'_k), \qquad x'_0 = x_0 \qquad (k \in \mathbf{N}_0).$$

(iv) Obtain the approximation $x'_k = x_0 \sum_{0\le j<k} r_0^j$, for $k \in \mathbf{N}_0$, which converges much slower than the one in part (iii).

(v) Generalize the preceding results (i) – (iii) to the case of $\mathrm{End}(\mathbf{R}^n)$ and $Y \in \mathrm{Aut}(\mathbf{R}^n)$. Use Exercise 2.45.(i) to verify that Newton's iteration in this case takes the form $X_{k+1} = 2X_k - X_k Y X_k$.

Exercise 2.49 (Sequel to Exercise 2.45). Suppose $A : \mathbf{R} \to \mathrm{End}^+(\mathbf{R}^n)$ is differentiable.

(i) Show that $DA(t) \in \mathrm{Lin}\,(\mathbf{R}, \mathrm{End}^+(\mathbf{R}^n)) \simeq \mathrm{End}^+(\mathbf{R}^n)$, for all $t \in \mathbf{R}$.

Suppose that $DA(t)$ is positive definite, for all $t \in \mathbf{R}$.

(ii) Prove that $A(t) - A(t')$ is positive definite, for all $t > t'$ in $\mathbf{R}$.

(iii) Now suppose that $A(t) \in \mathrm{Aut}(\mathbf{R}^n)$, for all $t \in \mathbf{R}$. By means of Exercise 2.45.(i) show that $A(t)^{-1} - A(t')^{-1}$ is negative definite, for all $t > t'$ in $\mathbf{R}$.

Finally, assume that A_0 and $B_0 \in \mathrm{End}^+(\mathbf{R}^n)$ and that $A_0 - B_0$ is positive definite. Also assume A_0 and $B_0 \in \mathrm{Aut}(\mathbf{R}^n)$.

(iv) Verify that $A_0^{-1} - B_0^{-1}$ is negative definite by considering $t \mapsto B_0 + t(A_0 - B_0)$.

Exercise 2.50 (Action of $\mathbf{R}^n$ on $C^\infty(\mathbf{R}^n)$). We recall the *action* of $\mathbf{R}^n$ on itself by translation, given by

$$T_a(x) = a + x \qquad (a,\ x \in \mathbf{R}^n).$$

Obviously, $T_a T_{a'} = T_{a+a'}$. We now define an action of $\mathbf{R}^n$ on $C^\infty(\mathbf{R}^n)$ by assigning to $a \in \mathbf{R}^n$ and $f \in C^\infty(\mathbf{R}^n)$ the function $T_a f \in C^\infty(\mathbf{R}^n)$ given by

$$(T_a f)(x) = f(x - a) \qquad (x \in \mathbf{R}^n).$$

Here we use the rule: "value of the transformed function at the transformed point equals value of the original function at the original point".

(i) Verify that we then have a *group action*, that is, for a and $a' \in \mathbf{R}^n$ we have

$$(T_a T_{a'}) f = T_a (T_{a'} f) \qquad (f \in C^\infty(\mathbf{R}^n)).$$

For every $a \in \mathbf{R}^n$ we introduce t_a by

$$t_a : C^\infty(\mathbf{R}^n) \to C^\infty(\mathbf{R}^n) \qquad \text{with} \qquad t_a f = \frac{d}{d\alpha}\Big|_{\alpha=0} T_{\alpha a} f.$$

(ii) Prove that $t_a f(x) = \langle\, a,\ -\operatorname{grad} f(x)\,\rangle \qquad (a \in \mathbf{R}^n,\ f \in C^\infty(\mathbf{R}^n))$.
Hint: For every $x \in \mathbf{R}^n$ we find

$$t_a f(x) = \frac{d}{d\alpha}\Big|_{\alpha=0} f(T_{-\alpha a} x) = Df(x) \frac{d}{d\alpha}\Big|_{\alpha=0} (x - \alpha a).$$

(iii) Demonstrate that from part (ii) follows, by means of Taylor expansion with respect to the variable $\alpha \in \mathbf{R}$,

$$T_{\alpha a} f = f + \langle \alpha a, -\operatorname{grad} f \rangle + \mathcal{O}(\alpha^2), \quad \alpha \to 0 \qquad (f \in C^\infty(\mathbf{R}^n)).$$

Background. In view of this result, the operator $-\operatorname{grad}$ is also referred to as the *infinitesimal generator* of the action of $\mathbf{R}^n$ on $C^\infty(\mathbf{R}^n)$.

Exercise 2.51 (Characteristic polynomial and Theorem of Hamilton–Cayley – sequel to Exercises 2.44 and 2.45). Denote the *characteristic polynomial* of $A \in \operatorname{End}(\mathbf{R}^n)$ by χ_A, that is

$$\chi_A(\lambda) = \det(\lambda I - A) =: \sum_{0 \le j \le n} (-1)^j \sigma_j(A) \lambda^{n-j}.$$

Then $\sigma_0(A) = 1$ and $\sigma_n(A) = \det A$. In this exercise we derive a recursion formula for the functions $\sigma_j : \operatorname{End}(\mathbf{R}^n) \to \mathbf{R}$ by means of differential calculus.

(i) $\det A$ is a homogeneous function of degree n in the matrix coefficients of A. Using this fact, show by means of Taylor expansion of $\det : \operatorname{End}(\mathbf{R}^n) \to \mathbf{R}$ at I that

$$\det(I + tA) = \sum_{0 \le j \le n} \frac{t^j}{j!} \det{}^{(j)}(I)(A^j) \qquad (A \in \operatorname{End}(\mathbf{R}^n), t \in \mathbf{R}).$$

Deduce

$$\chi_A(\lambda) = \sum_{0 \le j \le n} (-1)^j \frac{\lambda^{n-j}}{j!} \det{}^{(j)}(I)(A^j), \qquad \sigma_j(A) = \frac{\det{}^{(j)}(I)(A^j)}{j!}.$$

We define $C : \operatorname{End}(\mathbf{R}^n) \to \operatorname{End}(\mathbf{R}^n)$ by $C(A) = A^\sharp$, the complementary matrix of A, and we recall the mapping $F : \operatorname{Aut}(\mathbf{R}^n) \to \operatorname{Aut}(\mathbf{R}^n)$ with $F(A) = A^{-1}$ from Exercise 2.45. On $\operatorname{Aut}(\mathbf{R}^n)$ Cramer's rule from Formula (2.6) then can be written in the form $C = \det \cdot F$.

(ii) Apply Leibniz' rule, part (i) and Exercise 2.45.(vi) to this identity in order to get, for $0 \le k \le n$ and $A \in \operatorname{Aut}(\mathbf{R}^n)$,

$$\begin{aligned} \frac{1}{k!} C^{(k)}(I)(A^k) &= \sum_{0 \le j \le k} \frac{\det{}^{(j)}(I)(A^j)}{j!} \frac{1}{(k-j)!} F^{(k-j)}(I)(A^{k-j}) \\ &= \sum_{0 \le j \le k} (-1)^{k-j} \sigma_j(A) A^{k-j}. \end{aligned}$$

The coefficients of $C(A) \in \operatorname{End}(\mathbf{R}^n)$ are homogeneous polynomials of degree $n-1$ in those of A itself. Using this, deduce

$$C(A) = \frac{1}{(n-1)!} C^{(n-1)}(I)(A^{n-1}) = \sum_{0 \le j < n} (-1)^{n-1-j} \sigma_j(A) A^{n-1-j}.$$

(iii) Next apply Cramer's rule $AC(A) = \sigma_n(A)I$ to obtain the following *Theorem of Hamilton–Cayley*, which asserts that $A \in \mathrm{Aut}(\mathbf{R}^n)$ is annihilated by its own characteristic polynomial χ_A,

$$0 = \chi_A(A) = \sum_{0 \le j \le n} (-1)^j \sigma_j(A) A^{n-j} \qquad (A \in \mathrm{Aut}(\mathbf{R}^n)).$$

Using that $\mathrm{Aut}(\mathbf{R}^n)$ is dense in $\mathrm{End}(\mathbf{R}^n)$ show that the Theorem of Hamilton–Cayley is actually valid for all $A \in \mathrm{End}(\mathbf{R}^n)$.

(iv) From Exercise 2.44.(i) we know that $\det^{(1)} = \mathrm{tr} \circ C$. Differentiate this identity $k-1$ times at I and use Lemma 2.4.7 to obtain

$$\det{}^{(k)}(I)(A^k) = \mathrm{tr}\,(C^{(k-1)}(I)(A^{k-1})A) \qquad (A \in \mathrm{End}(\mathbf{R}^n)).$$

Here we employed once more the density of $\mathrm{Aut}(\mathbf{R}^n)$ in $\mathrm{End}(\mathbf{R}^n)$ to derive the equality on $\mathrm{End}(\mathbf{R}^n)$. By means of (i) and (ii) deduce the following recursion formula for the coefficients $\sigma_k(A)$ of χ_A:

$$k\,\sigma_k(A) = (-1)^{k-1} \sum_{0 \le j < k} (-1)^j \sigma_j(A)\,\mathrm{tr}(A^{k-j}) \qquad (A \in \mathrm{End}(\mathbf{R}^n)).$$

In particular,

$$\begin{aligned} \chi_A(\lambda) &= \lambda^n - \mathrm{tr}(A)\lambda^{n-1} + (\,\mathrm{tr}(A)^2 - \mathrm{tr}(A^2))\frac{\lambda^{n-2}}{2!} \\ &\quad -(\,\mathrm{tr}(A)^3 - 3\,\mathrm{tr}(A)\,\mathrm{tr}(A^2) + 2\,\mathrm{tr}(A^3))\frac{\lambda^{n-3}}{3!} + \cdots . \end{aligned}$$

Exercise 2.52 (Newton's Binomial Theorem, Leibniz' rule and Multinomial Theorem – needed for Exercises 2.53, 2.55 and 7.22).

(i) Prove the following identities, for x and $y \in \mathbf{R}^n$, a multi-index $\alpha \in \mathbf{N}_0^n$, and f and $g \in C^{|\alpha|}(\mathbf{R}^n)$, respectively

$$\frac{(x+y)^\alpha}{\alpha!} = \sum_{\beta \le \alpha} \frac{x^\beta}{\beta!}\,\frac{y^{\alpha-\beta}}{(\alpha-\beta)!}, \qquad \frac{D^\alpha(fg)}{\alpha!} = \sum_{\beta \le \alpha} \frac{D^\beta f}{\beta!}\,\frac{D^{\alpha-\beta} g}{(\alpha-\beta)!}.$$

Here the summation is over all multi-indices $\beta \in \mathbf{N}_0^n$ satisfying $\beta_i \le \alpha_i$, for every $1 \le i \le n$.
Hint: The former identity follows by Taylor's formula applied to $y \mapsto \frac{y^\alpha}{\alpha!}$, while the latter is obtained by computation in two different ways of the coefficient of the term of exponent α in the Taylor polynomial of order $|\alpha|$ of fg.

(ii) Derive by application of Taylor's formula to $x \mapsto \frac{\langle x,y\rangle^k}{k!}$ at $x = 0$, for all x and $y \in \mathbf{R}^n$ and $k \in \mathbf{N}_0$

$$\frac{\langle x, y\rangle^k}{k!} = \sum_{|\alpha|=k} \frac{x^\alpha y^\alpha}{\alpha!}, \qquad \text{in particular} \qquad \frac{(\sum_{1\le j\le n} x_j)^k}{k!} = \sum_{|\alpha|=k} \frac{x^\alpha}{\alpha!}.$$

The latter identity is called the *Multinomial Theorem.*

Exercise 2.53 (Symbol of differential operator – sequel to Exercise 2.52 – needed for Exercise 2.54). Let $k \in \mathbf{N}$. Given a function $\sigma : \mathbf{R}^n \times \mathbf{R}^n \to \mathbf{R}$ of the form

$$\sigma(x, \xi) = \sum_{|\alpha|\le k} p_\alpha(x)\xi^\alpha \qquad \text{with} \qquad p_\alpha \in C(\mathbf{R}^n),$$

we define the *linear partial differential operator*

$$P : C^k(\mathbf{R}^n) \to C(\mathbf{R}^n) \qquad \text{by} \qquad Pf(x) = P(x, D)f(x) = \sum_{|\alpha|\le k} p_\alpha(x)D^\alpha f(x).$$

The function σ is called the *total symbol* σ_P of the linear partial differential operator P. Conversely, we can recover σ_P from P as follows. For $\xi \in \mathbf{R}^n$ write $e^{\langle\cdot,\xi\rangle} \in C^\infty(\mathbf{R}^n)$ for the function $x \mapsto e^{\langle x,\xi\rangle}$.

(i) Verify $\sigma_P(x, \xi) = e^{-\langle x,\xi\rangle}(Pe^{\langle\cdot,\xi\rangle})(x)$.

Now let $l \in \mathbf{N}$ and let $\sigma_Q : \mathbf{R}^n \times \mathbf{R}^n \to \mathbf{R}$ be given by $\sigma_Q(x, \xi) = \sum_{|\beta|\le l} q_\beta(x)\xi^\beta$.

(ii) Using Leibniz' rule from Exercise 2.52.(i) show that

$$P(x, D)Q(x, D) = \sum_{|\gamma|\le k}\sum_{|\alpha|\le k} \frac{\alpha!}{(\alpha-\gamma)!} p_\alpha(x)\frac{1}{\gamma!}\sum_{|\beta|\le l} D^\gamma q_\beta(x)D^{\alpha-\gamma+\beta},$$

and deduce that the composition PQ is a linear partial differential operator too.

(iii) Denote by D_x and D_ξ differentiation with respect to the variables x and $\xi \in \mathbf{R}^n$, respectively. Using $D_\xi^\gamma \xi^\alpha = \frac{\alpha!}{(\alpha-\gamma)!}\xi^{\alpha-\gamma}$, deduce from (ii) that the total symbol σ_{PQ} of PQ is given by

$$\sigma_{PQ}(x, \xi) = \sum_{|\gamma|\le k} D_\xi^\gamma\Big(\sum_{|\alpha|\le k} p_\alpha(x)\xi^\alpha\Big)\frac{1}{\gamma!}D_x^\gamma\Big(\sum_{|\beta|\le l} q_\beta(x)\xi^\beta\Big).$$

Verify

$$\sigma_{PQ} = \sum_{|\gamma|\le k} \sigma_P^{(\gamma)}\frac{1}{\gamma!}D_x^\gamma \sigma_Q, \qquad \text{where} \qquad \sigma_P^{(\gamma)}(x, \xi) = D_\xi^\gamma \sigma_P(x, \xi).$$

Background. The coefficient functions p_α and the monomials ξ^α in the total symbol σ_P commute, whereas this is not the case for the p_α and the differential operators D^α in the differential operator P. Furthermore, $\sigma_{PQ} = \sigma_P\sigma_Q+$ additional terms, which as polynomials in ξ are of degree less than $k+l$. This shows that in general the mapping $P \mapsto \sigma_P$ from the space of linear partial differential operators to the space of functions: $\mathbf{R}^n \times \mathbf{R}^n \to \mathbf{R}$ is not multiplicative for the usual composition of operators and multiplication of functions, respectively. Because in general the differential operators P and Q do not commute, it is of interest to study their commutator $[\,P, Q\,] := PQ - QP$ (compare with Exercise 2.41).

(iv) Writing $(j) = (0, \dots, 0, j, 0, \dots, 0) \in \mathbf{N}_0^n$, show

$$\sigma_{[\,P,Q\,]} = \sigma_{PQ} - \sigma_{QP} \equiv \sum_{1\le j\le n} (D_\xi^{(j)}\sigma_P D_x^{(j)}\sigma_Q - D_\xi^{(j)}\sigma_Q D_x^{(j)}\sigma_P)$$

modulo terms of degree less than $k+l-1$ in ξ. Show that the sum above, evaluated at (x, ξ), equals

$$\sum_{1\le j\le n} \Big(\frac{\partial\sigma_P}{\partial\xi_j}\frac{\partial\sigma_Q}{\partial x_j} - \frac{\partial\sigma_P}{\partial x_j}\frac{\partial\sigma_Q}{\partial\xi_j}\Big)(x,\xi) = \{\sigma_p, \sigma_Q\}(x,\xi),$$

where $\{\sigma_p, \sigma_Q\} : \mathbf{R}^n \times \mathbf{R}^n \to \mathbf{R}$ denote the *Poisson brackets* of the total symbols σ_P and σ_Q. See Exercise 8.46.(xii) for more details.

Exercise 2.54 (Formula of Leibniz–Hörmander – sequel to Exercise 2.53 – needed for Exercise 2.55). Let the notation be as in the exercise. In particular, let σ_P be the total symbol of the linear partial differential operator P. Now suppose $g \in C^k(\mathbf{R}^n)$ and consider the mapping

$$Q : C^k(\mathbf{R}^n) \to C(\mathbf{R}^n) \qquad \text{given by} \qquad Q(f) = P(fg).$$

We shall prove that Q again is a linear partial differential operator. Indeed, use $e^{-\langle x,\xi\rangle} D_j(e^{\langle\cdot,\xi\rangle}g)(x) = (\xi_j + D_j)g(x)$, for $1 \le j \le n$, to show

$$\sigma_Q(x,\xi) = e^{-\langle x,\xi\rangle}(Pe^{\langle\cdot,\xi\rangle}g)(x) = P(x, \xi + D)g(x).$$

Note that the variable ξ and the differentiation D with respect to the variable x do commute; consequently, $P(x, \xi + D)$, which is a polynomial of degree k in its second argument, can be expanded formally. Taylor's formula applied to the partial function $\xi \mapsto P(x, \xi)$ gives an identity of polynomials. Use this to find

$$P(x, \xi + D) = \sum_{|\alpha|\le k} \frac{1}{\alpha!} P^{(\alpha)}(x,\xi) D^\alpha \qquad \text{with} \qquad P^{(\alpha)}(x,\xi) = D_\xi^\alpha \sigma_P(x,\xi).$$

Conclude

$$\sigma_Q(x,\xi) = \sum_{|\alpha|\le k} \frac{1}{\alpha!} P^{(\alpha)}(x,\xi) D^\alpha g(x) = \sum_{|\alpha|\le k} \frac{1}{\alpha!} D^\alpha g(x) P^{(\alpha)}(x,\xi).$$

This proves that Q is a linear partial differential operator with symbol σ_Q. Furthermore, deduce the following *formula of Leibniz–Hörmander*:

$$P(fg)(x) = Q(x, D)f(x) = \sum_{|\alpha|\le k} P^{(\alpha)}(x, D)f(x)\frac{1}{\alpha!}D^\alpha g(x).$$

Exercise 2.55 (Another proof of formula of Leibniz–Hörmander – sequel to Exercises 2.52 and 2.54). Let $a \in \mathbf{R}^n$ be arbitrary and introduce the monomials $g_\alpha(x) = (x-a)^\alpha$, for $\alpha \in \mathbf{N}_0^n$.

(i) Verify

$$D^\gamma g_\alpha(a) = \begin{cases} \alpha!, & \gamma = \alpha; \\ 0, & \gamma \neq \alpha. \end{cases}$$

Now let $P(x, D) = \sum_{|\beta|\le k} p_\beta(x)D^\beta$ be a linear partial differential operator, as in Exercise 2.53.

(ii) Deduce, in the notation from Exercise 2.54, for all $\alpha \in \mathbf{N}_0^n$,

$$P^{(\alpha)}(x, \xi) = D_\xi^\alpha \sigma_P(x, \xi) = \sum_{|\beta|\le k} p_\beta(x)\frac{\beta!}{(\beta-\alpha)!}\xi^{\beta-\alpha}.$$

(iii) Use Leibniz' rule from Exercise 2.52.(ii) and parts (i) and (ii) to show, for $f \in C^k(\mathbf{R}^n)$ and $\alpha \in \mathbf{N}_0^n$,

$$\begin{aligned} P(x, D)(fg_\alpha)(a) &= \sum_{|\beta|\le k} p_\beta(x)D^\beta(g_\alpha f)(a) \\ &= \sum_{|\beta|\le k} p_\beta(x)\frac{\beta!}{(\beta-\alpha)!}D^{\beta-\alpha}f(a) = P^{(\alpha)}(x, D)f(a). \end{aligned}$$

(iv) Finally, employ Taylor expansion of g at a of order k to obtain the formula of Leibniz–Hörmander from Exercise 2.54.

Exercise 2.56 (Necessary condition for extremum). Let U be a convex open subset of $\mathbf{R}^n$ and let $a \in U$ be a critical point for $f \in C^2(U, \mathbf{R})$. Show that $Hf(a)$ being positive, and negative, semidefinite is a necessary condition for f having a relative minimum, and maximum, respectively, at a.

Exercise 2.57. Define $f : \mathbf{R}^2 \to \mathbf{R}$ by $f(x) = 16x_1^2 - 24x_1x_2 + 9x_2^2 - 30x_1 - 40x_2$. Discuss f along the same lines as in Example 2.9.9.

Exercise 2.58. Define $f : \mathbf{R}^2 \to \mathbf{R}$ by $f(x) = 2x_2^2 - x_1(x_1 - 1)^2$. Show that f has a relative minimum at $(\frac{1}{3}, 0)$ and a saddle point at $(1, 0)$.

Exercise 2.59. Define $f : \mathbf{R}^2 \to \mathbf{R}$ by $f(x) = x_1 x_2 e^{-\frac{1}{2}\|x\|^2}$. Prove that f has a saddle point at the origin, and local maxima at $\pm(1, 1)$ as well as local minima at $\pm(1, -1)$. Verify $\lim_{\|x\|\to\infty} f(x) = 0$. Show that, actually, the local extrema are absolute.

Exercise 2.60. Define $f : \mathbf{R}^2 \to \mathbf{R}$ by $f(x) = 3x_1x_2 - x_1^3 - x_2^3$. Show that $(1, 1)$ and 0 are the only critical points of f, and that f has a local maximum at $(1, 1)$ and a saddle point at 0.

Exercise 2.61.

(i) Let $f : \mathbf{R} \to \mathbf{R}$ be continuous, and suppose that f achieves a local maximum at only one point and is unbounded from above. Show that f achieves a local minimum somewhere.

(ii) Define $f : \mathbf{R}^2 \to \mathbf{R}$ by $g(x) = x_1^2 + x_2^2(1 + x_1)^3$. Show that 0 is the only critical point of f and that f attains a local minimum there. Prove that f is unbounded both from above and below.

(iii) Define $f : \mathbf{R}^2 \to \mathbf{R}$ by $f(x) = 3x_1e^{x_2} - x_1^3 - e^{3x_2}$. Show that $(1, 0)$ is the only critical point of f and that f attains a local maximum there. Prove that f is unbounded both from above and below.

Background. For constructing f in (iii), start with the function from Exercise 2.60 and push the saddle point to infinity.

Exercise 2.62 (Morse function). A function $f \in C^2(\mathbf{R}^n)$ is called a *Morse function* if all of its critical points are nondegenerate.

(i) Show that every compact subset in $\mathbf{R}^n$ contains only finitely many critical points for a given Morse function f.

(ii) Show that $\{0\} \cup \{\, x \in \mathbf{R}^n \mid \|x\| = 1 \,\}$ is the set of critical points of $f(x) = 2\|x\|^2 - \|x\|^4$. Deduce from (i) that f is not a Morse function.

Background. Generically a function is a Morse function; if not, a small perturbation will change any given function into a Morse function, at least locally. Thus, the function $f(x) = x^3$ on $\mathbf{R}$ is not Morse, as its critical point at 0 is degenerate; but the neighboring function $x \mapsto x^3 - 3\epsilon^2 x$ for $\epsilon \in \mathbf{R} \setminus \{0\}$ is a Morse function, because its two critical points are at $\pm\epsilon$ and are both nondegenerate. We speak of this as a *Morsification* of the function f.

Exercise 2.63 (Gram's matrix and Cauchy–Schwarz inequality). Let $v_1, \dots, v_d$ be vectors in $\mathbf{R}^n$ and let $G(v_1, \dots, v_d) \in \mathrm{Mat}^+(d, \mathbf{R})$ be Gram's matrix associated with the vectors $v_1, \dots, v_d$ as in Formula (2.4).

(i) Verify that $G(v_1, \dots, v_d)$ is positive semidefinite and conclude that we have $\det G(v_1, \dots, v_d) \geq 0$, which is known as the *generalized Cauchy–Schwarz inequality*.

(ii) Let $d \leq n$. Prove that $v_1, \dots, v_d$ are linearly dependent if and only if $\det G(v_1, \dots, v_d) = 0$.

(iii) Apply (i) to vectors v_1 and v_2 in $\mathbf{R}^n$. Note that one obtains the Cauchy–Schwarz inequality $|\langle v_1, v_2\rangle| \leq \|v_1\|\,\|v_2\|$, which justifies the name given in (i).

(iv) Prove that for every triplet of vectors v_1, v_2 and v_3 in $\mathbf{R}^n$

$$\Big(\frac{\langle v_1, v_2\rangle}{\|v_1\|\,\|v_2\|}\Big)^2 + \Big(\frac{\langle v_2, v_3\rangle}{\|v_2\|\,\|v_3\|}\Big)^2 + \Big(\frac{\langle v_3, v_1\rangle}{\|v_3\|\,\|v_1\|}\Big)^2$$
$$\leq 1 + 2\frac{\langle v_1, v_2\rangle\,\langle v_2, v_3\rangle\langle v_3, v_1\rangle}{\|v_1\|^2\,\|v_2\|^2\,\|v_3\|^2}.$$

Exercise 2.64. Let $A \in \mathrm{End}^+(\mathbf{R}^n)$. Suppose that $z = x + iy \in \mathbf{C}^n$ is an eigenvector corresponding to the eigenvalue $\lambda = \mu + i\nu \in \mathbf{C}$, thus $Az = \lambda z$. Show that $Ax = \mu x - \nu y$ and $Ay = \nu x + \mu y$, and deduce

$$0 = \langle x, Ay\rangle - \langle Ax, y\rangle = \nu(\|x\|^2 + \|y\|^2).$$

Now verify the fact known from the Spectral Theorem 2.9.3 that the eigenvalues of A are real, with corresponding eigenvectors in $\mathbf{R}^n$.

Exercise 2.65 (Another proof of Spectral Theorem – sequel to Exercise 1.1). The following proof of Theorem 2.9.3 is not as efficient as the one in Section 2.9; on the other hand, it utilizes many techniques that have been studied in the current chapter. The notation is as in the theorem and its proof; in particular, $g : \mathbf{R}^n \to \mathbf{R}$ is given by $g(x) = \langle Ax, x\rangle$.

(i) Prove that there exists $a \in S^{n-1}$ such that $g(x) \leq g(a)$, for all $x \in S^{n-1}$.

Define $L = \mathbf{R}a$ and recall from Exercise 1.1 the notation $L^\perp$ for the orthocomplement of L.

(ii) Select $h \in S^{n-1} \cap L^\perp$. Show that $\gamma_h(t) = \frac{1}{\sqrt{1+t^2}}(a + th)$, for $t \in \mathbf{R}$, defines a mapping $\gamma_h : \mathbf{R} \to S^{n-1}$.

(iii) Deduce from (i) and (ii) that $g \circ \gamma_h : \mathbf{R} \to \mathbf{R}$ has a critical point at 0, and conclude $D(g \circ \gamma_h)(0) = 0$.

(iv) Prove $D\gamma_h(t) = (1+t^2)^{-\frac{3}{2}}(h - ta)$, for $t \in \mathbf{R}$. Verify by means of the chain rule that, for $t \in \mathbf{R}$,

$$D(g \circ \gamma_h)(t) = 2(1+t^2)^{-2}((1-t^2)\langle Aa, h\rangle + t(\langle Ah, h\rangle - \langle Aa, a\rangle)).$$

(v) Conclude from (iii) and (iv) that $\langle Aa, h\rangle = 0$, for all $h \in L^{\perp}$. Using Exercise 1.1.(iv) deduce that $Aa \in (L^{\perp})^{\perp} = L = \mathbf{R}a$, that is, there exists $\lambda \in \mathbf{R}$ with $Aa = \lambda a$.

(vi) Show that $A(L^{\perp}) \subset L^{\perp}$ and that A acting as a linear operator in $L^{\perp}$ is again self-adjoint. Now deduce the assertion of the theorem by mathematical induction over $n \in \mathbf{N}$.

Exercise 2.66 (Another proof of Lemma 2.1.1). Let $A \in \mathrm{Lin}(\mathbf{R}^n, \mathbf{R}^p)$. Use the notation from Example 2.9.6 and Theorem 2.9.3; in particular, denote by $\lambda_1, \dots, \lambda_n$ the eigenvalues of A^tA, each eigenvalue being repeated a number of times equal to its multiplicity.

(i) Show that $\lambda_j \geq 0$, for $1 \leq j \leq n$.

(ii) Using (i) show, for $h \in \mathbf{R}^n \setminus \{0\}$,

$$\frac{\|Ah\|^2}{\|h\|^2} = \frac{\langle A^tAh, h\rangle}{\|h\|^2} \leq \max\{\,\lambda_j \mid 1 \leq j \leq n\,\} \leq \sum_{1\leq j\leq n} \lambda_j = \mathrm{tr}(A^tA).$$

Conclude that the operator norm $\|\cdot\|$ satisfies

$$\|A\|^2 = \max\{\,\lambda_j \mid 1 \leq j \leq n\,\} \qquad \text{and} \qquad \|A\| \leq \|A\|_{\mathrm{Eucl}} \leq \sqrt{n}\,\|A\|.$$

(iii) Prove that $\|A\| = 1$ and $\|A\|_{\mathrm{Eucl}} = \sqrt{2}$, if

$$A = \begin{pmatrix} \cos\alpha & -\sin\alpha \\ \sin\alpha & \cos\alpha \end{pmatrix} \qquad (\alpha \in \mathbf{R}).$$

Exercise 2.67 (Minimax principle). Let the notation be as in the Spectral Theorem 2.9.3. In particular, suppose the eigenvalues of A to be enumerated in decreasing order, each eigenvalue being repeated a number of times equal to its multiplicity; thus: $\lambda_1 \geq \lambda_2 \geq \cdots$, with corresponding orthonormal eigenvectors $a_1, a_2, \dots$ in $\mathbf{R}^n$.

(i) Let $y \in \mathbf{R}^n$ be arbitrary. Show that there exist α_1 and $\alpha_2 \in \mathbf{R}$ with $\langle \alpha_1 a_1 + \alpha_2 a_2, y \rangle = 0$ and

$$\langle A(\alpha_1 a_1 + \alpha_2 a_2), \alpha_1 a_1 + \alpha_2 a_2 \rangle = \lambda_1 \alpha_1^2 + \lambda_2 \alpha_2^2 \geq \lambda_2 \| \alpha_1 a_1 + \alpha_2 a_2 \|^2.$$

Deduce, for the Rayleigh quotient R for A,

$$\lambda_2 \leq \max\{\, R(x) \mid x \in \mathbf{R}^n \setminus \{0\} \text{ with } \langle x, y \rangle = 0 \,\}.$$

On the other hand, by taking $y = a_1$, verify that the maximum is precisely λ_2. Conclude that

$$\lambda_2 = \min_{y \in \mathbf{R}^n} \max\{\, R(x) \mid x \in \mathbf{R}^n \setminus \{0\} \text{ with } \langle x, y \rangle = 0 \,\}.$$

(ii) Now prove by mathematical induction over $k \in \mathbf{N}$ that λ_{k+1} is given by

$$\min_{y_1, \dots, y_k \in \mathbf{R}^n} \max\{\, R(x) \mid x \in \mathbf{R}^n \setminus \{0\} \text{ with } \langle x, y_1 \rangle = \cdots = \langle x, y_k \rangle = 0 \,\}.$$

Background. This characterization of the eigenvalues by the minimax principle, and the algorithms based on it, do not require knowledge of the corresponding eigenvectors.

Exercise 2.68 (Polar decomposition of automorphism). Every $A \in \mathrm{Aut}(\mathbf{R}^n)$ can be represented in the form

$$A = KP, \qquad K \in \mathbf{O}(\mathbf{R}^n), \qquad P \in \mathrm{End}^+(\mathbf{R}^n) \cap \mathrm{Aut}(\mathbf{R}^n) \text{ positive definite},$$

and this decomposition is unique. We call P the *polar* or *radial* part of A. Prove this decomposition by noting that $A^t A \in \mathrm{End}^+(\mathbf{R}^n)$ is positive definite and has to be equal to $PK^tKP = P^2$. Verify that there exists a basis $(a_1, \dots, a_n)$ for $\mathbf{R}^n$ such that $A^t A a_j = \lambda_j^2 a_j$, where $\lambda_j > 0$, for $1 \leq j \leq n$. Next define P by $Pa_j = \lambda_j a_j$, for $1 \leq j \leq n$. Then P is uniquely determined by $A^t A$ and thus by A. Finally, define $K = AP^{-1}$ and verify that $K \in \mathbf{O}(\mathbf{R}^n)$.
Background. Similarly one finds a decomposition $A = P'K'$, which generalizes the polar decomposition $x = r(\cos\alpha, \sin\alpha)$ of $x \in \mathbf{R}^2 \setminus \{0\}$. Note that $\mathbf{O}(\mathbf{R}^n)$ is a subgroup of $\mathrm{Aut}(\mathbf{R}^n)$, whereas the collection of self-adjoint $P \in \mathrm{End}^+(\mathbf{R}^n) \cap \mathrm{Aut}(\mathbf{R}^n)$ is not a subgroup (the product of two self-adjoint operators is self-adjoint if and only if the operators commute). The polar decomposition is also known as the *Cartan decomposition*.

Exercise 2.69. Define $f : \mathbf{R}_+ \times \mathbf{R} \to \mathbf{R}$ by $f(x,t) = \frac{2x^2 t}{(x^2+t^2)^2}$. Show that $\int_0^1 f(x,t)\, dt = \frac{1}{1+x^2}$ and deduce

$$\lim_{x \to 0} \int_0^1 f(x,t)\, dt \neq \int_0^1 \lim_{x \to 0} f(x,t)\, dt.$$

Show that the conditions of the Continuity Theorem 2.10.2 are not satisfied.

Exercise 2.70 (Method of least squares). Let $I = [0, 1]$. Suppose we want to approximate on I the continuous function $f : I \to \mathbf{R}$ by an affine function of the form $t \mapsto \lambda t + c$. The *method of least squares* would require that λ and c be chosen to minimize the integral

$$\int_0^1 (\lambda t + c - f(t))^2\, dt.$$

Show $\lambda = 6 \int_0^1 (2t - 1) f(t)\, dt$ and $c = 2 \int_0^1 (2 - 3t) f(t)\, dt$.

Exercise 2.71 (Sequel to Exercise 0.1). Prove

$$F(x) := \int_0^{\frac{\pi}{2}} \log(1 + x\cos^2 t)\, dt = \pi \log\Big(\frac{1+\sqrt{1+x}}{2}\Big) \qquad (x > -1).$$

Hint: Differentiate F, use the substitution $t = \arctan u$ from Exercise 0.1.(ii), and deduce $F'(x) = \frac{\pi}{2}\left(\frac{1}{x} - \frac{1}{x\sqrt{1+x}}\right)$.

Exercise 2.72 (Sequel to Exercise 0.1).

(i) Prove $\int_0^\pi \log(1 - 2r\cos\alpha + r^2)\, d\alpha = 0$, for $|r| < 1$.
Hint: Differentiate with respect to r and use the substitution $\alpha = 2\arctan t$ from Exercise 0.1.(i).

(ii) Set $e_1 = (1, 0)$ and $x(r, \alpha) = r(\cos\alpha, \sin\alpha) \in \mathbf{R}^2$. Deduce that we have $\int_0^\pi \log \|e_1 - x(r,\alpha)\|\, d\alpha = 0$ for $|r| < 1$. Interpret this result geometrically.

(iii) The integral in (i) also can be computed along the lines of Exercise 0.18.(i). In fact, antidifferentiation of the geometric series $(1 - z)^{-1} = \sum_{k\in\mathbf{N}_0} z^k$ yields the power series $-\log(1-z) = \sum_{k\in\mathbf{N}} \frac{z^k}{k}$ which converges for $z \in \mathbf{C}$, $|z| \le 1$, $z \ne 1$. Substitute $z = re^{i\alpha}$, use De Moivre's formula, and take the real parts; this results in

$$\log(1 - 2r\cos\alpha + r^2) = -2\sum_{k\in\mathbf{N}} \frac{r^k}{k}\cos k\alpha.$$

Now interchange integration and summation.

(iv) Deduce from (i) that $\int_0^\pi \log(1 - 2r\cos\alpha + r^2)\, d\alpha = 2\pi\log|r|$, for $|r| > 1$.

Exercise 2.73 ($\int_{\mathbf{R}} e^{-x^2}\, dx = \sqrt{\pi}$ – needed for Exercise 2.87). Define $f : \mathbf{R} \to \mathbf{R}$ by

$$f(a) = \int_0^1 \frac{e^{-a^2(1+t^2)}}{1+t^2}\, dt.$$

(i) Prove by a substitution of variables that $f'(a) = -2e^{-a^2} \int_0^a e^{-x^2}\,dx$, for $a \in \mathbf{R}$.

Define $g : \mathbf{R} \to \mathbf{R}$ by $g(a) = f(a) + \left(\int_0^a e^{-x^2}\,dx\right)^2$.

(ii) Prove that $g' = 0$ on $\mathbf{R}$; then conclude that $g(a) = \frac{\pi}{4}$, for all $a \in \mathbf{R}$.

(iii) Prove that $0 \le f(a) \le e^{-a^2}$, for all $a \in \mathbf{R}$, and go on to show that (compare with Example 6.10.8 and Exercises 6.41 and 6.50.(i))

$$\int_{\mathbf{R}} e^{-x^2}\,dx = \sqrt{\pi}.$$

Background. The motivation for the arguments above will become apparent in Exercise 6.15.

Exercise 2.74 (Needed for Exercises 2.75 and 8.34). Let $f : \mathbf{R}^2 \to \mathbf{R}$ be a C^1 function that vanishes outside a bounded set in $\mathbf{R}^2$, and let $h : \mathbf{R} \to \mathbf{R}$ be a C^1 function. Then

$$D\left(\int_{-\infty}^{h(x)} f(x,t)\,dt\right) = f(x, h(x))\,h'(x) + \int_{-\infty}^{h(x)} D_1 f(x,t)\,dt \qquad (x \in \mathbf{R}).$$

Hint: Define $F : \mathbf{R}^2 \to \mathbf{R}$ by

$$F(y,z) = \int_{-\infty}^{z} f(y,t)\,dt.$$

Verify that F is a function with continuous partial derivative $D_1 F : \mathbf{R}^2 \to \mathbf{R}$ given by

$$D_1 F(y,z) = \int_{-\infty}^{z} D_1 f(y,t)\,dt.$$

On account of the Fundamental Theorem of Integral Calculus on $\mathbf{R}$, the function $D_2 F$ exists and is continuous, because $D_2 F(y,z) = f(y,z)$. Conclude that F is differentiable. Verify that the derivative of the mapping $\mathbf{R} \to \mathbf{R}$ with $x \mapsto F(x, h(x))$ is given by

$$\begin{aligned} DF(x,h(x)) &= \left(D_1 F(x,h(x))\ \ D_2 F(x,h(x))\right)\begin{pmatrix} 1 \\ h'(x) \end{pmatrix} \\ &= D_1 F(x,h(x)) + D_2 F(x,h(x))\,h'(x). \end{aligned}$$

Combination of these arguments now yields the assertion.

Exercise 2.75 (Repeated antidifferentiation and Taylor's formula – sequel to Exercise 2.74 – needed for Exercises 6.105 and 6.108). Let $f \in C(\mathbf{R})$ and $k \in \mathbf{N}$. Define $D^0 f = f$, and

$$D^{-k} f(x) = \int_0^x \frac{(x-t)^{k-1}}{(k-1)!} f(t)\,dt.$$

(i) Prove that $(D^{-k} f)' = D^{-k+1} f$ by means of Exercise 2.74, and deduce from the Fundamental Theorem of Integral Calculus on $\mathbf{R}$ that

$$D^{-k} f(x) = \int_0^x D^{-k+1} f(t)\,dt \qquad (k \in \mathbf{N}).$$

Show (compare with Exercise 2.81)

$$(\star) \qquad \int_0^x \frac{(x-t)^{k-1}}{(k-1)!} f(t)\,dt = \int_0^x \Big(\int_0^{t_1} \cdots \Big(\int_0^{t_{k-1}} f(t_k)\,dt_k \Big) \cdots dt_2 \Big)\,dt_1.$$

Background. The negative exponent in $D^{-k} f$ is explained by the fact that it is a k-fold **anti**derivative of f. Furthermore, in this fashion the notation is compatible with the one in Exercise 6.105.

In the light of the Fundamental Theorem of Integral Calculus on $\mathbf{R}$ the right–hand side of $(\star)$ is seen to be a k-fold antiderivative, say g, of f. Evidently, therefore, a k-fold antiderivative can also be calculated by one integration.

(ii) Prove that

$$g^{(j)}(0) = 0 \qquad (0 \le j < k).$$

Now assume $f \in C^{k+1}(\mathbf{R})$ with $k \in \mathbf{N}_0$. Accordingly

$$h(x) := \int_0^x \frac{(x-t)^k}{k!} f^{(k+1)}(t)\,dt$$

is a $(k+1)$-fold antiderivative of $f^{(k+1)}$, with $h^{(j)}(0) = 0$, for $0 \le j \le k$. On the other hand, f is also an $(k+1)$-fold antiderivative of $f^{(k+1)}$.

(iii) Show, for $x \in \mathbf{R}$,

$$\begin{aligned} f(x) - \sum_{0 \le j \le k} \frac{f^{(j)}(0)}{j!} x^j &= \int_0^x \frac{(x-t)^k}{k!} f^{(k+1)}(t)\,dt \\ &= \frac{x^{k+1}}{k!} \int_0^1 (1-t)^k f^{(k+1)}(tx)\,dt, \end{aligned}$$

Taylor's formula for f at 0, with the integral formula for the k-th remainder (see Lemma 2.8.2).

Exercise 2.76 (Higher-order generalization of Hadamard's Lemma). Let U be a convex open subset of $\mathbf{R}^n$ and $f \in C^\infty(U, \mathbf{R}^p)$. Under these conditions a higher-order generalization of Hadamard's Lemma 2.2.7 is as follows. For every $k \in \mathbf{N}$ there exists $\phi_k \in C^\infty(U, \mathrm{Lin}^k(\mathbf{R}^n, \mathbf{R}^p))$ such that

$$(\star) \qquad f(x) = \sum_{0 \le j < k} \frac{1}{j!} D^j f(a)((x-a)^j) + \phi_k(x)((x-a)^k),$$

$$(\star\star) \qquad \phi_k(a) = \frac{1}{k!} D^k f(a).$$

We will prove this result in three steps.

(i) Let x and $a \in U$. Applying the Fundamental Theorem of Integral Calculus on $\mathbf{R}$ to the line segment $L(a, x) \subset U$ and the mapping $t \mapsto f(x_t)$ with $x_t = a + t(x - a)$ (see Formula (2.15)) and using the continuity of Df, deduce

$$f(x) = f(a) + \int_0^1 \frac{d}{dt} f(x_t)\, dt = f(a) + \int_0^1 Df(x_t)\, dt\, (x-a)$$

$$=: f(a) + \phi(x)(x-a).$$

Derive $\phi \in C^\infty(U, \mathrm{Lin}(\mathbf{R}, \mathbf{R}^p))$ on account of the differentiability of the integral with respect to the parameter $x \in U$. Differentiate the identity above with respect to x at $x = a$ to find $\phi(a) = Df(a)$.

Note that this proof of Hadamard's Lemma is different from the one given in Lemma 2.2.7.

(ii) Prove $(\star)$ by mathematical induction over $k \in \mathbf{N}$. For $k = 1$ the result follows from part (i). Suppose the result has been obtained up to order k, and apply part (i) to $\phi_k \in C^\infty(U, \mathrm{Lin}^k(\mathbf{R}^n, \mathbf{R}^p))$. On the strength of Lemma 2.7.4 we obtain $\phi_{k+1} \in \mathbf{C}^\infty(U, \mathrm{Lin}^{k+1}(\mathbf{R}^n, \mathbf{R}^p))$ satisfying, for x near a,

$$\phi_k(x) = \phi_k(a) + \phi_{k+1}(x)(x-a) = \frac{1}{k!} D^k f(a) + \phi_{k+1}(x)(x-a).$$

(iii) For verifying $(\star\star)$, replace k by $k+1$ in $(\star)$ and differentiate the identity thus obtained $k+1$ times with respect to x at $x = a$. We find $D^{k+1} f(a) = (k+1)!\, \phi_{k+1}(a)$.

(iv) In order to obtain an estimate for the remainder term, apply Corollary 2.7.5 and verify

$$\|\phi_k(x)((x-a)^k)\| = \mathcal{O}(\|x-a\|^k), \qquad x \to a.$$

Background. It seems not to be possible to prove the higher-order generalization of Hadamard's Lemma without the integral representation used in part (i). Indeed, in (iii) we need the differentiability of $\phi_{k+1}(x)$ as such, without it being applied to the $(k+1)$-fold product of the vector $x - a$ with itself.

Exercise 2.77. Calculate

$$\int_{-\frac{\pi}{2}}^{\frac{\pi}{2}} \int_1^2 e^{x_2} \sin\left(\frac{x_1}{x_2}\right) dx_2\, dx_1.$$

Exercise 2.78 (Interchanging order of integration implies integration by parts). Consider $I = [\,a, b\,]$ and $F \in C(I^2)$. Prove

$$\int_a^b \int_a^x F(x, y)\, dy\, dx = \int_a^b \int_y^b F(x, y)\, dx\, dy.$$

Let f and $g \in C^1(I)$ and note $f(x) = f(a) + \int_a^x f'(y)\, dy$ and $g(y) = g(b) + \int_a^y g'(x)\, dx$, for x and $y \in I$. Now deduce

$$\int_a^b f(x)g'(x)\, dx = f(a)g(a) - f(b)g(b) - \int_a^b f'(y)g(y)\, dy.$$

Exercise 2.79 (Sequel to Exercise 0.1 – needed for Exercise 6.50). Show by means of the substitution $\alpha = 2\arctan t$ from Exercise 0.1.(i) that

$$\int_0^\pi \frac{d\alpha}{1 + p\cos\alpha} = \frac{\pi}{\sqrt{1 - p^2}} \qquad (|p| < 1).$$

Prove by changing the order of integration that

$$\int_0^\pi \frac{\log(1 + x\cos\alpha)}{\cos\alpha}\, d\alpha = \pi \arcsin x \qquad (|x| < 1).$$

Exercise 2.80 (Sequel to Exercise 0.1). Let $0 < a < 1$ and define $f : [\,0, \frac{\pi}{2}\,] \times [\,0, 1\,] \to \mathbf{R}$ by $f(x, y) = \frac{1}{1 - a^2y^2\sin^2 x}$. Substitute $x = \arctan t$ in the first integral, as in Exercise 0.1.(ii), in order to show

$$\int_0^{\frac{\pi}{2}} f(x, y)\, dx = \frac{\pi}{2\sqrt{1 - a^2y^2}}, \qquad \int_0^1 f(x, y)\, dy = \frac{1}{2a\sin x} \log\frac{1 + a\sin x}{1 - a\sin x}.$$

Deduce

$$\int_0^{\frac{\pi}{2}} \frac{1}{\sin x} \log\frac{1 + a\sin x}{1 - a\sin x}\, dx = \pi \arcsin a.$$

Exercise 2.81 (Exercise 2.75 revisited). Let $f \in C(\mathbf{R})$. Prove, for all $x \in \mathbf{R}$ and $k \in \mathbf{N}$ (compare with Exercise 2.75)

$$\int_0^x \Big(\int_0^{t_1} \cdots \Big(\int_0^{t_{k-1}} f(t_k)\, dt_k \Big) \cdots dt_2 \Big)\, dt_1 = \int_0^x \frac{(x-t)^{k-1}}{(k-1)!} f(t)\, dt.$$

Hint: If the choice is made to start from the left–hand side, use mathematical induction, and apply

$$\int_0^x \Big(\int_0^{t_1} \cdots dt \Big)\, dt_1 = \int_0^x \Big(\int_t^x \cdots dt_1 \Big)\, dt.$$

If the choice is to start from the right–hand side, use integration by parts.

Exercise 2.82. Using Example 2.10.14 and integration by parts show, for all $x \in \mathbf{R}$,

$$\int_{\mathbf{R}_+} \frac{1 - \cos xt}{t^2}\, dt = \int_{\mathbf{R}_+} \frac{\sin^2 xt}{t^2}\, dt = \frac{\pi}{2}|x|.$$

Deduce, using $\sin^2 2t = 4(\sin^2 t - \sin^4 t)$, and integration by parts, respectively,

$$\int_{\mathbf{R}_+} \frac{\sin^4 t}{t^2}\, dt = \frac{\pi}{4}, \qquad \int_{\mathbf{R}_+} \frac{\sin^4 t}{t^4}\, dt = \frac{\pi}{3}.$$

Exercise 2.83 (Sequel to Exercise 0.15). The notation is as in that exercise. Deduce from part (iv) by means of differentiation that

$$\int_{\mathbf{R}_+} \frac{x^{p-1} \log x}{x+1}\, dx = -\frac{\pi^2 \cos(\pi p)}{\sin^2(\pi p)} \qquad (0 < p < 1).$$

Exercise 2.84 (Sequel to Exercise 2.73 – needed for Exercise 6.83). Define $F : \mathbf{R}_+ \to \mathbf{R}$ by

$$F(\xi) = \int_{\mathbf{R}_+} e^{-\frac{1}{2}x^2} \cos(\xi x)\, dx.$$

By differentiating F obtain $F' + \xi F = 0$, and by using Exercise 2.73 deduce $F(\xi) = \sqrt{\frac{\pi}{2}} e^{-\frac{1}{2}\xi^2}$, for $\xi \in \mathbf{R}$. Conclude that

$$\sqrt{\frac{\pi}{2}}\, \xi e^{-\frac{1}{2}\xi^2} = \int_{\mathbf{R}_+} x e^{-\frac{1}{2}x^2} \sin(\xi x)\, dx \qquad (\xi \in \mathbf{R}).$$

Exercise 2.85 (Laplace's integrals). Define $F : \mathbf{R}_+ \to \mathbf{R}$ by

$$F(x) = \int_{\mathbf{R}_+} \frac{\sin xt}{t(1+t^2)}\, dt.$$

By differentiating F twice, obtain

$$F(x) - F''(x) = \int_{\mathbf{R}_+} \frac{\sin xt}{t}\, dt = \frac{\pi}{2} \qquad (x \in \mathbf{R}_+).$$

Since $\lim_{x\downarrow 0} F(x) = 0$ and F is bounded, deduce $F(x) = \frac{\pi}{2}(1-e^{-x})$. Conclude, on differentiation, that we have the following, known as *Laplace's integrals* (compare with Exercise 6.99.(iii))

$$\int_{\mathbf{R}_+} \frac{\cos xt}{1+t^2}\, dt = \int_{\mathbf{R}_+} \frac{t \sin xt}{1+t^2}\, dt = \frac{\pi}{2} e^{-x} \qquad (x \in \mathbf{R}_+).$$

Exercise 2.86. Prove for $x \geq 0$

$$\int_{\mathbf{R}_+} \frac{\log(1+x^2t^2)}{1+t^2}\, dt = \pi \log(1+x).$$

Hint: Use $|\frac{2x}{x^2-1}(\frac{1}{1+t^2} - \frac{1}{1+x^2t^2})| \leq \frac{4}{1+t^2}$, for all $x \geq \sqrt{2}$.

Exercise 2.87 (Sequel to Exercise 2.73 – needed for Exercises 6.53, 6.97 and 6.99).

(i) Using Exercise 2.73 prove for $x \geq 0$

$$I(x) := \int_{\mathbf{R}_+} e^{-t^2 - \frac{x^2}{t^2}}\, dt = \frac{1}{2}\sqrt{\pi}\, e^{-2x}.$$

We now derive this result in another way.

(ii) Prove $I(x) = \sqrt{x} \int_{\mathbf{R}_+} e^{-x(t^2+t^{-2})}\, dt$, for all $x \in \mathbf{R}_+$.

(iii) Write $\mathbf{R}_+$ as the union of $]\,0, 1\,]$ and $]\,1, \infty\,[$. Replace t by t^{-1} on $]\,0, 1\,]$ and conclude, for all $x \in \mathbf{R}_+$,

$$I(x) = \sqrt{x} \int_1^\infty (1+t^{-2}) e^{-x(t^2+t^{-2})}\, dt.$$

(iv) Prove by the substitution $u = t - t^{-1}$ that $I(x) = \sqrt{x} \int_{\mathbf{R}_+} e^{-x(u^2+2)}\, du$, for all $x \in \mathbf{R}_+$; and conclude that the identity in (i) holds.

(v) Let a and $b \in \mathbf{R}_+$. Prove

$$\int_{\mathbf{R}_+} \frac{1}{\sqrt{t}} e^{-a^2 t - \frac{b^2}{t}}\, dt = \sqrt{\pi}\,\frac{e^{-2ab}}{a}, \qquad \int_{\mathbf{R}_+} \frac{1}{t\sqrt{t}} e^{-a^2 t - \frac{b^2}{t}}\, dt = \sqrt{\pi}\,\frac{e^{-2ab}}{b}.$$

In particular, show

$$e^{-x} = \frac{1}{\sqrt{\pi}} \int_{\mathbf{R}_+} \frac{e^{-t}}{\sqrt{t}} e^{-\frac{x^2}{4t}}\, dt \qquad (x \geq 0).$$

Exercise 2.88. For the verification of

$$\int_0^1 \frac{\arctan t}{t\sqrt{1-t^2}}\, dt = \frac{\pi}{2}\log(1+\sqrt{2})$$

(note that the integrand is singular at $t = 1$) consider

$$F(x) = \int_0^1 \frac{\arctan xt}{t\sqrt{1-t^2}}\, dt \qquad (x \geq 0).$$

Differentiation under the integral sign now yields

$$F'(x) = \int_0^1 \frac{1}{1+x^2t^2}\,\frac{1}{\sqrt{1-t^2}}\, dt,$$

and with the substitution $t = \frac{1}{\sqrt{1+u^2}}$ we compute the definite integral to be

$$F'(x) = \int_{\mathbf{R}_+} \frac{1}{u^2+1+x^2}\, du = \frac{\pi}{2}\,\frac{1}{\sqrt{1+x^2}}.$$

Integrating this identity we obtain a constant $c \in \mathbf{R}$ such that

$$F(x) = c + \frac{\pi}{2}\operatorname{arcsinh} x = \frac{\pi}{2}\log(x+\sqrt{1+x^2}) + c.$$

From $F(0) = 0$ we get $c = 0$, which implies the desired formula.

Exercise 2.89. Using interchange of integrations we can give another proof of Exercise 2.88. Verify

$$\frac{\arctan t}{t} = \int_0^1 \frac{1}{1+t^2y^2}\, dy \qquad (t \in \mathbf{R}).$$

With F as in Exercise 2.88, now deduce from the computations in that exercise,

$$\begin{aligned} F(1) &= \int_0^1 \frac{1}{\sqrt{1-t^2}} \int_0^1 \frac{1}{1+t^2y^2}\, dy\, dt = \int_0^1\int_0^1 \frac{1}{(1+t^2y^2)\sqrt{1-t^2}}\, dt\, dy \\ &= \frac{\pi}{2}\int_0^1 \frac{1}{\sqrt{1+y^2}}\, dy = \frac{\pi}{2}\log(1+\sqrt{2}). \end{aligned}$$

Exercise 2.90 (Airy function – needed for Exercise 6.61). In the theory of light an important role is played by the *Airy function*

$$\mathrm{Ai} : \mathbf{R} \to \mathbf{C}, \qquad \mathrm{Ai}(x) = \frac{1}{2\pi} \int_{\mathbf{R}} e^{i(\frac{1}{3}t^3 + xt)}\, dt = \frac{1}{\pi} \int_{\mathbf{R}_+} \cos(\frac{1}{3}t^3 + xt)\, dt,$$

which satisfies *Airy's differential equation*

$$u''(x) - x\, u(x) = 0.$$

Although the integrand of $\mathrm{Ai}(x)$ does not die away as $|t| \to \infty$, its increasingly rapid oscillations induce convergence of the integral. However, the integral diverges when x lies in $\mathbf{C} \setminus \mathbf{R}$. (The Airy function is defined as the inverse Fourier transform of the function $t \mapsto e^{i\frac{1}{3}t^3}$, see Theorem 6.11.6.)

Define

$$f_\pm : \mathbf{C} \times \mathbf{R}^2 \to \mathbf{C} \qquad \text{by} \qquad f_\pm(z, x, t) = e^{\pm i(\frac{1}{3}z^3t^3 + zxt)},$$

and consider, for $z \in U_\pm = \{\, z \in \mathbf{C} \mid 0 < \pm \arg z < \frac{\pi}{3} \,\}$ and $x \in \mathbf{R}$,

$$F_\pm(z, x) = z \int_{\mathbf{R}_+} f_\pm(z, x, t)\, dt.$$

(i) Verify $2\pi\, \mathrm{Ai}(x) = F_-(1, x) + F_+(1, x)$. Further show $\lim_{t\to\infty} f_\pm(z, x, t) = 0$, for all $(z, x) \in U_\pm \times \mathbf{R}$ and that $f_\pm(z, x, t)$ remains bounded for $t \to \infty$, if $(z, x) \in K_\pm \times \mathbf{R}$; here $K_\pm = \overline{U}_\pm$, the closure of $U_\pm$. In particular, $1 \in K_\pm$.

(ii) In order to prove convergence of the $F_\pm(z, x)$, show, for $z \in \mathbf{C}$ near 1,

$$\int_{1+|x|}^{\infty} f_\pm(z, x, t)\, dt = \left[\frac{f_\pm(z, x, t)}{\pm i(z^3t^2 + zx)} \right]_{1+|x|}^{\infty} + \int_{1+|x|}^{\infty} \frac{2z^3 t}{\pm i(z^3t^2 + zx)^2} f_\pm(z, x, t)\, dt.$$

Using (i), prove that the boundary term converges and the integral converges absolutely and uniformly for $z \in K_\pm$ near 1 according to De la Vallée-Poussin's test. Deduce that $F_\pm(z, x)$ is a continuous function of $z \in K_\pm$ near 1.

(iii) Use $z\frac{\partial f_\pm}{\partial z}(z, x, t) = t\frac{\partial f_\pm}{\partial t}(z, x, t)$ to prove, for $z \in U_\pm$,

$$\begin{aligned}\frac{\partial F_\pm}{\partial z}(z, x) &= \int_{\mathbf{R}_+} f_\pm(z, x, t)\, dt + \int_{\mathbf{R}_+} t \frac{\partial f_\pm}{\partial t}(z, x, t)\, dt \\ &= \int_{\mathbf{R}_+} f_\pm(z, x, t)\, dt + \big[\, t f_\pm(z, x, t) \,\big]_0^\infty - \int_{\mathbf{R}_+} f_\pm(z, x, t)\, dt = 0.\end{aligned}$$

Conclude that $F_\pm(z, x)$ is a constant function of $z \in U_\pm$. Using (ii) now conclude that $z \mapsto F_\pm(z, x)$ actually is a constant function on $K_\pm$. (This may also be proved on the basis of Cauchy's Integral Theorem 8.3.11.)

(iv) Now, take $z = z_\pm := e^{\pm i\frac{\pi}{6}}$ and deduce from (iii)

$$\begin{aligned} F_\pm(1,x) &= F_\pm(z_\pm, x) = z_\pm \int_{\mathbf{R}_+} e^{-\frac{1}{3}t^3 \pm i z_\pm x t}\, dt \\ &= e^{\pm i\frac{\pi}{6}} \int_{\mathbf{R}_+} e^{-(\frac{1}{3}t^3 + \frac{1}{2}xt)}\, e^{\pm\frac{1}{2}\sqrt{3}\, i\, x t}\, dt. \end{aligned}$$

Conclude that $F_\pm(1,\cdot) : \mathbf{R} \to \mathbf{C}$ is a C^∞ function and deduce that $\mathrm{Ai} : \mathbf{R} \to \mathbf{C}$ is a C^∞ function.

(v) Show, for $x \in \mathbf{R}$,

$$\begin{aligned} \frac{\partial^2 F_\pm}{\partial x^2}(z_+, x) - x F_\pm(z_+, x) &= \pm i \int_{\mathbf{R}} (-t^2 \pm i z_\pm x) f_\pm(z_+, x, t)\, dt \\ &= \pm i \int_{\mathbf{R}_+} \frac{\partial f_\pm}{\partial t}(z_\pm, x, t)\, dt = \mp i. \end{aligned}$$

Finally deduce that the Airy function satisfies Airy's differential equation.

(vi) Deduce from (iv) that we have, with Γ denoting the Gamma function from Exercise 6.50 (see Exercise 6.61 for simplified expressions)

$$\begin{aligned} \mathrm{Ai}(0) &= \frac{1}{2\pi 3^{\frac{1}{6}}} \int_{\mathbf{R}_+} e^{-u} u^{-\frac{2}{3}}\, du = \frac{\Gamma(\frac{1}{3})}{2\pi 3^{\frac{1}{6}}}, \\ \mathrm{Ai}'(0) &= -\frac{3^{\frac{1}{6}}}{2\pi} \int_{\mathbf{R}_+} e^{-u} u^{-\frac{1}{3}}\, du = -\frac{3^{\frac{1}{6}}\Gamma(\frac{2}{3})}{2\pi}. \end{aligned}$$

(vii) Expand $e^{\pm i z_\pm x t}$ in part (iv) in a power series to show

$$F_\pm(1,x) = z_\pm \sum_{k\in\mathbf{N}_0} 3^{\frac{k-2}{3}} \Gamma\Big(\frac{k+1}{3}\Big) \frac{(\pm i z_\pm x)^k}{k!}.$$

Using $\Gamma(t+1) = t\Gamma(t)$ deduce the following power series expansion, which converges for all $x \in \mathbf{R}$,

$$\begin{aligned} \mathrm{Ai}(x) &= \mathrm{Ai}(0)\Big(1 + \frac{1}{3!}x^3 + \frac{1\cdot 4}{6!}x^6 + \frac{1\cdot 4\cdot 7}{9!}x^9 + \cdots\Big) \\ &\quad + \mathrm{Ai}'(0)\Big(x + \frac{2}{4!}x^4 + \frac{2\cdot 5}{7!}x^7 + \frac{2\cdot 5\cdot 8}{10!}x^{10} + \cdots\Big). \end{aligned}$$

Of course, this series also can be obtained by substitution of a power series $\sum_{k\in\mathbf{N}_0} a_k x^k$ in the differential equation and solving the recurrence relations.

(viii) Repeating the integration by parts in part (ii) prove $\mathrm{Ai}(x) = \mathcal{O}(x^{-k})$, $\quad x \to \infty$, for all $k \in \mathbf{N}$.

Exercises for Chapter 3

Exercise 3.1 (Needed for Exercise 6.59). Define $\Psi : \mathbf{R}^2_+ \to \mathbf{R}^2_+$ by

$$\Psi(y) = (y_1 + y_2, \frac{y_1}{y_2}).$$

Prove that Ψ is a C^∞ diffeomorphism, with inverse $\Phi : \mathbf{R}^2_+ \to \mathbf{R}^2_+$ given by $\Phi(x) = \frac{x_1}{x_2+1}(x_2, 1)$. Show that for $\Psi(y) = x$

$$\det D\Psi(y) = -\frac{y_1 + y_2}{y_2^2} = -\frac{(x_2 + 1)^2}{x_1}.$$

Exercise 3.2 (Needed for Exercises 3.19 and 6.64). Define $\Psi : \mathbf{R}^2 \to \mathbf{R}^2$ by

$$\Psi(y) = y_2(y_1(1, 0) + (1 - y_1)(0, 1)) = (y_1 y_2,\ (1 - y_1)y_2).$$

(i) Prove that $\det D\Psi(y) = y_2$, for $y \in \mathbf{R}^2$.

(ii) Prove that $\Psi :]\,0, 1\,[^2 \to \Delta^2$ is a C^∞ diffeomorphism, if we define $\Delta^2 \subset \mathbf{R}^2$ as the triangular open set

$$\Delta^2 = \{\, x \in \mathbf{R}^2_+ \mid x_1 + x_2 < 1 \,\}.$$

Exercise 3.3 (Needed for Exercise 3.22). Let $\Phi : \mathbf{R}^2 \to \mathbf{R}^2$ be defined by $\Phi(x) = (x_1 + x_2,\ x_1 - x_2)$.

(i) Prove that Φ is a C^∞ diffeomorphism.

(ii) Determine $\Phi(U)$ if U is the triangular open set given by

$$U = \{\, (y_1 y_2,\ y_1(1 - y_2)) \in \mathbf{R}^2 \mid (y_1, y_2) \in\ ]\,0, 1\,[\ \times\ ]\,0, 1\,[\,\}.$$

Exercise 3.4. Let $\Phi : \mathbf{R}^2 \to \mathbf{R}^2$ be defined by $\Phi(x) = e^{x_1}(\cos x_2,\ \sin x_2)$.

(i) Determine $\operatorname{im}(\Phi)$.

(ii) Prove that for every $x \in \mathbf{R}^2$ there exists a neighborhood U in $\mathbf{R}^2$ such that $\Phi : U \to \Phi(U)$ is a C^∞ diffeomorphism, but that Φ is not injective on all of $\mathbf{R}^2$.

(iii) Let $x = (0, \frac{\pi}{3})$, $y = \Phi(x)$, and let Ψ be the continuous inverse of Φ, defined in a neighborhood of y, such that $\Psi(y) = x$. Give an explicit formula for Ψ.

Exercise 3.5. Let $U = \,]-\pi,\ \pi\,[\ \times \mathbf{R}_+ \subset \mathbf{R}^2$, and define $\Phi : U \to \mathbf{R}^2$ by

$$\Phi(x) = (\cos x_1 \cosh x_2,\ \sin x_1 \sinh x_2).$$

(i) Prove that $\det D\Phi(x) = -(\sin^2 x_1 \cosh^2 x_2 + \cos^2 x_1 \sinh^2 x_2)$ for $x \in U$, and verify that Φ is regular on U.

(ii) Show that in general under Φ the lines $x_1 = a$ and $x_2 = b$ are mapped to hyperbolae and ellipses, respectively. Test whether Φ is injective.

(iii) Determine $\mathrm{im}(\Phi)$.
Hint: Assume $y \in \mathbf{R}^2 \setminus \{\, (y_1,\ 0) \mid y_1 \leq 0 \,\}$. Then prove that

$$\lim_{x_2 \to +\infty} \Big(\frac{y_1}{\cosh x_2}\Big)^2 + \Big(\frac{y_2}{\sinh x_2}\Big)^2 = 0,$$

$$\lim_{x_2 \downarrow 0} \Big(\frac{y_1}{\cosh x_2}\Big)^2 + \Big(\frac{y_2}{\sinh x_2}\Big)^2 = \infty.$$

Conclude, by application of the Intermediate Value Theorem 1.9.5, that $x_2 \in \mathbf{R}_+$ exists such that, for some suitable $x_1 \in \,]-\pi, \pi\,[$, one has

$$\frac{y_1}{\cosh x_2} = \cos x_1, \qquad \frac{y_2}{\sinh x_2} = \sin x_1.$$

Exercise 3.6 (Cylindrical coordinates – needed for Exercise 3.8). Define $\Psi : \mathbf{R}^3 \to \mathbf{R}^3$ by

$$\Psi(r, \alpha, x_3) = (r \cos\alpha,\ r \sin\alpha,\ x_3).$$

(i) Prove that Ψ is surjective, but not injective.

(ii) Find the points in $\mathbf{R}^3$ where Ψ is regular.

(iii) Determine a maximal open set V in $\mathbf{R}^3$ such that $\Psi : V \to \Psi(V)$ is a C^∞ diffeomorphism.

Background. The element $(r, \alpha, x_3) \in V$ consists of the *cylindrical coordinates* of $x = \Psi(r, \alpha, x_3)$.

Exercise 3.7 (Spherical coordinates – needed for Exercise 3.8). Define $\Psi : \mathbf{R}^3 \to \mathbf{R}^3$ by

$$\Psi(r, \alpha, \theta) = r(\cos\alpha \cos\theta,\ \sin\alpha \cos\theta,\ \sin\theta).$$

(i) Draw the images under Ψ of the planes r = constant, α = constant, and θ = constant, respectively, and those of the lines (α, θ), (r, θ) and (r, α) = constant, respectively.

(ii) Prove that Ψ is surjective, but not injective.

(iii) Prove

$$\det D\Psi(r, \alpha, \theta) = r^2 \cos\theta,$$

and determine the points $y = (r, \alpha, \theta) \in \mathbf{R}^3$ such that the mapping Ψ is regular at y.

(iv) Let $V = \mathbf{R}_+ \times \,]-\pi, \pi\,[\, \times \,]-\frac{\pi}{2}, \frac{\pi}{2}\,[\, \subset \mathbf{R}^3$. Prove that $\Psi : V \to \mathbf{R}^3$ is injective and determine $U := \Psi(V) \subset \mathbf{R}^3$. Prove that $\Psi : V \to U$ is a C^∞ diffeomorphism.

(v) Define $\Phi : \mathbf{R}^3 \setminus \{\, x \in \mathbf{R}^3 \mid x_1 \leq 0,\ x_2 = 0 \,\} \to \mathbf{R}^3$ by

$$\Phi(x) = \left(\|x\|,\ 2\arctan\left(\frac{x_2}{x_1 + \sqrt{x_1^2 + x_2^2}} \right),\ \arcsin\left(\frac{x_3}{\|x\|} \right) \right).$$

Prove that $\operatorname{im} \Phi = V$ and that $\Phi : U \to V$ is a C^∞ diffeomorphism. Verify that $\Phi : U \to V$ is the inverse of $\Psi : V \to U$.

Background. The element $(r, \alpha, \theta) \in V$ consists of the *spherical coordinates* of $x = \Psi(r, \alpha, \theta) \in U$.

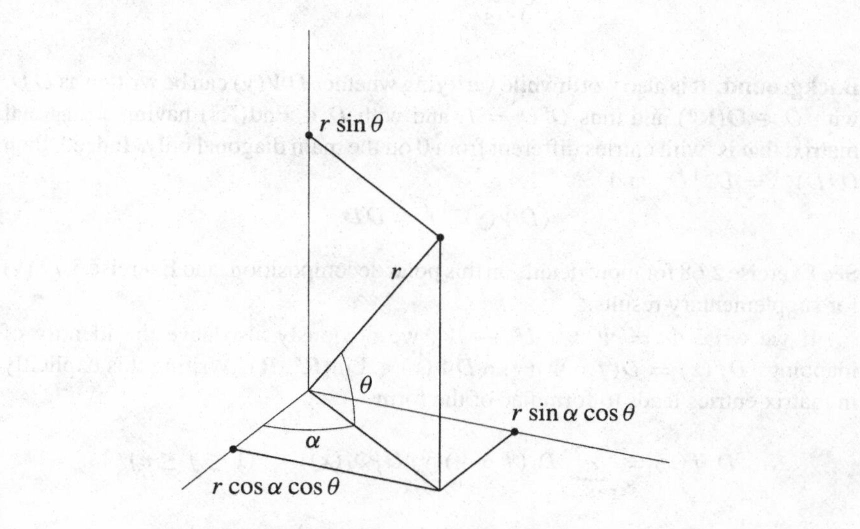

Illustration for Exercise 3.7: Spherical coordinates

Exercise 3.8 (Partial derivatives in arbitrary coordinates. Laplacian in cylindrical and spherical coordinates – sequel to Exercises 3.6 and 3.7 – needed for Exercises 3.9, 3.12, 5.79, 6.49, 6.101, 7.61 and 8.46). Let $\Psi : V \to U$ be a C^2 diffeomorphism between open subsets V and U in $\mathbf{R}^n$, and let $f : U \to \mathbf{R}$ be a C^2 function.

(i) Prove that $f \circ \Psi : V \to \mathbf{R}$ also is a C^2 function.

(ii) Let $\Psi(y) = x$, with $y \in V$ and $x \in U$. The chain rule then gives the identity of mappings $D(f \circ \Psi)(y) = Df(\Psi(y)) \circ D\Psi(y) \in \mathrm{Lin}(\mathbf{R}^n, \mathbf{R})$. Conclude that

$$(\star) \qquad (Df) \circ \Psi(y) = D(f \circ \Psi)(y) \circ D\Psi(y)^{-1}.$$

Use this and the identity $Df(x)^t = \mathrm{grad}\, f(x)$ to derive, by taking transposes, the following identity of mappings $V \to \mathbf{R}^n$:

$$(\mathrm{grad}\, f) \circ \Psi(y) = (D\Psi(y)^{-1})^t \,\mathrm{grad}(f \circ \Psi)(y) \qquad (y \in V).$$

Write

$$(D\Psi(y)^{-1})^t =: {}^tD\Psi(y)^{-1} = \big(\psi_{jk}(y)\big)_{1 \le j,k \le n},$$

and conclude that then

$$(D_j f) \circ \Psi = \Big(\sum_{1 \le k \le n} \psi_{jk}\, D_k\Big)(f \circ \Psi) \qquad (1 \le j \le n).$$

Background. It is also worthwhile verifying whether $D\Psi(y)$ can be written as OD, with $O \in \mathbf{O}(\mathbf{R}^n)$ and thus $O^t O = I$, and with $D \in \mathrm{End}(\mathbf{R}^n)$ having a diagonal matrix, that is, with entries different from 0 on the main diagonal only. Indeed, then $(OD)^{-1} = D^{-1}O^t$, and

$$(D\Psi(y)^{-1})^t = OD^{-1}.$$

See Exercise 2.68 for more details on this polar decomposition, and Exercise 5.79.(v) for supplementary results.

If we write $\Phi = \Psi^{-1} : U \to V$, we obviously also have the identity of mappings $Df(x) = D(f \circ \Psi)(y) \circ D\Phi(x) \in \mathrm{Lin}(\mathbf{R}^n, \mathbf{R})$. Writing this explicitly in matrix entries leads to formulae of the form

$$D_j f(x) = \sum_{1 \le k \le n} D_k(f \circ \Psi)(y)\, D_j \Phi_k(x) \qquad (1 \le j \le n).$$

This method could suggest that the inverse Φ of the coordinate transformation Ψ is explicitly calculated, and then partially differentiated. The treatment in part (ii) also calculates a (transpose) inverse, but of a **matrix**, namely $D\Psi(y)$, and not of Ψ itself.

(iii) Now let $\Psi : \mathbf{R}^3 \to \mathbf{R}^3$ be the substitution of cylindrical coordinates from Exercise 3.6, that is, $x = (r\cos\alpha,\ r\sin\alpha,\ x_3) = \Psi(r, \alpha, x_3) = \Psi(y)$. Assume that $y \in \mathbf{R}^3$ is a point at which Ψ is regular. Then prove that

$$D_1 f(x) = D_1 f(\Psi(y)) = \cos\alpha\, \frac{\partial(f\circ\Psi)}{\partial r}(y) - \frac{\sin\alpha}{r}\,\frac{\partial(f\circ\Psi)}{\partial\alpha}(y);$$

and derive analogous formulae for $D_2 f(x)$ and $D_3 f(x)$.

(iv) Now let $\Psi : \mathbf{R}^3 \to \mathbf{R}^3$ be the substitution of spherical coordinates $x = \Psi(y) = \Psi(r, \alpha, \theta)$ from Exercise 3.7, and assume that $y \in \mathbf{R}^3$ is a point at which Ψ is regular. Express $(D_j f)\circ\Psi(y) = D_j f(x)$, for $1 \le j \le 3$, in terms of $D_k(f\circ\Psi)(r, \alpha, \theta)$ with $1 \le k \le 3$.
Hint: We have

$$D\Psi(y) = \begin{pmatrix} \cos\alpha\cos\theta & -\sin\alpha & -\cos\alpha\sin\theta \\ \sin\alpha\cos\theta & \cos\alpha & -\sin\alpha\sin\theta \\ \sin\theta & 0 & \cos\theta \end{pmatrix}\begin{pmatrix} 1 & 0 & 0 \\ 0 & r\cos\theta & 0 \\ 0 & 0 & r \end{pmatrix}.$$

Show that the first matrix on the right–hand side is orthogonal. Therefore $(\operatorname{grad} f)\circ\Psi$ equals

$$\begin{pmatrix} \cos\alpha\cos\theta & -\sin\alpha & -\cos\alpha\sin\theta \\ \sin\alpha\cos\theta & \cos\alpha & -\sin\alpha\sin\theta \\ \sin\theta & 0 & \cos\theta \end{pmatrix}\begin{pmatrix} \dfrac{\partial(f\circ\Psi)}{\partial r} \\ \dfrac{1}{r\cos\theta}\dfrac{\partial(f\circ\Psi)}{\partial\alpha} \\ \dfrac{1}{r}\dfrac{\partial(f\circ\Psi)}{\partial\theta} \end{pmatrix}.$$

Background. The column vectors in this matrix, e_r, e_α and e_θ, say, form an orthonormal basis in $\mathbf{R}^3$, and point in the direction of the gradient of $x \mapsto r(x)$, $x \mapsto \alpha(x)$, $x \mapsto \theta(x)$, respectively. And furthermore, e_r has the same direction as x, whereas e_α is perpendicular to x and the x_3-axis.

The *Laplace operator* or *Laplacian* Δ, acting on C^2 functions $g : \mathbf{R}^3 \to \mathbf{R}$, is defined by

$$\Delta g = D_1^2 g + D_2^2 g + D_3^2 g.$$

(v) **(Laplacian in cylindrical coordinates).** Verify that in cylindrical coordinates y one has that $(D_1^2 f)\circ\Psi(y)$ equals, for $x = \Psi(y)$,

$$\begin{aligned} D_1(D_1 f)(x) &= \cos\alpha\,\frac{\partial}{\partial r}((D_1 f)\circ\Psi)(y) - \frac{\sin\alpha}{r}\,\frac{\partial}{\partial\alpha}((D_1 f)\circ\Psi)(y) \\ &= \cos\alpha\frac{\partial}{\partial r}\Big(\cos\alpha\,\frac{\partial(f\circ\Psi)}{\partial r}(y) - \frac{\sin\alpha}{r}\,\frac{\partial(f\circ\Psi)}{\partial\alpha}(y)\Big) \\ &\quad -\frac{\sin\alpha}{r}\,\frac{\partial}{\partial\alpha}\Big(\cos\alpha\,\frac{\partial(f\circ\Psi)}{\partial r}(y) - \frac{\sin\alpha}{r}\,\frac{\partial(f\circ\Psi)}{\partial\alpha}(y)\Big). \end{aligned}$$

Now prove

$$(\Delta f)\circ\Psi = \left(\frac{1}{r^2}\left(\left(r\frac{\partial}{\partial r}\right)^2 + \frac{\partial^2}{\partial\alpha^2}\right) + \frac{\partial^2}{\partial x_3^2}\right)(f\circ\Psi).$$

(vi) **(Laplacian in spherical coordinates).** Verify that this is given by the formula

$$(\Delta f)\circ\Psi = \frac{1}{r^2}\left(\frac{\partial}{\partial r}\left(r^2\frac{\partial}{\partial r}\right) + \frac{1}{\cos^2\theta}\left(\frac{\partial^2}{\partial\alpha^2} + \left(\cos\theta\frac{\partial}{\partial\theta}\right)^2\right)\right)(f\circ\Psi).$$

Background. In the Exercises 3.16, 7.60 and 7.61 we will encounter formulae for Δ on U in arbitrary coordinates in V. Our proofs in the last two cases require the theory of integration.

Exercise 3.9 (Angular momentum, Casimir, and Euler operators in spherical coordinates – sequel to Exercises 2.41 and 3.8 – needed for Exercises 3.17, 5.60 and 6.68). We employ the notations from the Exercises 2.41 and 3.8.

(i) Prove by means of Exercise 3.8.(iv) that for the substitution $x = \Psi(r,\alpha,\theta) \in \mathbf{R}^3$ of spherical coordinates one has

$$\begin{aligned}(L_1 f)\circ\Psi &= \left(\cos\alpha\tan\theta\frac{\partial}{\partial\alpha} - \sin\alpha\frac{\partial}{\partial\theta}\right)(f\circ\Psi),\\ (L_2 f)\circ\Psi &= \left(\sin\alpha\tan\theta\frac{\partial}{\partial\alpha} + \cos\alpha\frac{\partial}{\partial\theta}\right)(f\circ\Psi),\\ (L_3 f)\circ\Psi &= -\frac{\partial}{\partial\alpha}(f\circ\Psi),\end{aligned}$$

where L is the *angular momentum operator*.

(ii) Show that the following assertions concerning $f \in C^\infty(\mathbf{R}^3)$ are equivalent. First, the equation $Lf = 0$ is satisfied. And second, $f\circ\Psi$ is independent of α and θ, that is, f is a function of the distance to the origin.

(iii) Conclude that (see Exercise 2.41 for the definitions of H, X and Y)

$$\begin{aligned}(Hf)\circ\Psi &= -2i\frac{\partial}{\partial\alpha}(f\circ\Psi) &&=: H^*(f\circ\Psi),\\ (Xf)\circ\Psi &= e^{i\alpha}\left(\tan\theta\frac{\partial}{\partial\alpha} + i\frac{\partial}{\partial\theta}\right)(f\circ\Psi) &&=: X^*(f\circ\Psi),\\ (Yf)\circ\Psi &= -e^{-i\alpha}\left(\tan\theta\frac{\partial}{\partial\alpha} - i\frac{\partial}{\partial\theta}\right)(f\circ\Psi) = &&=: Y^*(f\circ\Psi).\end{aligned}$$

(iv) Show for the *Euler operator* E (see Exercise 2.41)

$$(Ef) \circ \Psi = \Big(r\frac{\partial}{\partial r}\Big)(f \circ \Psi) =: E^*(f \circ \Psi),$$

$$(E(E+1)f) \circ \Psi = \frac{\partial}{\partial r}\Big(r^2\frac{\partial}{\partial r}\Big)(f \circ \Psi).$$

(v) Verify for the *Casimir operator* C (see Exercise 2.41)

$$(Cf) \circ \Psi = \frac{1}{\cos^2\theta}\Big(\frac{\partial^2}{\partial\alpha^2} + \Big(\cos\theta\frac{\partial}{\partial\theta}\Big)^2\Big)(f \circ \Psi) := C^*(f \circ \Psi).$$

(vi) Prove for the *Laplace operator* Δ (compare with Exercise 2.41.(v))

$$(\Delta f) \circ \Psi = \frac{1}{r^2}(-C^* + E^*(E^*+1))(f \circ \Psi).$$

Exercise 3.10 (Sequel to Exercises 0.7 and 3.8 – needed for Exercise 7.68). Let $\Omega \subset \mathbf{R}^n$ be an open set. A C^2 function $f : \Omega \to \mathbf{R}$ is said to be *harmonic* (on Ω) if $\Delta f = 0$.

(i) Let $x^0 \in \mathbf{R}^3$. Prove that every harmonic function $x \mapsto f(x)$ on $\mathbf{R}^3 \setminus \{x^0\}$ for which $f(x)$ only depends on $\|x - x^0\|$ has the following form:

$$x \mapsto \frac{a}{\|x - x^0\|} + b \qquad (a,\, b \in \mathbf{R}).$$

(ii) Prove that every harmonic function f on $\mathbf{R}^3 \setminus \{x_3\text{-axis}\}$ for which $f(x)$ only depends on the "latitude" $\theta(x)$ of x has the following form:

$$x \mapsto a \log\tan\Big(\frac{\theta(x)}{2} + \frac{\pi}{4}\Big) + b \qquad (a,\, b \in \mathbf{R}).$$

Hint: See Exercise 0.7.

(iii) Let $\Omega = \mathbf{R}^n \setminus \{0\}$. Suppose that $f \in C^2(\mathbf{R}^n)$ is invariant under linear isometries, that is, $f(Ax) = f(x)$, for every $x \in \mathbf{R}^n$ and every $A \in \mathbf{O}(\mathbf{R}^n)$. Verify that there exists $f_0 \in C^2(\mathbf{R})$ satisfying $f(x) = f_0(\|x\|)$, for all $x \in \mathbf{R}^n$. Prove

$$\Delta f(x) = f_0''(\|x\|) + \frac{n-1}{\|x\|}f_0'(\|x\|) = \frac{1}{\|x\|^{n-1}}g'(\|x\|) \qquad (x \in \Omega),$$

with $g(r) = r^{n-1} f_0'(r)$, for $r > 0$. Conclude that for every harmonic function f on Ω that is invariant under linear isometries there exist numbers a and $b \in \mathbf{R}$ with (see also Exercise 2.30)

$$f(x) = \begin{cases} a \log \|x\| + b, & \text{if } n = 2; \\ \dfrac{a}{\|x\|^{n-2}} + b, & \text{if } n \neq 2. \end{cases}$$

Exercise 3.11 (Confocal coordinates – needed for Exercise 5.8). Introduce $f : \mathbf{R}_+ \times U \to \mathbf{R}$ by

$$U = \mathbf{R}_+^2 \quad \text{and} \quad f(y; x) = y^2 - y(x_1^2 + x_2^2 + 1) + x_1^2 \qquad (y \in \mathbf{R}_+,\ x \in U).$$

(i) Prove by means of the Intermediate Value Theorem 1.9.5 that, for every $x \in U$, there exist numbers $y_1 = y_1(x)$ and $y_2 = y_2(x)$ in $\mathbf{R}$ such that

$$f(y_1; x) = f(y_2; x) = 0 \quad \text{and} \quad 0 < y_1 < 1 < y_2.$$

Further define, in the notation of (i), $\Phi : U \to V$ by

$$V = \{\, y \in \mathbf{R}_+^2 \mid y_1 < 1 < y_2 \,\}, \qquad \Phi(x) = (y_1(x), y_2(x)).$$

(ii) Demonstrate that $\Phi : U \to V$ is a bijective mapping, by proving that the inverse $\Psi : V \to U$ of Φ is given by

$$\Psi(y) = \big(\sqrt{y_1 y_2},\ \sqrt{(1 - y_1)(y_2 - 1)}\big).$$

Hint: Consider the quadratic polynomial function $Y \mapsto (Y - y_1)(Y - y_2)$, for given $y \in \mathbf{R}^2$.

(iii) Prove that $\Psi : V \to U$ is a C^∞ mapping. Verify that if $\Psi(y) = x$,

$$\det D\Psi(y) = \frac{y_2 - y_1}{4 x_1 x_2}.$$

Use this to deduce that both $\Psi : V \to U$ and $\Phi : U \to V$ are C^∞ diffeomorphisms.

For each $y \in I = \{\, y \in \mathbf{R}_+ \mid y \neq 1 \,\}$ we now introduce $g_y : \mathbf{R}^2 \to \mathbf{R}$ and $V_y \subset \mathbf{R}^2$ by

$$g_y(x) = \frac{x_1^2}{y} + \frac{x_2^2}{y - 1} - 1 \quad \text{and} \quad V_y = g_y^{-1}(\{0\}), \quad \text{respectively.}$$

Note that V_y is a hyperbola or an ellipse, for $0 < y < 1$ or $1 < y$, respectively.

(iv) Show $x \in V_y$ if and only if $f(y; x) = 0$.

(v) Prove by means of (ii) that every $x \in U$ lies on precisely one hyperbola and also on precisely one ellipse from the collection $\{ V_y \mid y \in I \}$. And conversely, that for every $y \in V$ the hyperbola V_{y_1} and the ellipse V_{y_2} have precisely one point of intersection, namely $x = \Psi(y)$, in U.

Background. The last property in (v) is remarkable, and is related to the fact that all conics V_y possess the same foci, namely at the points $(\pm 1, 0)$. The element $y = (y_1(x), y_2(x)) \in V$ consists of the *confocal coordinates* of the point $x \in U$.

Exercise 3.12 (Moving frame, and gradient in arbitrary coordinates – sequel to Exercise 3.8 – needed for Exercises 3.16 and 7.60). Let $\Psi : V \to U$ be a C^2 diffeomorphism between open subsets V and U in $\mathbf{R}^n$. Then $(D_1\Psi(y), \dots, D_n\Psi(y))$ is a basis for $\mathbf{R}^n$ for every $y \in V$; the mapping

$$y \mapsto (D_1\Psi(y), \dots, D_n\Psi(y)) : V \to (\mathbf{R}^n)^n$$

is called the *moving frame* on V determined by Ψ. Define the C^1 mapping $G : V \to \mathbf{GL}(n, \mathbf{R})$ by

$$G(y) = D\Psi(y)^t \circ D\Psi(y) = (\langle D_j\Psi(y), D_k\Psi(y) \rangle)_{1 \le j,k \le n} =: (g_{jk}(y))_{1 \le j,k \le n}.$$

In other words, $G(y)$ is Gram's matrix associated with the set of basis vectors $D_1\Psi(y), \dots, D_n\Psi(y)$ in $\mathbf{R}^n$. Write the inverse matrix of $G(y)$ as

$$G(y)^{-1} = (g^{jk}(y))_{1 \le j,k \le n} \qquad (y \in V).$$

(i) Prove that $\det G(y) = (\det D\Psi(y))^2$, for all $y \in V$.

Therefore we can introduce the C^1 function

$$\sqrt{g} : V \to \mathbf{R} \qquad \text{by} \qquad \sqrt{g}(y) = \sqrt{\det G(y)} = |\det D\Psi(y)|.$$

Furthermore, let $f \in C^1(U)$ and consider $\operatorname{grad} f : U \to \mathbf{R}^n$. We now wish to express the mapping $(\operatorname{grad} f) \circ \Psi : V \to \mathbf{R}^n$ in terms of derivatives of $f \circ \Psi$ and of the moving frame on V determined by Ψ.

(ii) Using Exercise 3.8.(ii), demonstrate the following equality of vectors in $\mathbf{R}^n$:

$$(\operatorname{grad} f) \circ \Psi(y) = D\Psi(y) G(y)^{-1} \operatorname{grad}(f \circ \Psi)(y) \qquad (y \in V).$$

(iii) Verify the following equality on V of $\mathbf{R}^n$-valued mappings:

$$(\operatorname{grad} f) \circ \Psi = \sum_{1 \le j,k \le n} g^{jk} D_k(f \circ \Psi)\, D_j\Psi.$$

Exercise 3.13 (Divergence of vector field – needed for Exercise 3.14). Let U be an open subset of $\mathbf{R}^n$ and suppose $f : U \to \mathbf{R}^n$ to be a C^1 mapping; in this exercise we refer to f as a *vector field*. We define the *divergence* of the vector field f, notation: div f, as the function $\mathbf{R}^n \to \mathbf{R}$ with (see Definition 7.8.3 for more details)

$$\operatorname{div} f(x) = \operatorname{tr} Df(x) = \sum_{1 \le i \le n} D_i f_i(x).$$

As usual, tr denotes the trace of an element in $\operatorname{End}(\mathbf{R}^n)$. Next, let $h \in C^1(U)$. Using $(hf)_i = hf_i$ for $1 \le i \le n$, show, for $x \in U$,

$$\operatorname{div}(hf)(x) = Dh(x)\, f(x) + h(x) \operatorname{div} f(x) = (h \operatorname{div} f)(x) + \langle \operatorname{grad} h,\, f\rangle(x).$$

Exercise 3.14 (Divergence in arbitrary coordinates – sequel to Exercises 2.1, 2.44, 3.12 and 3.13 – needed for Exercises 3.15, 3.16, 7.60, 8.40 and 8.41). Let U and V be open subsets of $\mathbf{R}^n$, and let $\Psi : V \to U$ be a C^2 diffeomorphism. Next, assume $f : U \to \mathbf{R}^n$ to be a C^1 mapping; in this exercise we refer to f as a *vector field*. Then we define the C^1 vector field $\Psi^* f$, the *pullback* of f under Ψ (see Exercise 3.15), by

$$\Psi^* f : V \to \mathbf{R}^n \qquad \text{with} \qquad \Psi^* f(y) = D\Psi(y)^{-1}(f \circ \Psi)(y) \qquad (y \in V).$$

Denote the component functions of $\Psi^* f$ by $f^{(i)} \in C^1(V, \mathbf{R})$, for $1 \le i \le n$; thus, if $(e_1, \dots, e_n)$ is the standard basis for $\mathbf{R}^n$,

$$\Psi^* f = \sum_{1 \le i \le n} f^{(i)} e_i.$$

(i) Verify the identity of vectors in $\mathbf{R}^n$:

$$f \circ \Psi(y) = \sum_{1 \le i \le n} f^{(i)}(y)\, D_i\Psi(y) \qquad (y \in V).$$

In other words, the $f^{(i)}$ are the component functions of $f \circ \Psi : V \to \mathbf{R}^n$ with respect to the moving frame on V determined by Ψ in the terminology of Exercise 3.12.

The formula for $(\operatorname{div} f) \circ \Psi : V \to \mathbf{R}$, the divergence of f on U in the new coordinates in V, in terms of the pullback of f under Ψ or, what amounts to the same, the component functions of f with respect to the moving frame determined by Ψ, now reads, with $\sqrt{g}$ as in Exercise 3.12,

$$(\operatorname{div} f) \circ \Psi = \frac{1}{\sqrt{g}} \operatorname{div}(\sqrt{g}\, \Psi^* f) = \frac{1}{\sqrt{g}} \sum_{1 \le i \le n} D_i(\sqrt{g}\, f^{(i)}) : V \to \mathbf{R}.$$

We derive this identity in the following three steps; see Exercises 7.60 and 8.40 for other proofs.

(ii) Differentiate the equality $f \circ \Psi = D\Psi \cdot \Psi^* f : V \to \mathbf{R}^n$ to obtain, for $y \in V$ and $h \in \mathbf{R}^n$,

$$\begin{aligned} Df(\Psi(y))\, D\Psi(y)h &= D^2\Psi(y)(h, \Psi^* f(y)) + D\Psi(y)\, D(\Psi^* f)(y)h \\ &= D^2\Psi(y)(\Psi^* f(y), h) + D\Psi(y)\, D(\Psi^* f)(y)h, \end{aligned}$$

according to Theorem 2.7.9. Deduce the following identity of transformations in $\mathrm{End}(\mathbf{R}^n)$:

$$\begin{aligned} (Df) \circ \Psi(y) &= D^2\Psi(y)(\Psi^* f(y))\, D\Psi(y)^{-1} \\ &\quad + D\Psi(y)\, D(\Psi^* f)(y)\, D\Psi(y)^{-1}. \end{aligned}$$

By taking traces and using Exercise 2.1.(i) verify, for $y \in V$,

$$(\mathrm{div}\, f) \circ \Psi(y) = \mathrm{div}(\Psi^* f)(y) + \mathrm{tr}\,(D\Psi(y)^{-1} D^2\Psi(y)\Psi^* f(y)).$$

(iii) Show by means of the chain rule and ($\star\star$) in Exercise 2.44.(i), for $y \in V$ and $h \in \mathbf{R}^n$,

$$\begin{aligned} D(\det D\Psi)(y)h &= (D \det)(D\Psi(y))\, D^2\Psi(y)h \\ &= \det D\Psi(y)\ \mathrm{tr}\,(D\Psi(y)^{-1} D^2\Psi(y)h). \end{aligned}$$

Conclude, for $y \in V$, that

$$\begin{aligned} \mathrm{tr}\,(D\Psi(y)^{-1} D^2\Psi(y)\Psi^* f(y)) &= \frac{1}{\det D\Psi(y)} D(\det D\Psi)(y)\Psi^* f(y) \\ &= \frac{1}{\sqrt{g}(y)} D(\sqrt{g})(y)\, \Psi^* f(y). \end{aligned}$$

(iv) Use parts (ii) and (iii) and Exercise 3.13 to prove, for $y \in V$,

$$\begin{aligned} (\mathrm{div}\, f) \circ \Psi(y) &= \mathrm{div}(\Psi^* f)(y) + \frac{1}{\sqrt{g}(y)} D(\sqrt{g})(y)\, \Psi^* f(y) \\ &= \Big(\frac{1}{\sqrt{g}}\, \mathrm{div}(\sqrt{g}\, \Psi^* f)\Big)(y). \end{aligned}$$

Exercise 3.15 (Pullback and pushforward of vector field under diffeomorphism – sequel to Exercise 3.14 – needed for Exercise 8.46). (Compare with Exercise 5.79.) Pullback of objects (functions, vector fields, etc.) is associated with substitution of variables in these objects. Indeed, let U be an open subset of $\mathbf{R}^n$ and consider a vector field $f : U \to \mathbf{R}^n$ and the corresponding *ordinary differential equation* $\frac{dx}{dt}(t) = f(x(t))$ for a C^1 mapping $x : \mathbf{R} \to U$. Let V be an open set in $\mathbf{R}^n$ and let $\Psi : V \to U$ be a C^1 diffeomorphism.

(i) Prove that the substitution of variables $x = \Psi(y)$ leads to the following equation, where $\Psi^* f$ is as in Exercise 3.14:

$$D\Psi(y(t))\,\frac{dy}{dt}(t) = f \circ \Psi(y(t)), \qquad \text{thus} \qquad \frac{dy}{dt}(t) = \Psi^* f(y(t)).$$

(ii) Verify that the formula for the divergence in Exercise 3.14 may be written as

$$\Psi^* \circ \operatorname{div} = \frac{1}{\sqrt{g}} \operatorname{div}(\sqrt{g} \cdot \Psi^*).$$

Now let $\Phi : U \to V$ be a C^1 diffeomorphism. For every vector field $f : U \to \mathbf{R}^n$ we define the vector field $\Phi_* f$, the *pushforward* of f under Φ, by $\Phi_* f = (\Phi^{-1})^* f$.

(iii) Prove that $\Phi_* f : V \to \mathbf{R}^n$ with

$$\Phi_* f(y) = D\Phi(\Phi^{-1}(y))(f \circ \Phi^{-1})(y) = (D\Phi \cdot f) \circ \Phi^{-1}(y).$$

Exercise 3.16 (Laplacian in arbitrary coordinates – sequel to Exercises 3.12 and 3.14 – needed for Exercise 7.61). Let U be an open subset of $\mathbf{R}^n$ and suppose $f : U \to \mathbf{R}$ to be a C^2 function. We define the *Laplace operator*, or *Laplacian*, acting on f via

$$\Delta f = \operatorname{div}(\operatorname{grad} f) = \sum_{1 \le j \le n} D_j^2 f : U \to \mathbf{R}.$$

Now suppose that V is an open subset of $\mathbf{R}^n$ and that $\Psi : V \to U$ is a C^2 diffeomorphism. Then we have the following equality of mappings $V \to \mathbf{R}$ for Δ on U in the new coordinates in V:

$$\begin{aligned}(\Delta f) \circ \Psi &= \frac{1}{\sqrt{g}} \operatorname{div}(\sqrt{g}\, G^{-1} \operatorname{grad}(f \circ \Psi)) \\ &= \frac{1}{\sqrt{g}} \sum_{1 \le i, j \le n} D_i(\sqrt{g}\, g^{ij}\, D_j)(f \circ \Psi) \\ &= \Big(\sum_{1 \le i, j \le n} g^{ij}\, D_i D_j + \sum_{1 \le j \le n} \Big(\frac{1}{\sqrt{g}} \sum_{1 \le i \le n} D_i(\sqrt{g}\, g^{ij}) \Big) D_j \Big)(f \circ \Psi).\end{aligned}$$

We derive this identity in the following two steps, see Exercise 7.61 for another proof.

(i) Using Exercise 3.12.(ii) prove that the pullback of the vector field $\operatorname{grad} f : U \to \mathbf{R}^n$ under Ψ is given by

$$\Psi^*(\operatorname{grad} f) : V \to \mathbf{R}^n, \qquad \Psi^*(\operatorname{grad} f)(y) = G(y)^{-1} \operatorname{grad}(f \circ \Psi)(y).$$

(ii) Next apply Exercise 3.14.(iv) to obtain

$$(\Delta f)\circ\Psi = \operatorname{div}(\operatorname{grad} f)\circ\Psi = \frac{1}{\sqrt{g}}\operatorname{div}(\sqrt{g}G^{-1}\operatorname{grad}(f\circ\Psi)).$$

Phrased differently, $\Psi^*\circ\Delta = \frac{1}{\sqrt{g}}\operatorname{div}(\sqrt{g}\cdot G^{-1}\cdot\operatorname{grad}\circ\Psi^*)$.

The diffeomorphism Ψ is said to be *orthogonal* if $G(y)$ is a diagonal matrix, for all $y\in V$.

(iii) Verify that for an orthogonal Ψ the formula for $(\Delta f)\circ\Psi$ takes the following form:

$$(\star)\qquad (\Delta f)\circ\Psi = \frac{1}{\sqrt{g}}\sum_{1\le j\le n} D_j\left(\frac{\prod_{i\ne j}\|D_i\Psi\|}{\|D_j\Psi\|}D_j\right)(f\circ\Psi).$$

(iv) Verify that the condition of orthogonality of Ψ does **not** imply that $D\Psi(y)$, for $y\in V$, is an orthogonal matrix. Prove that the formula in Exercise 2.39.(v) is a special case of $(\star)$. Show that the transition to polar coordinates in $\mathbf{R}^2$, or to spherical coordinates in $\mathbf{R}^3$, respectively, is an orthogonal diffeomorphism. Also calculate Δ in polar and spherical coordinates, respectively, making use of $(\star)$; compare with Exercise 3.8.(v) and (vi). See also Exercise 5.13.

Background. Note that $D\Psi(y)$ enters the formula for $(\Delta f)\circ\Psi(y)$ only through its polar part, in the terminology of Exercise 2.68.

Exercise 3.17 (Casimir operator and spherical functions – sequel to Exercises 0.4 and 3.9 – needed for Exercises 5.61 and 6.68). We use the notations from Exercises 0.4 and 3.9. Let $l\in\mathbf{N}_0$ and $m\in\mathbf{Z}$ with $|m|\le l$. Define the *spherical function*

$$Y_l^m : V := \,]-\pi,\pi\,[\,\times\,\left]-\frac{\pi}{2},\frac{\pi}{2}\right[\,\to\mathbf{C}\qquad\text{by}\qquad Y_l^m(\alpha,\theta) = e^{im\alpha}P_l^{|m|}(\sin\theta).$$

Let $\mathcal{Y}_l$ be the $(2l+1)$-dimensional linear space over $\mathbf{C}$ of functions spanned by the Y_l^m, for $|m|\le l$. Through the linear isomorphism

$$\iota : f\mapsto \iota f\qquad\text{with}\qquad f(\alpha,\theta) = (\iota f)(\cos\alpha\cos\theta,\ \sin\alpha\cos\theta,\ \sin\theta),$$

continuous functions f on V are identified with continuous functions ιf on the unit sphere S^2 in $\mathbf{R}^3$. In particular, ιY_l^0 is a function on S^2 invariant under rotations about the x_3-axis. The zeros of ιY_l^0 divide S^2 up into parallel zones, and for that reason Y_l^0 is called a *zonal spherical function.*

(i) Prove by means of Exercise 0.4.(iii) that the differential operator C^* from Exercise 3.9.(v) acts on $\mathcal{Y}_l$ as the linear operator of multiplication by the scalar $l(l+1)$, that is, as a linear operator with a single eigenvalue, namely $l(l+1)$, with multiplicity $2l+1$. To do so, verify that

$$C^*\, Y_l^m = l(l+1)\, Y_l^m \qquad (|m| \leq l).$$

(ii) Prove by means of Exercises 3.9.(iii) and 0.4.(iv) that $\mathcal{Y}_l$ is invariant under the action of the differential operators H^*, X^* and Y^*. To do so, define $Y_l^{l+1} = Y_l^{-l-1} = 0$, and prove by means of Exercise 0.4.(iv), for all $|m| \leq l$,

$$H^*\, Y_l^m = 2m\, Y_l^m, \qquad X^*\, Y_l^m = i\, Y_l^{m+1},$$
$$Y^*\, Y_l^m = -i\,(l+m)(l-m+1) Y_l^{m-1}.$$

Another description of this result is as follows. Define $V_j = \frac{1}{j!}(Y^*)^j\, Y_l^l \in \mathcal{Y}_l$, for $0 \leq j \leq 2l$.

(iii) Show that

$$V_j = (-i)^j \frac{(2l)!}{(2l-j)!} Y_l^{l-j} \qquad (0 \leq j \leq 2l).$$

Then prove, for $0 \leq j \leq 2l$,

$$(\star) \qquad \begin{aligned} H^*\, V_j &= (2l - 2j)\, V_j, \qquad X^*\, V_j = (2l + 1 - j)\, V_{j-1}, \\ Y^*\, V_j &= (j+1) V_{j+1}. \end{aligned}$$

See Exercise 5.62.(v) for the matrices of X^* and Y^* with respect to the basis $(V_0, \dots, V_{2l})$, if l in that exercise is replaced by $2l$.

(iv) Conclude by means of Exercise 3.9.(vi) that the functions

$$\Psi(r, \alpha, \theta) \mapsto r^l\, Y_l^m(\alpha, \theta) \qquad \text{and} \qquad \Psi(r, \alpha, \theta) \mapsto r^{-l-1}\, Y_l^m(\alpha, \theta)$$

are harmonic (see Exercise 3.10) on the dense open subset $\operatorname{im}(\Psi) \subset \mathbf{R}^3$. For this reason the functions ιY_l^m are also called *spherical harmonic functions*, for they are restrictions to the two-sphere of harmonic functions on $\mathbf{R}^3$.

Background. In *quantum physics* this is the theory of the *(orbital) angular momentum operator* associated with a *particle without spin* in a rotationally symmetric force field (see also Exercise 5.61). Here l is called the *orbital angular momentum quantum number* and m the *magnetic quantum number*. The fact that the space $\mathcal{Y}_l$ of spherical functions is $(2l+1)$-dimensional constitutes the mathematical basis of the (elementary) theory of the *Zeeman effect*: the splitting of an atomic spectral line into an odd number of lines upon application of an external magnetic field along the x_3-axis.

Exercise 3.18 (Spherical coordinates in $\mathbf{R}^n$ – needed for Exercises 5.13 and 7.21). If x in $\mathbf{R}^n$ satisfies $x_1^2 + \cdots + x_n^2 = r^2$ with $r \geq 0$, then there exists a unique element $\theta_{n-2} \in [-\frac{\pi}{2}, \frac{\pi}{2}]$ such that

$$\sqrt{x_1^2 + \cdots + x_{n-1}^2} = r \cos\theta_{n-2}, \qquad x_n = r \sin\theta_{n-2}.$$

(i) Prove that repeated application of this argument leads to a C^∞ mapping

$$\Psi : [\,0, \infty\,[\, \times [\,-\pi, \pi\,] \times \left[-\frac{\pi}{2}, \frac{\pi}{2}\right]^{n-2} \to \mathbf{R}^n,$$

given by

$$\Psi : \begin{pmatrix} r \\ \alpha \\ \theta_1 \\ \theta_2 \\ \vdots \\ \theta_{n-2} \end{pmatrix} \mapsto \begin{pmatrix} r\cos\alpha\,\cos\theta_1\,\cos\theta_2 \cdots \cos\theta_{n-3}\cos\theta_{n-2} \\ r\sin\alpha\,\cos\theta_1\,\cos\theta_2 \cdots \cos\theta_{n-3}\cos\theta_{n-2} \\ r\sin\theta_1\,\cos\theta_2 \cdots \cos\theta_{n-3}\cos\theta_{n-2} \\ \vdots \\ r\sin\theta_{n-3}\,\cos\theta_{n-2} \\ r\sin\theta_{n-2} \end{pmatrix}.$$

(ii) Prove that $\Psi : V \to U$ is a bijective C^∞ mapping, if

$$V = \mathbf{R}_+ \times \,]-\pi, \pi\,[\, \times \left]-\frac{\pi}{2}, \frac{\pi}{2}\right[^{n-2},$$

$$U = \mathbf{R}^n \setminus \{\, x \in \mathbf{R}^n \mid x_1 \leq 0,\ x_2 = 0 \,\}.$$

(iii) Prove, for $(r, \alpha, \theta_1, \theta_2, \ldots, \theta_{n-2}) \in V$,

$$\det D\Psi(r, \alpha, \theta_1, \theta_2, \ldots, \theta_{n-2}) = r^{n-1} \cos\theta_1 \cos^2\theta_2 \cdots \cos^{n-2}\theta_{n-2}.$$

Hint: In the bottom row of the determinant only the first and the last entries differ from 0. Therefore expand according to this row; the result is $(-1)^{n-1}\sin\theta_{n-2}\det A_{n1} + r\cos\theta_{n-2}\det A_{nn}$. Use the multilinearity of $\det A_{n1}$ and $\det A_{nn}$ to take out factors $\cos\theta_{n-2}$ or $\sin\theta_{n-2}$, take the last column in A_{n1} to the front position, and complete the proof by mathematical induction on $n \in \mathbf{N}$.

(iv) Now prove that $\Psi : V \to U$ is a C^∞ diffeomorphism.

We want to add another derivation of the formula in (iii). For this purpose, the C^∞ mapping

$$f : U \times V \to \mathbf{R}^n,$$

is defined by, for $x \in U$ and $y = (r, \alpha, \theta_1, \theta_2, \dots, \theta_{n-2}) \in V$,

$$\begin{aligned}
f_1(x; y) &= x_1^2 - \quad r^2 \cos^2\alpha \cos^2\theta_1 \cos^2\theta_2 \cdots \cos^2\theta_{n-2},\\
f_2(x; y) &= x_1^2 + x_2^2 - \quad r^2 \cos^2\theta_1 \cos^2\theta_2 \cdots \cos^2\theta_{n-2},\\
\vdots \quad &\quad \vdots \qquad\qquad\qquad \vdots\\
f_{n-1}(x; y) &= x_1^2 + x_2^2 + \cdots + x_{n-1}^2 - \quad r^2 \cos^2\theta_{n-2},\\
f_n(x; y) &= x_1^2 + x_2^2 + \cdots + x_{n-1}^2 + x_n^2 - \quad r^2.
\end{aligned}$$

(v) Prove $\det \frac{\partial f}{\partial y}(x; y)$ and $\det \frac{\partial f}{\partial x}(x; y)$ are given by, respectively,

$$(-1)^n 2^n r^{2n-1} \cos\alpha \sin\alpha \cos^3\theta_1 \sin\theta_1 \cos^5\theta_2 \sin\theta_2 \cdots \cos^{2n-3}\theta_{n-2} \sin\theta_{n-2},$$

$$2^n r^n \cos\alpha \sin\alpha \cos^2\theta_1 \sin\theta_1 \cos^3\theta_2 \sin\theta_2 \cdots \cos^{n-1}\theta_{n-2} \sin\theta_{n-2}.$$

Conclude that the formula in (iii) follows.

Background. In Exercise 5.13 we present a third way to find the formula in (iii).

Exercise 3.19 (Standard $(n+1)$-tope – sequel to Exercise 3.2 – needed for Exercises 6.65 and 7.27). Let Δ^n be the following open set in $\mathbf{R}^n$, bounded by $n+1$ hypersurfaces:

$$\Delta^n = \{\, x \in \mathbf{R}^n_+ \mid \sum_{1 \le j \le n} x_j < 1 \,\}.$$

The *standard* $(n+1)$*-tope* is defined as the closure of Δ^n in $\mathbf{R}^n$. This terminology has the following origin. A *polygon* (ἡ γωνία = angle) is a set in $\mathbf{R}^2$ bounded by a finite number of lines, a *polyhedron* (ἡ ἕδρα = basis) is a set in $\mathbf{R}^3$ bounded by a finite number of planes, and a *polytope* (ὁ τόπος = place) is a set in $\mathbf{R}^n$ bounded by a finite number of hyperplanes. In particular, the standard 3-tope is a rectangular triangle and the standard 4-tope a rectangular *tetrahedron* (τετρα = four, in compounds).

(i) Prove that $\Psi : \,]0, 1[^n \to \Delta^n$ is a C^∞ diffeomorphism, if $\Psi(y) = x$, where x is given by

$$x_j = (1 - y_{j-1}) \prod_{j \le k \le n} y_k \qquad (1 \le j \le n,\ y_0 = 0).$$

Hint: For $x \in \Delta^n$ one has $y_j \in \,]0, 1[$, for $1 \le j \le n$, if

$$\begin{aligned}
y_n &= x_1 + \cdots \cdots + x_n\\
y_{n-1}\, y_n &= x_1 + \cdots \cdots + x_{n-1}\\
y_{n-2}\, y_{n-1}\, y_n &= x_1 + \cdots \cdots + x_{n-2}\\
\vdots \quad &\quad \vdots \quad \vdots\\
y_1\, y_2 \cdots \cdots y_{n-1}\, y_n &= x_1.
\end{aligned}$$

The result above may also be obtained by generalization to $\mathbf{R}^n$ of the geometrical construction from Exercise 3.2 using induction.

(ii) Let $y \in \mathbf{R}^n$, $Y_1 = 1 \in \mathbf{R}$, and assume that $Y_{n-1} \in \mathbf{R}^{n-1}$ has been defined. Next, let $\widetilde{Y}_n = (Y_{n-1}, 0) \in \mathbf{R}^n$, let $e_n \in \mathbf{R}^n$ be the n-th unit vector, and assume

$$Y_n = y_{n-1}\,\widetilde{Y}_n + (1 - y_{n-1})\,e_n, \qquad X_n = y_n Y_n.$$

Verify that one then has $X_n = x = \Psi(y)$.

(iii) Prove

$$\det D\Psi(y) = y_2\, y_3^2 \cdots y_{n-1}^{n-2}\, y_n^{n-1} \qquad (y \in\,]\,0, 1\,[^n).$$

Hint: To each row in $D\Psi(y)$ add all other rows above it.

Exercise 3.20 (Diffeomorphism of triangle onto square – sequel to Exercise 0.3 – needed for Exercises 6.39 and 6.40). Define $\Psi : V \to \mathbf{R}^2$ by

$$V = \{\, y \in \mathbf{R}_+^2 \mid y_1 + y_2 < \frac{\pi}{2}\,\}, \qquad \Psi(y) = \left(\frac{\sin y_1}{\cos y_2}, \frac{\sin y_2}{\cos y_1}\right).$$

(i) Prove that $\Psi : V \to U$ is a C^∞ diffeomorphism when $U = \,]\,0, 1\,[^2 \subset \mathbf{R}^2$. **Hint:** For $y \in V$ we have $0 < y_1 < \frac{\pi}{2} - y_2 < \frac{\pi}{2}$; and therefore

$$0 < \sin y_1 < \sin\left(\frac{\pi}{2} - y_2\right) = \cos y_2.$$

This enables us to conclude that $\Psi(y) \in U$. Conversely, given $x \in U$, the solution $y \in \mathbf{R}^2$ of $\Psi(y) = x$ is given by

$$y = \left(\arctan x_1 \sqrt{\frac{1 - x_2^2}{1 - x_1^2}},\ \arctan x_2 \sqrt{\frac{1 - x_1^2}{1 - x_2^2}}\right).$$

From $x_1 x_2 < 1$ it follows that

$$x_2 \sqrt{\frac{1 - x_1^2}{1 - x_2^2}} < \frac{1}{x_1}\sqrt{\frac{1 - x_1^2}{1 - x_2^2}} = \left(x_1 \sqrt{\frac{1 - x_2^2}{1 - x_1^2}}\right)^{-1}.$$

Because of the identity $\arctan t + \arctan \frac{1}{t} = \frac{\pi}{2}$ from Exercise 0.3 we get $y \in V$.

(ii) Prove that $\det D\Psi(y) = 1 - \Psi_1(y)^2 \Psi_2(y)^2 > 0$, for all $y \in V$.

(iii) Generalize the above for the mapping $\Psi : V \to\ ]\,0, 1\,[^n \subset \mathbf{R}^n$, where

$$V = \{\, y \in \mathbf{R}^n_+ \mid y_1 + y_2 < \frac{\pi}{2},\ y_2 + y_3 < \frac{\pi}{2},\ \dots,\ y_n + y_1 < \frac{\pi}{2}\,\},$$

$$\Psi(y) = \Big(\frac{\sin y_1}{\cos y_2},\ \frac{\sin y_2}{\cos y_3},\ \dots,\ \frac{\sin y_n}{\cos y_1}\Big).$$

Prove that $\det D\Psi(y) = 1 - (-1)^n \Psi_1(y)^2 \cdots \Psi_n(y)^2 > 0$, for all $y \in V$.
Hint: Consider, for prescribed $x \in\ ]\,0, 1\,[^n$, the affine function $\chi : \mathbf{R} \to \mathbf{R}$ given by

$$\chi(t) = x_n^2(1 - x_1^2(1 - \cdots (1 - x_{n-1}^2(1 - t)) \cdots)).$$

This χ maps the set $]\,0, 1\,[$ into itself; and more in particular, in such a way as to be distance-decreasing. Consequently there exists a unique $t_0 \in\ ]\,0, 1\,[$ with $\chi(t_0) = t_0$. Choose $y \in V$ with

$$\sin^2 y_n = t_0, \qquad \sin^2 y_j = x_j^2(1 - \sin^2 y_{j+1}) \quad (n > j \geq 1).$$

Then $\Psi(y) = x$ if and only if $\sin^2 y_n = x_n^2(1 - \sin^2 y_1)$, and this follows from $\chi(\sin^2 y_n) = \sin^2 y_n$.

Exercise 3.21. Consider the mapping

$$\Phi_{A,a} : \mathbf{R}^2 \to \mathbf{R}^2 \qquad \text{given by} \qquad \Phi_{A,a}(x) = (\,\langle x, x\,\rangle,\ \langle Ax, x\rangle + \langle a, x\rangle\,),$$

where $A \in \mathrm{End}^+(\mathbf{R}^2)$ and $a \in \mathbf{R}^2$. Demonstrate the existence of O in $\mathbf{O}(\mathbf{R}^2)$ such that $\Phi_{A,a} = \Phi_{B,b} \circ O$, where $B \in \mathrm{End}(\mathbf{R}^2)$ has a diagonal matrix and $b \in \mathbf{R}^2$. Prove that there are the following possibilities for the set of singular points of $\Phi_{A,a}$:

(i) $\mathbf{R}^2$;

(ii) a straight line through the origin;

(iii) the union of two straight lines intersecting at right angles, one of which at least runs through the origin;

(iv) a hyperbola with mutually perpendicular asymptotes, one branch of which runs through the origin.

Try to formulate the conditions for A and a which lead to these various cases.

Exercise 3.22 (Wave equation in one spatial variable – sequel to Exercise 3.3 – needed for Exercise 8.33). Define $\Psi : \mathbf{R}^2 \to \mathbf{R}^2$ by $\Psi(y) = \frac{1}{2}(y_1 + y_2,\ y_1 - y_2) = (x, t)$. From Exercise 3.3 we know that Ψ is a C^∞ diffeomorphism. Assume that $u : \mathbf{R}^2 \to \mathbf{R}$ is a C^2 function satisfying the *wave equation*

$$\frac{\partial^2 u}{\partial t^2}(x, t) = \frac{\partial^2 u}{\partial x^2}(x, t).$$

(i) Prove that $u \circ \Psi$ satisfies the equation $D_1 D_2(u \circ \Psi)(y) = 0$.
Hint: Compare with Exercise 3.8.(ii) and (v).

This means that $D_2(u \circ \Psi)(y)$ is independent of y_1, that is, $D_2(u \circ \Psi)(y) = F'_-(y_2)$, for a C^2 function $F_- : \mathbf{R} \to \mathbf{R}$.

(ii) Conclude that $u(x,t) = (u \circ \Psi)(y) = F_+(y_1) + F_-(y_2) = F_+(x+t) + F_-(x-t)$.

Next, assume that u satisfies the *initial conditions*

$$u(x,0) = f(x); \qquad \frac{\partial u}{\partial t}(x,0) = g(x) \qquad (x \in \mathbf{R}).$$

Here f and $g : \mathbf{R} \to \mathbf{R}$ are prescribed C^2 and C^1 functions, respectively.

(iii) Prove

$$F'_+(x) + F'_-(x) = f'(x); \qquad F'_+(x) - F'_-(x) = g(x) \qquad (x \in \mathbf{R}),$$

and conclude that the solution u of the wave equation satisfying the initial conditions is given by

$$u(x,t) = \frac{1}{2}(f(x+t) + f(x-t)) + \frac{1}{2}\int_{x-t}^{x+t} g(y)\,dy,$$

which is known as *D'Alembert's formula*.

Exercise 3.23 (Another proof of Proposition 3.2.3 – sequel to Exercise 2.37). Suppose that U and V are open subsets of $\mathbf{R}^n$ both containing 0. Consider $\Phi \in C^1(U,V)$ satisfying $\Phi(0) = 0$ and $D\Phi(0) = I$. Set $\Xi = I - \Phi \in C^1(U, \mathbf{R}^n)$.

(i) Show that we may assume, by shrinking U if necessary, that $D\Phi(x) \in \mathrm{Aut}(\mathbf{R}^n)$, for all $x \in U$, and that Ξ is Lipschitz continuous with a Lipschitz constant $\le \frac{1}{2}$.

(ii) By means of the reverse triangle inequality from Lemma 1.1.7.(iii) show, for x and $x' \in U$,

$$\|\Phi(x) - \Phi(x')\| = \|x - x' - (\Xi(x) - \Xi(x'))\| \ge \frac{1}{2}\|x - x'\|.$$

Select $\delta > 0$ such that $K := \{\, x \in \mathbf{R}^n \mid \|x\| \le 4\delta \,\} \subset U$ and $V_0 := \{\, y \in \mathbf{R}^n \mid \|y\| < \delta \,\} \subset V$. Then K is a compact subset of U with a nonempty interior.

(iii) Show that $\|\Phi(x)\| \ge 2\delta$, for $x \in \partial K$.

(iv) Using Exercise 2.37 show that, given $y \in V_0$, we can find $x \in K$ with $\Phi(x) = y$.

(v) Show that this $x \in K$ is unique.

(vi) Use parts (iv) and (v) in order to give a proof of Proposition 3.2.3 without recourse to the Contraction Lemma 1.7.2.

Exercise 3.24 (Lipschitz continuity of inverse – sequel to Exercises 1.27 and 2.37 – needed for Exercises 3.25 and 3.26). Suppose that $\Phi \in C^1(\mathbf{R}^n, \mathbf{R}^n)$ is regular everywhere and that there exists $k > 0$ such that $\|\Phi(x) - \Phi(x')\| \geq k\|x - x'\|$, for all x and $x' \in \mathbf{R}^n$. Under these stronger conditions we can draw a conclusion better than the one in Exercise 1.27, specifically, that Φ is surjective onto $\mathbf{R}^n$. In fact, in four steps we shall prove that Φ is a C^1 diffeomorphism.

(i) Prove that Φ is an open mapping.

(ii) On the strength of Exercise 1.27.(i) conclude that Φ is an injective and closed mapping.

(iii) Deduce from (i) and (ii) and the connectedness of $\mathbf{R}^n$ that Φ is surjective.

(iv) Using Proposition 3.2.2 show that Φ is a C^1 diffeomorphism.

The surjectivity of Φ can be proved without appeal to a topological argument as in part (iii).

(v) Let $y \in \mathbf{R}^n$ be arbitrary and consider $f : \mathbf{R}^n \to \mathbf{R}$ given by $f(x) = \|\Phi(x) - y\|$. Prove that $\|\Phi(x)\|$ goes to infinity, and therefore $f(x)$ as well, as $\|x\|$ goes to infinity. Conclude that $\{ x \in \mathbf{R}^n \mid f(x) \leq \delta \}$ is compact, for every $\delta > 0$, and nonempty if δ is sufficiently large. Now apply Exercise 2.37.

Exercise 3.25 (Criterion for bijectivity – sequel to Exercise 3.24). Suppose that $\Phi \in C^1(\mathbf{R}^n, \mathbf{R}^n)$ and that there exists $k > 0$ such that

$$\langle D\Phi(x)h, h\rangle \geq k\|h\|^2 \qquad (x \in \mathbf{R}^n,\ h \in \mathbf{R}^n).$$

Then Φ is a C^1 diffeomorphism, since it satisfies the conditions of Exercise 3.24; indeed:

(i) Show that Φ is regular everywhere.

(ii) Prove that $\langle \Phi(x) - \Phi(x'), x - x'\rangle \geq k\|x - x'\|^2$, and using the Cauchy–Schwarz inequality obtain that $\|\Phi(x) - \Phi(x')\| \geq k\|x - x'\|$, for all x and $x' \in \mathbf{R}^n$.
Hint: Apply the Mean Value Theorem on $\mathbf{R}$ to $f : [0, 1] \to \mathbf{R}^n$ with $f(t) = \langle \Phi(x_t), x - x'\rangle$. Here x_t is as in Formula (2.15).

Background. The uniformity in the condition on Φ can not be weakened if we insist on surjectivity, as can be seen from the example of $\arctan : \mathbf{R} \to \left]-\frac{\pi}{2}, \frac{\pi}{2}\right[$. Furthermore, compare with Exercise 2.26.

Exercise 3.26 (Perturbation of identity – sequel to Exercises 2.24 and 3.24). Let $\Xi : \mathbf{R}^n \to \mathbf{R}^n$ be a C^1 mapping that is Lipschitz continuous with Lipschitz constant $0 < m < 1$. Then $\Phi : \mathbf{R}^n \to \mathbf{R}^n$ defined by $\Phi(x) = x - \Xi(x)$ is a C^1 diffeomorphism. This is a consequence of Exercise 3.24.

(i) Using Exercise 2.24 show that Φ is regular everywhere.

(ii) Verify that the estimate from Exercise 3.24 is satisfied.

Define $\Phi' : \mathbf{R}^n \times \mathbf{R}^n \to \mathbf{R}^n \times \mathbf{R}^n$ by $\Phi'(x, y) = (x - \Xi(y), y - \Xi(x))$.

(iii) Prove that Φ' is a C^1 diffeomorphism.

Exercise 3.27 (Needed for Exercise 6.22). Suppose $f : \mathbf{R}^n \to \mathbf{R}^n$ is a C^1 mapping and, for each $t \in \mathbf{R}$, consider $\Phi_t : \mathbf{R}^n \to \mathbf{R}^n$ given by $\Phi_t(x) = x - t\,f(x)$. Let $K \subset \mathbf{R}^n$ be compact.

(i) Show that $\Phi_t : K \to \mathbf{R}^n$ is injective if t is sufficiently small.
Hint: Use Corollary 2.5.5 to find some Lipschitz constant $c > 0$ for $f|_K$. Next, consider any t with $k := c\,|t| < 1$ (note the analogy with Exercise 3.26).

Suppose in addition that f satisfies $\|f(x)\| = 1$ if $\|x\| = 1$, that $\langle f(x), x\rangle = 0$, and that $f(rx) = rf(x)$, for $x \in \mathbf{R}^n$ and $r \geq 0$.

(ii) For $n \in 2\mathbf{N}$, show that $f(x) = (-x_2, x_1, -x_4, x_3, \dots, -x_{2n}, x_{2n-1})$ is an example of a mapping f satisfying the conditions above.

For $r > 0$, define $S(r) = \{\, x \in \mathbf{R}^n \mid \|x\| = r \,\}$, the sphere in $\mathbf{R}^n$ of center 0 and radius r.

(iii) Prove that Φ_t is a surjection from $S(r)$ onto $S(r\sqrt{1+t^2})$, for $r \geq 0$, if t is sufficiently small.
Hint: By homogeneity it is sufficient to consider the case of $r = 1$. Define $K = \{\, x \in \mathbf{R}^n \mid \frac{1}{2} \leq \|x\| \leq \frac{3}{2} \,\}$ and choose t small enough so that $k = c|t| < \frac{1}{3}$. Then, for $x_0 \in S(1)$, the auxiliary mapping $x \mapsto x_0 + t\,f(x)$ maps K into itself, since $\|tf(x)\| < \frac{1}{2}$, and it is a contraction with contraction factor k. Next, use the Contraction Lemma to find a unique solution $x \in K$ for $\Phi_t(x) = x_0$. Finally, multiply x and x_0 by $\sqrt{1+t^2}$.

Let $0 < a < b$ and let A be the open annulus $\{\, x \in \mathbf{R}^n \mid a < \|x\| < b \,\}$.

(iv) Use the Global Inverse Function Theorem to prove that $\Phi_t : A \to \sqrt{1+t^2}\,A$ is a C^1 diffeomorphism if t is sufficiently small.

Exercise 3.28 (Another proof of the Implicit Function Theorem 3.5.1 – sequel to Exercise 2.6). The notation is as in the theorem. The details are left for the reader to check.

(i) **Verification of conditions of Exercise 2.6.** Write $A = D_x f(x^0; y^0)^{-1} \in \mathrm{Aut}(\mathbf{R}^n)$. Take $0 < \epsilon < 1$ arbitrary; considering that W is open and $D_x f$ continuous at (x^0, y^0), there exist numbers $\delta > 0$ and $\eta > 0$, possibly requiring further specification, such that for $x \in V(x^0; \delta)$ and $y \in V(y^0; \eta)$,

$$(\star) \qquad (x, y) \in W \qquad \text{and} \qquad \|I - A \circ D_x f(x; y)\|_{\mathrm{Eucl}} \le \epsilon.$$

Let $y \in V(y^0; \eta)$ be fixed for the moment. We now wish to apply Exercise 2.6, with

$$V = V(x^0; \delta), \qquad V \ni x \mapsto f(x; y), \qquad F(x) = x - Af(x; y).$$

We make use of the fact that f is differentiable with respect to x, to prove that F is a contraction. Indeed, because the derivative of F is given by $I - A \circ D_x f(x; y)$, application of the estimates (2.17) and $(\star)$ yields, for every $x, x' \in V$,

$$\|F(x) - F(x')\| \le \sup_{\xi \in V} \|I - A \circ D_x f(\xi; y)\|_{\mathrm{Eucl}} \|x - x'\| \le \epsilon \|x - x'\|.$$

Because $Af(x^0; y^0) = 0$, the continuity of $y \mapsto Af(x^0; y)$ implies that

$$\|Af(x^0; y)\| \le (1 - \epsilon)\, \delta,$$

if necessary by taking η smaller still.

(ii) **Continuity of ψ at y^0.** The existence and uniqueness of $x = \psi(y)$ now follow from application of Exercise 2.6; the exercise also yields the estimate

$$\|x - x^0\| \le \frac{1}{1 - \epsilon} \|Af(x^0; y)\| = \frac{1}{1 - \epsilon} \|A(f(x^0; y) - f(x^0; y^0))\|.$$

On account of f being differentiable with respect to y we then find a constant $K > 0$ such that for $y \in V(y^0; \eta)$,

$$(\star\star) \qquad \|\psi(y) - \psi(y^0)\| = \|x - x^0\| \le K \|y - y^0\|.$$

This proves the continuity of ψ at y^0, but not the differentiability.

(iii) **Differentiability of ψ at y^0.** For this we use Formula (3.3). We obtain

$$f(x; y) = D_x f(x^0; y^0)(x - x^0) + D_y f(x^0; y^0)(y - y^0) + R(x, y).$$

Given any $\zeta > 0$ we can now arrange that for all $x \in V(x^0; \delta')$ and $y \in V(y^0; \eta')$,

$$\|R(x, y)\| \le \zeta (\|x - x^0\|^2 + \|y - y^0\|^2)^{1/2},$$

if necessary by choosing $\delta' < \delta$ and $\eta' < \eta$. In particular, with $x = \psi(y)$ we obtain

$$\begin{aligned} \psi(y) - \psi(y^0) = x - x^0 &= -D_x f(x^0; y^0)^{-1} \circ D_y f(x^0; y^0)(y - y^0) \\ &\quad + \widetilde{R}(\psi(y), y), \end{aligned}$$

where, on the basis of the estimate (⋆⋆), for all $y \in V(y^0; \eta')$,

$$\|\widetilde{R}(\psi(y),\, y)\| = \| - A \circ R(\psi(y),\, y)\| \le \zeta\, \|A\|_{\text{Eucl}}\, (K^2+1)^{1/2}\, \|y - y^0\|.$$

This proves that ψ is differentiable at y^0 and that the formula for $D\psi(y)$ is correct at y^0.

(iv) **Differentiability of ψ at** y. Because $y \mapsto D_x f(\psi(y); y)$ is continuous at y^0 we can, using Lemma 2.1.2 and taking η smaller if necessary, arrange that, for all $y \in V(y^0; \eta)$,

$$(\psi(y),\, y) \in W, \qquad f(\psi(y); y) = 0, \qquad D_x f(\psi(y); y) \in \operatorname{Aut}(\mathbf{R}^n).$$

Application of part (iii) yields the differentiability of ψ at y, with the desired formula for $D\psi(y)$.

(v) ψ **belongs to $\mathbf{C}^{\mathbf{k}}$**. That $\psi \in C^k$ can be proved by induction on k.

Exercise 3.29 (Another proof of the Local Inverse Function Theorem 3.2.4). The Implicit Function Theorem 3.5.1 was proved by means of the Local Inverse Function Theorem 3.2.4. Show that, in turn, the latter theorem can be obtained as a corollary of the former. Note that combination of this result with Exercise 3.28 leads to another proof of the Local Inverse Function Theorem.

Exercise 3.30. Let $f : \mathbf{R}^3 \to \mathbf{R}^2$ be defined by

$$f(x; y) = (x_1^3 + x_2^3 - 3ax_1x_2,\; yx_1 - x_2).$$

Let $V = \mathbf{R} \setminus \{-1\}$ and let $\psi : V \to \mathbf{R}^2$ be defined by

$$\psi(y) = \frac{3ay}{y^3+1}(1,\, y).$$

(i) Prove that $f(\psi(y); y) = 0$, for all $y \in V$.

(ii) Determine the $y \in V$ for which $D_x f(\psi(y); y) \in \operatorname{Aut}(\mathbf{R}^2)$.

Exercise 3.31. Consider the equation $\sin(x^2 + y) - 2x = 0$ for $x \in \mathbf{R}$ with $y \in \mathbf{R}$ as a parameter.

(i) Prove the existence of neighborhoods V and U of 0 in $\mathbf{R}$ such that for every $y \in V$ there exists a unique solution $x = \psi(y) \in U$. Prove that ψ is a C^∞ mapping on V, and that $\psi'(0) = \frac{1}{2}$.

We will also use the following notations: $x = \psi(y)$, $x' = \psi'(y)$ and $x'' = \psi''(y)$.

(ii) Prove $2(x\cos(x^2+y) - 1)\,x' + \cos(x^2+y) = 0$, for all $y \in V$; and again calculate $\psi'(0)$.

(iii) Show, for all $y \in V$,

$$(x\cos(x^2+y) - 1)x'' + (\cos(x^2+y) - 4x^3)(x')^2 - 4x^2\,x' - x = 0;$$

and prove that $\psi''(0) = \frac{1}{4}$.

Remark. Apparently p, the Taylor polynomial of order 2 of ψ at 0, is given by $p(y) = \frac{1}{2}y + \frac{1}{8}y^2$. Approximating the function sin in the original equation by its Taylor polynomial of order 1 at 0, we find the problem $x^2 - 2x + y = 0$. One verifies that $p(y)$, modulo terms that are $\mathcal{O}(y^3)$, $y \to 0$, is a solution of the latter problem (see Exercise 0.11.(v)).

Exercise 3.32 (Kepler's equation – needed for Exercise 6.67). This equation reads, for unknown $x \in \mathbf{R}$ with $y \in \mathbf{R}^2$ as a parameter:

$$(\star) \qquad x = y_1 + y_2 \sin x.$$

It plays an important role in the theory of planetary motion. There x represents the eccentric anomaly, which is a variable describing the position of a planet in its elliptical orbit, y_1 is proportional to time and y_2 is the eccentricity of the ellipse. Newton used his iteration method from Exercise 2.47 to obtain approximate values for x. We have $|y_2| < 1$, hence we set $V = \{ y \in \mathbf{R}^2 \mid |y_2| < 1 \}$.

(i) Prove that $(\star)$ has a unique solution $x = \psi(y)$ if y belongs to the open subset V of $\mathbf{R}^2$. Show that $\psi : V \to \mathbf{R}$ is a C^∞ function.

(ii) Verify $\psi(\pi, y_2) = \pi$, for $|y_2| < 1$; and also $\psi(y_1 + 2\pi, y_2) = \psi(y) + 2\pi$, for $y \in V$.

(iii) Show that $\psi(-y_1, y_2) = -\psi(y_1, y_2)$; in particular, $\psi(0, y_2) = 0$ for $|y_2| < 1$. More generally, prove $D_1^k\psi(0, y_2) = 0$, for $k \in 2\mathbf{N}_0$ and $|y_2| < 1$.

(iv) Compute $D_1\psi(y)$, for $y \in V$.

(v) Let $y_2 = 1$. Prove that there exists a unique function $\xi : \mathbf{R} \to \mathbf{R}$ such that $x = \xi(y_1)$ is a solution of equation $(\star)$. Show that $\xi(0) = 0$ and that ξ is continuous. Verify

$$y_1 = \frac{x^3}{6}(1 + \mathcal{O}(x^2)), \quad x \to 0, \qquad \text{and deduce} \qquad \lim_{y_1 \to 0} \frac{\xi(y_1)}{(6y_1)^{1/3}} = 1.$$

Prove that this implies that ξ is not differentiable at 0.

Exercise 3.33. Consider the following equation for $x \in \mathbf{R}$ with $y \in \mathbf{R}^2$ as a parameter:

$$(\star) \qquad x^3 y_1 + x^2 y_1 y_2 + x + y_1^2 y_2 = 0.$$

(i) Prove that there are neighborhoods V of $(-1, 1)$ in $\mathbf{R}^2$ and U of 1 in $\mathbf{R}$ such that for every $y \in V$ the equation $(\star)$ has a unique solution $x = \psi(y) \in U$. Prove that the mapping $\psi : V \to \mathbf{R}$ given by $y \mapsto x = \psi(y)$ is a C^∞ mapping.

(ii) Calculate the partial derivatives $D_1\psi(-1, 1)$, $D_2\psi(-1, 1)$ and finally also $D_1 D_2\psi(-1, 1)$.

(iii) Prove that there do not exist neighborhoods V as above and U' of -1 in $\mathbf{R}$ such that for every $y \in V$ the equation ($\star$) has a **unique** solution $x = x(y) \in U'$. **Hint:** Explicitly determine the three solutions of equation ($\star$) in the special case $y_1 = -1$.

Exercise 3.34. Consider the following equations for $x \in \mathbf{R}^2$ with $y \in \mathbf{R}^2$ as a parameter

$$e^{x_1+x_2} + y_1 y_2 = 2 \qquad \text{and} \qquad x_1^2 y_1 + x_2 y_2^2 = 0.$$

(i) Prove that there are neighborhoods V of $(1, 1)$ in $\mathbf{R}^2$ and U of $(1, -1)$ in $\mathbf{R}^2$ such that for every $y \in V$ there exists a unique solution $x = \psi(y) \in U$. Prove that $\psi : V \to \mathbf{R}^2$ is a C^∞ mapping.

(ii) Find $D_1\psi_1$, $D_2\psi_1$, $D_1\psi_2$, $D_2\psi_2$ and $D_1 D_2\psi_1$, all at the point $y = (1, 1)$.

Exercise 3.35.

(i) Show that $x \in \mathbf{R}^2$ in a neighborhood of $0 \in \mathbf{R}^2$ can uniquely be solved in terms of $y \in \mathbf{R}^2$ in a neighborhood of $0 \in \mathbf{R}^2$, from the equations

$$e^{x_1} + e^{x_2} + e^{y_1} + e^{y_2} = 4 \qquad \text{and} \qquad e^{x_1} + e^{2x_2} + e^{3y_1} + e^{4y_2} = 4.$$

(ii) Likewise, show that $x \in \mathbf{R}^2$ in a neighborhood of $0 \in \mathbf{R}^2$ can uniquely be solved in terms of $y \in \mathbf{R}^2$ in a neighborhood of $0 \in \mathbf{R}^2$, from the equations

$$e^{x_1} + e^{x_2} + e^{y_1} + e^{y_2} = 4 + x_1 \qquad \text{and} \qquad e^{x_1} + e^{2x_2} + e^{3y_1} + e^{4y_2} = 4 + x_2.$$

(iii) Calculate $D_2 x_1$ and $D_2^2 x_1$ for the function $y \mapsto x_1(y)$, at the point $y = 0$.

Exercise 3.36. Let $f : \mathbf{R} \to \mathbf{R}$ be a continuous function and consider $b > 0$ such that

$$f(0) \neq -1 \qquad \text{and} \qquad \int_0^b f(t)\, dt = 0.$$

Prove that the equation

$$\int_x^a f(t)\, dt = x,$$

for a in $\mathbf{R}$ sufficiently close to b, possesses a unique solution $x = x(a) \in \mathbf{R}$ with $x(a)$ near 0. Prove that $a \mapsto x(a)$ is a C^1 mapping, and calculate $x'(b)$. What goes wrong if $f(t) = -1 + 2t$ and $b = 1$?

Exercise 3.37 (Addition to Application 3.6.A). The notation is as in that application. We now deal with the case $n = 3$ in more detail. Dividing by a_3 and applying a translation (to make the coefficient of x^2 disappear) we may assume that

$$f(x) = f_{a,b}(x) = x^3 - 3ax + 2b \qquad ((a,\ b) \in \mathbf{R}^2).$$

(i) Prove that

$$S = \{\, (t^2, t^3) \in \mathbf{R}^2 \mid t \in \mathbf{R} \,\}$$

is the collection S of the $(a, b) \in \mathbf{R}^2$ such that $f_{a,b}(x)$ has a double root, and sketch S.
Hint: write $f(x) = (x - s)(x - t)^2$.

(ii) Determine the regions G_1 and G_3, respectively, of the (a, b) in $\mathbf{R}^2 \setminus S$ where $f_{a,b}$ has one and three real zeros, respectively.

(iii) Let $c \in \mathbf{R}$. Prove that the points $(a, b) \in \mathbf{R}^2$ with $f_{a,b}(c) = 0$ lie on the straight line with the equation $3ca - 2b = c^3$. Determine the other two zeros of $f_{a,b}(x)$, for (a, b) on this line. Prove that this line intersects the set S transversally once, and is once tangent to it (unless $c = 0$). (Note that $3ct^2 - 2t^3 = c^3$ implies $(t - c)^2(2t + c) = 0$). In G_3, three such lines run through each point; explain why. Draw a few of these lines.

(iv) Draw the set

$$\{\, (a, b, x) \in \mathbf{R}^3 \mid x^3 - 3ax + 2b = 0 \,\},$$

by constructing the cross-section with the plane a = constant for $a < 0$, $a = 0$ and $a > 0$. Also draw in these planes a few vertical lines (a and b both constant). Do this, for given a, by taking b in each of the intervals where $(a, b) \in G_1$ or $(a, b) \in G_3$, respectively, plus the special points b for which $a, b \in S$.

Exercise 3.38 (Addition to Application 3.6.E). The notation is as in that application.

(i) Determine $\psi''(0)$.

(ii) Let $A = \begin{pmatrix} 1 & 1 \\ 0 & 1 \end{pmatrix}$. Prove that there exist no neighborhood V of 0 in $\mathbf{R}$ for which there is a differentiable mapping $\xi : V \to M$ such that

$$\xi(0) = \begin{pmatrix} 1 & 0 \\ 0 & -1 \end{pmatrix} \qquad \text{and} \qquad \xi(t)^2 + tA\xi(t) = I \qquad (t \in V).$$

Hint: Suppose that such V and ξ do exist. What would this imply for $\xi'(0)$?

Exercise 3.39. Define $f : \mathbf{R}^2 \to \mathbf{R}$ by $f(x) = 2x_1^3 - 3x_1^2 + 2x_2^3 + 3x_2^2$. Note that $f(x)$ is divisible by $x_1 + x_2$.

(i) Find the four critical points of f in $\mathbf{R}^2$. Show that f has precisely one local minimum and one local maximum on $\mathbf{R}^2$.

(ii) Let $N = \{ x \in \mathbf{R}^2 \mid f(x) = 0 \}$. Find the points of N that do not possess a neighborhood in $\mathbf{R}^2$ where one can obtain, from the equation $f(x) = 0$, either x_2 as a function of x_1, or x_1 as a function of x_2. Sketch N.

Exercise 3.40. If $x_1^2 + x_2^2 + x_3^2 + x_4^2 = 1$ and $x_1^3 + x_2^3 + x_3^3 + x_4^3 = 0$, then $D_1 x_3$ is not uniquely determined. We now wish to consider two of the variables as functions of the other two; specifically, x_3 as a dependent and x_1 as an independent variable. Then either x_2 or x_4 may be chosen as the other independent variable. Calculate $D_1 x_3$ for both cases.

Exercise 3.41 (Simple eigenvalues and corresponding eigenvectors are C^∞ functions of matrix entries – sequel to Exercise 2.21). Consider the $n + 1$ equations

$$(\star) \qquad Ax = \lambda x \qquad \text{and} \qquad \langle x, x \rangle = 1,$$

in the $n + 1$ unknowns x, λ, with $x \in \mathbf{R}^n$ and $\lambda \in \mathbf{R}$, where $A \in \mathrm{Mat}(n, \mathbf{R})$ is a parameter. Assume that x^0, λ^0 and A^0 satisfy $(\star)$, and that λ^0 is a simple root of the characteristic polynomial $\lambda \mapsto \det(\lambda I - A^0)$ of A^0.

(i) Prove that there are numbers $\eta > 0$ and $\delta > 0$ such that, for every $A \in \mathrm{Mat}(n, \mathbf{R})$ with $\|A - A^0\|_{\mathrm{Eucl}} < \eta$, there exist unique $x \in \mathbf{R}^n$ and $\lambda \in \mathbf{R}$ with

$$\|x - x^0\| < \delta, \qquad |\lambda - \lambda^0| < \delta \qquad \text{and} \qquad x,\ \lambda,\ A \text{ satisfy } (\star).$$

Hint: Define $f : \mathbf{R}^n \times \mathbf{R} \times \mathrm{Mat}(n, \mathbf{R}) \to \mathbf{R}^n \times \mathbf{R}$ by

$$f(x, \lambda; A) = (Ax - \lambda x,\ \langle x, x \rangle - 1).$$

Apply Exercise 2.21.(iv) to conclude that

$$\det \frac{\partial f}{\partial(x, \lambda)}(x^0, \lambda^0; A^0) = 2 \lim_{\lambda \to \lambda^0} \frac{\det(A^0 - \lambda I)}{\lambda^0 - \lambda};$$

and now use the fact that λ^0 is a simple zero.

(ii) Prove that these x and λ are C^∞ functions of the matrix entries of A.

Exercise 3.42. Let $p(x) = \sum_{0 \le k \le n} a_k x^k$, with $a_k \in \mathbf{R}$, $a_n = 1$. Assume that the polynomial p has real roots c_k with $1 \le k \le n$ only, and that these are simple. Let $\epsilon \in \mathbf{R}$ and $m \in \mathbf{N}_0$, and define

$$p_\epsilon(x) = p(x; \epsilon) = p(x) - \epsilon x^m.$$

Then consider the polynomial equation $p_\epsilon(x) = 0$ for x, with ϵ as a parameter.

(i) Prove that for ϵ chosen sufficiently close to 0, the polynomial p_ϵ also has n simple roots $c_k(\epsilon) \in \mathbf{R}$, with $c_k(\epsilon)$ near c_k for $1 \le k \le n$.

(ii) Prove that the mappings $\epsilon \mapsto c_k(\epsilon)$ are differentiable, with

$$c_k'(0) = \frac{c_k^m}{p'(c_k)} \qquad (1 \le k \le n).$$

(iii) If $m \le n$ and ϵ sufficiently close to 0, this determines all roots of p_ϵ. Why?

Next, assume that $m > n$ and that $\epsilon \ne 0$. Define $y = y(x, \epsilon)$ by $y = \delta x$ with $\delta = \epsilon^{1/(m-n)}$.

(iv) Check that $p_\epsilon(x) = 0$ implies that y satisfies the equation

$$y^m = y^n + \sum_{0 \le k < n} \delta^{n-k} a_k\, y^k.$$

(v) Demonstrate that in this case, for $\epsilon \ne 0$ but sufficiently close to 0, the polynomial p_ϵ has $m - n$ additional roots $c_k(\epsilon)$, for $n + 1 \le k \le m$, of the following form:

$$c_k(\epsilon) = \epsilon^{1/(n-m)} e^{2\pi i k/(m-n)} + b_k(\epsilon),$$

with $b_k(\epsilon)$ bounded for $\epsilon \to 0$, $\epsilon \ne 0$. This characterizes the other roots of p_ϵ.

Exercise 3.43 (Cardioid). This is the curve $C \subset \mathbf{R}^2$ described in polar coordinates (r, α) for $\mathbf{R}^2$ (i.e. $(x, y) = r(\cos\alpha, \sin\alpha)$) by the equation $r = 2(1 + \cos\alpha)$.

(i) Prove $C = \{\, (x, y) \in \mathbf{R}^2 \mid 2\sqrt{x^2 + y^2} = x^2 + y^2 - 2x \,\}$.

One has in fact

$$C = \{\, (x, y) \in \mathbf{R}^2 \mid x^4 - 4x^3 + 2y^2x^2 - 4y^2x + y^4 - 4y^2 = 0 \,\}.$$

Because this is a quartic equation in x with coefficients in y, it allows an explicit solution for x in terms of y. Answer the following questions, without solving the quartic equation.

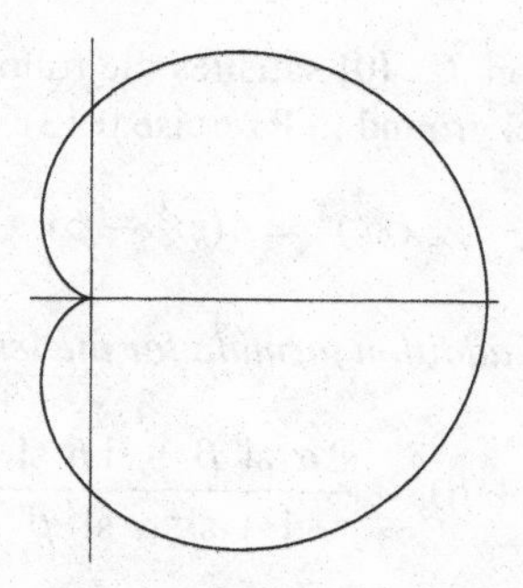

Illustration for Exercise 3.43: Cardioid

(i) Let $D = \,]-\frac{3}{2}\sqrt{3}, \frac{3}{2}\sqrt{3}\,[\, \subset \mathbf{R}$. Prove that a function $\psi : D \to \mathbf{R}$ exists such that $(\psi(y), y) \in C$, for all $y \in D$.

(ii) For $y \in D \setminus \{0\}$, express the derivative $\psi'(y)$ in terms of y and $x = \psi(y)$.

(iii) Determine the internal extrema of ψ on D.

Exercise 3.44 (Addition formula for lemniscatic sine – needed for Exercise 7.5). The mapping

$$a : [-1, 1] \to \mathbf{R} \qquad \text{given by} \qquad a(x) = \int_0^x \frac{1}{\sqrt{1-t^4}}\, dt$$

arises in the computation of the length of the lemniscate (see Exercise 7.5.(i)). Note that $a(1) =: \frac{\varpi}{2}$ is well-defined as the improper integral converges. We have $\varpi = 2.622\,057\,554\,292 \cdots$. Set $I = \,]-1, 1\,[$ and $J = \,]-\frac{\varpi}{2}, \frac{\varpi}{2}\,[$.

(i) Verify that $a : I \to J$ is a C^∞ bijection.

The function $\mathrm{sl} : J \to I$, the *lemniscatic sine*, is defined as the inverse mapping of a, thus

$$\alpha = \int_0^{\mathrm{sl}\,\alpha} \frac{1}{\sqrt{1-t^4}}\, dt \qquad (\alpha \in J).$$

Many properties of the lemniscatic sine are similar to those of the ordinary sine, for instance $\mathrm{sl}\,\frac{\varpi}{2} = 1$.

(ii) Show that sl is differentiable on J and prove

$$\mathrm{sl}' = \sqrt{1 - \mathrm{sl}^4}, \qquad \mathrm{sl}'' = -2\,\mathrm{sl}^3.$$

(iii) Prove that $\wp := \mathrm{sl}^{-2}$ on $J \setminus \{0\}$ satisfies the following differential equation (compare with the Background in Exercise 0.13):

$$(\wp')^2 = 4(\wp^3 - \wp).$$

We now prove the following *addition formula for the lemniscatic sine*, valid for α, β and $\alpha + \beta \in J$,

$$\mathrm{sl}(\alpha + \beta) = \frac{\mathrm{sl}\,\alpha\ \mathrm{sl}'\,\beta + \mathrm{sl}\,\beta\ \mathrm{sl}'\,\alpha}{1 + \mathrm{sl}^2\,\alpha\ \mathrm{sl}^2\,\beta}.$$

To this end consider

$$f : I \times I^2 \to \mathbf{R} \qquad \text{with} \qquad f(x; y, z) = \frac{x\sqrt{1-y^4} + y\sqrt{1-x^4}}{1 + x^2y^2} - z.$$

(iv) Prove the existence of η, ζ and $\delta > 0$ such that for all y with $|y| < \eta$ and z with $|z| < \zeta$ there is $x = \psi(y) \in \mathbf{R}$ with $|x| < \delta$ and $f(x; y, z) = 0$. Verify that ψ is a C^∞ mapping on $]-\eta, \eta[\, \times\,]-\zeta, \zeta[$.

From now on we treat z as a constant and consider ψ as a function of y alone, with a slight abuse of notation.

(v) Show that the derivative of the mapping $\psi :]-\eta, \eta[\to \mathbf{R}$ is given by

$$\psi'(y) = -\frac{\sqrt{1-\psi(y)^4}}{\sqrt{1-y^4}}, \qquad \text{that is} \qquad \frac{\psi'(y)}{\sqrt{1-\psi(y)^4}} + \frac{1}{\sqrt{1-y^4}} = 0.$$

Hint:

$$D_x f(x; y, z) = \frac{1}{\sqrt{1-x^4}} \frac{\sqrt{1-x^4}\sqrt{1-y^4}(1-x^2y^2) - 2xy(x^2+y^2)}{(1+x^2y^2)^2}.$$

(vi) Using the Fundamental Theorem of Integral Calculus and the equality $\psi(0) = z$ deduce

$$\int_z^{\psi(y)} \frac{1}{\sqrt{1-t^4}}\,dt + \int_0^y \frac{1}{\sqrt{1-t^4}}\,dt = 0,$$

in other words

$$\int_0^x \frac{1}{\sqrt{1-t^4}}\,dt + \int_0^y \frac{1}{\sqrt{1-t^4}}\,dt = \int_0^z \frac{1}{\sqrt{1-t^4}}\,dt,$$

$$z = \frac{x\sqrt{1-y^4} + y\sqrt{1-x^4}}{1 + x^2y^2}.$$

Applying (ii), verify that the addition formula for the lemniscatic sine is a direct consequence. Note that in Exercise 0.10.(iv) we consider the special case of $x = y$.

For a second proof of the addition formula introduce new variables $\gamma = \frac{\alpha+\beta}{2}$ and $\delta = \frac{\alpha-\beta}{2}$, and denote the right–hand side of the addition formula by $g(\gamma, \delta)$.

(vii) Use (ii) to verify $\frac{\partial g}{\partial \delta}(\gamma, \delta) = 0$ and conclude that $g(\gamma, \delta) = g(\gamma, \gamma)$ for all γ and δ. The addition formula now follows from $g(\gamma, \gamma) = \mathrm{sl}\, 2\gamma = \mathrm{sl}(\alpha + \beta)$.

(viii) Deduce the following division formula for the lemniscatic sine:

$$\mathrm{sl}\,\frac{\alpha}{2} = \sqrt{\frac{\sqrt{1+\mathrm{sl}^2\,\alpha} - 1}{\sqrt{1-\mathrm{sl}^2\,\alpha} + 1}} \qquad (0 \le \alpha \le \frac{\varpi}{2}).$$

Conclude $\mathrm{sl}\,\frac{\varpi}{4} = \sqrt{\sqrt{2} - 1}$ (compare with Exercise 0.10.(iv)).

Hint: Write $h(\alpha) = \sqrt{1 - \mathrm{sl}^2\,\alpha}$ and use the same technique as in part (vii) to verify

$$h(\alpha + \beta) = \frac{h(\alpha)h(\beta) - \mathrm{sl}\,\alpha\, \mathrm{sl}\,\beta \sqrt{1+\mathrm{sl}^2\,\alpha}\sqrt{1+\mathrm{sl}^2\,\beta}}{1 + \mathrm{sl}^2\,\alpha\,\mathrm{sl}^2\,\beta}.$$

Set $a = \mathrm{sl}\,\frac{\alpha}{2}$ and deduce

$$h(\alpha) = \frac{1 - a^2 - a^2(1+a^2)}{1+a^4} = \frac{1 - 2a^2 - a^4}{1+a^4}.$$

Finally solve for a^2.

Exercise 3.45 (Discriminant locus). In applications of the Implicit Function Theorem 3.5.1 it is often interesting to study the subcollection in the parameter space consisting of the points $y \in \mathbf{R}^p$ where the condition of the theorem that

$$D_x f(x; y) \in \mathrm{Aut}(\mathbf{R}^n)$$

is not met. That is, those $y \in \mathbf{R}^p$ for which there exists at least one $x \in \mathbf{R}^n$ such that $(x, y) \in U$, $f(x; y) = 0$, and $\det D_x f(x; y) = 0$. This set is called the *discriminant locus* or the *bifurcation set*.

We now do this for the general quadratic equation

$$f(x; y) = y_2 x^2 + y_1 x + y_0 \qquad (y_i \in \mathbf{R},\ 0 \le i \le 2,\ y_2 \ne 0).$$

Prove that the discriminant locus D is given by

$$D = \{\, y \in \mathbf{R}^3 \mid y_1^2 - 4y_0 y_2 = 0 \,\}.$$

Show that D is a conical surface by substituting $y_0 = \frac{1}{2}(z_2 + z_0)$, $y_1 = z_1$, $y_2 = \frac{1}{2}(z_2 - z_0)$. Indeed, we find

$$D = \{\, z \in \mathbf{R}^3 \mid z_0^2 + z_1^2 - z_2^2 = 0 \,\}.$$

The complement in $\mathbf{R}^3$ of the cone D consists of two disjoint open sets, D_+ and D_-, made up of points $y \in \mathbf{R}^3$ for which $y_1^2 - 4y_0 y_2 > 0$ and < 0, respectively. What we find is that the equation $f(x; y) = 0$ has two distinct, one, or no solution $x \in \mathbf{R}$ depending on whether $y \in \mathbf{R}^3$ lies in D_+, D or D_-.

Exercise 3.46. In *thermodynamics* one studies real variables P, V and T obeying a relation of the form $f(P, V, T) = 0$, for a C^1 function $f : \mathbf{R}^3 \to \mathbf{R}$. This relation is used to consider one of the variables, say P, as a function of another, say T, while keeping the third variable, V, constant. The notation $\frac{\partial P}{\partial T}\big|_V$ is used for the derivative of P with respect to T under constant V. Prove

$$\frac{\partial P}{\partial T}\bigg|_V \frac{\partial T}{\partial V}\bigg|_P \frac{\partial V}{\partial P}\bigg|_T = -1,$$

and try to formulate conditions on f for this formula to hold.

Exercise 3.47 (Newtonian mechanics). Let $x(t) \in \mathbf{R}^n$ be the position at time $t \in \mathbf{R}$ of a classical point particle with n degrees of freedom, and assume that $t \mapsto x(t)$ is a C^2 curve in $\mathbf{R}^n$. According to Newton the motion of such a particle is described by the following second-order differential equation for x:

$$m\, x''(t) = -\operatorname{grad} V(x(t)) \qquad (t \in \mathbf{R}).$$

Here $m > 0$ is the *mass* of the particle and $V : \mathbf{R}^n \to \mathbf{R}$ a C^1 function representing the *potential energy* of the particle. We define the *kinetic energy* $T : \mathbf{R}^n \to \mathbf{R}$ and the *total energy* $E : \mathbf{R}^n \times \mathbf{R}^n \to \mathbf{R}$, respectively, by

$$T(v) = \frac{1}{2} m \|v\|^2 \qquad \text{and} \qquad E(x, v) = T(v) + V(x) \qquad (v,\ x \in \mathbf{R}^n).$$

(i) Prove that *conservation of energy*, that is, $t \mapsto E(x(t),\ x'(t))$ being a constant function (equal to $E \in \mathbf{R}$, say) on $\mathbf{R}$ is equivalent to Newton's differential equation.

(ii) From now on suppose $n = 1$. Verify that in this case x also satisfies a first-order differential equation

$$(\star) \qquad x'(t) = \pm\Big(\frac{2}{m}(E - V(x(t)))\Big)^{1/2},$$

where one sign only applies. Conclude that x is not uniquely determined if $t \mapsto x(t)$ is required to be a C^1 function satisfying $(\star)$.
Hint: Consider the situation where $x'(t_0) = 0$, at some time t_0.

(iii) Now assume $x'(t) > 0$, for $t_0 \le t \le t_1$. Prove that then $V(x) < E$, for $x(t_0) \le x \le x(t_1)$, and that

$$t = \int_{x(t_0)}^{x(t)} \Big(\frac{2}{m}(E - V(y))\Big)^{-1/2} dy.$$

Prove by means of this result that $x(t)$ is uniquely determined as a function of $x(t_0)$ and E.

Further assume that $a < b$; $V(a) = V(b) = E$; $V(x) < E$, for $a < x < b$; $V'(a) < 0$, $V'(b) > 0$ (that is, if $x(t_0) = a$, $x(t_1) = b$, then $x'(t_0) = x'(t_1) = 0$; $x''(t_0) > 0$; $x''(t_1) < 0$).

(iv) Give a proof that then

$$T := 2 \int_a^b \left(\frac{2}{m}(E - V(x))\right)^{-1/2} dx < \infty.$$

Hint: Use, for example, Taylor expansion of $x'(t)$ in powers of $t - t_0$.

(v) Prove the following. Each motion $t \mapsto x(t)$ with total energy E which for some t finds itself in $[\,a, b\,]$ has the property that $x(t) \in [\,a, b\,]$ for all t, and, in addition, has a unique continuation for all $t \in \mathbf{R}$ (being a solution of Newton's differential equation). This solution is periodic with period T, i.e., $x(t + T) = x(t)$, for all $t \in \mathbf{R}$.

Exercise 3.48 (Fundamental Theorem of Algebra – sequel to Exercises 1.7 and 1.25). This theorem asserts that for every nonconstant polynomial function $p : \mathbf{C} \to \mathbf{C}$ there is a $z \in \mathbf{C}$ with $p(z) = 0$ (see Example 8.11.5 and Exercise 8.13 for other proofs).

(i) Verify that the theorem follows from $\mathrm{im}(p) = \mathbf{C}$.

(ii) Using Exercise 1.25 show that $\mathrm{im}(p)$ is a closed subset of $\mathbf{C}$.

(iii) Note that the derivative p' possesses at most finitely many zeros, and conclude by applying the Local Inverse Function Theorem 3.7.1 on $\mathbf{C}$ that $\mathrm{im}(p)$ has a nonempty interior.

(iv) Assume that w belongs to the boundary of $\mathrm{im}(p)$. Then by (ii) there exists a $z \in \mathbf{C}$ with $p(z) = w$. Prove by contradiction, using the Local Inverse Function Theorem on $\mathbf{C}$, that $p'(z) = 0$. Conclude that the boundary of $\mathrm{im}(p)$ is a finite set.

(v) Prove by means of Exercise 1.7 that $\mathrm{im}(p) = \mathbf{C}$.

Exercises for Chapter 4

Exercise 4.1. What is the number of degrees of freedom of a bicycle? Assume, for simplicity, that the frame is rigid, that the wheels can rotate about their axes, but cannot wobble, and that both the chain and the pedals are rigidly coupled to the rear wheel. What is the number of degrees of freedom when the bicycle rests on the ground with one or with two wheels, respectively?

Exercise 4.2. Let $V \subset \mathbf{R}^2$ be the set of points which in polar coordinates (r, α) for $\mathbf{R}^2$ satisfy the equation

$$r = \frac{6}{1 - 2\cos\alpha}.$$

Show that locally V satisfies: (i) Definition 4.1.3 of zero-set, (ii) Definition 4.1.2 of parametrized set, and (iii) Definition 4.2.1 of C^∞ submanifold in $\mathbf{R}^2$ of dimension 1. Sketch V.

Hint: (i) $3x_1^2 + 24x_1 - x_2^2 + 36 = 0$; (ii) $\phi(t) = (-4 + 2\cosh t,\ 2\sqrt{3}\sinh t)$, where $\cosh t = \frac{1}{2}(e^t + e^{-t})$ and $\sinh t = \frac{1}{2}(e^t - e^{-t})$.

Exercise 4.3. Prove that each of the following sets in $\mathbf{R}^2$ is a C^∞ submanifold in $\mathbf{R}^2$ of dimension 1:

$$\{\, x \in \mathbf{R}^2 \mid x_2 = x_1^3 \,\}, \qquad \{\, x \in \mathbf{R}^2 \mid x_1 = x_2^3 \,\}, \qquad \{\, x \in \mathbf{R}^2 \mid x_1 x_2 = 1 \,\};$$

but that this is not the case for the following subsets of $\mathbf{R}^2$:

$$\{\, x \in \mathbf{R}^2 \mid x_2 = |x_1| \,\}, \qquad \{\, x \in \mathbf{R}^2 \mid (x_1 x_2 - 1)(x_1^2 + x_2^2 - 2) = 0 \,\},$$

$$\{\, x \in \mathbf{R}^2 \mid x_2 = -x_1^2, \text{ for } x_1 \le 0;\ x_2 = x_1^2, \text{ for } x_1 \ge 0 \,\}.$$

Why is it that

$$\{\, x \in \mathbf{R}^2 \mid \|x\| < 1 \,\} \qquad \text{and} \qquad \{\, x \in \mathbf{R}^2 \mid |x_1| < 1,\ |x_2| < 1 \,\}$$

are submanifolds of $\mathbf{R}^2$, but not

$$\{\, x \in \mathbf{R}^2 \mid \|x\| \le 1 \,\}\ ?$$

Exercise 4.4 (Cycloid). (See also Example 5.3.6.) This is the curve $\phi : \mathbf{R} \to \mathbf{R}^2$ defined by

$$\phi(t) = (t - \sin t,\ 1 - \cos t).$$

(i) Prove that ϕ is an injective C^∞ mapping.

(ii) Prove that ϕ is an immersion at every point $t \in \mathbf{R}$ with $t \ne 2k\pi$, for $k \in \mathbf{Z}$.

Exercise 4.5. Consider the mapping $\phi : \mathbf{R}^2 \to \mathbf{R}^3$ defined by

$$\phi(\alpha, \theta) = (\cos\alpha\cos\theta,\ \sin\alpha\cos\theta,\ \sin\theta).$$

(i) Prove that ϕ is a surjection of $\mathbf{R}^2$ onto the unit sphere $S^2 \subset \mathbf{R}^3$ with $S^2 = \{ x \in \mathbf{R}^3 \mid \|x\| = 1 \}$. Describe the inverse image under ϕ of a point in S^2, paying special attention to the points $(0, 0, \pm 1)$.

(ii) Find the points in $\mathbf{R}^2$ where the mapping ϕ is regular.

(iii) Draw the images of the lines with the equations $\theta =$ constant and $\alpha =$ constant, respectively, for different values of θ and α, respectively.

(iv) Determine a maximal open set $D \subset \mathbf{R}^2$ such that $\phi : D \to S^2$ is injective. Is $\phi|_D$ regular on D? And surjective? Is $\phi : D \to \phi(D)$ an embedding?

Exercise 4.6 (Surfaces of revolution – needed for Exercises 5.32 and 6.25). Assume that C is a curve in $\mathbf{R}^3$ lying in the half-plane $\{ x \in \mathbf{R}^3 \mid x_1 > 0,\ x_2 = 0 \}$. Further assume that $C = \mathrm{im}(\gamma)$ for the C^k embedding $\gamma : I \to \mathbf{R}^3$ with

$$\gamma(s) = (\gamma_1(s), 0,\ \gamma_3(s)).$$

The *surface of revolution* V *formed by revolving* C *about the* x_3*-axis* is defined as $V = \mathrm{im}(\phi)$ for the mapping $\phi : I \times \mathbf{R} \to \mathbf{R}^3$ with

$$\phi(s, t) = (\gamma_1(s)\cos t,\ \gamma_1(s)\sin t,\ \gamma_3(s)).$$

(i) Prove that ϕ is injective if the domain of ϕ is suitably restricted, and determine a maximal domain with this property.

(ii) Prove that ϕ is a C^k immersion everywhere (it is useful to note that C does not intersect the x_3-axis).

(iii) Prove that $x = \phi(s, t) \in V$ implies that, for $t \neq (2k + 1)\pi$ with $k \in \mathbf{Z}$,

$$t = 2\arctan\left(\frac{x_2}{x_1 + \sqrt{x_1^2 + x_2^2}}\right),$$

while for t in a neighborhood of $(2k + 1)\pi$ with $k \in \mathbf{Z}$,

$$t = 2\,\mathrm{arccotan}\left(\frac{x_2}{-x_1 + \sqrt{x_1^2 + x_2^2}}\right).$$

Use the fact that $\phi|_D$, for a suitably chosen domain $D \subset \mathbf{R}^2$, is a C^k embedding. Prove that the surface of revolution V is a C^k submanifold in $\mathbf{R}^3$ of dimension 2.

(iv) What complications can arise in the arguments above if C does intersect the x_3-axis?

Apply the preceding construction to the *catenary* $C = \mathrm{im}(\gamma)$ given by

$$\gamma(s) = a(\cosh s, 0, s) \qquad (a > 0).$$

The resulting surface $V = \mathrm{im}(\phi)$ in $\mathbf{R}^3$ is called the *catenoid*

$$\phi(s, t) = a(\cosh s \cos t, \cosh s \sin t, s).$$

(v) Prove that $V = \{ x \in \mathbf{R}^3 \mid x_1^2 + x_2^2 - a^2 \cosh^2(\frac{x_3}{a}) = 0 \}$.

Exercise 4.7. Define the C^∞ functions f and $g : \mathbf{R} \to \mathbf{R}$ by

$$f(t) = t^3, \qquad \text{and} \qquad g(t) = \begin{cases} e^{-1/t^2}, & t > 0; \\ 0, & t = 0; \\ -e^{-1/t^2}, & t < 0. \end{cases}$$

(i) Prove that g is a C^1 function.

(ii) Prove that f and g are not regular everywhere.

(iii) Prove that f and g are both injective.

Define $h : \mathbf{R} \to \mathbf{R}^2$ by $h(t) = (g(t), |g(t)|)$.

(iv) Prove that h is a C^1 curve in $\mathbf{R}^2$ and that $\mathrm{im}(h) = \{ (x, |x|) \mid |x| < 1 \}$. Is there a contradiction between the last two assertions?

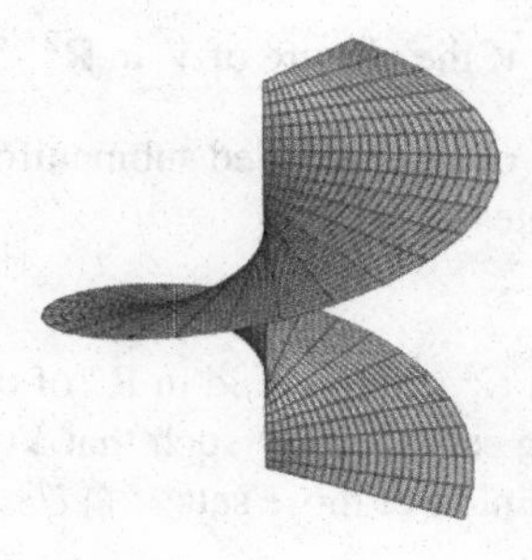

Illustration for Exercise 4.8: Helicoid

Exercise 4.8 (Helix and helicoid – needed for Exercises 5.32 and 7.3). (ἡ ἕλιξ = curl of hair.) The *spiral* or *helix* is the curve in $\mathbf{R}^3$ defined by

$$t \mapsto (\cos t,\ \sin t,\ at) \qquad (a \in \mathbf{R}_+,\ t \in \mathbf{R}).$$

The *helicoid* is the surface in $\mathbf{R}^3$ defined by

$$(s, t) \mapsto (s\cos t,\ s\sin t,\ at) \qquad ((s, t) \in \mathbf{R}_+ \times \mathbf{R}).$$

(i) Prove that the helix and the helicoid are C^∞ submanifolds in $\mathbf{R}^3$ of dimension 1 and 2, respectively. Verify that the helicoid is generated when through every point of the helix a line is drawn which intersects the x_3-axis and which runs parallel to the plane $x_3 = 0$.

(ii) Prove that a point x of the helicoid satisfies the equation

$$x_2 = x_1 \tan\left(\frac{x_3}{a}\right), \qquad \text{when} \qquad \frac{x_3}{a} \notin \left]\,(k+\frac{1}{4})\pi,\ (k+\frac{3}{4})\pi\,\right[\,;$$
$$x_1 = x_2 \cot\left(\frac{x_3}{a}\right), \qquad \text{when} \qquad \frac{x_3}{a} \in \left[\,(k+\frac{1}{4})\pi,\ (k+\frac{3}{4})\pi\,\right],$$

where $k \in \mathbf{Z}$.

Exercise 4.9 (Closure of embedded submanifold). Define $f : \mathbf{R}_+ \to \,]\,0, 1\,[$ by $f(t) = \frac{2}{\pi}\arctan t$, and consider the mapping $\phi : \mathbf{R}_+ \to \mathbf{R}^2$, given by $\phi(t) = f(t)(\cos t,\ \sin t)$.

(i) Show that ϕ is a C^∞ embedding and conclude that $V = \phi(\mathbf{R}_+)$ is a C^∞ submanifold in $\mathbf{R}^2$ of dimension 1.

(ii) Prove that V is closed in the open subset $U = \{\, x \in \mathbf{R}^2 \mid 0 < \|x\| < 1 \,\}$ of $\mathbf{R}^2$.

(iii) As usual, denote by $\overline{V}$ the closure of V in $\mathbf{R}^2$. Show that $\overline{V} \setminus V = \partial U$.

Background. The closure of an embedded submanifold may be a complicated set, as this example demonstrates.

Exercise 4.10. Let V be a C^k submanifold in $\mathbf{R}^n$ of dimension d. Prove that there exist countably many open sets U in $\mathbf{R}^n$ such that $V \cap U$ is as in Definition 4.2.1 and V is contained in the union of those sets $V \cap U$.

Exercise 4.11. Define $g : \mathbf{R}^2 \to \mathbf{R}$ by $g(x) = x_1^3 - x_2^3$.

(i) Prove that g is a surjective C^∞ function.

(ii) Prove that g is a C^∞ submersion at every point $x \in \mathbf{R}^2 \setminus \{0\}$.

(iii) Prove that for all $c \in \mathbf{R}$ the set $\{ x \in \mathbf{R}^2 \mid g(x) = c \}$ is a C^∞ submanifold in $\mathbf{R}^2$ of dimension 1.

Exercise 4.12. Define $g : \mathbf{R}^3 \to \mathbf{R}$ by $g(x) = x_1^2 + x_2^2 - x_3^2$.

(i) Prove that g is a surjective C^∞ function.

(ii) Prove that g is a submersion at every point $x \in \mathbf{R}^3 \setminus \{0\}$.

(iii) Prove that the two sheets of the cone

$$g^{-1}(\{0\}) \setminus \{0\} = \{ x \in \mathbf{R}^3 \setminus \{0\} \mid x_1^2 + x_2^2 = x_3^2 \}$$

form a C^∞ submanifold in $\mathbf{R}^3$ of dimension 2 (compare with Example 4.2.3).

Exercise 4.13. Consider the mappings $\phi_i : \mathbf{R}^2 \to \mathbf{R}^3$ $(i = 1, 2)$ defined by

$$\phi_1(y) = (ay_1 \cos y_2,\ by_1 \sin y_2,\ y_1^2),$$
$$\phi_2(y) = (a \sinh y_1 \cos y_2,\ b \sinh y_1 \sin y_2,\ c \cosh y_1),$$

with images equal to an *elliptic paraboloid* and a *hyperboloid of two sheets*, respectively; here $a, b, c > 0$.

(i) Establish at which points of $\mathbf{R}^2$ the mapping ϕ_i is regular.

(ii) Determine $V_i = \operatorname{im}(\phi_i)$, and prove that V_i is a C^∞ manifold in $\mathbf{R}^3$ of dimension 2.
Hint: Recall the Submersion Theorem 4.5.2.

Exercise 4.14 (Quadrics – needed for Exercises 5.6 and 5.12). A *nondegenerate quadric* in $\mathbf{R}^n$ is a set having the form

$$V = \{ x \in \mathbf{R}^n \mid \langle Ax, x \rangle + \langle b, x \rangle + c = 0 \},$$

where $A \in \operatorname{End}^+(\mathbf{R}^n) \cap \operatorname{Aut}(\mathbf{R}^n)$, $b \in \mathbf{R}^n$ and $c \in \mathbf{R}$. Introduce the discriminant $\Delta = \langle b, A^{-1}b \rangle - 4c \in \mathbf{R}$.

(i) Prove that V is an $(n-1)$-dimensional C^∞ submanifold of $\mathbf{R}^n$, provided $\Delta \neq 0$.

(ii) Suppose $\Delta = 0$. Prove that $-\frac{1}{2}A^{-1}b \in V$ and that $W = V \setminus \{ -\frac{1}{2}A^{-1}b \}$ always is an $(n-1)$-dimensional C^∞ submanifold of $\mathbf{R}^n$.

(iii) Now assume that A is positive definite. Use the substitution of variables $x = y - \frac{1}{2}A^{-1}b$ to prove the following. If $\Delta < 0$, then $V = \emptyset$; if $\Delta = 0$, then $V = \{ -\frac{1}{2}A^{-1}b \}$ and $W = \emptyset$; and if $\Delta > 0$, then V is an ellipsoid.

Exercise 4.15 (Needed for Exercise 4.16). Let $1 \le d \le n$, and let there be given vectors $a^{(1)}, \dots, a^{(d)} \in \mathbf{R}^n$ and numbers $r_1, \dots, r_d \in \mathbf{R}$. Let $x^0 \in \mathbf{R}^n$ such that $\|x^0 - a^{(i)}\| = r_i$, for $1 \le i \le d$. Prove by means of the Submersion Theorem 4.5.2 that at x^0

$$V = \{\, x \in \mathbf{R}^n \mid \|x - a^{(i)}\| = r_i \ (1 \le i \le d) \,\}$$

is a C^∞ manifold in $\mathbf{R}^n$ of codimension d, under the assumption that the vectors $x^0 - a^{(1)}, \dots, x^0 - a^{(d)}$ form a linearly independent system in $\mathbf{R}^n$.

Exercise 4.16 (Sequel to Exercise 4.15). For Exercise 4.15 it is also possible to give a fully explicit solution: by induction over $d \le n$ one can prove that in the case when the vectors $a^{(1)} - a^{(2)}, \dots, a^{(1)} - a^{(d)}$ are linearly independent, the set V equals

$$\{\, x \in H_d \mid \|x - b^{(d)}\|^2 = s_d \,\},$$

where $b^{(d)} \in H_d$, $s_d \in \mathbf{R}$, and

$$H_d = \{\, x \in \mathbf{R}^n \mid \langle x,\, a^{(1)} - a^{(i)} \rangle = c_j, \ \text{for all } i = 2, \dots, d \,\}$$

is a plane of dimension $n - d + 1$ in $\mathbf{R}^n$. If $s_d > 0$ we therefore obtain an $(n-d)$-dimensional sphere with radius $\sqrt{s_d}$ in H_d; if $s_d = 0$, a point; and if $s_d < 0$, the empty set. In particular: if $d = n$, V consists of two points at most. First try to verify the assertions for $d = 2$. Pay particular attention also to what may happen in the case $n = 3$.

Exercise 4.17 (Lines on hyperboloid of one sheet). Consider the *hyperboloid of one sheet* in $\mathbf{R}^3$

$$V = \{\, x \in \mathbf{R}^3 \mid \frac{x_1^2}{a^2} + \frac{x_2^2}{b^2} - \frac{x_3^2}{c^2} = 1 \,\} \qquad (a,\, b,\, c > 0).$$

Note that for $x \in V$,

$$\Big(\frac{x_1}{a} + \frac{x_3}{c}\Big)\Big(\frac{x_1}{a} - \frac{x_3}{c}\Big) = \Big(1 + \frac{x_2}{b}\Big)\Big(1 - \frac{x_2}{b}\Big).$$

(i) From this, show that through every point of V run two different straight lines that lie entirely on V, and that thus one finds two one-parameter families of straight lines lying entirely on V.

(ii) Prove that two different lines from one family do not intersect and are non-parallel.

(iii) Prove that two lines from different families either intersect or are parallel.

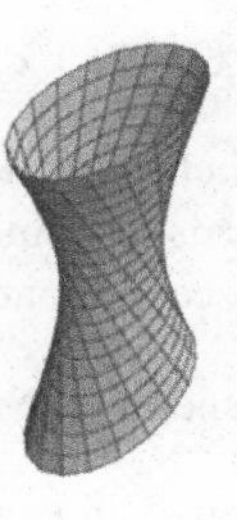

Illustration for Exercise 4.17: Lines on hyperboloid of one sheet

Exercise 4.18 (Needed for Exercise 5.5). Let $a, v : \mathbf{R} \to \mathbf{R}^n$ be C^k mappings for $k \in \mathbf{N}_\infty$, with $v(s) \neq 0$ for $s \in \mathbf{R}$. Then for every $s \in \mathbf{R}$ the mapping $t \mapsto a(s) + tv(s)$ defines a straight line in $\mathbf{R}^n$. The mapping $f : \mathbf{R}^2 \to \mathbf{R}^n$ with

$$f(s, t) = a(s) + tv(s),$$

is called a *one-parameter family of lines* in $\mathbf{R}^n$.

(i) Prove that f is a C^k mapping.

Henceforth assume that $n = 2$. Let $s_0 \in \mathbf{R}$ be fixed.

(ii) Assume the vector $v'(s_0)$ is not a multiple of $v(s_0)$. Prove that for all s sufficiently near s_0 there is a unique $t = t(s) \in \mathbf{R}$ such that (s, t) is a singular point of f, and that this t is C^{k-1}-dependent on s. Also prove that the derivative of $s \mapsto f(s, t(s))$ is a multiple of $v(s)$.

(iii) Next assume that $v'(s_0)$ is a multiple of $v(s_0)$. If $a'(s_0)$ is not a multiple of $v(s_0)$, then (s_0, t) is a regular point for f, for all $t \in \mathbf{R}$. If $a'(s_0)$ is a multiple of $v(s_0)$, then (s_0, t) is a singular point of f, for all $t \in \mathbf{R}$.

Exercise 4.19. Let there be given two open subsets $U, V \subset \mathbf{R}^n$, a C^1 diffeomorphism $\Phi : U \to V$ and two C^1 functions $g, h : V \to \mathbf{R}$. Assume $V = g^{-1}(0) \neq \emptyset$ and $h(x) \neq 0$, for all $x \in V$.

(i) Prove that g is submersive on V if and only if the composition $g \circ \Phi$ is submersive on U.

(ii) Prove that g is submersive on V if and only if the pointwise product $h\,g$ is submersive on V.

We consider a special case. The *cardioid* $C \subset \mathbf{R}^2$ is the curve described in polar coordinates (r, α) on $\mathbf{R}^2$ by the equation $r = 2(1 + \cos\alpha)$.

(iii) Prove $C = \{\, x \in \mathbf{R}^2 \mid 2\sqrt{x_1^2 + x_2^2} = x_1^2 + x_2^2 - 2x_1 \,\}$.

(iv) Show that $C = \{ x \in \mathbf{R}^2 \mid x_1^4 - 4x_1^3 + 2x_1^2x_2^2 - 4x_1x_2^2 + x_2^4 - 4x_2^2 = 0 \}$.

(v) Prove that the function $g : \mathbf{R}^2 \to \mathbf{R}$ given by $g(x) = x_1^4 - 4x_1^3 + 2x_1^2x_2^2 - 4x_1x_2^2 + x_2^4 - 4x_2^2$ is a submersion at all points of $C \setminus \{(0,0)\}$.
Hint: Use the preceding parts of this exercise.

We now consider the general case again.

(vi) In this situation, give an example where V is a submanifold of $\mathbf{R}^n$, but g is not submersive on V.
Hint: Take inspiration from the way in which part (ii) was answered.

Exercise 4.20. Show that an injective and proper immersion is a proper embedding.

Exercise 4.21. We define the position of a rigid body in $\mathbf{R}^3$ by giving the coordinates of three points in general position on that rigid body. Substantiate that the space of the positions of the rigid body is a C^∞ manifold in $\mathbf{R}^9$ of dimension 6.

Exercise 4.22 (Rotations in $\mathbf{R}^3$ and Euler's formula – sequel to Exercise 2.5 – needed for Exercises 5.58, 5.65 and 5.72). (Compare with Example 4.6.2.) Consider

$$\alpha \in \mathbf{R} \quad \text{with} \quad 0 \le \alpha \le \pi, \qquad \text{and} \qquad a \in \mathbf{R}^3 \quad \text{with} \quad \|a\| = 1.$$

Let $R_{\alpha,a} \in \mathbf{O}(\mathbf{R}^3)$ be the *rotation* in $\mathbf{R}^3$ with the following properties. $R_{\alpha,a}$ fixes a and maps the linear subspace N_a orthogonal to a into itself; in N_a, the effect of $R_{\alpha,a}$ is that of rotation by the angle α such that, if $0 < \alpha < \pi$, one has for all $y \in N_a$ that $\det(a \ \ y \ \ R_{\alpha,a}y) > 0$. Note that in Exercise 2.5 we proved that every rotation in $\mathbf{R}^3$ is of the form just described. Now let $x \in \mathbf{R}^3$ be arbitrarily chosen.

(i) Prove that the decomposition of x into the component along a and the component y perpendicular to a is given by (compare with Exercise 1.1.(ii))

$$x = \langle x, a \rangle \, a + y \qquad \text{with} \qquad y = x - \langle x, a \rangle \, a.$$

(ii) Show that $a \times x = a \times y$ is a vector of length $\|y\|$ which lies in N_a and which is perpendicular to y (see the Remark on linear algebra in Section 5.3 for the definition of the cross product $a \times x$ of a and x).

(iii) Prove $R_{\alpha,a}y = (\cos\alpha)\, y + (\sin\alpha)\, a \times x$, and conclude that we have the following, known as *Euler's formula*:

$$R_{\alpha,a}x = (1 - \cos\alpha)\langle a, x \rangle \, a + (\cos\alpha)\, x + (\sin\alpha)\, a \times x.$$

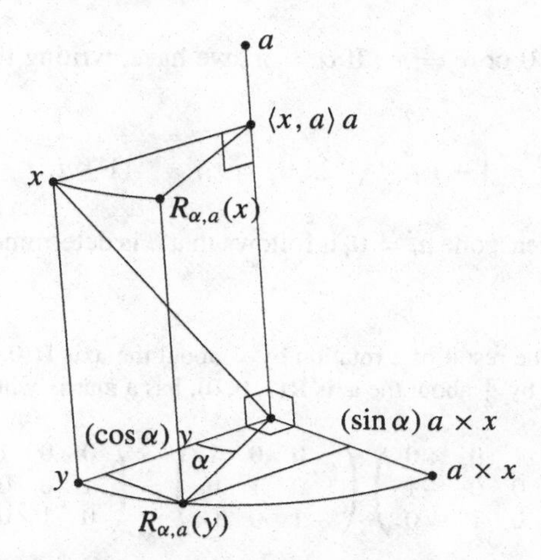

Illustration for Exercise 4.22: Rotation in $\mathbf{R}^3$

Verify that the matrix of $R_{\alpha,a}$ with respect to the standard basis for $\mathbf{R}^3$ is given by

$$\begin{pmatrix} \cos\alpha + \ \ a_1^2\, c(\alpha) & -a_3 \sin\alpha + a_1 a_2\, c(\alpha) & a_2 \sin\alpha + a_1 a_3\, c(\alpha) \\ a_3 \sin\alpha + a_1 a_2\, c(\alpha) & \cos\alpha + \ \ a_2^2\, c(\alpha) & -a_1 \sin\alpha + a_2 a_3\, c(\alpha) \\ -a_2 \sin\alpha + a_1 a_3\, c(\alpha) & a_1 \sin\alpha + a_2 a_3\, c(\alpha) & \cos\alpha + \ \ a_3^2\, c(\alpha) \end{pmatrix},$$

where $c(\alpha) = 1 - \cos\alpha$.

On geometrical grounds it is evident that

$$R_{\alpha,a} = R_{\beta,b} \qquad \Longleftrightarrow \qquad (\alpha = \beta = 0, \ a, \ b \text{ arbitrary})$$
$$\text{or} \quad (0 < \alpha = \beta < \pi, \ a = b) \quad \text{or} \quad (\alpha = \beta = \pi, \ a = \pm b).$$

We now ask how this result can be found from the matrix of $R_{\alpha,a}$. Recall the definition from Section 2.1 of the trace of a matrix.

(iv) Prove

$$\cos\alpha = \frac{1}{2}(-1 + \operatorname{tr} R_{\alpha,a}), \qquad R_{\alpha,a} - R_{\alpha,a}^t = 2(\sin\alpha)\, r_a,$$

where

$$r_a = \begin{pmatrix} 0 & -a_3 & a_2 \\ a_3 & 0 & -a_1 \\ -a_2 & a_1 & 0 \end{pmatrix}.$$

Using these formulae, show that $R_{\alpha,a}$ uniquely determines the *angle* $\alpha \in \mathbf{R}$ with $0 \le \alpha \le \pi$ and the *axis of rotation* $a \in \mathbf{R}^3$, unless $\sin\alpha = 0$, that is, unless $\alpha = 0$ or $\alpha = \pi$. If $\alpha = \pi$, we have, writing the matrix of $R_{\pi,a}$ as (r_{ij}),

$$2a_i^2 = 1 + r_{ii}, \qquad 2a_i a_j = r_{ij} \qquad (1 \le i, j \le 3,\ i \ne j).$$

Because at least one $a_i \ne 0$, it follows that a is determined by $R_{\pi,a}$, to within a factor ± 1.

Example. The result of a rotation by $\frac{\pi}{2}$ about the axis $\mathbf{R}(0, 1, 0)$, followed by a rotation by $\frac{\pi}{2}$ about the axis $\mathbf{R}(1, 0, 0)$, has a matrix which equals

$$\begin{pmatrix} 1 & 0 & 0 \\ 0 & 0 & -1 \\ 0 & 1 & 0 \end{pmatrix} \begin{pmatrix} 0 & 0 & 1 \\ 0 & 1 & 0 \\ -1 & 0 & 0 \end{pmatrix} = \begin{pmatrix} 0 & 0 & 1 \\ 1 & 0 & 0 \\ 0 & 1 & 0 \end{pmatrix}.$$

The last matrix is seen to arise from the rotation by $\alpha = \frac{2\pi}{3}$ about the axis $\mathbf{R}a = \mathbf{R}(1, 1, 1)$, since in that case $\cos\alpha = -\frac{1}{2}$, $2\sin\alpha = \sqrt{3}$, and

$$R_{\alpha,a} - R_{\alpha,a}^t = \sqrt{3} \begin{pmatrix} 0 & -\frac{1}{\sqrt{3}} & \frac{1}{\sqrt{3}} \\ \frac{1}{\sqrt{3}} & 0 & -\frac{1}{\sqrt{3}} \\ -\frac{1}{\sqrt{3}} & \frac{1}{\sqrt{3}} & 0 \end{pmatrix}.$$

The composition of these rotations in reverse order equals the rotation by $\alpha = \frac{2\pi}{3}$ about the axis $\mathbf{R}a = \mathbf{R}(1, 1, -1)$

Let $0 \ne w \in \mathbf{R}^3$ and write $\alpha = \|w\|$ and $a = \frac{1}{\|w\|} w$. Define the mapping

$$R : \{\, w \in \mathbf{R}^3 \mid 0 \le \|w\| \le \pi \,\} \to \mathrm{End}(\mathbf{R}^3),$$
$$R : w \mapsto R_w := R_{\alpha,a} \qquad (R_0 = I).$$

(v) Prove that R is differentiable at 0, with derivative

$$DR(0) \in \mathrm{Lin}\,(\mathbf{R}^3, \mathrm{End}(\mathbf{R}^3)) \qquad \text{given by} \qquad DR(0)h : x \mapsto h \times x.$$

(vi) Demonstrate that there exists an open neighborhood U of 0 in $\mathbf{R}^3$ such that $R(U)$ is a C^∞ submanifold in $\mathrm{End}(\mathbf{R}^3) \simeq \mathbf{R}^9$ of dimension 3.

Exercise 4.23 (Rotation group of $\mathbf{R}^n$ – needed for Exercise 5.58). (Compare with Exercise 4.22.) Let $\mathrm{A}(n, \mathbf{R})$ be the linear subspace in $\mathrm{Mat}(n, \mathbf{R})$ of the antisymmetric matrices $A \in \mathrm{Mat}(n, \mathbf{R})$, that is, those for which $A^t = -A$. Let $\mathbf{SO}(n, \mathbf{R})$ be the group of those $B \in \mathbf{O}(n, \mathbf{R})$ with $\det B = 1$ (see Example 4.6.2).

(i) Determine $\dim \mathrm{A}(n, \mathbf{R})$.

(ii) Prove that $\mathbf{SO}(n, \mathbf{R})$ is a C^∞ submanifold of $\mathrm{Mat}(n, \mathbf{R})$, and determine the dimension of $\mathbf{SO}(n, \mathbf{R})$.

(iii) Let $A \in \mathrm{A}(n, \mathbf{R})$. Prove that, for all $x, y \in \mathbf{R}^n$, the mapping (see Example 2.4.10)

$$t \mapsto \langle e^{tA}x, e^{tA}y \rangle$$

is a constant mapping; and conclude that $e^A \in \mathbf{O}(n, \mathbf{R})$.

(iv) Prove that $\exp : A \mapsto e^A$ is a mapping $\mathrm{A}(n, \mathbf{R}) \to \mathbf{SO}(n, \mathbf{R})$; in other words e^A is a *rotation* in $\mathbf{R}^n$.
Hint: Consider the continuous function $[\,0, 1\,] \to \mathbf{R} \setminus \{0\}$ defined by $t \mapsto \det(e^{tA})$, or apply Exercise 2.44.(ii).

(v) Determine $D(\exp)(0)$, and prove that exp is a C^∞ diffeomorphism of a suitably chosen open neighborhood of 0 in $\mathrm{A}(n, \mathbf{R})$ onto an open neighborhood of I in $\mathbf{SO}(n, \mathbf{R})$.

(vi) Prove that exp is surjective (at least for $n = 2$ and $n = 3$), but that exp is not injective.

The curve $t \mapsto e^{tA}$ is called a *one-parameter group of rotations* in $\mathbf{R}^n$, A is called the corresponding *infinitesimal rotation* or *infinitesimal generator*, and $\mathbf{SO}(n, \mathbf{R})$ is the *rotation group* of $\mathbf{R}^n$.

Exercise 4.24 (Special linear group – sequel to Exercise 2.44). Let the notation be as in Exercise 2.44.

(i) Prove, by means of Exercise 2.44.(i), that $\det : \mathrm{Mat}(n, \mathbf{R}) \to \mathbf{R}$ is singular, that is to say, nonsubmersive, in $A \in \mathrm{Mat}(n, \mathbf{R})$ if and only if $\operatorname{rank} A \le n-2$.
Hint: $\operatorname{rank} A \le r - 1$ if and only if every minor of A of order r is equal to 0.

(ii) Prove that $\mathbf{SL}(n, \mathbf{R}) = \{ A \in \mathrm{Mat}(n, \mathbf{R}) \mid \det A = 1 \}$, the *special linear group*, is a C^∞ submanifold in $\mathrm{Mat}(n, \mathbf{R})$ of codimension 1.

Exercise 4.25 (Stratification of $\mathrm{Mat}(p \times n, \mathbf{R})$ **by rank).** Let $r \in \mathbf{N}_0$ with $r \le r_0 = \min(p, n)$, and denote by M_r the subset of $\mathrm{Mat}(p \times n, \mathbf{R})$ formed by the matrices of rank r.

(i) Prove that M_r is a C^∞ submanifold in $\mathrm{Mat}(p \times n, \mathbf{R})$ of codimension given by $(p - r)(n - r)$, that is, $\dim M_r = r(n + p - r)$.
Hint: Start by considering the case where $A \in M_r$ is of the form

$$A = \begin{matrix} \\ r \\ p-r \end{matrix} \begin{matrix} \quad r \quad\;\; n-r \\ \left(\begin{array}{c|c} B & C \\ \hline D & E \end{array} \right) \end{matrix},$$

with $B \in \mathbf{GL}(r, \mathbf{R})$. Then multiply from the right by the following matrix in $\mathbf{GL}(n, \mathbf{R})$:

$$\begin{pmatrix} I & -B^{-1}C \\ 0 & I \end{pmatrix},$$

to prove that rank $A = r$ if and only if $E - DB^{-1}C = 0$. Deduce that M_r near A equals the graph of $(B, C, D) \mapsto DB^{-1}C$.

(ii) Show that $r \mapsto \dim M_r$ is monotonically increasing on $[\,0, r_0\,]$. Verify that $\dim M_{r_0} = pn$ and deduce that M_{r_0} is an open subset of $\mathrm{Mat}(p \times n, \mathbf{R})$.

(iii) Prove that the set $\{\, A \in \mathrm{Mat}(p \times n, \mathbf{R}) \mid \operatorname{rank} A \leq r \,\}$ is given by the vanishing of polynomial functions: all the minors of A of order $r + 1$ have to vanish. Deduce that this subset is closed in $\mathrm{Mat}(p \times n, \mathbf{R})$, and that the closure of M_r is the union of the $M_{r'}$ with $0 \leq r' \leq r$. Conclude that $\{\, A \in \mathrm{Mat}(p \times n, \mathbf{R}) \mid \operatorname{rank} A \geq r \,\}$ is open, because its complement is given by the condition rank $A \leq r - 1$. Let $A \in \mathrm{Mat}(p \times n, \mathbf{R})$, and show that rank $A' \geq$ rank A, for every $A' \in \mathrm{Mat}(p \times n, \mathbf{R})$ sufficiently close to A.

Background. $E - DB^{-1}C$ is called the *Schur complement* of B in A, and is denoted by $(A \,|\, B)$. We have $\det A = \det B \det(A \,|\, B)$. Above we obtained what is called a *stratification*: the "singular" object under consideration, which is here the closure of M_r, has the form of a finite union of *strata* each of which is a submanifold, with the closure of each of these strata being in turn composed of the stratum itself and strata of lower dimensions.

Exercise 4.26 (Hopf fibration and stereographic projection – needed for Exercises 5.68 and 5.70). Let $x \in \mathbf{R}^4$, and write $\alpha = \alpha(x) = x_1 + ix_4$ and $\beta = \beta(x) = x_3 + ix_2 \in \mathbf{C}$ (where $i = \sqrt{-1}$). This gives an identification of $\mathbf{R}^4$ with $\mathbf{C}^2$ via $x \leftrightarrow (\alpha(x), \beta(x))$. Define $g : \mathbf{R}^4 \to \mathbf{R}^3$ by (compare with the last column in the matrix in Exercise 5.66.(vi))

$$g(x) = g(\alpha, \beta) = \begin{pmatrix} 2\,\mathrm{Re}(\alpha\overline{\beta}) \\ 2\,\mathrm{Im}(\alpha\overline{\beta}) \\ |\alpha|^2 - |\beta|^2 \end{pmatrix} = \begin{pmatrix} 2(x_2x_4 + x_1x_3) \\ 2(x_3x_4 - x_1x_2) \\ x_1^2 - x_2^2 - x_3^2 + x_4^2 \end{pmatrix}.$$

(i) Prove that g is a submersion on $\mathbf{R}^4 \setminus \{0\}$.
Hint: We have

$$Dg(x) = 2 \begin{pmatrix} x_3 & x_4 & x_1 & x_2 \\ -x_2 & -x_1 & x_4 & x_3 \\ x_1 & -x_2 & -x_3 & x_4 \end{pmatrix}.$$

The row vectors in this matrix all have length $2\|x\|$ and are mutually orthogonal in $\mathbf{R}^4$. Or else, let M_j be the matrix obtained from $Dg(x)$ by deleting the column which does not contain x_j. Then $\det M_j = 8x_j\|x\|^2$, for $1 \leq j \leq 4$.

(ii) If $g(x) = c \in \mathbf{R}^3$, then

$$(\star) \qquad 2\alpha\overline{\beta} = c_1 + ic_2, \qquad |\alpha|^2 - |\beta|^2 = c_3.$$

Prove $\|g(x)\| = |\alpha|^2 + |\beta|^2 = \|x\|^2$. Deduce that the sphere in $\mathbf{R}^4$ of center 0 and radius $\sqrt{r}$ is mapped by g into the sphere in $\mathbf{R}^3$ of center 0 and radius $r \geq 0$.

Now assume that $\gamma = c_1 + ic_2 \in \mathbf{C}$ and $c_3 \in \mathbf{R}$ are given, and define $p_\pm = \frac{1}{2}(\|c\| \mp c_3)$.

(iii) Show that $p_\pm = p_\pm(c)$ is the unique number ≥ 0 such that $\frac{|\gamma|^2}{4p_\pm} - p_\pm = \pm c_3$. Verify that $(\alpha, \beta) \in \mathbf{C}^2$ satisfies $(\star)$ if and only if

$$(\star\star) \qquad |\alpha| = \sqrt{p_-}, \qquad |\beta| = \sqrt{p_+}, \qquad \left(\frac{\alpha}{\beta}\right)^{\pm 1} = \frac{c_1 \pm ic_2}{\|c\| \mp c_3}.$$

If $\gamma = 0$, we choose the sign in the third equation that makes sense. In other words, if $\gamma = 0$

$$|\alpha| = \sqrt{c_3}, \qquad \beta = 0, \qquad \text{if} \quad c_3 > 0;$$
$$\alpha = 0, \qquad |\beta| = \sqrt{-c_3}, \qquad \text{if} \quad c_3 < 0.$$

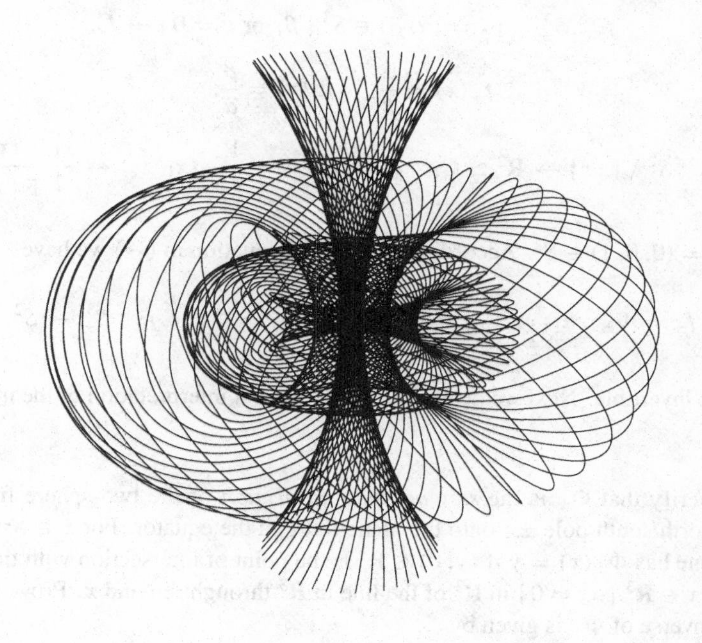

Illustration for Exercise 4.26: Hopf fibration
View by stereographic projection onto $\mathbf{R}^3$ of the Hopf fibration of $S^3 \subset \mathbf{R}^4$ by great circles. Depicted are the (partial) inverse images of points on three parallel circles in S^2

Identify the unit circle S^1 in $\mathbf{R}^2$ with $\{ z \in \mathbf{C} \mid |z| = 1 \}$, and define

$$\Psi_\pm : S^1 \times (\mathbf{R}^3 \setminus \{ (0, 0, c_3) \mid c_3 \gtreqless 0 \}) \to \mathbf{C}^2 \simeq \mathbf{R}^4 \qquad \text{by}$$

$$\Psi_\pm(z, c) = \frac{z}{\sqrt{2(\|c\| \mp c_3)}} \cdot \begin{cases} (c_1 + ic_2, \ \|c\| - c_3); \\ (\|c\| + c_3, \ c_1 - ic_2). \end{cases}$$

(iv) Verify that $\Psi_\pm$ is a mapping such that $g \circ \Psi_\pm$ is the projection onto the second factor in the Cartesian product (compare with the Submersion Theorem 4.5.2.(iv)).

Let $S^{n-1} = \{ x \in \mathbf{R}^n \mid \|x\| = 1 \}$, for $2 \leq n \leq 4$. The *Hopf mapping* $h : S^3 \to \mathbf{R}^3$ is defined to be the restriction of g to S^3.

(v) Derive from the previous parts that $h : S^3 \to S^2$ is surjective. Prove that the inverse image under h of a point in S^2 is a great circle on S^3.

Consider the mappings

$$f_\pm : S^3_\pm = \{ x = (\alpha, \beta) \in S^3 \mid \beta, \text{ or } \alpha \neq 0 \} \to \mathbf{C},$$

$$f_+(x) = \frac{\alpha}{\beta}, \qquad f_-(x) = \frac{\overline{\beta}}{\overline{\alpha}};$$

$$\Phi_\pm : S^2 \setminus \{\pm n\} \to \mathbf{R}^2 \simeq \mathbf{C}, \qquad \Phi_\pm(x) = \frac{1}{1 \mp x_3}(x_1, x_2) \leftrightarrow \frac{x_1 + ix_2}{1 \mp x_3}.$$

Here $n = (0, 0, 1) \in S^2$. According to the third equation in ($\star\star$) we have

$$f_\pm = \Phi_\pm \circ h|_{S^3_\pm}; \qquad \text{and thus} \qquad h|_{S^3_\pm} = \Phi_\pm^{-1} \circ f_\pm : S^3_\pm \to S^2,$$

if $\Phi_\pm$ is invertible. Next we determine a geometrical interpretation of the mapping $\Phi_\pm$.

(vi) Verify that $\Phi_\pm$ is the *stereographic projection* of the two-sphere from the north/south pole $\pm n$ onto the plane through the equator. For $x \in S^2 \setminus \{\pm n\}$ one has $\Phi_\pm(x) = y$ if $(y, 0) \in \mathbf{R}^3$ is the point of intersection with the plane $\{ x \in \mathbf{R}^3 \mid x_3 = 0 \}$ in $\mathbf{R}^3$ of the line in $\mathbf{R}^3$ through $\pm n$ and x. Prove that the inverse of $\Phi_\pm$ is given by

$$\Phi_\pm^{-1} : \mathbf{R}^2 \to S^2 \setminus \{n\}, \qquad \Phi_\pm^{-1}(y) = \frac{1}{\|y\|^2 + 1}(2y, \ \pm(\|y\|^2 - 1)) \in \mathbf{R}^3.$$

In particular, suppose $y \leftrightarrow y_1 + iy_2 = \frac{\alpha}{\beta}$ with $\alpha, \beta \in \mathbf{C}$, $\beta \neq 0$ and $|\alpha|^2 + |\beta|^2 = 1$. Verify that we recover the formulae for $c = \Phi^{-1}(y) \in S^2$ given in $(\star)$

$$c_1 + ic_2 = \frac{2y}{y\overline{y}+1} = 2\alpha\overline{\beta}, \qquad c_1 - ic_2 = \frac{2\overline{y}}{y\overline{y}+1} = 2\overline{\alpha}\beta,$$

$$c_3 = \frac{y\overline{y}-1}{y\overline{y}+1} = |\alpha|^2 - |\beta|^2.$$

Background. We have now obtained the *Hopf fibration* of S^3 into disjoint circles, all of which are isomorphic with S^1 and which are parametrized by points of S^2; the parameters of such a circle depend smoothly on the coordinates of the corresponding point in S^2. This fibration plays an important role in *algebraic topology*. See the Exercises 5.68 and 5.70 for other properties of the Hopf fibration.

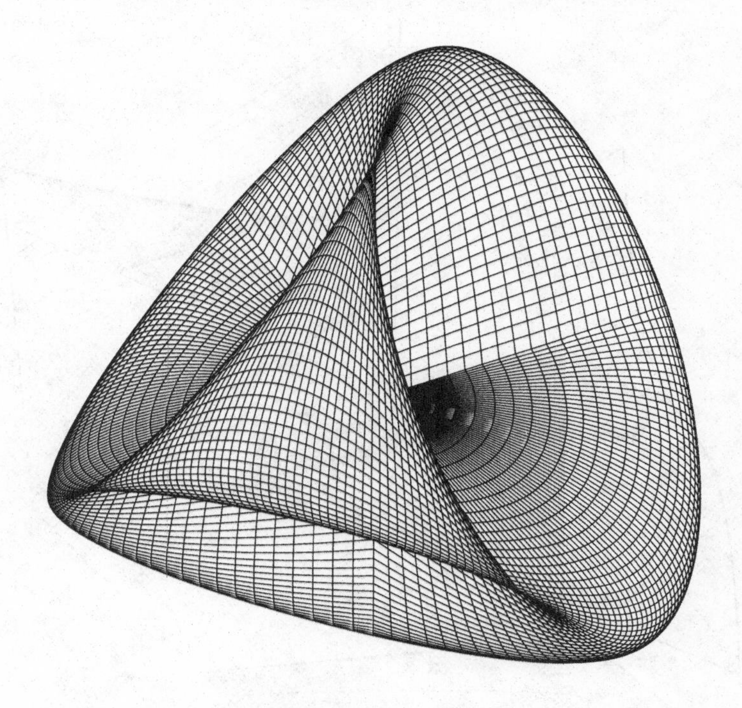

Illustration for Exercise 4.27: Steiner's Roman surface

Exercise 4.27 (Steiner's Roman surface – needed for Exercise 5.33). Let $\Psi : \mathbf{R}^3 \to \mathbf{R}^3$ be defined by

$$\Psi(y) = (y_2y_3,\ y_3y_1,\ y_1y_2).$$

The image V under Ψ of the unit sphere S^2 in $\mathbf{R}^3$ is called *Steiner's Roman surface*.

(i) Prove that V is contained in the set

$$W := \{ x \in \mathbf{R}^3 \mid g(x) := x_2^2x_3^2 + x_3^2x_1^2 + x_1^2x_2^2 - x_1x_2x_3 = 0 \}.$$

(ii) If $x \in W$ and $d := \sqrt{x_2^2x_3^2 + x_3^2x_1^2 + x_1^2x_2^2} \neq 0$, then $x \in V$. Prove this.
Hint: Let $y_1 = \frac{x_2x_3}{d}$, etc.

(iii) Prove that W is the union of V and the intervals $\left]-\infty, -\frac{1}{2}\right[$ and $\left]\frac{1}{2}, \infty\right[$ along each of the three coordinate axes.

(iv) Find the points in W where g is not a submersion.

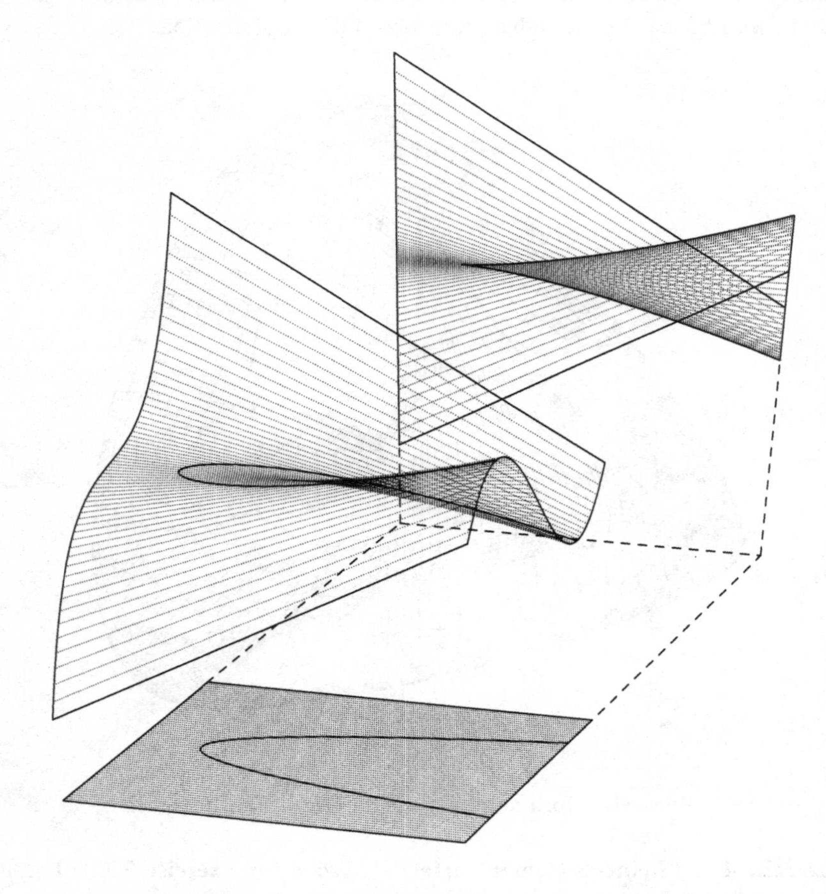

Illustration for Exercise 4.28: Cayley's surface
For clarity of presentation a left–handed coordinate system has been used

Exercise 4.28 (Cayley's surface). Let the cubic surface V in $\mathbf{R}^3$ be given as the zero-set of the function $g : \mathbf{R}^3 \to \mathbf{R}$ with

$$g(x) = x_2^3 - x_1x_2 - x_3.$$

(i) Prove that V is a C^∞ submanifold in $\mathbf{R}^3$ of dimension 2.

(ii) Prove that $\phi : \mathbf{R}^2 \to \mathbf{R}^3$ is a C^∞ parametrization of V by $\mathbf{R}^2$, if $\phi(y) = (y_1, y_2, y_2^3 - y_1y_2)$.

Let $\pi : \mathbf{R}^3 \to \mathbf{R}^3$ be the orthogonal projection with $\pi(x) = (x_1, 0, x_3)$; and define $\Xi = \pi \circ \phi$ with

$$\Xi : \mathbf{R}^2 \simeq \mathbf{R}^2 \times \{0\} \to \mathbf{R} \times \{0\} \times \mathbf{R} \simeq \mathbf{R}^2, \qquad \Xi(x_1, x_2, 0) = (x_1, 0, x_2^3 - x_1x_2).$$

(iii) Prove that the set $S \subset \mathbf{R}^2 \times \{0\}$ consisting of the singular points for Ξ, equals the parabola $\{ (x_1, x_2, 0) \in \mathbf{R}^3 \mid x_1 = 3x_2^2 \}$.

(iv) $\Xi(S) \subset \mathbf{R} \times \{0\} \times \mathbf{R}$ is the *semicubic parabola* $\{ (x_1, 0, x_3) \in \mathbf{R}^3 \mid 4x_1^3 = 27x_3^2 \}$. Prove this.

(v) Investigate whether $\Xi(S)$ is a submanifold in $\mathbf{R}^2$ of dimension 1.

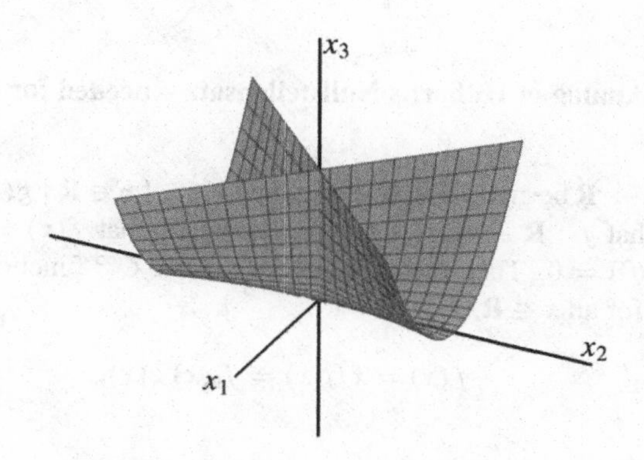

Illustration for Exercise 4.29: Whitney's umbrella

Exercise 4.29 (Whitney's umbrella). Let $g : \mathbf{R}^3 \to \mathbf{R}$ and $V \subset \mathbf{R}^3$ be defined by

$$g(x) = x_1^2 - x_3x_2^2, \qquad V = \{x \in \mathbf{R}^3 \mid g(x) = 0\}.$$

Further define

$$S^- = \{ (0, 0, x_3) \in \mathbf{R}^3 \mid x_3 < 0 \}, \qquad S^+ = \{ (0, 0, x_3) \in \mathbf{R}^3 \mid x_3 \geq 0 \}.$$

(i) Prove that V is a C^∞ submanifold in $\mathbf{R}^3$ at each point $x \in V \setminus (S^- \cup S^+)$, and determine the dimension of V at each of these points x.

(ii) Verify that the x_2-axis and the x_3-axis are both contained in V. Also prove that V is a C^∞ submanifold in $\mathbf{R}^3$ of dimension 1 at every point $x \in S^-$.

(iii) Intersect V with the plane $\{ (x_1, y_1, x_3) \in \mathbf{R}^3 \mid (x_1, x_3) \in \mathbf{R}^2 \}$, for every $y_1 \in \mathbf{R}$, and show that $V \setminus S^- = \mathrm{im}(\phi)$, where $\phi : \mathbf{R}^2 \to \mathbf{R}^3$ is defined by $\phi(y) = (y_1 y_2,\ y_1,\ y_2^2)$.

(iv) Prove that $\phi : \mathbf{R}^2 \setminus \{ (0, y_2) \in \mathbf{R}^2 \mid y_2 \in \mathbf{R} \} \to \mathbf{R}^3 \setminus S^+$ is an embedding.

(v) Prove that V is not a C^0 submanifold in $\mathbf{R}^3$ at the point x if $x \in S^+$.
Hint: Intersect V with the plane $\{ (x_1, x_2, x_3) \in \mathbf{R}^3 \mid (x_1, x_2) \in \mathbf{R}^2 \}$.

Exercise 4.30 (Submersions are locally determined by one image point). Let x_i in $\mathbf{R}^n$ and $g_i : \mathbf{R}^n \to \mathbf{R}^{n-d}$ be C^1 mappings, for $i = 1, 2$. Assume that $g_1(x_1) = g_2(x_2)$ and that g_i is a submersion at x_i, for $i = 1, 2$. Prove that there exist open neighborhoods U_i of x_i, for $i = 1, 2$, in $\mathbf{R}^n$ and a C^1 diffeomorphism $\Phi : U_1 \to U_2$ such that

$$\Phi(x_1) = x_2 \qquad \text{and} \qquad g_1 = g_2 \circ \Phi.$$

Exercise 4.31 (Analog of Hilbert's Nullstellensatz – needed for Exercises 4.32 and 5.74).

(i) Let $g : \mathbf{R} \to \mathbf{R}$ be given by $g(x) = x$; then $V := \{ x \in \mathbf{R} \mid g(x) = 0 \} = \{0\}$. Assume that $f : \mathbf{R} \to \mathbf{R}$ is a C^∞ function such that $f(x) = 0$, for $x \in V$, that is, $f(0) = 0$. Then prove that there exists a C^∞ function $f_1 : \mathbf{R} \to \mathbf{R}$ such that for all $x \in \mathbf{R}$,

$$f(x) = x f_1(x) = f_1(x)\, g(x).$$

Hint: $f(x) - f(0) = \int_0^1 \frac{df}{dt}(tx)\, dt = x \int_0^1 f'(tx)\, dt$.

(ii) Let $U_0 \subset \mathbf{R}^n$ be an open subset, let $g : U_0 \to \mathbf{R}^{n-d}$ be a C^∞ submersion, and define $V := \{ x \in U_0 \mid g(x) = 0 \}$. Assume that $f : U_0 \to \mathbf{R}$ is a C^∞ function such that $f(x) = 0$, for all $x \in V$. Then prove that, for every $x^0 \in U_0$, there exist a neighborhood U of x^0 in U_0, and C^∞ functions $f_i : U \to \mathbf{R}$, with $1 \le i \le n - d$, such that for all $x \in U$,

$$f(x) = f_1(x) g_1(x) + \cdots + f_{n-d}(x) g_{n-d}(x).$$

Hint: Let $x = \Psi(y, c)$ be the substitution of variables from assertion (iv) of the Submersion Theorem 4.5.2, with $(y, c) \in W := \Phi(U) \subset \mathbf{R}^d \times \mathbf{R}^{n-d}$. In particular then, for all $(y, c) \in W$, one has $g_i(\Psi(y, c)) = c_i$, for $1 \le i \le n - d$. It follows that for every C^∞ function $h : W \to \mathbf{R}$,

$$\begin{aligned} h(y, c) - h(y, 0) &= \int_0^1 \frac{dh}{dt}(y, tc)\, dt = \sum_{1 \le i \le n-d} c_i \int_0^1 D_{d+i} h(y, tc)\, dt \\ &= \sum_{1 \le i \le n-d} g_i(\Psi(y, c))\, h_i(y, c). \end{aligned}$$

Reformulation of (ii). The C^∞ functions $f : U_0 \to \mathbf{R}$ form a ring R under the operations of pointwise addition and multiplication. The collection of the $f \in R$ with the property $f|_V = 0$ forms an ideal I in R. The result above asserts that the ideal I is generated by the functions $g_1, \dots, g_{n-d}$, in a local sense at least.

(iii) Generally speaking, the functions f_i from (ii) are not uniquely determined. Verify this, taking the example where $g : \mathbf{R}^3 \to \mathbf{R}^2$ is given by $g(x) = (x_1, x_2)$ and $f(x) := x_1x_3 + x_2^2$. Note $f(x) = (x_3 - x_2)g_1(x) + (x_1 + x_2)g_2(x)$.

(iv) Let $V := \{ x \in \mathbf{R}^2 \mid g(x) := x_2^2 = 0 \}$. The function $f(x) = x_2$ vanishes for every $x \in V$. Even so, there do not exist for every $x \in \mathbf{R}^2$ a neighborhood U of x and a C^∞ function $f_1 : U \to \mathbf{R}$ such that for all $x \in U$ we have $f(x) = f_1(x)g(x)$. Verify that this is not in contradiction with the assertion from (ii).

Background. Note that in the situation of part (iv) there does exist a C^∞ function f_1 such that $f^2(x) = f_1(x)g(x)$, for all $x \in \mathbf{R}^2$. In the context of **polynomial functions** one has, in the case when the vectors grad $g_i(x)$, for $1 \le i \le n - d$ and $x \in V$, are not known to be linearly independent, the following, known as *Hilbert's Nullstellensatz.*[1] Write $\mathbf{C}[x]$ for the ring of polynomials in the variables $x_1, \dots, x_n$ with coefficients in $\mathbf{C}$. Assume that $f, g_1, \dots, g_c \in \mathbf{C}[x]$ and let $V := \{ x \in \mathbf{C}^n \mid g_i(x) = 0 \ (1 \le i \le c) \}$. If $f|_V = 0$, then there exists a number $m \in \mathbf{N}$ such that f^m belongs to the ideal in $\mathbf{C}[x]$ generated by $g_1, \dots, g_c$.

Exercise 4.32 (Analog of de l'Hôpital's rule – sequel to Exercise 4.31). We use the notation and the results of Exercise 4.31.(ii). We consider the special case where $\dim V = n - 1$ and will now investigate to what extent the C^∞ submersion $g : U_0 \to \mathbf{R}$ near $x^0 \in V$ is determined by the set $V = \{ x \in U_0 \mid g(x) = 0 \}$. Assume that $\widetilde{g} : U_0 \to \mathbf{R}$ is a C^∞ submersion for which also $V = \{ x \in U_0 \mid \widetilde{g}(x) = 0 \}$. Then we have that $\widetilde{g}(x) = 0$, for all $x \in V$. Hence we find a neighborhood U of x^0 in U_0 and a C^∞ function $f : U \to \mathbf{R}$ such that for all $x \in U$,

$$\widetilde{g}(x) = f(x)g(x).$$

[1] A proof can be found on p. 380 in Lang, S.: *Algebra.* Addison-Wesley Publishing Company, Reading 1993, or Peter May, J.: Munshi's proof of the Nullstellensatz. Amer. Math. Monthly 110 (2003), 133 – 140.

It is evident that f restricted to U does not equal 0 outside V.

(i) Prove that $\operatorname{grad} \widetilde{g}(x) = f(x) \operatorname{grad} g(x)$, for all $x \in V$, and conclude that $f(x) \neq 0$, for all $x \in U$. Check the analogy of this result with de l'Hôpital's rule (in that case, g is not required to be differentiable on V).

(ii) Prove by means of Proposition 1.9.8.(iv) that the sign of f is constant on the connected components of $V \cap U$.

Exercise 4.33 (Functional dependence – needed for Exercise 4.34). (See also Exercises 4.35 and 6.37.)

(i) Define $f : \mathbf{R}^2 \supset\!\rightarrow \mathbf{R}^2$ by

$$f(x) = \Big(\frac{x_1 + x_2}{1 - x_1 x_2},\ \arctan x_1 + \arctan x_2\Big) \qquad (x_1 x_2 \neq 1).$$

Prove that the rank of $Df(x)$ is constant and equals 1. Show that $\tan(f_2(x)) = f_1(x)$.

In a situation like the one above f_1 is said to be *functionally dependent* on f_2. We shall now study such situations.

Let $U_0 \subset \mathbf{R}^n$ be an open subset, and let $f : U_0 \to \mathbf{R}^p$ be a C^k mapping, for $k \in \mathbf{N}_\infty$. We assume that the rank of $Df(x)$ is constant and equals r, for all $x \in U_0$. We have already encountered the cases $r = n$, of an immersion, and $r = p$, of a submersion. Therefore we assume that $r < \min(n, p)$. Let $x^0 \in U_0$ be fixed.

(ii) Show that there exists a neighborhood U of x^0 in $\mathbf{R}^n$ and that the coordinates of $\mathbf{R}^p$ can be permuted, such that $\pi \circ f$ is a submersion on U, where $\pi : \mathbf{R}^p \to \mathbf{R}^r$ is the projection onto the first r coordinates.

Now consider a level set $N(c) = \{x \in U \mid \pi(f(x)) = c\} \subset \mathbf{R}^n$. If this is a C^k submanifold, of dimension $n - r$, choose a C^k parametrization $\phi_c : D \to \mathbf{R}^n$ for it, where $D \subset \mathbf{R}^{n-r}$ is open.

(iii) Calculate $D(\pi \circ f \circ \phi_c)(y)$, for a $y \in D$. Next, calculate $D(f \circ \phi_c)(y)$, using the fact that $D\pi(f(x))$ is injective on the image of $Df(x)$, for $x \in U$. (Why is this true?) Conclude that f is constant on $N(c)$.

(iv) Shrink U, if necessary, for the preceding to apply, and show that π is injective on $f(U)$.

This suggests that $f(U)$ is a submanifold in $\mathbf{R}^p$ of dimension r, with $\pi : f(U) \to \mathbf{R}^r$ as its coordinatization. To actually demonstrate this, we may have to shrink U, in order to find, by application of the Submersion Theorem 4.5.2, a C^k diffeomorphism $\Psi : V \to \mathbf{R}^n$ with V open in $\mathbf{R}^n$, such that

$$\pi \circ f \circ \Psi(x) = (x_1, \dots, x_r) \qquad (x \in V).$$

Then π^{-1} can locally be written as $f \circ \Psi \circ \lambda$, with λ affine linear; consequently, it is certain to be a C^k mapping.

(v) Verify the above, and conclude that $f(U)$ is a C^k submanifold in $\mathbf{R}^p$ of dimension r, and that the components $f_{r+1}(x), \dots, f_p(x)$ of $f(x) \in \mathbf{R}^p$, for $x \in U$, can be expressed in terms of $f_1(x), \dots, f_r(x)$, by means of C^k mappings.

Exercise 4.34 (Sequel to Exercise 4.33 – needed for Exercise 7.5).

(i) Let $f : \mathbf{R}_+ \to \mathbf{R}$ be a C^1 function such that $f(1) = 0$ and $f'(x) = \frac{1}{x}$. Define $F : \mathbf{R}_+^2 \to \mathbf{R}^2$ by

$$F(x, y) = (f(x) + f(y),\ xy) \qquad (x,\ y \in \mathbf{R}_+).$$

Verify that $DF(x, y)$ has constant rank equal to 1 and conclude by Exercise 4.33.(v) that a C^1 function $g : \mathbf{R}_+ \to \mathbf{R}$ exists such that $f(x) + f(y) = g(xy)$. Substitute $y = 1$, and prove that f satisfies the following *functional equation* (see Exercise 0.2.(v)):

$$f(x) + f(y) = f(xy) \qquad (x,\ y \in \mathbf{R}_+).$$

(ii) Let there be given a C^1 function $f : \mathbf{R} \to \mathbf{R}$ with $f(0) = 0$ and $f'(x) = \frac{1}{1+x^2}$. Prove in similar fashion as in (i) that f satisfies the equation (see Exercises 0.2.(vii) and 4.33.(i))

$$f(x) + f(y) = f\Big(\frac{x + y}{1 - xy}\Big) \qquad (x,\ y \in \mathbf{R},\ xy \neq 1).$$

(iii) **(Addition formula for lemniscatic sine).** (See Exercises 0.10, 3.44 and 7.5 for background information.) Define $f : [\,0, 1\,[\, \to \mathbf{R}$ by

$$f(x) = \int_0^x \frac{dt}{\sqrt{1 - t^4}}.$$

Prove, for x and $y \in [\,0, 1\,[$ sufficiently small,

$$f(x) + f(y) = f(a(x, y)) \qquad \text{with} \qquad a(x, y) = \frac{x\sqrt{1 - y^4} + y\sqrt{1 - x^4}}{1 + x^2y^2}.$$

Hint: $$D_1 a(x, y) = \frac{\sqrt{1 - x^4}\sqrt{1 - y^4}(1 - x^2y^2) - 2xy(x^2 + y^2)}{\sqrt{1 - x^4}(1 + x^2y^2)^2}.$$

Exercise 4.35 (Rank Theorem). For completeness we now give a direct proof of the principal result from Exercise 4.33. The details are left for the reader to check. Observe that the result is a generalization of the Rank Lemma 4.2.7, and also of the Immersion Theorem 4.3.1 and the Submersion Theorem 4.5.2. However, contrary to the case of immersions or submersions, the case of constant rank $r < \min(n, p)$ in the Rank Theorem below does not occur generically, see Exercise 4.25.(ii) and (iii).

(Rank Theorem). Let $U_0 \subset \mathbf{R}^n$ be an open subset, and let $f : U_0 \to \mathbf{R}^p$ be a C^k mapping, for $k \in \mathbf{N}_\infty$. Assume that $Df(x) \in \mathrm{Lin}(\mathbf{R}^n, \mathbf{R}^p)$ has constant rank equal to $r \leq \min(n, p)$, for all $x \in U_0$. Then there exist, for every $x^0 \in U_0$, neighborhoods U of x^0 in U_0 and W of $f(x^0)$ in $\mathbf{R}^p$, respectively, and C^k diffeomorphisms $\Phi : W \to \mathbf{R}^p$ and $\Psi : V \to U$ with V an open set in $\mathbf{R}^n$, respectively, such that for all $y = (y_1, \dots, y_n) \in V$,

$$\Phi \circ f \circ \Psi(y_1, \dots, y_n) = (y_1, \dots, y_r, 0, \dots, 0) \in \mathbf{R}^p.$$

Proof. We write $x = (x_1, \dots, x_n)$ for the coordinates in $\mathbf{R}^n$, and $v = (v_1, \dots, v_p)$ for those in $\mathbf{R}^p$. As in the proof of the Submersion Theorem 4.5.2 we assume (this may require prior permutation of the coordinates of $\mathbf{R}^n$ and $\mathbf{R}^p$) that for all $x \in U_0$,

$$(D_j f_i(x))_{1 \leq i, j \leq r}$$

is the matrix of an operator in $\mathrm{Aut}(\mathbf{R}^r)$. Note that in the proof of the Submersion Theorem $z = (x_{d+1}, \dots, x_n) \in \mathbf{R}^{n-d}$ plays the role that is here filled by $(x_1, \dots, x_r)$. As in the proof of that theorem we define $\Psi^{-1} : U_0 \to \mathbf{R}^n$ by $\Psi^{-1}(x) = y$, with

$$y_i = \begin{cases} f_i(x), & 1 \leq i \leq r; \\ x_i, & r < i \leq n. \end{cases}$$

Then, for all $x \in U_0$,

$$D\Psi^{-1}(x) = \begin{array}{c} \\ r \\ n-r \end{array} \begin{array}{c} \begin{array}{cc} r & n-r \end{array} \\ \left(\begin{array}{c|c} D_j f_i(x) & \star \\ \hline 0 & I_{n-r} \end{array} \right) \end{array}.$$

Applying the Local Inverse Function Theorem 3.2.4 we find open neighborhoods U of x^0 in U_0 and V of $y^0 = \Psi^{-1}(x^0)$ in $\mathbf{R}^n$ such that $\Psi^{-1}|_U : U \to V$ is a C^k diffeomorphism.

Next, we define C^k functions $g_i : V \to \mathbf{R}$ by

$$g_i(y_1, \dots, y_n) = g_i(y) = f_i \circ \Psi(y) \qquad (r < i \leq p, \ y \in V).$$

Then, for $y = \Psi^{-1}(x) \in V$,

$$f \circ \Psi(y) = f(x) = (y_1, \dots, y_r, g_{r+1}(y), \dots, g_p(y)).$$

Next we study the functions g_i, for $r < i \leq p$. In fact, we obtain, for $y \in V$,

$$D(f \circ \Psi)(y) = \begin{array}{c} r \\ \\ p-r \\ \\ \end{array} \begin{array}{c} \begin{array}{cc} r & n-r \end{array} \\ \left(\begin{array}{c|ccc} I_r & & 0 & \\ \hline \star & D_{r+1} g_{r+1}(y) & \cdots & D_n g_{r+1}(y) \\ \vdots & \vdots & & \vdots \\ \star & D_{r+1} g_p(y) & \cdots & D_n g_p(y) \end{array} \right) \end{array}.$$

Since $D(f \circ \Psi)(y) = Df(x) \circ D\Psi(y)$, the matrix above also has constant rank equal to r, for all $y \in V$. But then the coefficient functions in the bottom right part of the matrix must identically vanish on V. In effect, therefore, we have

$$g_i(y_1, \dots, y_n) = g_i(y_1, \dots, y_r) \qquad (r < i \le p,\ y \in V).$$

Finally, let $p_r : \mathbf{R}^p \to \mathbf{R}^r$ be the projection $(v_1, \dots, v_p) \mapsto (v_1, \dots, v_r)$ onto the first r coordinates. Similarly to the proof of the Immersion Theorem we define $\Phi : \mathrm{dom}(\Phi) \to \mathbf{R}^p$ by $\mathrm{dom}(\Phi) = p_r(V) \times \mathbf{R}^{p-r}$ and

$$\Phi(v) = w, \qquad \text{with} \qquad w_i = \begin{cases} v_i, & 1 \le i \le r; \\ v_i - g_i(v_1, \dots, v_r), & r < i \le p. \end{cases}$$

Then, for $v \in \mathrm{dom}(\Phi)$,

$$D\Phi(v) = \begin{matrix} \\ r \\ p-r \end{matrix} \begin{matrix} \begin{matrix} r & p-r \end{matrix} \\ \left(\begin{array}{c|c} I_r & 0 \\ \hline \star & I_{p-r} \end{array} \right) \end{matrix}.$$

Because $f(x^0) = (y_1^0, \dots, y_r^0, g_{r+1}(y^0), \dots, g_p(y^0)) \in \mathrm{dom}(\Phi)$, it follows by the Local Inverse Function Theorem 3.2.4 that there exists an open neighborhood W of $f(x^0)$ in $\mathbf{R}^p$ such that $\Phi|_W : W \to \Phi(W)$ is a C^k diffeomorphism. By shrinking, if need be, the neighborhood U we can arrange that $f(U) \subset W$. But then, for all $y \in V$,

$$\begin{aligned} \Phi \circ f \circ \Psi(y_1, \dots, y_n) &= \Phi(y_1, \dots, y_r, g_{r+1}(y_1, \dots, y_r), \dots, g_p(y_1, \dots, y_r)) \\ &= (y_1, \dots, y_r, 0, \dots, 0). \end{aligned}$$

❑

Background. Since $\Phi \circ f \circ \Psi$ is the composition of the submersion $(y_1, \dots, y_n) \mapsto (y_1, \dots, y_r)$ and the immersion $(y_1, \dots, y_r) \mapsto (y_1, \dots, y_r, 0, \dots, 0)$, the mapping f of constant rank sometimes is called a *subimmersion*.

Exercises for Chapter 5

Exercise 5.1. Let V be the manifold from Exercise 4.2. Determine the geometric tangent space of V at $(-2, 0)$ in three ways, by successively considering V as a zero-set, a parametrized set and a graph.

Exercise 5.2. Let V be the hyperboloid of two sheets from Exercise 4.13. Determine the geometric tangent space of V at an arbitrary point of V in three ways, by successively considering V as a zero-set, a parametrized set and a graph.

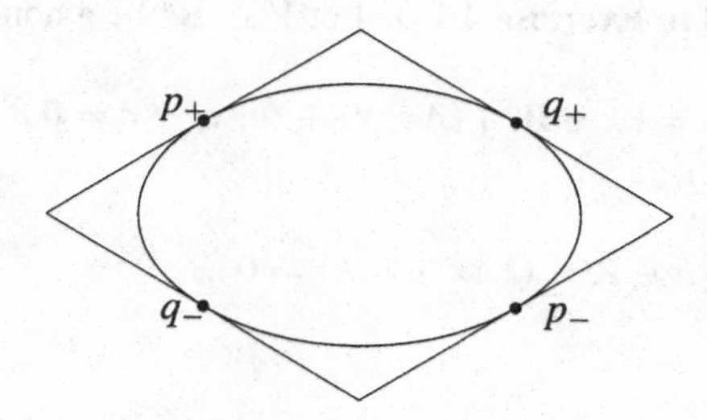

Illustration for Exercise 5.3

Exercise 5.3. Let there be the ellipse $V = \{\, x \in \mathbf{R}^2 \mid \frac{x_1^2}{a^2} + \frac{x_2^2}{b^2} = 1 \,\}$ $(a, b > 0)$ and a point $p_+ = (x_1^0, x_2^0)$ in V.

(i) Prove that there exists a circumscribed parallelogram $P \subset \mathbf{R}^2$ such that $P \cap V = \{\, p_\pm, q_\pm \,\}$, where $p_- = -p_+$, and the points p_+, q_+, p_-, q_- are the midpoints of the sides of P, respectively. Prove

$$q_\pm = \pm\Big(\frac{a}{b}\, x_2^0,\ -\frac{b}{a}\, x_1^0\Big).$$

(ii) Show that the mapping $\Phi : V \to V$ with $\Phi(p_+) = q_+$, for all $p_+ \in V$, satisfies $\Phi^4 = I$.

Exercise 5.4. Let $V \subset \mathbf{R}^2$ be defined by $V = \{\, x \in \mathbf{R}^2 \mid x_1^3 - x_1^2 = x_2^2 - x_2 \,\}$.

(i) Prove that V is a C^∞ submanifold in $\mathbf{R}^2$ of dimension 1.

(ii) Prove that the tangent line of V at $p_0 = (0, 0)$ has precisely one other point of intersection with the manifold V, namely $p_1 = (1, 0)$. Repeat this procedure with p_i in the role of p_{i-1}, for $i \in \mathbf{N}$, and prove $p_4 = p_0$.

Exercise 5.5 (Sequel to Exercise 4.18). The set $V = \{ (s, \frac{1}{2}s^2 - 1) \in \mathbf{R}^2 \mid s \in \mathbf{R} \}$ is a parabola. Prove that the one-parameter family of lines in $\mathbf{R}^2$

$$f(s, t) = (s, \frac{1}{2}s^2 - 1) + t(-s, 1)$$

consists entirely of straight lines through points $x \in V$ that are orthogonal to $T_x V$. Determine the singular points (s, t) of the mapping $f : \mathbf{R}^2 \to \mathbf{R}^2$ and the corresponding singular values $f(s, t)$. Draw a sketch of this one-parameter family and indicate the set of singular values. Discuss the relevance of Exercise 4.18.

Exercise 5.6 (Sequel to Exercise 4.14). Let $V \subset \mathbf{R}^n$ be a nondegenerate quadric given by

$$V = \{ x \in \mathbf{R}^n \mid \langle Ax, x \rangle + \langle b, x \rangle + c = 0 \}.$$

Let $x \in W = V \setminus \{-\frac{1}{2}A^{-1}b\}$.

(i) Prove $T_x V = \{ h \in \mathbf{R}^n \mid \langle 2Ax + b, h \rangle = 0 \}$.

(ii) Prove

$$\begin{aligned} x + T_x V &= \{ h \in \mathbf{R}^n \mid \langle 2Ax + b, x - h \rangle = 0 \} \\ &= \{ h \in \mathbf{R}^n \mid \langle Ax, h \rangle + \tfrac{1}{2}\langle b, x + h \rangle + c = 0 \}. \end{aligned}$$

Exercise 5.7. Let $V = \{ x \in \mathbf{R}^3 \mid \frac{x_1^2}{a^2} + \frac{x_2^2}{b^2} + \frac{x_3^2}{c^2} = 1 \}$, where $a, b, c > 0$.

(i) Prove (see also Exercise 5.6.(ii))

$$h \in x + T_x V \qquad \Longleftrightarrow \qquad \frac{x_1 h_1}{a^2} + \frac{x_2 h_2}{b^2} + \frac{x_3 h_3}{c^2} = 1.$$

Let $p(x)$ be the orthogonal projection in $\mathbf{R}^3$ of the origin in $\mathbf{R}^3$ onto $x + T_x V$.

(ii) Prove

$$p(x) = \left(\frac{x_1^2}{a^4} + \frac{x_2^2}{b^4} + \frac{x_3^2}{c^4} \right)^{-1} \left(\frac{x_1}{a^2}, \frac{x_2}{b^2}, \frac{x_3}{c^2} \right).$$

(iii) Show that $P = \{ p(x) \mid x \in V \}$ is also described by

$$P = \{ y \in \mathbf{R}^3 \mid \|y\|^4 = a^2 y_1^2 + b^2 y_2^2 + c^2 y_3^2 \} \setminus \{(0, 0, 0)\}.$$

Exercise 5.8 (Sequel to Exercise 3.11). Let the notation be that of part (v) of that exercise. For $x = \Psi(y) \in U$, show that the hyperbola V_{y_1} and the ellipse V_{y_2} are orthogonal at the point x.

Exercise 5.9. Let $a_i \neq 0$ for $1 \leq i \leq 3$ and consider the three C^∞ surfaces in $\mathbf{R}^3$ defined by

$$\{ x \in \mathbf{R}^3 \mid \|x\|^2 = a_i x_i \} \qquad (1 \leq i \leq 3).$$

Prove that these surfaces intersect mutually orthogonally, that is, the normal vectors to the tangent planes at the points of intersection are mutually orthogonal.

Exercise 5.10. Let $B^n = \{ x \in \mathbf{R}^n \mid \|x\| \leq 1 \}$ be the closed unit ball in $\mathbf{R}^n$ and let $S^{n-1} = \{ x \in \mathbf{R}^n \mid \|x\| = 1 \}$ be the unit sphere in $\mathbf{R}^n$.

(i) Show that S^{n-1} is a C^∞ submanifold in $\mathbf{R}^n$ of dimension $n-1$.

(ii) Let V be a C^1 submanifold of $\mathbf{R}^n$ entirely contained in B^n and assume that $V \cap S^{n-1} = \{x\}$. Prove that $T_x V \subset T_x S^{n-1}$.
Hint: Let $v \in T_x V$. Then there exists a differentiable curve $\gamma : I \to V$ with $\gamma(t_0) = x$ and $\gamma'(t_0) = v$. Verify that $t \mapsto \|\gamma(t)\|^2$ has a maximum at t_0, and conclude $\langle x, v \rangle = 0$.

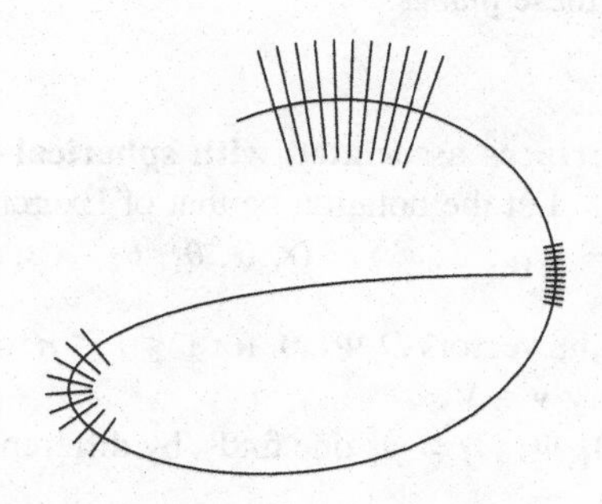

Illustration for Exercise 5.11: Tubular neighborhood

Exercise 5.11 (Local existence of tubular neighborhood of (hyper)surface – needed for Exercise 7.35). Here we use the notation from Example 5.3.3. In particular, $D_0 \subset \mathbf{R}^2$ is an open set, $\phi : D_0 \to \mathbf{R}^3$ a C^k embedding for $k \geq 2$, and $V = \phi(D_0)$. For $x = \phi(y) \in V$, let

$$n(x) = D_1\phi(y) \times D_2\phi(y).$$

Define $\Psi : \mathbf{R} \times D_0 \to \mathbf{R}^3$ by $\Psi(t, y) = \phi(y) + t\, n(\phi(y))$.

(i) Prove that, for every $y \in D_0$,

$$\det D\Psi(0, y) = \|n \circ \phi(y)\|^2.$$

Conclude that for every $y \in D_0$ there exist a number $\delta > 0$ and an open neighborhood D of y in D_0 such that $\Psi : W := \,]-\delta, \delta[\, \times D \to U := \Psi(W)$ is a C^{k-1} diffeomorphism onto the open neighborhood U of $x = \phi(y)$ in $\mathbf{R}^3$.

Interpretation: For every $x \in \phi(D) \subset V$ an open neighborhood U of x in $\mathbf{R}^3$ exists such that the line segments orthogonal to $\phi(D)$ of length 2δ, and having as centers the points from $\phi(D) \cap U$, are disjoint and together fill the entire neighborhood U. Through every $u \in U$ a unique line runs orthogonal to $\phi(D)$.

(ii) Use the preceding to prove that there exists a C^{k-1} submersion $g : U \to \mathbf{R}$ such that $\phi(D) \cap U = N(g, 0)$ (compare with Theorem 4.7.1.(iii)).

(iii) Use Example 5.3.11 to generalize the above in the case where V is a C^k hypersurface in $\mathbf{R}^n$.

Exercise 5.12 (Sequel to Exercise 4.14). Let K_i be the quadric in $\mathbf{R}^3$ given by the equation $x_1^2 + x_2^2 - x_3^2 = i$, with $i = 1, -1$. Then K_i is a C^∞ submanifold in $\mathbf{R}^3$ of dimension 2 according to Exercise 4.14. Find all planes in $\mathbf{R}^3$ that do not have transverse intersection with K_i; these are tangent planes. Also determine the cross-section of K_i with these planes.

Exercise 5.13 (Hypersurfaces associated with spherical coordinates in $\mathbf{R}^n$ – sequel to Exercise 3.18). Let the notation be that of Exercise 3.18; in particular, write $x = \Psi(y)$ with $y = (y_1, \dots, y_n) = (r, \alpha, \theta_1, \theta_2, \dots, \theta_{n-2}) \in V$.

(i) Prove that the column vectors $D_i\Psi(y)$, for $1 \le i \le n$, of $D\Psi(y)$ are pairwise orthogonal for every $y \in V$.
Hint: From $\langle \Psi(y), \Psi(y) \rangle = y_1^2$ one finds, by differentiation with respect to y_j,

$$(\star) \qquad \langle \Psi(y), D_j\Psi(y) \rangle = 0 \qquad (2 \le j \le n).$$

Since $\Psi(y) = y_1 D_1\Psi(y)$, this implies

$$\langle D_1\Psi(y), D_j\Psi(y) \rangle = 0 \qquad (2 \le j \le n).$$

Likewise, $(\star)$ gives that $\langle \Psi(y), D_iD_j\Psi(y) \rangle + \langle D_i\Psi(y), D_j\Psi(y) \rangle = 0$, for $1 \le i < j \le n$. We know that $D_i\Psi(y)$ can be written as $\cos y_j$ times a vector independent of y_j, if $i < j$; and therefore $D_jD_i\Psi(y) = D_iD_j\Psi(y)$ is a scalar multiple of $D_i\Psi(y)$. Accordingly, using $(\star)$ we find

$$\langle D_i\Psi(y), D_j\Psi(y) \rangle = 0 \qquad (2 \le i < j \le n).$$

(ii) Demonstrate, for $y \in V$,

$$\|D_1\Psi(y)\| = 1, \qquad \|D_i\Psi(y)\| = r\cos\theta_{i-1}\cdots\cos\theta_{n-2} \qquad (2 \le i \le n).$$

Hint: Check

$$(\star\star) \qquad \Psi_j(y) = \Psi_j(y_1, y_j, \dots, y_n) \qquad (1 \le j \le n),$$

$$(\star\star\star) \qquad \sum_{1 \le j \le i} \Psi_j(y)^2 = y_1^2 \cos^2 y_{i+1} \cdots \cos^2 y_n \qquad (1 \le i \le n).$$

Let $2 \le i \le n$. Then $(\star\star)$ implies that

$$D_i\Psi(y)^t = (D_i\Psi_1(y), \dots, D_i\Psi_i(y), 0, \dots, 0),$$

and $(\star\star\star)$ yields $\Psi_1 D_i\Psi_1 + \cdots + \Psi_i D_i\Psi_i = 0$. Show that $D_i^2\Psi_j = -\Psi_j$, for $1 \le j \le i$, and thereby prove $(D_i\Psi_1)^2 + \cdots + (D_i\Psi_i)^2 - \Psi_1^2 - \cdots - \Psi_i^2 = 0$.

(iii) Using the preceding parts, prove the formula from Exercise 3.18.(iii)

$$\det D\Psi(y) = r^{n-1} \cos\theta_1 \cos^2\theta_2 \cdots \cos^{n-2}\theta_{n-2}.$$

(iv) Given $y^0 \in V$, the hypersurfaces $\{\, \Psi(y) \in \mathbf{R}^n \mid y \in V,\ y_j = y_j^0 \,\}$, for $1 \le j \le n$, all go through $\Psi(y^0)$. Prove by means of part (i) that these hypersurfaces in $\mathbf{R}^n$ are mutually orthogonal at the point $\Psi(y^0)$.

Exercise 5.14. Let V be a C^k submanifold in $\mathbf{R}^n$ of dimension d, and let $x^0 \in V$. Let $\psi : \mathbf{R}^{n-d} \to \mathbf{R}^n$ be a C^k mapping with $\psi(0) = 0$. Define $f : V \times \mathbf{R}^{n-d} \to \mathbf{R}^n$ by $f(x, y) = x + \psi(y)$. Then prove the equivalence of the following assertions.

(i) There exist open neighborhoods U of x^0 in $\mathbf{R}^n$ and W of 0 in $\mathbf{R}^{n-d}$ such that $f|_{(V\cap U)\times W}$ is a C^k diffeomorphism onto a neighborhood of x^0 in $\mathbf{R}^n$.

(ii) $\operatorname{im} D\psi(0) \cap T_{x^0}V = \{0\}$.

Exercise 5.15. Suppose that V is a C^k submanifold in $\mathbf{R}^n$ of dimension d, consider $A \in \operatorname{Lin}(\mathbf{R}^n, \mathbf{R}^d)$ and let $x \in V$. Prove that the following assertions are equivalent.

(i) There exists an open neighborhood U of x in $\mathbf{R}^n$ such that $A|_{V\cap U}$ is a C^k diffeomorphism from $V \cap U$ onto an open neighborhood of Ax in $\mathbf{R}^d$.

(ii) $\ker A \cap T_xV = \{0\}$.

Prove that assertion (ii) implies that A is surjective.

Exercise 5.16 (Implicit Function Theorem in geometric form). We consider the Implicit Function Theorem 3.5.1 and some of its relations to manifolds and tangent spaces. Let the notation and the assumptions be as in the theorem. Furthermore, we write $v^0 = (x^0, y^0)$ and $V_0 = \{\, (x, y) \in U \times V \mid f(x; y) = 0 \,\}$

(i) Show that V_0 is a submanifold in $\mathbf{R}^n \times \mathbf{R}^p$ of dimension p. Prove that

$$\begin{aligned} T_{v^0} V_0 &= \{\, (\xi, \eta) \in \mathbf{R}^n \times \mathbf{R}^p \mid D_x f(v^0)\xi + D_y f(v^0)\eta = 0 \,\} \\ &= \{\, (\xi, \eta) \in \mathbf{R}^n \times \mathbf{R}^p \mid \xi = D\psi(y^0)\eta \,\}. \end{aligned}$$

Denote by $\pi : \mathbf{R}^n \times \mathbf{R}^p \to \mathbf{R}^p$ the projection along $\mathbf{R}^n$ onto $\mathbf{R}^p$, satisfying $\pi(\xi, \eta) = \eta$.

(ii) Show that $\pi|_{T_{v^0} V_0} \in \mathrm{Lin}(T_{v^0} V_0, \mathbf{R}^p)$ is a bijection. Conversely, if this mapping is given to be a bijection, prove that the condition $D_x f(v^0) \in \mathrm{Aut}(\mathbf{R}^n)$ from the Implicit Function Theorem is satisfied.

Exercise 5.17 (Tangent bundle of submanifold). Let $k \in \mathbf{N}$, let V be a C^k submanifold in $\mathbf{R}^n$ of dimension d, let $x \in V$ and $v \in T_x V$. Then $(x, v) \in \mathbf{R}^{2n}$ is said to be a *(geometric) tangent vector* of V. Define TV, the *tangent bundle* of V, as the subset of $\mathbf{R}^{2n}$ consisting of all (geometric) tangent vectors of V.

(i) Use the local description of $V \cap U = N(g, 0)$ according to Theorem 4.7.1.(iii) and consider the C^{k-1} mapping

$$G : U \times \mathbf{R}^n \to \mathbf{R}^{n-d} \times \mathbf{R}^{n-d} \quad \text{with} \quad G(x, v) = \begin{pmatrix} g(x) \\ Dg(x)v \end{pmatrix}.$$

Verify that $(x, v) \in U \times \mathbf{R}^n$ belongs to TV if and only if $G(x, v) = 0$.

(ii) Prove that $DG(x, v) \in \mathrm{Lin}(\mathbf{R}^{2n}, \mathbf{R}^{2n-2d})$ is given by

$$DG(x, v)(\xi, \eta) = \begin{pmatrix} Dg(x)\xi \\ D^2 g(x)(\xi, v) + Dg(x)\eta \end{pmatrix} \qquad ((\xi, \eta) \in \mathbf{R}^{2n}).$$

Deduce that G is a C^{k-1} submersion and apply the Submersion Theorem 4.5.2 to show that TV is a C^{k-1} submanifold in $\mathbf{R}^{2n}$ of dimension $2d$.

(iii) Denote by $\pi : TV \to V$ the projection of the bundle onto the base space V given by $\pi(x, v) = x$. Show that $\pi^{-1}(\{x\}) = T_x V$, for all $x \in V$, that is, the tangent spaces are the fibres of the projection.

Background. One may think of the tangent bundle of V as the disjoint union of the tangent spaces at all points of V. The reason for taking the disjoint union of the tangent spaces is the following: although all the geometric tangent spaces $x + T_x V$ are affinely isomorphic with $\mathbf{R}^d$, geometrically they might differ from each other although they might intersect, and we do not wish to water this fact down by including into the union just one point from among the points lying in each of these different tangent spaces.

(iv) Denote by S^{n-1} the unit sphere in $\mathbf{R}^n$. Verify that the *unit tangent bundle* of S^{n-1} is given by

$$\{ (x, v) \in \mathbf{R}^n \times \mathbf{R}^n \mid \langle x, x \rangle = \langle v, v \rangle = 1, \ \langle x, v \rangle = 0 \}.$$

Background. In fact, the unit tangent bundle of S^2 can be identified with $\mathbf{SO}(3, \mathbf{R})$ (see Exercises 2.5 and 5.58) via the mapping $(x, v) \mapsto (x \ v \ x \times v) \in \mathbf{SO}(3, \mathbf{R})$. This shows that the unit tangent bundle of S^2 has a nontrivial structure.

Exercise 5.18 (Lemniscate). Let $c = (\frac{1}{\sqrt{2}}, 0) \in \mathbf{R}^2$. We define the *lemniscate* L (lemniscatus = adorned with ribbons) by (see also Example 6.6.4)

$$L = \{ x \in \mathbf{R}^2 \mid \|x - c\| \, \|x + c\| = \|c\|^2 \}.$$

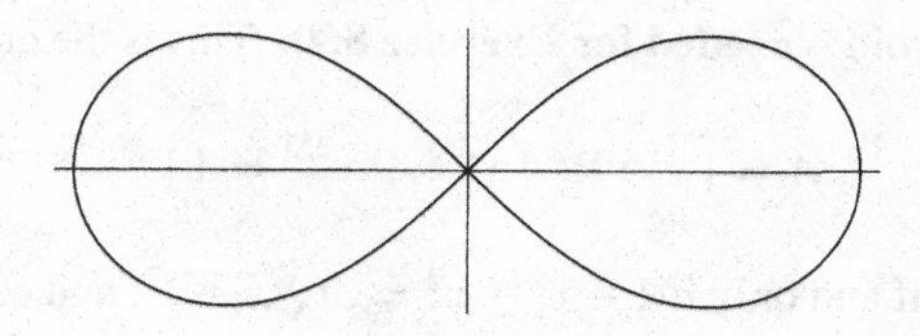

Illustration for Exercise 5.18: Lemniscate

(i) Show $L = \{ x \in \mathbf{R}^2 \mid g(x) = 0 \}$ where $g : \mathbf{R}^2 \to \mathbf{R}$ is given by $g(x) = \|x\|^4 - x_1^2 + x_2^2$. Deduce $L \subset \{ x \in \mathbf{R}^2 \mid \|x\| \leq 1 \}$.

(ii) For every point $x \in L \setminus \{O\}$ with $O = (0, 0)$, prove that L is a C^∞ submanifold in $\mathbf{R}^2$ at x of dimension 1.

(iii) Show that, with respect to polar coordinates (r, α) for $\mathbf{R}^2$,

$$L = \{ (r, \alpha) \in [\, 0, 1\,] \times \Big[-\frac{\pi}{4}, \frac{\pi}{4} \Big] \mid r^2 = \cos 2\alpha \}.$$

(iv) From $r^4 = x_1^2 - x_2^2$ and $r^2 = x_1^2 + x_2^2$ for $x \in L$ derive

$$L = \{ \frac{1}{\sqrt{2}} r(\sqrt{1 + r^2}, \pm\sqrt{1 - r^2}) \mid -1 \leq r \leq 1 \}.$$

Prove that in a neighborhood of O the set L is the union of two C^∞ submanifolds in $\mathbf{R}^2$ of dimension 1 which intersect at O. Calculate the (smaller) angle at O between these submanifolds.

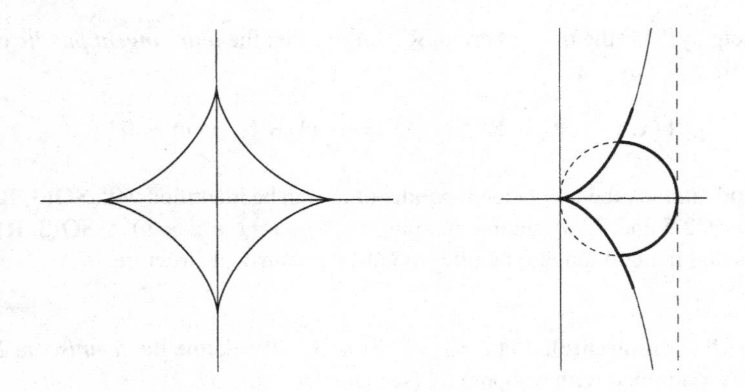

Illustration for Exercise 5.19: Astroid, and for Exercise 5.20: Cissoid

Exercise 5.19 (Astroid – needed for Exercise 8.4). This is the curve A defined by

$$A = \{ x \in \mathbf{R}^2 \mid x_1^{2/3} + x_2^{2/3} = 1 \}.$$

(i) Prove $x \in A$ if and only if $1 - x_1^2 - x_2^2 = 3(\sqrt[3]{x_1 x_2})^2$; and conclude $\|x\| \leq 1$, if $x \in A$.
Hint: Verify

$$\left(x_1^{2/3} + x_2^{2/3}\right)^3 = x_1^2 + x_2^2 + (1 - x_1^2 - x_2^2)\left(x_1^{2/3} + x_2^{2/3}\right).$$

This gives

$$\left(x_1^{2/3} + x_2^{2/3}\right)^3 - \left(x_1^{2/3} + x_2^{2/3}\right) = (x_1^2 + x_2^2)\left(1 - x_1^{2/3} - x_2^{2/3}\right).$$

Now factorize the left–hand side, and then show

$$\left(x_1^{2/3} + x_2^{2/3} - 1\right)\left(x_1^2 + x_2^2 + \left(x_1^{2/3} + x_2^{2/3}\right)^2 + x_1^{2/3} + x_2^{2/3}\right) = 0.$$

(ii) Prove that $A \setminus \{ (-1, 0) \} = \operatorname{im} \phi$, where $\phi : \,]-\pi, \pi[\, \to \mathbf{R}^2$ is given by $\phi(t) = (\cos^3 t,\ \sin^3 t)$.

Define $D = \,]-\pi, \pi[\, \setminus \{ -\frac{\pi}{2},\ 0,\ \frac{\pi}{2} \} \subset \mathbf{R}$ and

$$S = \{ (1, 0),\ (0, 1),\ (0, -1),\ (-1, 0) \} \subset A.$$

(iii) Verify that $\phi : D \to \mathbf{R}^2$ is a C^∞ embedding. Demonstrate that A is a C^∞ submanifold in $\mathbf{R}^2$ of dimension 1, except at the points of S.

(iv) Verify that the slope of the tangent line of A at $\phi(t)$ equals $-\tan t$, if $t \neq -\frac{\pi}{2}, \frac{\pi}{2}$. Now prove by Taylor expansion of $\phi(t)$ for $t \to 0$ and by symmetry arguments that A has an ordinary cusp at the points of S. Conclude that A is not a C^1 submanifold in $\mathbf{R}^2$ of dimension 1 at the points of S.

Exercise 5.20 (Diocles' cissoid). (ὁ κισσός = ivy.) Let $0 \le y < \sqrt{2}$. Then the vertical line in $\mathbf{R}^2$ through the point $(2 - y^2, 0)$ has a unique point of intersection, say $\chi(y) \in \mathbf{R}^2$, with the circular arc $\{x \in \mathbf{R}^2 \mid \|x - (1, 0)\| = 1,\ x_2 \ge 0\}$; and $\chi(y) \neq (0, 0)$. The straight line through $(0, 0)$ and $\chi(y)$ intersects the vertical line through $(y^2, 0)$ at a unique point, say $\phi(y) \in \mathbf{R}^2$.

(i) Verify that

$$\phi(y) = \left(y^2, \frac{y^3}{\sqrt{2 - y^2}}\right) \qquad (0 \le y < \sqrt{2}).$$

(ii) Show that the mapping $\phi : \,]0, \sqrt{2}[\, \to \mathbf{R}^2$ given in (i) is a C^∞ embedding.

Diocles' cissoid is the set $V \subset \mathbf{R}^2$ defined by

$$V = \left\{ \left(y^2, \frac{y^3}{\sqrt{2 - y^2}}\right) \in \mathbf{R}^2 \;\middle|\; -\sqrt{2} < y < \sqrt{2} \right\}.$$

(iii) Show that $V \setminus \{(0, 0)\}$ is a C^∞ submanifold in $\mathbf{R}^2$ of dimension 1.

(iv) Prove $V = g^{-1}(\{0\})$, with $g(x) = x_1^3 + (x_1 - 2)x_2^2$.

(v) Show that $g : \mathbf{R}^2 \to \mathbf{R}$ is surjective, and that, for every $c \in \mathbf{R} \setminus \{0\}$, the level set $g^{-1}(\{c\})$ is a C^∞ submanifold in $\mathbf{R}^2$ of dimension 1.

(vi) Prove by Taylor expansion that V has an ordinary cusp at the point $(0, 0)$, and deduce that at the point $(0, 0)$ the set V is not a C^1 submanifold in $\mathbf{R}^2$ of dimension 1.

Exercise 5.21 (Conchoid and trisection of angle). (ἡ κόγχη = shell.) A *Nicomedes conchoid* is a curve K in $\mathbf{R}^2$ defined as follows. Let $O = (0, 0) \in \mathbf{R}^2$. Let $d > 0$ be arbitrary. Choose a point A on the line $x_1 = 1$. Then the line through A and O contains two points, say P_A and P'_A, whose distance to A equals d. Define $K = K_d$ as the collection of these points P_A and P'_A, for all possible A. Define the C^∞ mapping $g_d : \mathbf{R}^2 \to \mathbf{R}$ by

$$g_d(x) = x_1^4 + x_1^2 x_2^2 - 2x_1^3 - 2x_1 x_2^2 + (1 - d^2)x_1^2 + x_2^2.$$

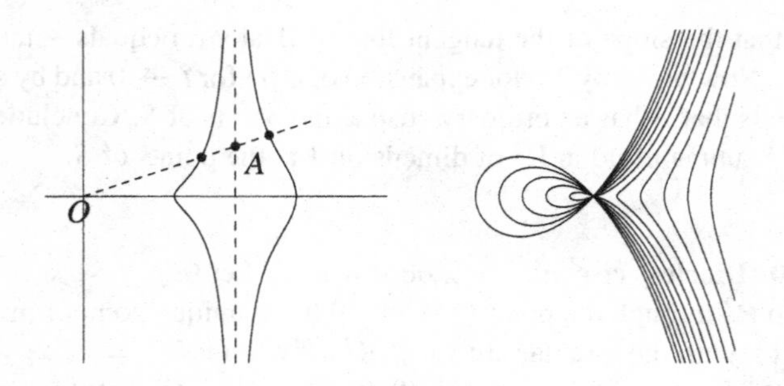

Illustration for Exercise 5.21: Conchoid, and a family of branches

(i) Prove

$$\{ x \in \mathbf{R}^2 \mid g_d(x) = 0 \} = \begin{cases} K_d \cup \{O\}, & \text{if } 0 < d < 1; \\ K_d, & \text{if } 1 \le d < \infty. \end{cases}$$

(ii) Prove the following. For all $d > 0$, the mapping g_d is not a submersion at O. If we assume $0 < d < 1$, then g_d is a submersion at every point of K_d, and K_d is a C^∞ submanifold in $\mathbf{R}^2$ of dimension 1. If $d \ge 1$, then g_d is a submersion at every point of $K_d \setminus \{O\}$.

(iii) Prove that $\mathrm{im}(\phi_d) \subset K_d$, if the C^∞ mapping $\phi_d : \mathbf{R} \to \mathbf{R}^2$ is given by

$$\phi_d(s) = \frac{-d + \cosh s}{\cosh s}(1, \sinh s).$$

Hint: Take the distance, say t, between O and A as a parameter in the description of K_d, and then substitute $t = \cosh s$.

(iv) Show that

$$\phi_d'(s) = \frac{1}{\cosh^2 s}(d \sinh s, -d + \cosh^3 s).$$

Also prove the following. The mapping ϕ_d is an immersion if $d \ne 1$. Furthermore, ϕ_1 is an immersion on $\mathbf{R} \setminus \{0\}$, while ϕ_1 is not an immersion at 0.

(v) Prove that, for $d > 1$, the curve K_d intersects with itself at O. Show by Taylor expansion that K_1 has an ordinary cusp at O.

Background. The *trisection* of an angle can be performed by means of a conchoid. Let α be the angle between the line segment OA and the positive part of the x_1-axis (see illustration). Assume one is given the conchoid K with $d = 2|OA|$. Let B be the point of intersection of the line through A parallel to the x_1-axis and that part of K which is furthest from the origin. Then the angle β between OB and the positive part of the x_1-axis satisfies $\beta = \frac{1}{3}\alpha$.

Indeed, let C be the point of intersection of OB and the line $x_1 = 1$, and let D be the midpoint of the line segment BC. In view of the properties of K one then has $|AO| = |DC|$. One readily verifies $|DA| = |DB|$. This leads to $|DC| = |DA|$, and so $|AO| = |AD|$. Because $\angle(ODA) = 2\beta$, it follows that $\angle(AOD) = 2\beta$, and $\alpha = 2\beta + \beta = 3\beta$.

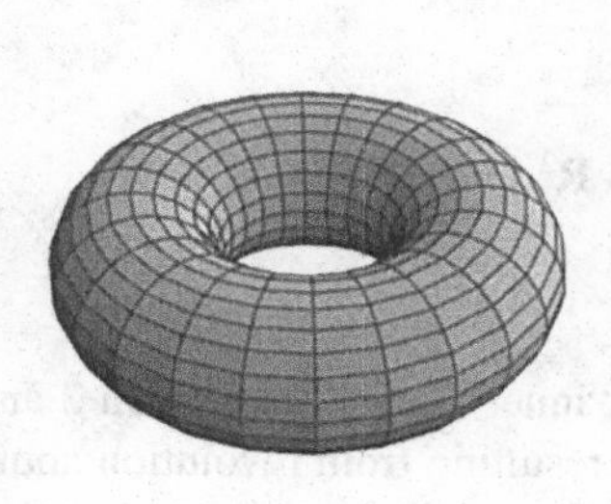

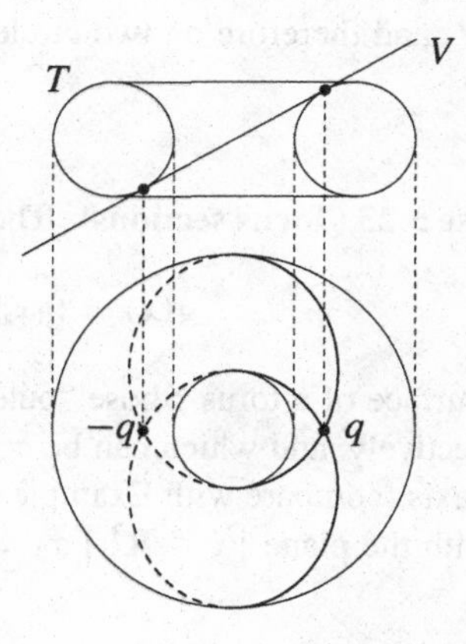

Illustration for Exercise 5.22: Villarceau's circles

Exercise 5.22 (Villarceau's circles). A tangent plane V of a 2-dimensional C^k manifold T with $k \in \mathbf{N}_\infty$ in $\mathbf{R}^3$ is said to be *bitangent* if V is tangent to T at precisely two different points. In particular, if V is bitangent to the toroidal surface T (see Example 4.4.3) given by

$$x = ((2 + \cos\theta)\cos\alpha,\ (2 + \cos\theta)\sin\alpha,\ \sin\theta) \qquad (-\pi \le \alpha,\ \theta \le \pi),$$

the intersection of V and T consists of two circles intersecting exactly at the two tangent points. (Note that the two tangent planes of T parallel to the plane $x_3 = 0$ are each tangent to T along a circle; in particular they are not bitangent.)

Indeed, let $p \in T$ and assume that $V = T_pT$ is bitangent. In view of the rotational symmetry of T we may assume p in the quadrant $x_1 < 0$, $x_3 < 0$ and in the plane $x_2 = 0$. If we now intersect V with the plane $x_2 = 0$, it follows, again by symmetry, that the resulting tangent line L in this intersection must be tangent to T; at a point q, in $x_1 > 0$, $x_3 > 0$ and $x_2 = 0$. But this implies $0 \in L$. We have $q = (2 + \cos\theta,\ 0,\ \sin\theta)$, for some $\theta \in\ ]0, \pi[$ yet to be determined. Show that $\theta = \frac{2\pi}{3}$ and prove that the tangent plane V is given by the condition $x_1 = \sqrt{3}\,x_3$.

The intersection $V \cap T$ of the tangent plane V and the toroidal surface T now consists of those $x \in T$ for which $x_1 = \sqrt{3}\,x_3$. Deduce

$$x \in V \cap T \quad \Longleftrightarrow \quad x_1 = \sqrt{3}\sin\theta, \qquad x_2 = \pm(1 + 2\cos\theta), \qquad x_3 = \sin\theta.$$

But then

$$x_1^2 + (x_2 \mp 1)^2 + x_3^2 = 3\sin^2\theta + 4\cos^2\theta + \sin^2\theta = 4;$$

such x therefore lie on the sphere of center $(0,\ \pm 1,\ 0)$ and radius 2, but also in the plane V, and therefore on two circles intersecting at the points $q = (\frac{3}{2},\ 0,\ \frac{1}{2}\sqrt{3})$ and $-q$.

Exercise 5.23 (Torus sections). The zero-set T in $\mathbf{R}^3$ of $g : \mathbf{R}^3 \to \mathbf{R}$ with

$$g(x) = (\|x\|^2 - 5)^2 + 16(x_1^2 - 1)$$

is the surface of a torus whose "outer circle" and "inner circle" are of radii 3 and 1, respectively, and which can be conceived of as resulting from revolution about the x_1-axis (compare with Example 4.6.3). Let T_c, for $c \in \mathbf{R}$, be the cross-section of T with the plane $\{\, x \in \mathbf{R}^3 \mid x_3 = c \,\}$, orthogonal to the x_3-axis. Inspection of

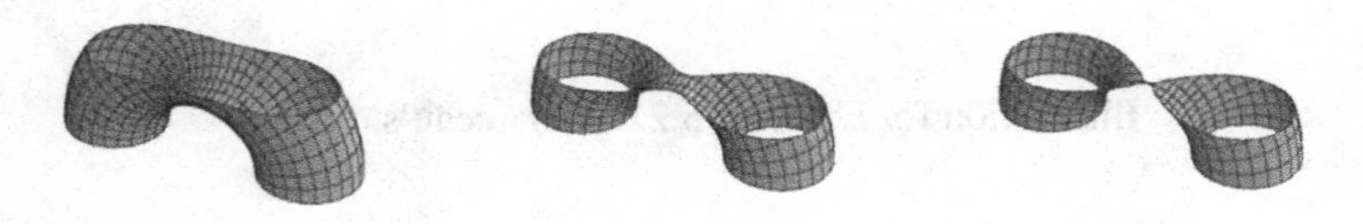

Illustration for Exercise 5.23: Torus sections

the illustration reveals that, for $c \geq 0$, the curve T_c successively takes the following forms: $\emptyset$ for $c > 3$, point for $c = 3$, elongated and changing into 8-shaped for $3 > c > 1$, 8-shaped for $c = 1$, two oval shapes for $1 > c > 0$, two circles for $c = 0$. We will now prove some of these assertions.

(i) Prove that $(x, c) \in T_c$ if and only if $x \in \mathbf{R}^2$ satisfies $g_c(x) = 0$, where

$$g_c(x) = (\|x\|^2 + c^2 - 5)^2 + 16(x_1^2 - 1).$$

Show that $x \in V_c := \{\, x \in \mathbf{R}^2 \mid g_c(x) = 0 \,\}$ implies

$$|x_1| \leq 1 \quad \text{and} \quad 1 - c^2 \leq \|x\|^2 \leq 9 - c^2.$$

Conclude that $V_c = \emptyset$, for $c > 3$; and $V_3 = \{0\}$.

(ii) Verify that $0 \in V_c$ with $c \geq 0$ implies $c = 3$ or $c = 1$; prove that in these cases g_c is nonsubmersive at $0 \in \mathbf{R}^2$. Verify that g_c is submersive at every point of $V_c \setminus \{0\}$, for all $c \geq 0$. Show that V_c is a C^∞ submanifold in $\mathbf{R}^2$ of dimension 1, for every c with $3 > c > 1$ or $1 > c \geq 0$.

(iii) Define

$$\phi_\pm : \left]-\sqrt{2},\, \sqrt{2}\right[\to \mathbf{R}^2 \qquad \text{by} \qquad \phi_\pm(t) = t(\sqrt{2 - t^2},\, \pm\sqrt{2 + t^2}).$$

By intersecting V_1 with circles around 0 of radius $2t$, demonstrate that $V_1 = \mathrm{im}(\phi_+) \cup \mathrm{im}(\phi_-)$. Conclude that V_1 intersects with itself at 0 at an angle of $\frac{\pi}{2}$ radians.

(iv) Draw a single sketch showing, in a neighborhood of $0 \in \mathbf{R}^2$, the three manifolds V_c, for c slightly larger than 1, for $c = 1$, and for c slightly smaller than 1. Prove that, for c near 1, the V_c qualitatively have the properties sketched. **Hint:** With t in a suitable neighborhood of 0, substitute $\|x\|^2 = 4t^2\sqrt{1-d}$, where $5 - c^2 = 4\sqrt{1-d}$. One therefore has d in a neighborhood of 0. One finds

$$\begin{aligned} x_1^2 &= d + 2(1-d)t^2 - (1-d)t^4, \\ x_2^2 &= -d + 2(2\sqrt{1-d} - 1 + d)t^2 + (1-d)t^4. \end{aligned}$$

Now determine, in particular, for which c one has $x_1^2 \lesseqqgtr x_2^2$ for $x \in V_c$, and establish whether the V_c go through 0.

Exercise 5.24 (Conic section in polar coordinates – needed for Exercises 5.53 and 6.28). Let $t \mapsto x(t)$ be a differentiable curve in $\mathbf{R}^2$ and let $f_\pm \in \mathbf{R}^2$ be two distinct fixed points, called the *foci*. Suppose the angle between $x(t) - f_-$ and $-x'(t)$ equals the angle between $x(t) - f_+$ and $x'(t)$, for all $t \in \mathbf{R}$ (angle of incidence equals angle of reflection).

(i) Prove

$$\sum_\pm \frac{\langle x(t) - f_\pm,\, x'(t)\rangle}{\|x(t) - f_\pm\|} = 0; \qquad \text{hence} \qquad \Big(\sum_\pm \|x(t) - f_\pm\|\Big)' = 0,$$

on account of Example 2.4.8. Deduce $\sum_\pm \|x(t) - f_\pm\| = 2a$, for some $a \geq \frac{1}{2}\|f_+ - f_-\| > 0$ and all $t \in \mathbf{R}$.

(ii) By a translation, if necessary, we may assume $f_- = 0$. For $a > 0$ and $f_+ \in \mathbf{R}^2$ with $\|f_+\| \neq 2a$, we define the two sets C_+ and C_- by $C_\pm = \{x \in \mathbf{R}^2 \mid \pm\|x - f_+\| = 2a - \|x\|\}$. Then

$$C := \{x \in \mathbf{R}^2 \mid \|x\| = \langle x, \epsilon\rangle + d\} = \begin{cases} C_+, & d > 0; \\ C_-, & d < 0. \end{cases}$$

Here we introduced the *eccentricity vector* $\epsilon = \frac{1}{2a} f_+ \in \mathbf{R}^2$ and $d = \frac{4a^2 - \|f_+\|^2}{4a} \in \mathbf{R}$. Next, define the numbers $e \geq 0$, the *eccentricity* of C, and $c > 0$ and $b \geq 0$ by

$$\|\epsilon\| = e = \frac{c}{a} = \frac{\sqrt{a^2 \mp b^2}}{a} \qquad \text{as} \qquad e \lesseqgtr 1.$$

Deduce that $\|f_+\| = 2c$, in other words, the distance between the foci equals $2c$, and that $d = a(1 - e^2) = \pm \frac{b^2}{a}$ as $e \lesseqgtr 1$. Conversely, for $e \neq 1$,

$$a = \frac{d}{1 - e^2}, \qquad b = \frac{\pm d}{\sqrt{\pm(1 - e^2)}} \qquad \text{as} \qquad e \lesseqgtr 1.$$

Consider $x \in \mathbf{R}^2$ perpendicular to f_+ of length d. Prove that $x \in C$ if $d > 0$, and $f_+ + x \in C$ if $d < 0$. The chord of C perpendicular to ϵ of length $2d$ is called the *latus rectum* of C.

(iii) Use part (ii) to show that the substitution of variables $x = y + a\epsilon$ turns the equation $\|x\|^2 = (\langle x, \epsilon \rangle + d)^2$ into

$$\|y\|^2 - \langle y, \epsilon \rangle^2 = \pm b^2, \qquad \text{that is} \qquad \frac{y_1^2}{a^2} \pm \frac{y_2^2}{b^2} = 1, \qquad \text{as} \qquad e \lesseqgtr 1.$$

In the latter equation ϵ is assumed to be along the y_1-axis; this may require a rotation about the origin. Furthermore, it is the equation of a conic section with ϵ along the line connecting the foci. Deduce that C equals the ellipse C_+ with *semimajor axis* a and *semiminor axis* b if $0 < e < 1$, a parabola if $e = 1$, and the branch C_- nearest to f_+ of the hyperbola with *transverse axis* a and *conjugate axis* b if $e > 1$. (For the other branch, which is nearest to 0, of the hyperbola, consider $-d > 0$; this corresponds to the choice of $a < 0$, and an element x in that branch satisfies $\|x - f_+\| - \|x\| = 2a$.)

(iv) Finally, introduce polar coordinates (r, α) by $r = \|x\|$ and $\cos\alpha = -\frac{\langle x, \epsilon \rangle}{\|x\| \, \|\epsilon\|} = -\frac{\langle x, \epsilon \rangle}{re}$ (this choice makes the *periapse*, that is, the point of C nearest to the origin correspond with $\alpha = 0$). Show that the equation for C takes the form $r = \frac{d}{1 + e\cos\alpha}$, compare with Exercise 4.2.

Exercise 5.25 (Heron's formula and sine and cosine rule – needed for Exercise 5.39). Let $\Delta \subset \mathbf{R}^2$ be a triangle of area O whose sides have lengths A_i, for $1 \leq i \leq 3$, respectively. We then write $2S = \sum_{1 \leq i \leq 3} A_i$ for the perimeter of Δ.

(i) Prove the following, known as *Heron's formula*:

$$O^2 = S(S - A_1)(S - A_2)(S - A_3).$$

Hint: We may assume the vertices of Δ to be given by the vectors $a_3 =$

0, a_1 and $a_2 \in \mathbf{R}^2$. Then $2\,O = \det(a_1\ a_2)$ (a triangle is one half of a parallelogram), and so

$$4\,O^2 = \begin{vmatrix} \|a_1\|^2 & \langle a_2, a_1 \rangle \\ \langle a_1, a_2 \rangle & \|a_2\|^2 \end{vmatrix}$$

$$= (\|a_1\|\|a_2\| + \langle a_1, a_2 \rangle)(\|a_1\|\|a_2\| - \langle a_1, a_2 \rangle)$$

$$= \frac{1}{4}((\|a_1\| + \|a_2\|)^2 - \|a_1 - a_2\|^2)(\|a_1 - a_2\|^2 - (\|a_1\| - \|a_2\|)^2).$$

Now write $\|a_1\| = A_2$, $\|a_2\| = A_1$ and $\|a_1 - a_2\| = A_3$.

(ii) Note that the first identity above is equivalent to $2\,O = A_1 A_2 \sin \angle(a_1, a_2)$, that is, O is half the height times the length of the base. Denote by α_i the angle of Δ opposite A_i, for $1 \le i \le 3$. Deduce the following *sine rule*, and the *cosine rule*, respectively, for Δ and $1 \le i \le 3$:

$$\frac{\sin \alpha_i}{A_i} = \frac{2O}{\prod_{1 \le i \le 3} A_i}, \qquad A_i^2 = A_{i+1}^2 + A_{i+2}^2 - 2A_{i+1}A_{i+2}\cos\alpha_i.$$

In the latter formula the indices $1 \le i \le 3$ are taken modulo 3.

Exercise 5.26 (Cauchy–Schwarz inequality in $\mathbf{R}^n$, Grassmann's, Jacobi's and Lagrange's identities in $\mathbf{R}^3$ – needed for Exercises 5.27, 5.58, 5.59, 5.65, 5.66 and 5.67).

(i) Show, for v and $w \in \mathbf{R}^n$ (compare with the Remark on linear algebra in Example 5.3.11 and see Exercise 7.1.(ii) for another proof),

$$\langle v, w\rangle^2 + \sum_{1 \le i < j \le n} (v_i w_j - v_j w_i)^2 = \|v\|^2\,\|w\|^2.$$

Derive from this the Cauchy–Schwarz inequality $|\langle v, w\rangle| \le \|v\|\,\|w\|$, for v and $w \in \mathbf{R}^n$.

Assume v_1, v_2, v_3 and v_4 arbitrary vectors in $\mathbf{R}^3$.

(ii) Prove the following, known as *Grassmann's identity*, by writing out the components on both sides:

$$v_1 \times (v_2 \times v_3) = \langle v_3, v_1 \rangle\, v_2 - \langle v_1, v_2 \rangle\, v_3.$$

Background. Here is a rule to remember this formula. Note $v_1 \times (v_2 \times v_3)$ is perpendicular to $v_2 \times v_3$; hence, it is a linear combination $\lambda\, v_2 + \mu\, v_3$ of v_2 and v_3. Because it is perpendicular to v_1 too, we have $\lambda\langle v_1, v_2\rangle + \mu\langle v_3, v_1\rangle = 0$; and, indeed, this is satisfied by $\lambda = \langle v_3, v_1\rangle$ and $\mu = -\langle v_1, v_2\rangle$. It turns out to be tedious to convert this argument into a rigorous proof.

Instead we give another proof of Grassmann's identity. Let (e_1, e_2, e_3) denote the standard basis in $\mathbf{R}^3$. Given any two different indices i and j, let k denote the third index different from i and j. Then $e_i \times e_j = \det(e_i\ e_j\ e_k)\, e_k$. We obtain, for any two distinct indices i and j,

$$\langle e_i, e_j \times (v_2 \times v_3)\rangle = \langle e_i \times e_j, v_2 \times v_3\rangle = \det(e_i\ e_j\ e_k)\langle e_k, v_2 \times v_3\rangle = v_{2i}\, v_{3j} - v_{2j}\, v_{3i}.$$

Obviously this is also valid if $i = j$. Therefore i can be chosen arbitrarily, and thus we find, for all j,

$$e_j \times (v_2 \times v_3) = v_{3j}\, v_2 - v_{2j}\, v_3 = \langle v_3, e_j\rangle v_2 - \langle e_j, v_2\rangle v_3.$$

Grassmann's identity now follows by linearity.

(iii) Prove the following, known as *Jacobi's identity*:

$$v_1 \times (v_2 \times v_3) + v_2 \times (v_3 \times v_1) + v_3 \times (v_1 \times v_2) = 0.$$

For an alternative proof of Jacobi's identity, show that

$$(v_1, v_2, v_3, v_4) \mapsto \langle\, v_4 \times v_1,\ v_2 \times v_3\,\rangle + \langle\, v_4 \times v_2,\ v_3 \times v_1\,\rangle + \langle\, v_4 \times v_3,\ v_1 \times v_2\,\rangle$$

is an antisymmetric 4-linear form on $\mathbf{R}^3$, which implies that it is identically 0.

Background. An algebra for which multiplication is anticommutative and which satisfies Jacobi's identity is known as a *Lie algebra*. Obviously $\mathbf{R}^3$, regarded as a vector space over $\mathbf{R}$, and further endowed with the cross multiplication of vectors, is a Lie algebra.

(iv) Demonstrate

$$(v_1 \times v_2) \times (v_3 \times v_4) = \det(v_4\, v_1\, v_2)\, v_3 - \det(v_1\, v_2\, v_3)\, v_4,$$

and also the following, known as *Lagrange's identity*:

$$\langle v_1 \times v_2, v_3 \times v_4\rangle = \langle v_1, v_3\rangle\, \langle v_2, v_4\rangle - \langle v_1, v_4\rangle\, \langle v_2, v_3\rangle = \begin{vmatrix} \langle v_1, v_3\rangle & \langle v_1, v_4\rangle \\ \langle v_2, v_3\rangle & \langle v_2, v_4\rangle \end{vmatrix}.$$

Verify that the identity $\langle v_1, v_2\rangle^2 + \|v_1 \times v_2\|^2 = \|v_1\|^2\, \|v_2\|^2$ is a special case of Lagrange's identity.

(v) Alternatively, verify that Lagrange's identity follows from Formula 5.3 by use of the polarization identity from Lemma 1.1.5.(iii). Next use

$$\langle v_1 \times v_2, v_3 \times v_4\rangle = \langle v_1,\ v_2 \times (v_3 \times v_4)\rangle$$

to obtain Grassmann's identity in a way which does not depend on writing out the components.

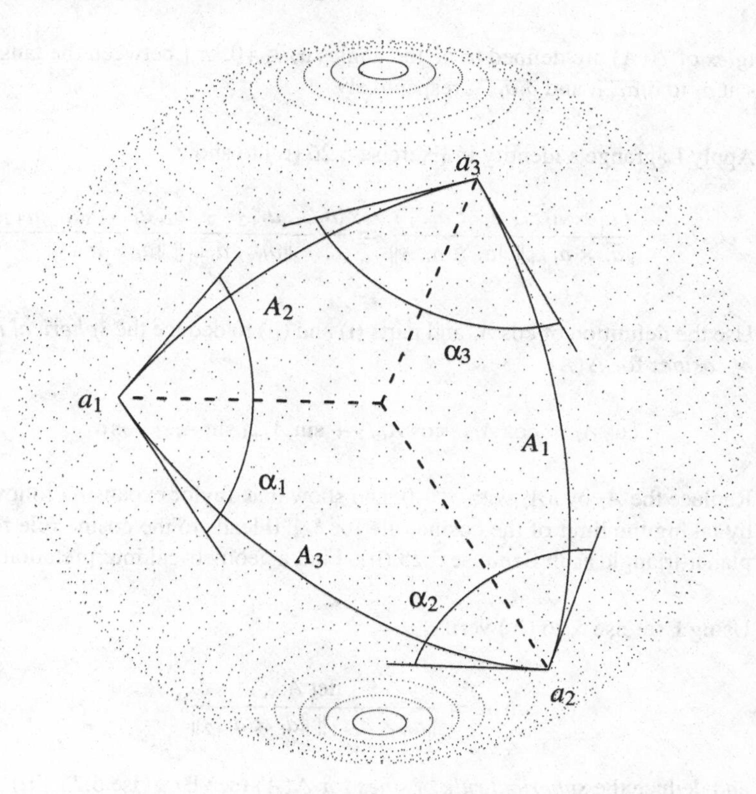

Illustration for Exercise 5.27: Spherical trigonometry

Exercise 5.27 (Spherical trigonometry – sequel to Exercise 5.26 – needed for Exercises 5.65, 5.66, 5.67 and 5.72). Let $S^2 = \{ x \in \mathbf{R}^3 \mid \|x\| = 1 \}$. A *great circle* on S^2 is the intersection of S^2 with a two-dimensional linear subspace of $\mathbf{R}^3$. Two points a_1 and $a_2 \in S^2$ with $a_1 \neq \pm a_2$ determine a great circle on S^2, the *segment* $\overline{a_1 a_2}$ is the shortest arc of this great circle between a_1 and a_2. Now let a_1, a_2 and $a_3 \in S^2$, let $A = (a_1\ a_2\ a_3) \in \operatorname{Mat}(3, \mathbf{R})$ be the matrix having a_1, a_2 and a_3 in its columns, and suppose $\det A > 0$. The *spherical triangle* $\Delta(A)$ is the union of the segments $\overline{a_i a_{i+1}}$ for $1 \leq i \leq 3$, here the indices $1 \leq i \leq 3$ are taken modulo 3. The length of $\overline{a_{i+1} a_{i+2}}$ is defined to be

$$A_i \in \,]\,0, \pi\,[\qquad \text{satisfying} \qquad \cos A_i = \langle a_{i+1}, a_{i+2} \rangle \qquad (1 \leq i \leq 3).$$

The A_i are called the *sides* of $\Delta(A)$.

(i) Show

$$\sin A_i = \|a_{i+1} \times a_{i+2}\|.$$

The *angles* of $\Delta(A)$ are defined to be the angles $\alpha_i \in\ ]0, \pi[$ between the tangent vectors at a_i to $\overline{a_i a_{i+1}}$, and $\overline{a_i a_{i+2}}$, respectively.

(ii) Apply Lagrange's identity in Exercise 5.26.(iv) to show

$$\cos\alpha_i = \frac{\langle a_i \times a_{i+1},\ a_i \times a_{i+2} \rangle}{\|a_i \times a_{i+1}\|\ \|a_i \times a_{i+2}\|} = \frac{\langle a_{i+1}, a_{i+2}\rangle - \langle a_i, a_{i+1}\rangle \langle a_i,\ a_{i+2}\rangle}{\|a_i \times a_{i+1}\|\ \|a_i \times a_{i+2}\|}.$$

(iii) Use the definition of $\cos A_i$ and parts (i) and (ii) to deduce the *spherical rule of cosines* for $\Delta(A)$

$$\cos A_i = \cos A_{i+1} \cos A_{i+2} + \sin A_{i+1} \sin A_{i+2} \cos\alpha_i.$$

Replace the A_i by δA_i with $\delta > 0$, and show that Taylor expansion followed by taking the limit of the cosine rule for $\delta \downarrow 0$ leads to the cosine rule for a planar triangle as in Exercise 5.25.(ii). Find a geometrical interpretation.

(iv) Using Exercise 5.26.(iv) verify

$$\sin\alpha_i = \frac{\det A}{\|a_i \times a_{i+1}\|\ \|a_i \times a_{i+2}\|},$$

and deduce the *spherical rule of sines* for $\Delta(A)$ (see Exercise 5.25.(ii))

$$\frac{\sin\alpha_i}{\sin A_i} = \frac{\det A}{\prod_{1\le i\le 3} \sin A_i} = \frac{\det A}{\prod_{1\le i\le 3} \|a_i \times a_{i+1}\|}.$$

The triple $(\widehat{a_1}, \widehat{a_2}, \widehat{a_3})$ of vectors in $\mathbf{R}^3$, not necessarily belonging to S^2, is called the *dual basis* to (a_1, a_2, a_3) if

$$\langle \widehat{a_i}, a_j \rangle = \delta_{ij}.$$

After normalization of the $\widehat{a_i}$ we obtain $a_i' \in S^2$. Then $A' = (a_1' a_2' a_3') \in \mathrm{Mat}(3, \mathbf{R})$ determines the *polar triangle* $\Delta'(A) := \Delta(A')$ associated with A.

(v) Prove $\Delta'(A') = \Delta(A)$.

(vi) Show

$$\widehat{a_i} = \frac{1}{\det A} a_{i+1} \times a_{i+2}, \qquad a_i' = \frac{1}{\|a_{i+1} \times a_{i+2}\|} a_{i+1} \times a_{i+2}.$$

From Exercise 5.26.(iv) deduce $\widehat{a_i} \times \widehat{a_{i+1}} = \frac{1}{\det A}\, a_{i+2}$. Use (ii) to verify

$$\begin{aligned}
\cos A_i' &= \frac{\langle\, a_{i+2} \times a_i,\, a_i \times a_{i+1}\,\rangle}{\|a_i \times a_{i+1}\|\, \|a_i \times a_{i+2}\|} = -\cos\alpha_i &&= \cos(\pi - \alpha_i),\\
\cos\alpha_i' &= \frac{\langle\, \widehat{a_i} \times \widehat{a_{i+1}},\, \widehat{a_i} \times \widehat{a_{i+2}}\,\rangle}{\|\widehat{a_i} \times \widehat{a_{i+1}}\|\, \|\widehat{a_i} \times \widehat{a_{i+2}}\|} = -\langle\, a_{i+2},\, a_{i+1}\,\rangle &&= \cos(\pi - A_i).
\end{aligned}$$

Deduce the following relation between the sides and angles of a spherical triangle and of its polar:

$$\alpha_i + A_i' = \alpha_i' + A_i = \pi.$$

(vii) Apply (iv) in the case of $\Delta'(A)$ and use (vi) to obtain $\sin A_i \sin\alpha_{i+1} \sin\alpha_{i+2} = \det A'$. Next show

$$\frac{\sin\alpha_i}{\sin A_i} = \frac{\det A'}{\det A}.$$

(viii) Apply the spherical rule of cosines to $\Delta'(A)$ and use (vi) to obtain

$$\cos\alpha_i = \sin\alpha_{i+1} \sin\alpha_{i+2} \cos A_i - \cos\alpha_{i+1} \cos\alpha_{i+2}.$$

(ix) Deduce from the spherical rule of cosines that

$$\cos(A_{i+1} + A_{i+2}) = \cos A_{i+1} \cos A_{i+2} - \sin A_{i+1} \sin A_{i+2} < \cos A_i$$

and obtain $\sum_{1 \le i \le 3} A_i < 2\pi$. Use (vi) to show

$$\sum_{1 \le i \le 3} \alpha_i > \pi.$$

Background. Part (ix) proves that spherical trigonometry differs significantly from plane trigonometry, where we would have $\sum_{1 \le i \le 3} \alpha_i = \pi$. The *excess*

$$\sum_{1 \le i \le 3} \alpha_i - \pi$$

turns out to be the area of $\Delta(A)$, see Exercise 7.13.

As an application of the theory above we determine the length of the segment $\overline{a_1 a_2}$ for two points a_1 and $a_2 \in S^2$ with longitudes ϕ_1 and ϕ_2 and latitudes θ_1 and θ_2, respectively.

(x) The length is $|\theta_2-\theta_1|$ if a_1, a_2 and the north pole of S^2 are coplanar. Otherwise, choose $a_3 \in S^2$ equal to the north pole of S^2. By interchanging the roles of a_1 and a_2 if necessary, we may assume that $\det A > 0$. Show that $\Delta(A)$ satisfies $\alpha_3 = \phi_2 - \phi_1$, $A_1 = \frac{\pi}{2} - \theta_2$ and $A_2 = \frac{\pi}{2} - \theta_1$. Deduce that A_3, the side we are looking for, and the remaining angles α_1 and α_2 are given by

$$\cos A_3 = \sin\theta_1 \sin\theta_2 + \cos\theta_1 \cos\theta_2 \cos(\phi_2 - \phi_1) = \langle a_1, a_2\rangle;$$

$$\sin\alpha_i = \frac{\sin(\phi_2 - \phi_1)\cos\theta_{i+1}}{\sin A_3} = \frac{\det A}{\|a_1 \times a_2\|\,\|a_i \times e_3\|} \qquad (1 \le i \le 2).$$

Verify that the most northerly/southerly latitude θ_0 reached by the great circle determined by a_1 and a_2 is given by (see Exercise 5.43 for another proof)

$$\cos\theta_0 = \sin\alpha_i \cos\theta_i = \frac{\det A}{\|a_1 \times a_2\|} \qquad (1 \le i \le 2).$$

Exercise 5.28. Let $n \ge 3$. In this exercise the indices $1 \le j \le n$ are taken modulo n. Let $(e_1, \dots, e_n)$ be the standard basis in $\mathbf{R}^n$.

(i) Prove $e_1 \times e_2 \times \cdots \times e_{n-1} = (-1)^{n-1} e_n$.

(ii) Check that $e_1 \times \cdots \times e_{j-1} \times e_{j+1} \times \cdots \times e_n = (-1)^{j-1} e_j$.

(iii) Conclude

$$e_{j+1} \times \cdots \times e_n \times e_1 \times \cdots \times e_{j-1} = (-1)^{(j-1)(n-j+1)} e_j$$
$$= \begin{cases} e_j, & \text{if } n \text{ odd;} \\ (-1)^{j-1} e_j, & \text{if } n \text{ even.} \end{cases}$$

Exercise 5.29 (Stereographic projection). Let V be a C^1 submanifold in $\mathbf{R}^n$ of dimension d and let $\Phi : V \to \mathbf{R}^p$ be a C^1 mapping. Then Φ is said to be *conformal* or *angle-preserving* at $x \in V$ if there exists a constant $c = c(x) > 0$ such that for all $v, w \in T_x V$

$$\langle D\Phi(x)v,\ D\Phi(x)w \rangle = c(x)\,\langle v, w\rangle.$$

In addition, Φ is said to be *conformal* or *angle-preserving* if $x \mapsto c(x)$ is a C^1 mapping from V to $\mathbf{R}$.

(i) Prove that $d \le p$ is necessary for Φ to be a conformal mapping.

Let $S = \{ x \in \mathbf{R}^3 \mid x_1^2 + x_2^2 + (x_3 - 1)^2 = 1 \}$ and let $n = (0, 0, 2)$. Define the *stereographic projection* $\Phi : S \setminus \{n\} \to \mathbf{R}^2$ by $\Phi(x) = y$, where $(y, 0) \in \mathbf{R}^3$ is the point of intersection with the plane $x_3 = 0$ in $\mathbf{R}^3$ of the line in $\mathbf{R}^3$ through n and $x \in S \setminus \{n\}$.

(ii) Prove that Φ is a bijective C^∞ mapping for which

$$\Phi^{-1}(y) = \frac{2}{4 + \|y\|^2}(2y_1,\ 2y_2,\ \|y\|^2) \qquad (y \in \mathbf{R}^2).$$

Conclude that Φ is a C^∞ diffeomorphism.

(iii) Prove that circles on S are mapped onto circles or straight lines in $\mathbf{R}^2$, and vice versa.
Hint: A circle on S is the cross-section of a plane $V = \{ x \in \mathbf{R}^3 \mid \langle a, x\rangle = a_0 \}$, for $a \in \mathbf{R}^3$ and $a_0 \in \mathbf{R}$, and S. Verify

$$\Phi^{-1}(y) \in V \qquad \Longleftrightarrow \qquad \frac{1}{4}(2a_3 - a_0)\|y\|^2 + a_1 y_1 + a_2 y_2 - a_0 = 0.$$

Examine the consequences of $2a_3 - a_0 = 0$.

(iv) Prove that Φ is a conformal mapping.

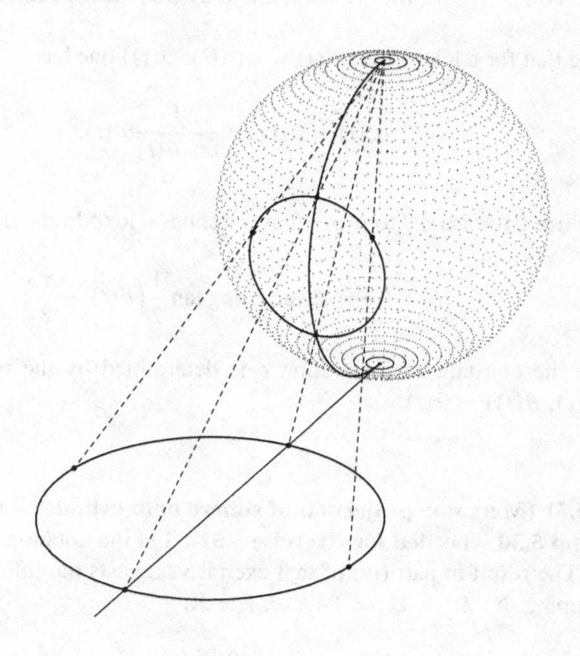

Illustration for Exercise 5.29: Stereographic projection

Exercise 5.30 (Course of constant compass heading – sequel to Exercise 0.7 – needed for Exercise 5.31). Let $D = \,]-\pi, \pi[\times]-\frac{\pi}{2}, \frac{\pi}{2}[\subset \mathbf{R}^2$ and let $\phi : D \to S^2 = \{ x \in \mathbf{R}^3 \mid \|x\| = 1 \}$ be the C^1 embedding from Exercise 4.5 with $\phi(\alpha, \theta) = (\cos\alpha\cos\theta, \ \sin\alpha\cos\theta, \ \sin\theta)$. Assume $\delta : I \to D$ with $\delta(t) = (\alpha(t), \theta(t))$ is a C^1 curve, then $\gamma := \phi \circ \delta : I \to S^2$ is a C^1 curve on S^2. Let $t \in I$ and let

$$\mu : \theta \mapsto (\cos\alpha(t)\cos\theta, \ \sin\alpha(t)\cos\theta, \ \sin\theta)$$

be the meridian on S^2 given by $\alpha = \alpha(t)$ and $\gamma(t) \in \operatorname{im}(\mu)$.

(i) Prove $\langle \gamma'(t), \mu'(\theta(t)) \rangle = \theta'(t)$.

(ii) Show that the angle β at the point $\gamma(t)$ between the curve γ and the meridian μ is given by

$$\cos\beta = \frac{\theta'(t)}{\sqrt{\cos^2\theta(t)\,\alpha'(t)^2 + \theta'(t)^2}}.$$

By definition, a *loxodrome* is a curve γ on S^2 which makes a constant angle β with all meridians on S^2 (λοξός means skew, and ὁ δρόμος means course).

(iii) Prove that for a loxodrome $\gamma(t) = \phi(\alpha(t), \theta(t))$ one has

$$\alpha'(t) = \pm\tan\beta \, \frac{1}{\cos\theta(t)} \theta'(t).$$

(iv) Now use Exercise 0.7 to prove that γ defines a loxodrome if

$$\alpha(t) + c = \pm\tan\beta \, \log\tan\frac{1}{2}\Big(\theta(t) + \frac{\pi}{2}\Big).$$

Here the constant of integration c is determined by the requirement that $\phi(\alpha(t), \theta(t)) = \gamma(t)$.

Exercise 5.31 (Mercator projection of sphere onto cylinder – sequel to Exercises 0.7 and 5.30 – needed for Exercise 5.51). Let the notation be that of Exercise 5.30. The result in part (iv) of that exercise suggests the following definition for the mapping $\Phi : D \to E := \,]-\pi, \pi[\times \mathbf{R}$:

$$\Phi(\alpha, \theta) = \Big(\alpha, \ \log\tan\Big(\frac{\theta}{2} + \frac{\pi}{4}\Big)\Big) =: (\alpha, \tau).$$

(i) Prove that $\det D\Phi(\alpha, \theta) = \frac{1}{\cos\theta} \neq 0$, and that $\Phi : D \to E$ is a C^1 diffeomorphism.

(ii) Use the formulae from Exercise 0.7 to prove

$$\sin\theta = \tanh\tau, \qquad \cos\theta = \frac{1}{\cosh\tau}.$$

Here $\cosh\tau = \frac{1}{2}(e^\tau + e^{-\tau})$, $\sinh\tau = \frac{1}{2}(e^\tau - e^{-\tau})$, $\tanh\tau = \frac{\sinh\tau}{\cosh\tau}$.

Now define the C^1 parametrization

$$\psi : E \to S^2 \qquad \text{by} \qquad \psi(\alpha, \tau) = \Big(\frac{\cos\alpha}{\cosh\tau}, \frac{\sin\alpha}{\cosh\tau}, \tanh\tau\Big).$$

The inverse $\psi^{-1} : S^2 \to E$ is said to be the *Mercator projection* of S^2 onto $]-\pi, \pi[\, \times \mathbf{R}$. Note that in the (α, τ)-coordinates for S^2 a loxodrome is given by the **linear** equation

$$\alpha + c = \pm\tau \tan\beta.$$

(iii) Verify that the Mercator projection: $S^2 \setminus \{x \in \mathbf{R}^3 \mid x_1 \leq 0,\ x_2 = 0\} \to\,]-\pi, \pi[\, \times \mathbf{R}$ is given by (see Exercise 3.7.(v))

$$x \mapsto \Big(2\arctan\Big(\frac{x_2}{x_1 + \sqrt{x_1^2 + x_2^2}}\Big),\ \frac{1}{2}\log\Big(\frac{1 + x_3}{1 - x_3}\Big)\Big).$$

(iv) The stereographic projection of S^2 onto the cylinder $\{x \in \mathbf{R}^3 \mid x_1^2 + x_2^2 = 1\} \simeq\,]-\pi, \pi] \times \mathbf{R}$ assigns to a point $x \in S^2$ the nearest point of intersection with the cylinder of the straight line through x and the origin (compare with Exercise 5.29). Show that the Mercator projection is **not** the same as this stereographic projection. Yet more exactly, prove the following assertion. If $x \mapsto (\alpha, \xi) \in\,]-\pi, \pi] \times \mathbf{R}$ is the stereographic projection of x, then

$$x \mapsto (\alpha,\ \log(\sqrt{\xi^2 + 1} + \xi))$$

is the Mercator projection of x. The Mercator projection, therefore, is not a projection in the strict sense of the word projection.

(v) A slightly simpler construction for finding the Mercator coordinates (α, τ) of the point $x = (\cos\alpha\cos\theta,\ \sin\alpha\cos\theta,\ \sin\theta)$ is as follows. Using Exercise 0.7 verify that $r(\cos\alpha, \sin\alpha, 0)$ with $r = \tan(\frac{\theta}{2} + \frac{\pi}{4})$ is the stereographic projection of x from $(0, 0, 1)$ onto the equatorial plane $\{x \in \mathbf{R}^3 \mid x_3 = 0\}$. Thus $x \mapsto (\alpha, \log r)$ is the Mercator projection of x.

(vi) Prove

$$\Big\langle \frac{\partial\psi}{\partial\alpha}(\alpha, \tau),\ \frac{\partial\psi}{\partial\alpha}(\alpha, \tau)\Big\rangle = \Big\langle \frac{\partial\psi}{\partial\tau}(\alpha, \tau),\ \frac{\partial\psi}{\partial\tau}(\alpha, \tau)\Big\rangle = \frac{1}{\cosh^2\tau},$$

$$\Big\langle \frac{\partial\psi}{\partial\alpha}(\alpha, \tau),\ \frac{\partial\psi}{\partial\tau}(\alpha, \tau)\Big\rangle = 0.$$

(vii) Show from this that the angle between two curves ζ_1 and ζ_2 in E intersecting at $\zeta_1(t_1) = \zeta_2(t_2)$ equals the angle between the image curves $\psi \circ \zeta_1$ and $\psi \circ \zeta_2$ on S^2, which then intersect at $\psi \circ \zeta_1(t_1) = \psi \circ \zeta_2(t_2)$. In other words, the Mercator projection $\psi^{-1} : S^2 \to E$ is **conformal** or **angle-preserving**.
Hint: $(\psi \circ \zeta)'(t) = \frac{\partial \psi}{\partial \alpha}(\zeta(t))\,\alpha'(t) + \frac{\partial \psi}{\partial \tau}(\zeta(t))\,\tau'(t)$.

(viii) Now consider a triangle on S^2 whose sides are loxodromes which do not run through either the north or the south poles of S^2. Prove that the sum of the internal angles of that triangle equals π radians.

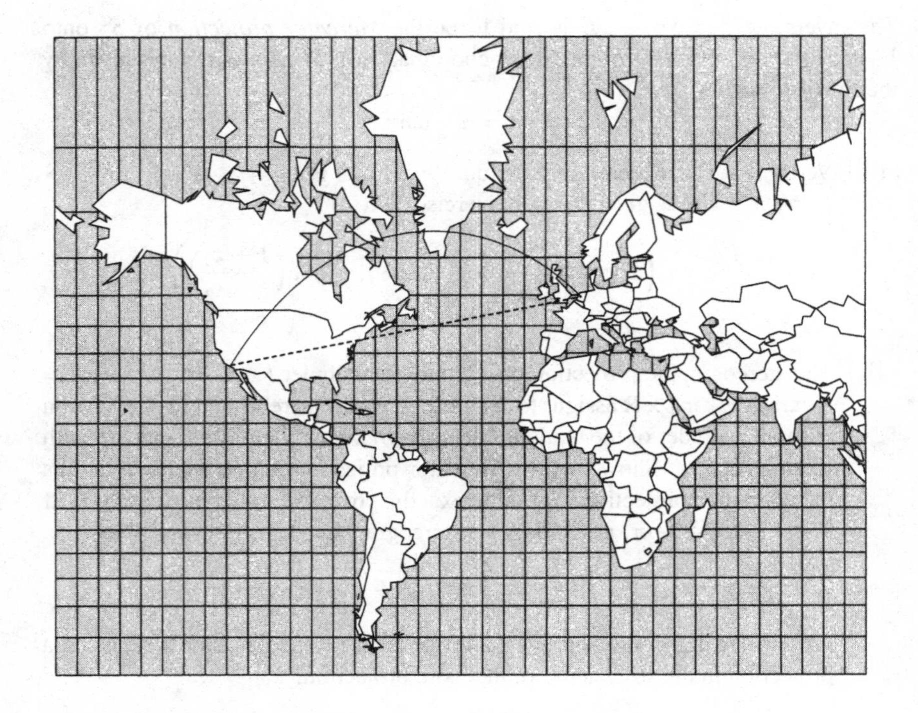

Illustration for Exercise 5.31: Loxodrome in Mercator projection
The dashed line is the loxodrome from Los Angeles, USA to Utrecht, The Netherlands, the solid line is the shortest curve along the Earth's surface connecting these two cities

Exercise 5.32 (Local isometry of catenoid and helicoid – sequel to Exercises 4.6 and 4.8). Let $D \subset \mathbf{R}^2$ be an open subset and assume that $\phi : D \to V$ and $\widetilde{\phi} : \widetilde{D} \to V$ are both C^1 parametrizations of surfaces V and $\widetilde{V}$, respectively, in $\mathbf{R}^3$. Assume the mapping

$$\Phi : V \to \widetilde{V} \qquad \text{given by} \qquad \Phi(x) = \widetilde{\phi} \circ \phi^{-1}(x)$$

is the restriction of a C^1 mapping $\Phi : \mathbf{R}^3 \to \mathbf{R}^3$. We say that V and $\widetilde{V}$ are *locally isometric* (under the mapping Φ) if for all $x \in V$

$$\langle D\Phi(x)v,\ D\Phi(x)w \rangle = \langle v, w \rangle \qquad (v,\ w \in T_x V).$$

(i) Prove that V and $\widetilde{V}$ are locally isometric under Φ if for all $y \in D$

$$\| D_j\phi(y) \| = \| D_j\widetilde{\phi}(y) \| \quad (j = 1, 2),$$
$$\langle D_1\phi(y),\ D_2\phi(y) \rangle = \langle D_1\widetilde{\phi}(y),\ D_2\widetilde{\phi}(y) \rangle.$$

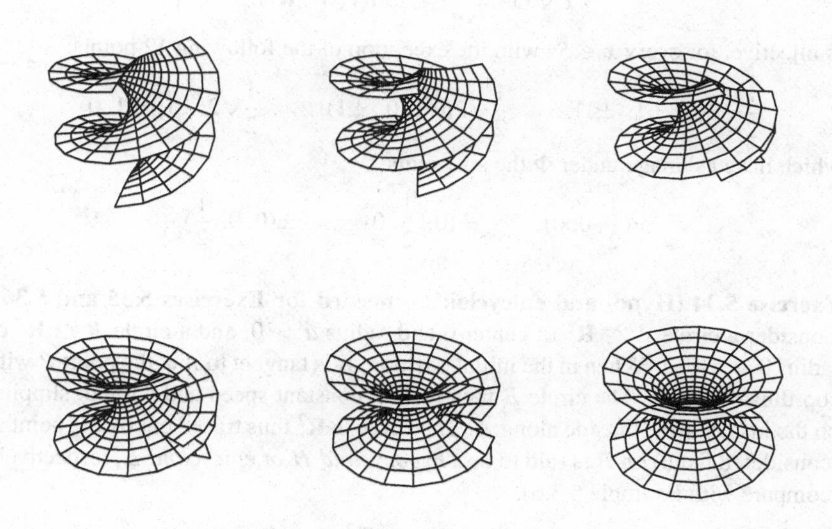

Illustration for Exercise 5.32: From helicoid to catenoid

We now recall the catenoid V from Exercise 4.6:

$$\phi : \mathbf{R}^2 \to V \qquad \text{with} \qquad \phi(s, t) = a(\cosh s \cos t,\ \cosh s \sin t,\ s),$$

and the helicoid $\widetilde{V}$ from Exercise 4.8:

$$\phi' : \mathbf{R}_+ \times \mathbf{R} \to \widetilde{V} \qquad \text{with} \qquad \phi'(s, t) = (s \cos t,\ s \sin t,\ at).$$

(ii) Verify that

$$\widetilde{\phi} : \mathbf{R}^2 \to \widetilde{V} \qquad \text{with} \qquad \widetilde{\phi}(s, t) = a(\sinh s \cos t,\ \sinh s \sin t,\ t)$$

is another C^1 parametrization of the helicoid.

(iii) Prove that the catenoid V and the helicoid $\widetilde{V}$ are locally isometric surfaces in $\mathbf{R}^3$.

Background. Let $a = 1$. Consider the part of the helicoid determined by a single revolution about $\ell = \{ (0, 0, t) \mid 0 \le t \le 2\pi \}$. Take the "warp" out of this surface, then wind the surface around the catenoid such that ℓ is bent along the circle $\{ (\cos t, \sin t, 0) \mid 0 \le t \le 2\pi \}$ on the catenoid. The surfaces can thus be made exactly to coincide without any stretching, shrinking or tearing.

Exercise 5.33 (Steiner's Roman surface – sequel to Exercise 4.27). Demonstrate that

$$D\Phi(x)|_{T_x S^2} \in \mathrm{Lin}(T_x S^2, \mathbf{R}^3)$$

is injective, for every $x \in S^2$ with the exception of the following 12 points:

$$\frac{1}{2}\sqrt{2}(0, \pm 1, \pm 1), \qquad \frac{1}{2}\sqrt{2}(\pm 1, 0, \pm 1), \qquad \frac{1}{2}\sqrt{2}(\pm 1, \pm 1, 0),$$

which have as image under Φ the six points

$$\pm(\frac{1}{2}, 0, 0), \qquad \pm(0, \frac{1}{2}, 0), \qquad \pm(0, 0, \frac{1}{2}).$$

Exercise 5.34 (Hypo- and epicycloids – needed for Exercises 5.35 and 5.36). Consider a circle $A \subset \mathbf{R}^2$ of center 0 and radius $a > 0$, and a circle $B \subset \mathbf{R}^2$ of radius $0 < b < a$. When in the initial position, B is tangent to A at the point P with coordinates $(a, 0)$. The circle B then rolls at constant speed and without slipping on the inside or the outside along A. The curve in $\mathbf{R}^2$ thus traced out by the point P (considered fixed) on B is said to be a *hypocycloid* H or *epicycloid* E, respectively (compare with Example 5.3.6).

(i) Prove that $H = \mathrm{im}(\phi)$, with $\phi : \mathbf{R} \to \mathbf{R}^2$ given by

$$\phi(\alpha) = \begin{pmatrix} (a-b)\cos\alpha + b\cos\left(\dfrac{a-b}{b}\alpha\right) \\ \\ (a-b)\sin\alpha - b\sin\left(\dfrac{a-b}{b}\alpha\right) \end{pmatrix};$$

and $E = \mathrm{im}(\psi)$, with

$$\psi(\alpha) = \begin{pmatrix} (a+b)\cos\alpha - b\cos\left(\dfrac{a+b}{b}\alpha\right) \\ \\ (a+b)\sin\alpha - b\sin\left(\dfrac{a+b}{b}\alpha\right) \end{pmatrix}.$$

Hint: Let M be the center of the circle B. Describe the motion of P in terms of the rotation of M about 0 and the rotation of P about M. With regard to the latter, note that only in the initial situation does the radius vector of M coincide with the positive direction of the x_1-axis.

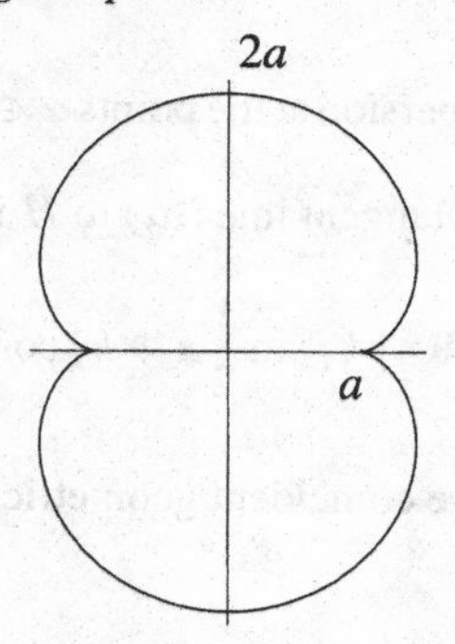

Illustration for Exercise 5.34.(ii): Nephroid

Note that the cardioid from Exercise 3.43 is an epicycloid for which $a = b$. An epicycloid for which $b = \frac{1}{2}a$ is said to be a *nephroid* (ὁ νεφρός = kidneys), and is given by

$$\nu(\alpha) = \frac{1}{2}a \begin{pmatrix} 3\cos\alpha - \cos 3\alpha \\ 3\sin\alpha - \sin 3\alpha \end{pmatrix}.$$

(ii) Prove that the nephroid is a C^∞ submanifold in $\mathbf{R}^2$ of dimension 1 at all of its points, with the exception of the points $(a, 0)$ and $(-a, 0)$; and that the curve has ordinary cusps at these points.
Hint: We have

$$\nu(\alpha) = \begin{pmatrix} a \\ 0 \end{pmatrix} + \begin{pmatrix} \frac{3}{2}a\alpha^2 + \mathcal{O}(\alpha^4) \\ 2a\alpha^3 + \mathcal{O}(\alpha^4) \end{pmatrix}, \qquad \alpha \to 0.$$

Exercise 5.35 (Steiner's hypocycloid – sequel to Exercise 5.34 – needed for Exercise 8.6). A special case of the hypocycloid H occurs when $b = \frac{1}{3}a$:

$$\phi(\alpha) = b \begin{pmatrix} 2\cos\alpha + \cos 2\alpha \\ 2\sin\alpha - \sin 2\alpha \end{pmatrix}.$$

Define $C_r = \{\, r(\cos\theta, \sin\theta) \in \mathbf{R}^2 \mid \theta \in \mathbf{R} \,\}$, for $r > 0$.

(i) Prove that $\|\phi(\alpha)\| = b\sqrt{5 + 4\cos 3\alpha}$. Conclude that there are no points of H outside C_a, nor inside C_b, while $H \cap C_a$ and $H \cap C_b$ are given by, respectively,

$$\{\,\phi(0),\ \phi(\frac{2}{3}\pi),\ \phi(\frac{4}{3}\pi)\,\}, \qquad \{\,\phi(\frac{1}{3}\pi),\ \phi(\pi),\ \phi(\frac{5}{3}\pi)\,\}.$$

(ii) Prove that ϕ is not an immersion at the points $\alpha \in \mathbf{R}$ for which $\phi(\alpha) \in H \cap C_a$.

(iii) Prove that the geometric tangent line $l(\alpha)$ to H in $\phi(\alpha) \in H \setminus C_a$ is given by

$$l(\alpha) = \{\, h \in \mathbf{R}^2 \mid h_1 \sin\frac{1}{2}\alpha + h_2 \cos\frac{1}{2}\alpha = b \sin\frac{3}{2}\alpha \,\}.$$

(iv) Show that H and C_b have coincident geometric tangent lines at the points of $H \cap C_b$.

We say that C_b is the *incircle* of H.

(v) Prove that $H \cap l(\alpha)$ consists of the three points

$$x^{(0)}(\alpha) := \phi(\alpha), \qquad x^{(1)}(\alpha) := \phi(\pi - \frac{1}{2}\alpha),$$
$$x^{(2)}(\alpha) := \phi(2\pi - \frac{1}{2}\alpha) = \phi(-\frac{1}{2}\alpha).$$

(vi) Establish for which $\alpha \in \mathbf{R}$ one has $x^{(i)}(\alpha) = x^{(j)}(\alpha)$, with $i, j \in \{0, 1, 2\}$. Conclude that H has ordinary cusps at the points of $H \cap C_a$, and corroborate this by Taylor expansion.

(vii) Prove that for all $\alpha \in \mathbf{R}$

$$\|x^{(1)}(\alpha) - x^{(2)}(\alpha)\| = 4b, \qquad \|\frac{1}{2}(x^{(1)}(\alpha) + x^{(2)}(\alpha))\| = b.$$

That is, the segment of $l(\alpha)$ lying between those points of intersection with H that are different from $\phi(\alpha)$ is of constant length $4b$, and the midpoint of that segment lies on the incircle of H.

(viii) Prove that for $\phi(\alpha) \in H \setminus C_a$ the geometric tangent lines at $x^{(1)}(\alpha)$ and $x^{(2)}(\alpha)$ are mutually orthogonal, intersecting at $-\frac{1}{2}(x^{(1)}(\alpha) + x^{(2)}(\alpha)) \in C_b$.

Exercise 5.36 (Elimination theory: equations for hypo- and epicycloid – sequel to Exercises 5.34 and 5.35 - Needed for Exercise 5.37). Let the notation be as in Exercise 5.34. Let C be a hypocycloid or epicycloid, that is, $C = \mathrm{im}(\phi)$ or $C = \mathrm{im}(\psi)$, and let $\beta = \frac{a \mp b}{b}\alpha$. Assume there exists $m \in \mathbf{N}$ such that $\beta = m\alpha$. Then the coordinates (x_1, x_2) of $x \in C \subset \mathbf{R}^2$ satisfy a **polynomial** equation; we now proceed to find the equation by means of the following *elimination theory*.

We commence by parametrizing the unit circle (save one point) without goniometric functions as follows. We choose a special point $a = (-1, 0)$ on the unit circle; then for every $t \in \mathbf{R}$ the line through a of slope t, given by $x_2 = t(x_1 + 1)$, intersects with the circle at precisely one other point. The quadratic equation

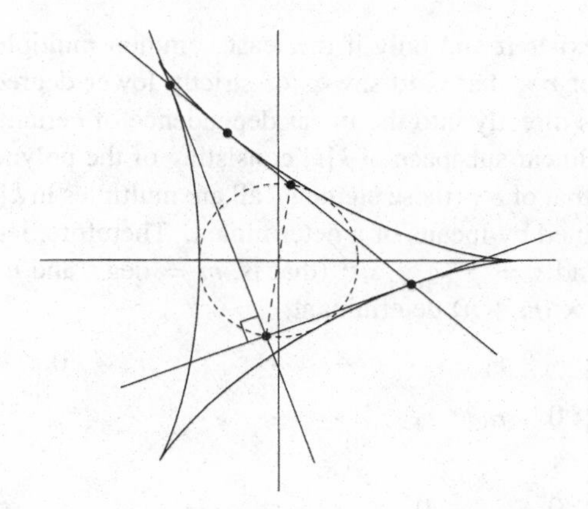

Illustration for Exercise 5.35: Steiner's hypocycloid

$x_1^2 + t^2(x_1+1)^2 - 1 = 0$ for the x_1-coordinate of such a point of intersection can be divided by a factor $x_1 + 1$ (corresponding to the point of intersection a), and we obtain the linear equation $t^2(x_1+1) + x_1 - 1 = 0$, with the solution $x_1 = \frac{1-t^2}{1+t^2}$, from which $x_2 = \frac{2t}{1+t^2}$. Accordingly, for every angle $\alpha \in \,]-\pi, \pi[$ there is a unique $t \in \mathbf{R}$ such that

$$\cos\alpha = \frac{1-t^2}{1+t^2} \qquad \text{and} \qquad \sin\alpha = \frac{2t}{1+t^2},$$

(in order to also obtain $\alpha = -\pi$ one might allow $t = \infty$). By means of this substitution and of $\cos\beta + i\sin\beta = (\cos\alpha + i\sin\alpha)^m$ we find a rational parametrization of the curve C, that is, polynomials $p_1, q_1, p_2, q_2 \in \mathbf{R}[t]$ such that the points of C (save one) form the image of the mapping $\mathbf{R} \to \mathbf{R}^2$ given by

$$t \mapsto \left(\frac{p_1}{q_1}, \frac{p_2}{q_2}\right).$$

For $i = 1, 2$ and $x_i \in \mathbf{R}$ fixed, the condition $p_i/q_i = x_i$ is of a polynomial nature in t, specifically $q_i x_i - p_i = 0$.

Now the condition $x \in C$ is equivalent to the condition that these two polynomials $q_i x_i - p_i \in \mathbf{R}[t]$ (whose coefficients depend on the point x) have a **common** zero. Because it can be shown by algebraic means that (for reasonable p_i, q_i such as here) substitution of nonreal complex values for t in $(p_1/q_1,\ p_2/q_2)$ cannot possibly lead to real pairs (x_1, x_2), this condition can be replaced by the (seemingly weaker) condition that the two polynomials have a common zero in $\mathbf{C}$, and this is equivalent to the condition that they possess a nontrivial common factor in $\mathbf{R}[t]$.

Elimination theory gives a condition under which two polynomials $r, s \in k[t]$ of given degree over an arbitrary field k possess a nontrivial common factor, which condition is itself a polynomial identity in the coefficients of r and s. Indeed, such a common factor exists if and only if the least common multiple of r and s is a nontrivial divisor of $r\,s$, that is to say, is of strictly lower degree. But the latter condition translates directly into the linear dependence of certain elements of the finite-dimensional linear subspace of $k[t]$ consisting of the polynomials of degree strictly lower than that of $r\,s$ (these elements all are multiples in $k[t]$ of r or s); and this can be ascertained by means of a determinant. Therefore, let the *resultant* of $r = \sum_{0\le i\le m} r_i t^i$ and $s = \sum_{0\le i\le n} s_i t^i$ (that is, $m = \deg r$ and $n = \deg s$) be the following $(m+n)\times(m+n)$ determinant:

$$\operatorname{Res}(r,s) = \begin{vmatrix} r_0 & r_1 & \cdots & \cdots & \cdots & \cdots & r_m & 0 & \cdots & 0 \\ 0 & r_0 & r_1 & \cdots & \cdots & \cdots & \cdots & r_m & \ddots & \vdots \\ \vdots & \ddots & \ddots & \ddots & & & & & \ddots & 0 \\ 0 & \cdots & 0 & r_0 & r_1 & \cdots & \cdots & \cdots & \cdots & r_m \\ s_0 & s_1 & \cdots & s_{n-1} & s_n & 0 & \cdots & \cdots & \cdots & 0 \\ 0 & s_0 & s_1 & \cdots & \cdots & s_n & \ddots & & & \vdots \\ \vdots & \ddots & \ddots & \ddots & & & \ddots & \ddots & & \vdots \\ \vdots & & \ddots & \ddots & \ddots & & & \ddots & \ddots & \vdots \\ \vdots & & & \ddots & \ddots & \ddots & & & \ddots & 0 \\ 0 & \cdots & \cdots & \cdots & 0 & s_0 & s_1 & \cdots & \cdots & s_n \end{vmatrix}.$$

Here the first n rows comprise the coefficients of r, the remaining m rows those of s. The polynomials r and s possess a nontrivial factor in $k[t]$ precisely when $\operatorname{Res}(r,s) = 0$. One can easily see that if the same formula is applied, but with either r or s a polynomial of too low a degree (that is, $r_m = 0$ or $s_n = 0$), the vanishing of the determinant remains equivalent to the existence of a common factor of r and s. If, however, both r and s are of too low degree, the determinant vanishes in any case; this might informally be interpreted as an indication for a "common zero at infinity". Thus the point "$t = \infty$" on C will occur as a solution of the relevant polynomial equation, although it was not included in the parametrization.

In the case considered above one has $r = q_1 x_1 - p_1$ and $s = q_2 x_2 - p_2$, and the coefficients r_i, s_j of r and s depend on the coordinates (x_1, x_2). If we allow these coordinates to vary, we may consider the said coefficients as polynomials of degree ≤ 1 in $\mathbf{R}[x_1, x_2]$, while $\operatorname{Res}(r,s)$ is a polynomial of total degree $\le n+m$ in $\mathbf{R}[x_1, x_2]$ whose zeros precisely form the curve C.

Now consider the specific case where C is Steiner's hypocycloid from Exercise 5.35, $(a = 3, b = 1)$; this is parametrized by

$$\phi(\alpha) = \begin{pmatrix} 2\cos\alpha + \cos 2\alpha \\ 2\sin\alpha - \sin 2\alpha \end{pmatrix}.$$

As indicated above, we perform the substitution of rational functions of t for the goniometric functions of α, to obtain the parametrization

$$t \mapsto \left(\frac{3 - 6t^2 - t^4}{(1+t^2)^2}, \frac{8t^3}{(1+t^2)^2} \right),$$

so that in this case $p_1 = 3 - 6t^2 - t^4$, $p_2 = 8t^3$, and $q_1 = q_2 = (1+t^2)^2$. Then, for a given point (x_1, x_2),

$$r = (x_1 - 3) + (2x_1 + 6)t^2 + (x_1 + 1)t^4 \qquad \text{and} \qquad s = x_2 + 2x_2t^2 - 8t^3 + x_2t^4,$$

which leads to the following expression for the resultant $\operatorname{Res}(r,\ s)$:

$$\begin{vmatrix} x_1 - 3 & 0 & 2x_1 + 6 & 0 & x_1 + 1 & 0 & 0 & 0 \\ 0 & x_1 - 3 & 0 & 2x_1 + 6 & 0 & x_1 + 1 & 0 & 0 \\ 0 & 0 & x_1 - 3 & 0 & 2x_1 + 6 & 0 & x_1 + 1 & 0 \\ 0 & 0 & 0 & x_1 - 3 & 0 & 2x_1 + 6 & 0 & x_1 + 1 \\ x_2 & 0 & 2x_2 & -8 & x_2 & 0 & 0 & 0 \\ 0 & x_2 & 0 & 2x_2 & -8 & x_2 & 0 & 0 \\ 0 & 0 & x_2 & 0 & 2x_2 & -8 & x_2 & 0 \\ 0 & 0 & 0 & x_2 & 0 & 2x_2 & -8 & x_2 \end{vmatrix}.$$

By virtue of its regular structure, this 8×8 determinant in $\mathbf{R}[x_1, x_2]$ is less difficult to calculate than would be expected in a general case: one can repeatedly use one collection of similar rows to perform elementary matrix operations on another such collection, and then conversely use the resulting collection to treat the first one, in a way analogous to the Euclidean algorithm. In this special case all pairs of coefficients of x_1 and x_2 in corresponding rows are equal, which will even make the total degree ≤ 4, as one can see. Although it is convenient to have the help of a computer in making the calculation, the result easily fits onto a single line:

$$\operatorname{Res}(r,\ s) = 2^{12}(x_1^4 + 2x_1^2x_2^2 + x_2^4 - 8x_1^3 + 24x_1x_2^2 + 18(x_1^2 + x_2^2) - 27).$$

It is an interesting exercise to check that this polynomial is invariant under reflection with respect to the x_1-axis and under rotation by $\frac{2\pi}{3}$ about the origin, that is, under the substitutions $(x_1, x_2) := (x_1, -x_2)$ and $(x_1, x_2) := (-\frac{1}{2}x_1 - \frac{\sqrt{3}}{2}x_2,\ \frac{\sqrt{3}}{2}x_1 - \frac{1}{2}x_2)$. It can in fact be shown that $x_1^2 + x_2^2$, $3x_1x_2^2 - x_1^3 = -x(x - \sqrt{3}y)(x + \sqrt{3}y)$ and $x_1^4 + 2x_1^2x_2^2 + x_2^4 = (x_1^2 + x_2^2)^2$, neglecting scalars, are the **only** homogeneous polynomials of degree 2, 3, and 4, respectively, that are invariant under these operations.

Exercise 5.37 (Discriminant of polynomial – sequel to Exercise 5.36). For a polynomial function $f(t) = \sum_{0 \leq i \leq n} a_i t^i$ with $a_n \neq 0$ and $a_i \in \mathbf{C}$, we define $\Delta(f)$, the *discriminant* of f, by

$$\operatorname{Res}(f,\ f') = (-1)^{\frac{n(n-1)}{2}} a_n \Delta(f),$$

where f' denotes the derivative of f (see Exercise 3.45). Prove that $\Delta(f) = 0$ is the necessary condition on the coefficients of f for f to have a multiple zero over $\mathbf{C}$. Show

$$\Delta(f) = \begin{cases} -4a_0a_2 + a_1^2, & n = 2; \\ -27a_0^2a_3^2 + 18a_0a_1a_2a_3 - 4a_0a_2^3 - 4a_1^3a_3 + a_1^2a_2^2, & n = 3. \end{cases}$$

$\Delta(f)$ is a homogeneous polynomial in the coefficients of f of degree $2n - 2$.

Exercise 5.38 (Envelopes and caustics). Let $V \subset \mathbf{R}$ be open and let $g : \mathbf{R}^2 \times V \to \mathbf{R}$ be a C^1 function. Define

$$g_y(x) = g(x; y) \quad (x \in \mathbf{R}^2, \ y \in V), \qquad N_y = \{ x \in \mathbf{R}^2 \mid g_y(x) = 0 \}.$$

Assume, for a start, $V = \mathbf{R}$ and $g(x; y) = \|x - (y, 0)\|^2 - 1$. The sets N_y, for $y \in \mathbf{R}$, then form a collection of circles of radius 1, all centered on the x_1-axis. The lines $x_2 = \pm 1$ always have the point $(y, \pm 1)$ in common with N_y and are tangent to N_y at that point. Note that

$$\{ x \in \mathbf{R}^2 \mid x_2 = \pm 1 \} = \{ x \in \mathbf{R}^2 \mid \exists\, y \in \mathbf{R} \text{ with } g(x; y) = 0 \text{ and } D_y g(x; y) = 0 \}.$$

We therefore define the *discriminant* or *envelope* E of the collection of zero-sets $\{ N_y \mid y \in V \}$ by

$$(\star) \qquad E = \{ x \in \mathbf{R}^2 \mid \exists\, y \in V \text{ with } G(x; y) = 0 \}, \qquad G(x; y) = \begin{pmatrix} g(x; y) \\ D_y g(x; y) \end{pmatrix}.$$

(i) Consider the collection of straight lines in $\mathbf{R}^2$ with the property that for each of them the distance between the points of intersection with the x_1-axis and the x_2-axis, respectively, equals 1. Prove that this collection is of the form

$$\{ N_y \mid 0 < y < 2\pi, \ y \neq \frac{\pi}{2}, \ \pi, \ \frac{3\pi}{2} \},$$
$$g(x; y) = x_1 \sin y + x_2 \cos y - \cos y \sin y.$$

Then prove

$$E = \{ \phi(y) := (\cos^3 y, \ \sin^3 y) \mid 0 < y < 2\pi, \ y \neq \frac{\pi}{2}, \ \pi, \ \frac{3\pi}{2} \}.$$

Show that (see also Exercise 5.19)

$$E \cup \{(\pm 1, 0), (0, \pm 1)\} = \{ x \in \mathbf{R}^2 \mid x_1^{2/3} + x_2^{2/3} = 1 \}.$$

We now study the set E in $(\star)$ in more general cases. We assume that the sets N_y all are C^1 submanifolds in $\mathbf{R}^2$ of dimension 1, and, in addition, that the conditions of the Implicit Function Theorem 3.5.1 are met; that is, we assume that for every $x \in E$ there exist an open neighborhood U of x in $\mathbf{R}^2$, an open set $D \subset \mathbf{R}$ and a C^1 mapping $\psi : D \to \mathbf{R}^2$ such that $G(\psi(y); y) = 0$, for $y \in D$, while

$$E \cap U = \{ \psi(y) \mid y \in D \}.$$

In particular then $g(\psi(y); y) = 0$, for all $y \in D$; and hence also

$$D_x g(\psi(y); y)\, \psi'(y) + D_y g(\psi(y); y) = 0 \qquad (y \in D).$$

(ii) Now prove

$$(\star\star) \qquad \psi(y) \in E \cap N_y; \qquad T_{\psi(y)}E = T_{\psi(y)}N_y \qquad (y \in D).$$

(iii) Next, show that the converse of $(\star\star)$ also holds. That is, $\psi : D \to \mathbf{R}^2$ is a C^1 mapping with $\operatorname{im}\psi \subset E$ if $\operatorname{im}\psi \subset \mathbf{R}^2$ satisfies $\psi(y) \in N_y$ and $T_{\psi(y)} \operatorname{im}\psi = T_{\psi(y)}N_y$.

Note that $(\star\star)$ highlights the geometric meaning of an envelope.

(iv) The ellipse in $\mathbf{R}^2$ centered at 0, of major axis 2 along the x_1-axis, and of minor axis $2b > 0$ along the x_2-axis, occurs as image under the mapping $\phi : y \mapsto (\cos y,\ b\sin y)$, for $y \in \mathbf{R}$. A circle of radius 1 and center on this ellipse therefore has the form

$$N_y = \{\, x \in \mathbf{R}^2 \mid \|x - (\cos y,\ b\sin y)\|^2 - 1 = 0 \,\} \qquad (y \in \mathbf{R}).$$

Prove that the envelope of the collection $\{\, N_y \mid y \in \mathbf{R} \,\}$ equals the union of the curve

$$y \mapsto \phi(y) + I(y)(b\cos y,\ \sin y), \qquad I(y) = (b^2\cos^2 y + \sin^2 y)^{-1/2};$$

and the *toroid*, defined by

$$y \mapsto \phi(y) - I(y)(b\cos y,\ \sin y).$$

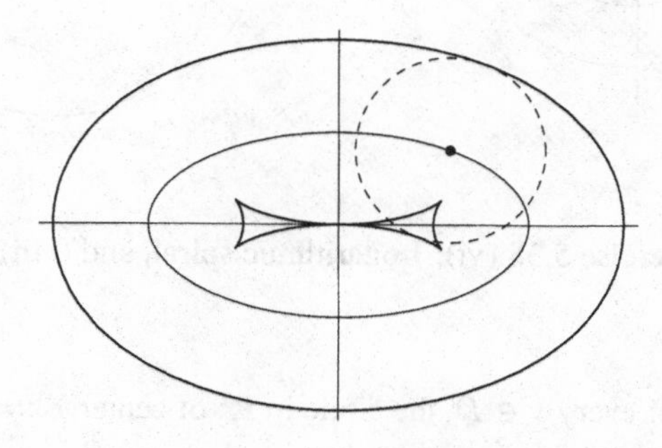

Illustration for Exercise 5.38.(iv): Toroid

Next, assume $C = \operatorname{im}(\phi) \subset \mathbf{R}^2$, where $\phi : D \to \mathbf{R}^2$ and $D \subset \mathbf{R}$ open, is a C^3 embedding.

(v) Then prove that, for every $y \in D$, the line N_y in $\mathbf{R}^2$ perpendicular to the curve C at the point $\phi(y)$ is given by

$$N_y = \{ x \in \mathbf{R}^2 \mid g(x; y) = \langle x - \phi(y), \phi'(y) \rangle = 0 \}.$$

We now define the *evolute* E of C as the envelope of the collection $\{ N_y \mid y \in D \}$. Assume $\det(\phi'(y)\ \phi''(y)) \neq 0$, for all $y \in D$. Prove that E then possesses the following C^1 parametrization:

$$y \mapsto x(y) = \phi(y) + \frac{\|\phi'(y)\|^2}{\det(\phi'(y)\ \phi''(y))} \begin{pmatrix} 0 & -1 \\ 1 & 0 \end{pmatrix} \phi'(y) : D \to \mathbf{R}^2.$$

(vi) A *logarithmic spiral* is a curve in $\mathbf{R}^2$ of the form

$$y \mapsto e^{ay}(\cos y, \sin y) \qquad (y \in \mathbf{R},\ a \in \mathbf{R}).$$

Prove that the logarithmic spiral with $a = 1$ has the following curve as its evolute:

$$y \mapsto e^y \Big(\cos\big(\frac{\pi}{2} - y\big), \sin\big(\frac{\pi}{2} - y\big) \Big) \qquad (y \in \mathbf{R}).$$

In other words, this logarithmic spiral is its own evolute.

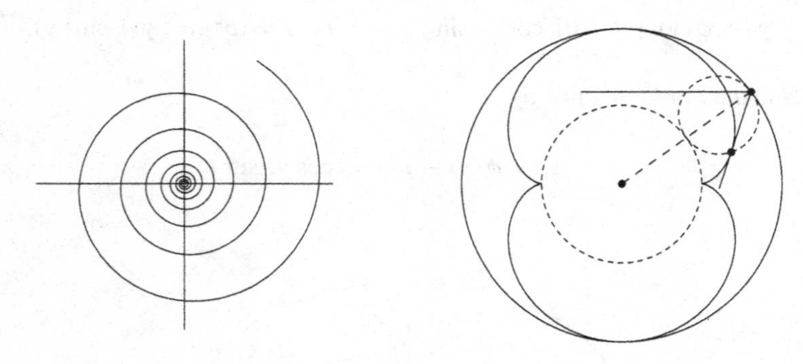

Illustration for Exercise 5.38.(vi): Logarithmic spiral, and (viii): Reflected light

(vii) Prove that, for every $y \in D$, the circle in $\mathbf{R}^2$ of center $Y := \phi(y) \in C$ which goes through the origin 0 is given by

$$N_y = \{ x \in \mathbf{R}^2 \mid \|x\|^2 - 2\langle x, \phi(y) \rangle = 0 \}.$$

Now define

$$E \quad \text{is the envelope of the collection} \quad \{ N_y \mid y \in D \}.$$

Check that the point $X := x \in \mathbf{R}^2$ lying on E and parametrized by y satisfies

$$x \in N_y \qquad \text{and} \qquad \langle x, \phi'(y) \rangle = 0.$$

That is, the line segments YO and YX are of equal length and the line segment OX is orthogonal to the tangent line at Y to C. Let $L_y \subset \mathbf{R}^2$ be the straight line through Y and X. Then the ultimate course of a light ray starting from O and reflected at Y by the curve C will be along a part of L_y. Therefore, the envelope of the collection $\{ L_y \mid y \in D \}$ is said to be the *caustic* of C relative to O (ἡ καῦσις = sun's heat, burning). Since the tangent line at X to E coincides with the tangent line at X to the circle N_y, the line L_y is orthogonal to E. Consequently, the caustic of C relative to O is the evolute of E.

(viii) Consider the circle in $\mathbf{R}^2$ of center 0 and radius $a > 0$, and the collection of straight lines in $\mathbf{R}^2$ parallel to the x_1-axis and at a distance $\leq a$ from that axis. Assume that these lines are reflected by the circle in such a way that the angle between the incident line and the normal equals the angle between the reflected line and the normal. Prove that a reflected line is of the form N_α, for $\alpha \in \mathbf{R}$, where

$$N_\alpha = \{ x \in \mathbf{R}^2 \mid x_2 - x_1 \tan 2\alpha + a \cos\alpha \, \tan 2\alpha - a \sin\alpha = 0 \}.$$

Prove that the envelope of the collection $\{ N_\alpha \mid \alpha \in \mathbf{R} \}$ is given by the curve

$$\alpha \mapsto \frac{1}{4} a \begin{pmatrix} 3\cos\alpha - \cos 3\alpha \\ 3\sin\alpha - \sin 3\alpha \end{pmatrix} = \frac{1}{2} a \begin{pmatrix} \cos\alpha(3 - 2\cos^2\alpha) \\ 2\sin^3\alpha \end{pmatrix}.$$

Background. Note that this is a nephroid from Exercise 5.34. The caustic of sunlight reflected by the inner wall of a teacup of radius a is traced out by a point on a circle of radius $\frac{1}{4}a$ when this circle rolls on the outside along a circle of radius $\frac{1}{2}a$.

Exercise 5.39 (Geometric–arithmetic and Kantorovich's inequalities – needed for Exercise 7.55).

(i) Show

$$n^n \prod_{1 \leq j \leq n} x_j^2 \leq \|x\|^{2n} \qquad (x \in \mathbf{R}^n).$$

(ii) Assume that $a_1, \dots, a_n$ in $\mathbf{R}$ are nonnegative. Prove the *geometric–arithmetic inequality*

$$\Big(\prod_{1 \leq j \leq n} a_j \Big)^{1/n} \leq a := \frac{1}{n} \sum_{1 \leq j \leq n} a_j.$$

Hint: Write $a_j = n a x_j^2$.

Let $\Delta \subset \mathbf{R}^2$ be a triangle of area O and perimeter l.

(iii) **(Sequel to Exercise 5.25).** Prove by means of this exercise and part (ii) the *isoperimetric inequality* for Δ:

$$O \leq \frac{l^2}{12\sqrt{3}},$$

where equality obtains if and only if Δ is equilateral.

Assume that $t_1, \dots, t_n \in \mathbf{R}$ are nonnegative and satisfy $t_1 + \cdots + t_n = 1$. Furthermore, let $0 < m < M$ and suppose $x_j \in \mathbf{R}$ satisfy $m \leq x_j \leq M$ for all $1 \leq j \leq n$. Then we have *Kantorovich's inequality*

$$\Big(\sum_{1 \leq j \leq n} t_j\, x_j\Big)\Big(\sum_{1 \leq j \leq n} t_j\, x_j^{-1}\Big) \leq \frac{(m+M)^2}{4m\,M},$$

which we shall prove in three steps.

(iv) Verify that the inequality remains unchanged if we replace every x_j by $\lambda\, x_j$ with $\lambda > 0$. Conclude we may assume $m\,M = 1$, and hence $0 < m < 1$.

(v) Determine the extrema of the function $x \mapsto x + x^{-1}$ on $I = [\,m,\ m^{-1}\,]$; next show $x + x^{-1} \leq m + M$ for $x \in I$, and deduce

$$\sum_{1 \leq j \leq n} t_j\, x_j + \sum_{1 \leq j \leq n} t_j\, x_j^{-1} \leq m + M.$$

(vi) Verify that Kantorovich's inequality follows from the geometric–arithmetic inequality.

Exercise 5.40. Let C be the cardioid from Exercise 3.43. Calculate the critical points of the restriction to C of the function $f(x, y) = y$, in other words find the points where the tangent line to C is horizontal.

Exercise 5.41 (Hölder's, Minkowski's and Young's inequalities – needed for Exercises 6.73 and 6.75). For $p \geq 1$ and $x \in \mathbf{R}^n$ we define

$$\|x\|_p = \Big(\sum_{1 \leq j \leq n} |x_j|^p\Big)^{1/p}.$$

Now assume that $p \geq 1$ and $q \geq 1$ satisfy

$$\frac{1}{p} + \frac{1}{q} = 1.$$

(i) Prove the following, known as *Hölder's inequality*:

$$|\langle x, y\rangle| \leq \|x\|_p \|y\|_q \qquad (x, y \in \mathbf{R}^n).$$

For $p = 2$, what well-known inequality does this turn into?
Hint: Let $f(x, y) = \sum_{1\leq j\leq n} |x_j||y_j| - \|x\|_p \|y\|_q$, and calculate the maximum of f for fixed $\|x\|_p$ and y.

(ii) Prove *Minkowski's inequality*

$$\|x + y\|_p \leq \|x\|_p + \|y\|_p \qquad (x, y \in \mathbf{R}^n).$$

Hint: Let $\phi(t) = \|x + ty\|_p - \|x\|_p - t\|y\|_p$, and use Hölder's inequality to determine the sign of $\phi'(t)$.

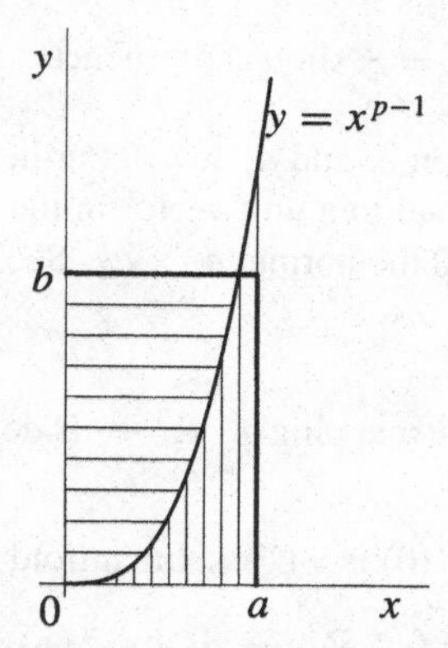

Illustration for Exercise 5.41: Another proof of Young's inequality
Because $(p - 1)(q - 1) = 1$, one has $x^{p-1} = y$ if and only if $x = y^{q-1}$

For the sake of completeness we point out another proof of Hölder's inequality, which shows it to be a consequence of *Young's inequality*

$$(\star) \qquad ab \leq \frac{a^p}{p} + \frac{b^q}{q},$$

valid for any pair of positive numbers a and b. Note that $(\star)$ is a generalization of the inequality $ab \leq \frac{1}{2}(a^2 + b^2)$, which derives from Exercise 5.39.(ii) or from $0 \leq (a - b)^2$. To prove $(\star)$, consider the function $f : \mathbf{R}_+ \to \mathbf{R}$, with $f(t) = p^{-1}t^p + q^{-1}t^{-q}$.

(iii) Prove that $f'(t) = 0$ implies $t = 1$, and that $f''(t) > 0$ for $t \in \mathbf{R}_+$.

(iv) Verify that $(\star)$ follows from $f(a^{\frac{1}{q}}b^{-\frac{1}{p}}) \geq f(1) = 1$.

(v) Prove Hölder's inequality by substitution into ($\star$) of $a = \frac{x_j}{\|x\|_p}$ and $b = \frac{y_j}{\|y\|_q}$ $(1 \leq j \leq n)$ and subsequent summation on j.

Exercise 5.42 (Luggage). On certain intercontinental flights the following luggage restrictions apply. Besides cabin luggage, a passenger may bring at most two pieces of luggage for transportation. For this luggage to be transported free of charge, the maximum allowable sum of all lengths, widths and heights is 270 cm, with no single length, width or height exceeding 159 cm. Calculate the maximal volume a passenger may take with him.

Exercise 5.43. Let a_1 and $a_2 \in S^2 = \{ x \in \mathbf{R}^3 \mid \|x\| = 1 \}$ be linearly independent vectors and denote by $L(a_1, a_2)$ the plane in $\mathbf{R}^3$ spanned by a_1 and a_2. If $x \in \mathbf{R}^3$ is constrained to belong to $S^2 \cap L(a_1, a_2)$, prove that the maximum, and minimum value, $x_{i\pm}$ for the coordinate function $x \mapsto x_i$, for $1 \leq i \leq 3$, is given by

$$x_{i\pm} = \pm \frac{\|e_i \times (a_1 \times a_2)\|}{\|a_1 \times a_2\|} = \pm \sin \beta_i, \qquad \text{hence} \qquad \cos \beta_i = \frac{|\det(a_1\ a_2\ e_i)|}{\|a_1 \times a_2\|}.$$

Here β_i is the angle between e_i and $a_1 \times a_2$. Furthermore, these extremal values for x_i are attained for x equal to a unit vector in the intersection of $L(a_1, a_2)$ and the plane spanned by e_i and the normal $a_1 \times a_2$. See Exercise 5.27.(x) for another derivation.

Exercise 5.44. Consider the mapping $g : \mathbf{R}^3 \to \mathbf{R}$ defined by $g(x) = x_1^3 - 3x_1x_2^2 - x_3$.

(i) Show that the set $g^{-1}(0)$ is a C^∞ submanifold in $\mathbf{R}^3$.

(ii) Define the function $f : \mathbf{R}^3 \to \mathbf{R}$ by $f(x) = x_3$. Show that under the constraint $g(x) = 0$ the function f does not have (local) extrema.

(iii) Let $B = \{ x \in \mathbf{R}^3 \mid g(x) = 0,\ x_1^2 + x_2^2 \leq 1 \}$. Prove that the function f has a maximum on B.

(iv) Assume that f attains its maximum on B at the point b. Then show $b_1^2 + b_2^2 = 1$.

(v) Using the method of Lagrange multipliers, calculate the points $b \in B$ where f has its maximum on B.

(vi) Define $\phi : \mathbf{R} \to \mathbf{R}^3$ by $\phi(\alpha) = (\cos \alpha,\ \sin \alpha,\ \cos 3\alpha)$. Prove that ϕ is an immersion.

(vii) Prove $\phi(\mathbf{R}) = \{ x \in \mathbf{R}^3 \mid g(x) = 0,\ x_1^2 + x_2^2 = 1 \}$.

(viii) Find the points in $\mathbf{R}$ where the function $f \circ \phi$ has a maximum, and compare this result with that of part (v).

Exercise 5.45. Let $U \subset \mathbf{R}^n$ be an open set, and let there be, for $i = 1, 2$, the function g_i in $C^1(U)$. Let $V_i = \{ x \in U \mid g_i(x) = 0 \}$ and grad $g_i(x) \neq 0$, for all $x \in V_i$. Assume $g_2|_{V_1}$ has its maximum at $x \in V_1$, while $g_2(x) = 0$. Prove that V_1 and V_2 are tangent to each other at the point x, that is, $x \in V_1 \cap V_2$ and $T_x V_1 = T_x V_2$.

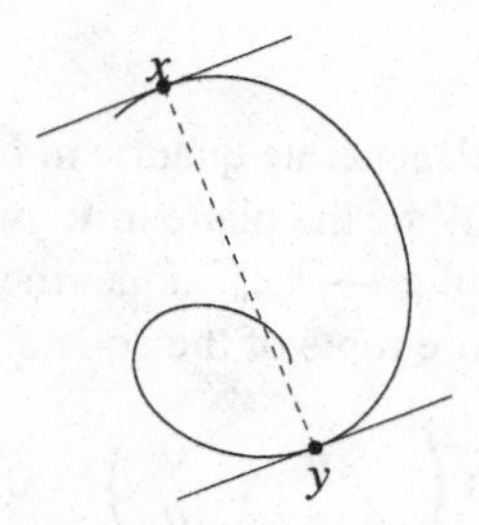

Illustration for Exercise 5.46: Diameter of submanifold

Exercise 5.46 (Diameter of submanifold). Let V be a C^1 submanifold in $\mathbf{R}^n$ of dimension $d > 0$. Define $\delta(V) \in [0, \infty]$, the *diameter of* V, by

$$\delta(V) = \sup\{ \|x - y\| \mid x,\ y \in V \}.$$

(i) Prove $\delta(V) > 0$. Prove that $\delta(V) < \infty$ if V is compact, and that in that case $x^0, y^0 \in V$ exist such that $\delta(V) = \|x^0 - y^0\|$.

(ii) Show that for all $x, y \in V$ with $\delta(V) = \|x - y\|$ we have $x - y \perp T_x V$ and $x - y \perp T_y V$.

(iii) Show that the diameter of the zero-set in $\mathbf{R}^3$ of $x \mapsto x_1^4 + x_2^4 + x_3^4 - 1$ equals $2\sqrt[4]{3}$.

(iv) Show that the diameter of the zero-set in $\mathbf{R}^3$ of $x \mapsto x_1^4 + 2x_2^4 + 3x_3^4 - 1$ equals $2\sqrt[4]{\frac{11}{6}}$.

Exercise 5.47 (Another proof of the Spectral Theorem 2.9.3). Let $A \in \mathrm{End}^+(\mathbf{R}^n)$ and define $f : \mathbf{R}^n \to \mathbf{R}$ by $f(x) = \langle Ax, x \rangle$.

(i) Use the compactness of the unit sphere S^{n-1} in $\mathbf{R}^n$ to show the existence of $a \in S^{n-1}$ such that $f(x) \leq f(a)$, for all $x \in S^{n-1}$.

(ii) Prove that there exists $\lambda \in \mathbf{R}$ with $Aa = \lambda a$, and $\lambda = f(a)$. In other words, the maximum of $f|_{S^{n-1}}$ is a real eigenvalue of A.

(iii) Verify that there exist an orthonormal basis $(a_1, \dots, a_n)$ of $\mathbf{R}^n$ and a vector $(\lambda_1, \dots, \lambda_n) \in \mathbf{R}^n$ such that $Aa_j = \lambda_j a_j$, for $1 \le j \le n$.

(iv) Use the preceding to find the apices of the conic with the equation

$$36x_1^2 + 96x_1x_2 + 64x_2^2 + 20x_1 - 15x_2 + 25 = 0.$$

Exercise 5.48. Let V be a nondegenerate quadric in $\mathbf{R}^3$ with equation $\langle Ax, x\rangle = 1$ where $A \in \mathrm{End}^+(\mathbf{R}^3)$, and let W be the plane in $\mathbf{R}^3$ with equation $\langle a, x\rangle = 0$ with $a \ne 0$. Prove that the extrema of $x \mapsto \|x\|^2$ under the constraint $x \in V \cap W$ equal λ_1^{-1} and λ_2^{-1}, where λ_1 and λ_2 are roots of the equation

$$\det \begin{pmatrix} a & 0 \\ A - \lambda I & a \end{pmatrix} = 0.$$

Show that this equation in λ is of degree ≤ 2, and give a geometric argument why the roots λ_1 and λ_2 are real and positive if the equation is quadratic.

Exercise 5.49 (Another proof of Hadamard's inequality and Iwasawa decomposition). Consider an arbitrary basis $(g_1, \dots, g_n)$ for $\mathbf{R}^n$; using the *Gram–Schmidt orthonormalization process* we obtain from this an orthonormal basis $(k_1, \dots, k_n)$ for $\mathbf{R}^n$ by means of

$$h_j := g_j - \sum_{1 \le i < j} \langle g_j, k_i \rangle k_i \in \mathbf{R}^n, \qquad k_j := \frac{1}{\|h_j\|} h_j \in \mathbf{R}^n \qquad (1 \le j \le n).$$

(Verify that, indeed, $h_j \ne 0$, for $1 \le j \le n$.) Thus there exist numbers $u_{1j}, \dots, u_{jj}$ in $\mathbf{R}$ with

$$g_j = \sum_{1 \le i \le j} u_{ij} k_i \qquad (1 \le j \le n).$$

For the matrix $G \in \mathbf{GL}(n, \mathbf{R})$ with the g_j, for $1 \le j \le n$, as column vectors this implies $G = KU$, where $U = (u_{ij}) \in \mathbf{GL}(n, \mathbf{R})$ is an upper triangular matrix and $K \in \mathbf{O}(n, \mathbf{R})$ the matrix which sends the standard unit basis $(e_1, \dots, e_n)$ for $\mathbf{R}^n$ into the orthonormal basis $(k_1, \dots, k_n)$. Further, U can be written as AN, where $A \in \mathbf{GL}(n, \mathbf{R})$ is a diagonal matrix with coefficients $a_{jj} = \|h_j\| > 0$, and $N \in \mathbf{GL}(n, \mathbf{R})$ an upper triangular matrix with $n_{jj} = 1$, for $1 \le j \le n$. Conclude that every $G \in \mathbf{GL}(n, \mathbf{R})$ can be written as $G = KAN$, this is called the *Iwasawa decomposition* of G.

The notation now is as in Example 5.5.3. So let $G \in \mathbf{GL}(n, \mathbf{R})$ be given, and write $G = KU$. If $u_j \in \mathbf{R}^n$ denotes the j-th column vector of U, we have $|u_{jj}| \le \|u_j\|$; whereas $G = KU$ implies $\|g_j\| = \|u_j\|$. Hence we obtain *Hadamard's inequality*

$$|\det G| = |\det U| = \prod_{1 \le j \le n} |u_{jj}| \le \prod_{1 \le j \le n} \|g_j\|.$$

Background. The Iwasawa decomposition $G = KAN$ is unique. Indeed, $G^t G = N^t A^2 N$. If $K_1 A_1 N_1 = K_2 A_2 N_2$, we therefore find $N_1^t A_1^2 N_1 = N_2^t A_2^2 N_2$. This implies

$$A_1^2 N := A_1^2 N_1 N_2^{-1} = (N_1^t)^{-1} N_2^t A_2^2 = (N^t)^{-1} A_2^2.$$

Since N and N^t are triangular in opposite direction they must be equal to I, and therefore $A_1 = A_2$, which implies $K_1 = K_2$.

Exercise 5.50. Let V be the torus from Example 4.6.3, let $a \in \mathbf{R}^3$ and let f_a be the linear function on $\mathbf{R}^3$ with $f_a(x) = \langle a, x \rangle$. Find the points of V where $f_a|_V$ has its maxima or minima, and examine how these points depend on $a \in \mathbf{R}^3$.
Hint: Consider the geometric approach.

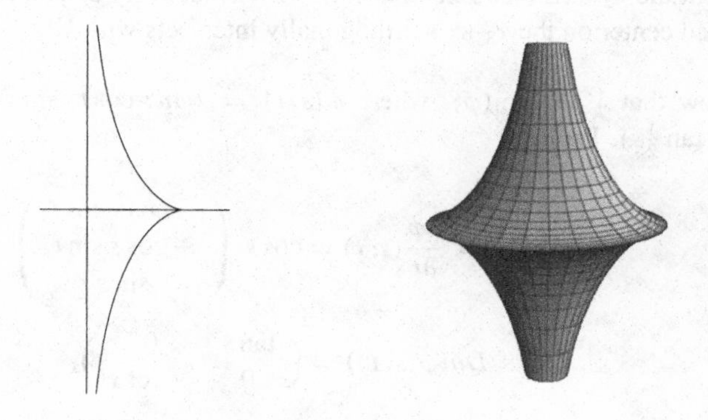

Illustration for Exercise 5.51 : Tractrix and pseudosphere

Exercise 5.51 (Tractrix and pseudosphere – sequel to Exercise 5.31 – needed for Exercises 6.48 and 7.19). A (point) walker w in the (x_1, x_3)-plane $\simeq \mathbf{R}^2$ pulls a (point) cart k along, by means of a rigid rod of length 1. Initially w is at $(0, 0)$ and k at $(1, 0)$; then w starts to walk up or down along the x_3-axis. The curve T in $\mathbf{R}^2$ traced out by k is known as the *tractrix* (tractare = to pull).

(i) Define $f : \,]0, 1] \to [0, \infty[$ by

$$f(x) = \int_x^1 \frac{\sqrt{1 - t^2}}{t} \, dt.$$

Prove that f is surjective, and $T = \{ (x, \pm f(x)) \mid x \in \,]0, 1] \}$.

(ii) Using the substitution $x = \cos \alpha$, prove

$$f(x) = \log(1 + \sqrt{1 - x^2}) - \log x - \sqrt{1 - x^2} \qquad (0 < x \leq 1).$$

Verify $T = \{\, (\sin s,\ \cos s + \log\tan\frac{s}{2}) \mid 0 < s < \pi \,\}$. Conclude

$$T = \Big\{ \Big(\cos\theta,\ -\sin\theta + \log\tan\Big(\frac{\theta}{2} + \frac{\pi}{4}\Big)\Big) \Bigm| -\frac{\pi}{2} < \theta < \frac{\pi}{2} \Big\}.$$

Deduce that the x_3-coordinate of w equals $\log\tan(\frac{\theta}{2} + \frac{\pi}{4})$ when the angle of the rod with the horizontal direction is θ. Note that this gives a geometrical construction of the Mercator coordinates from Exercise 5.31 of a point on S^2.

(iii) Show that $T \setminus \{(1, 0)\}$ is a C^∞ submanifold in $\mathbf{R}^2$ of dimension 1, and that T is not a C^∞ submanifold at $(1, 0)$.

(iv) Let V be the surface of revolution in $\mathbf{R}^3$ obtained by revolving $T \setminus \{(1, 0)\}$ about the x_3-axis (see Exercise 4.6). Prove that every sphere in $\mathbf{R}^3$ of radius 1 and center on the x_3-axis orthogonally intersects with V.

(v) Show that $V \subset \mathrm{im}(\phi)$, where $\phi(s, t) = (\sin s\cos t,\ \sin s\sin t,\ \cos s + \log\tan\frac{1}{2}s)$. Prove

$$\frac{\partial\phi}{\partial s}(s, t) \times \frac{\partial\phi}{\partial t}(s, t) = \cos s \begin{pmatrix} -\cos s\cos t \\ -\cos s\sin t \\ \sin s \end{pmatrix},$$

$$Dn(\phi(s, t)) = \begin{pmatrix} \tan s & 0 \\ 0 & -\cot s \end{pmatrix}.$$

Prove that the Gauss curvature of V equals -1 everywhere. Explain why V is called the *pseudosphere*.

Exercise 5.52 (Curvature of planar curve and evolute).

(i) Let $f : I \to \mathbf{R}$ be a C^2 function. Show that the planar curve in $\mathbf{R}^3$ parametrized by $t \mapsto (t, f(t), 0)$ has curvature

$$\kappa = \frac{|f''|}{(1 + f'^2)^{3/2}} : I \to [0, \infty[.$$

(ii) Let $\gamma : I \to \mathbf{R}^2$ be a C^2 embedding. Prove that the planar curve in $\mathbf{R}^3$ parametrized by $t \mapsto (\gamma_1(t), \gamma_2(t), 0)$ has curvature

$$\kappa = \frac{|\det(\gamma'\ \gamma'')|}{\|\gamma'\|^3} : I \to [0, \infty[.$$

(iii) Let the notation be as in Exercise 5.38.(v). Show that the parametrization of the evolute E also can be written as

$$y \mapsto \phi(y) + \frac{1}{\kappa(y)} N(y),$$

where $\kappa(y)$ denotes the curvature of $\mathrm{im}(\phi)$ at $\phi(y)$, and $N(y)$ the normal to $\mathrm{im}(\phi)$ at $\phi(y)$.

Exercise 5.53 (Curvature of ellipse or hyperbola – sequel to Exercise 5.24). We use the notation of that exercise. In particular, given $0 \neq \epsilon \in \mathbf{R}^2$ and $0 \neq d \in \mathbf{R}$, consider $C = \{ x \in \mathbf{R}^2 \mid \|x\| = \langle x, \epsilon \rangle + d \}$. Suppose that $\mathbf{R} \ni t \mapsto x(t) \in \mathbf{R}^2$ is a C^2 curve with image equal to C. In the following we often write x instead of $x(t)$ to simplify the notation, and similarly x' and x''.

(i) Prove by differentiation

$$\langle \epsilon - \frac{1}{\|x\|} x, x' \rangle = 0, \qquad \text{and deduce} \qquad \epsilon = \frac{1}{\|x\|} x + \frac{d}{l} J x',$$

with $J \in \mathbf{SO}(2, \mathbf{R})$ as in Lemma 8.1.5 and $l = \det(x \; x') = -\langle J x', x \rangle$.

(ii) From now on, assume C to be either an ellipse or a hyperbola. Applying part (i), the definition of C and Exercise 5.24.(ii), show

$$(\|x\| \, \|x'\|)^2 = \frac{l^2}{d^2} \|x\|^2 \Big(1 + e^2 - 2 \frac{\langle x, \epsilon \rangle}{\|x\|} \Big) = \pm \frac{l^2}{b^2} (2a\|x\| - \|x\|^2).$$

(iii) Differentiate the identity in part (i) once more in order to obtain $\frac{\|x \times x'\|^2}{\|x\|^3} = \frac{d}{l} \det(x' \; x'')$; in other words, $\det(x' \; x'') = \frac{l^3}{d\|x\|^3}$. Denoting by κ the curvature of C, conclude on account of Formula (5.17) and part (ii)

$$\kappa = \frac{|\det(x' \; x'')|}{\|x'\|^3} = \frac{1}{d} \Big(\frac{|\det(x \; x')|}{\|x\| \, \|x'\|} \Big)^3 = \frac{ab}{(\pm (2a\|x\| - \|x\|^2))^{\frac{3}{2}}} = \frac{b^4}{a^2 h^3}.$$

Here $h = \frac{b}{a}(\pm (2a\|x\| - \|x\|^2))^{\frac{1}{2}}$ is the distance between $x \in C$ and the point of intersection of the line perpendicular to C at x and of the focal axis $\mathbf{R}\epsilon$. The sign $\pm$ occurs as C is an ellipse or a hyperbola, respectively. For the last equality, suppose $\epsilon = (e, 0)$ and deduce $|x_1'| = \frac{l}{d} \frac{x_2}{\|x\|}$ from part (i) using that $J^2 = -I$.

Exercise 5.54 (Independence of curvature and torsion of parametrization). Prove directly that the formulae in (5.17) for the curvature and torsion of a curve in $\mathbf{R}^3$ are independent of the choice of the parametrization $\gamma : I \to \mathbf{R}^3$.

Hint: Let $\widetilde{\gamma} : \widetilde{I} \to \mathbf{R}^3$ be another parametrization and assume $\gamma(t) = \widetilde{\gamma}(\widetilde{t})$, with $t \in I$ and $\widetilde{t} \in \widetilde{I}$; then $t = (\gamma^{-1} \circ \widetilde{\gamma})(\widetilde{t}) =: \Psi(\widetilde{t})$. Application of the chain rule to $\widetilde{\gamma} = \gamma \circ \Psi$ on $\widetilde{I}$ therefore gives, for $\widetilde{t} \in \widetilde{I}$, $t = \Psi(\widetilde{t}) \in I$,

$$\begin{aligned}
\widetilde{\gamma}'(\widetilde{t}) &= \Psi'(\widetilde{t})\,\gamma'(t), \\
\widetilde{\gamma}''(\widetilde{t}) &= \Psi''(\widetilde{t})\,\gamma'(t) + \Psi'(\widetilde{t})^2\,\gamma''(t), \\
\widetilde{\gamma}'''(\widetilde{t}) &= \qquad\qquad\qquad \cdots + \Psi'(\widetilde{t})^3\,\gamma'''(t).
\end{aligned}$$

Exercise 5.55 (Projections of curve). Let the notation be as in Section 5.8. Suppose $\gamma : J \to \mathbf{R}^3$ to be a biregular C^4 parametrization by arc length s starting at $0 \in J$. Using a rotation in $\mathbf{R}^3$ we may assume that $T(0)$, $N(0)$, $B(0)$ coincide with the standard basis vectors e_1, e_2, e_3 in $\mathbf{R}^3$.

(i) Set $\kappa = \kappa(0)$, $\kappa' = \kappa'(0)$ and $\tau = \tau(0)$. By means of the formulae of Frenet–Serret prove

$$\gamma'(0) = \begin{pmatrix} 1 \\ 0 \\ 0 \end{pmatrix}, \qquad \gamma''(0) = \begin{pmatrix} 0 \\ \kappa \\ 0 \end{pmatrix}, \qquad \gamma'''(0) = \begin{pmatrix} -\kappa^2 \\ \kappa' \\ \kappa\tau \end{pmatrix}.$$

(ii) Applying a translation in $\mathbf{R}^3$ we can arrange that $\gamma(0) = 0$. Using Taylor expansion show

$$\begin{aligned}
\gamma(s) &= s\gamma'(0) + \frac{s^2}{2}\gamma''(0) + \frac{s^3}{6}\gamma'''(0) + \mathcal{O}(s^4) \\
&= \begin{pmatrix} s \qquad - \dfrac{\kappa^2}{6}s^3 \\ \dfrac{\kappa}{2}s^2 + \dfrac{\kappa'}{6}s^3 \\ \dfrac{\kappa\tau}{6}s^3 \end{pmatrix} + \mathcal{O}(s^4), \qquad s \to 0.
\end{aligned}$$

Deduce

$$\lim_{s\to 0} \frac{\gamma_2(s)}{\gamma_1(s)^2} = \frac{\kappa}{2}, \qquad \lim_{s\to 0} \frac{\gamma_3(s)^2}{\gamma_2(s)^3} = \frac{2\tau^2}{9\kappa}, \qquad \lim_{s\to 0} \frac{\gamma_3(s)}{\gamma_1(s)^3} = \frac{\kappa\tau}{6}.$$

(iii) Show that the orthogonal projections of $\mathrm{im}(\gamma)$ onto the following planes near the origin are approximated by the curves, respectively (here we suppose $\tau \neq 0$) (see the Illustration for Exercise 4.28):

osculating,	$x_2 = \dfrac{\kappa}{2}x_1^2,$	parabola;
normal,	$x_3^2 = \dfrac{2\tau^2}{9\kappa}x_2^3,$	semicubic parabola;
rectifying,	$x_3 = \dfrac{\kappa\tau}{6}x_1^3,$	cubic.

(iv) Prove that the parabola in the osculating plane in turn is approximated near the origin by the osculating circle with equation $x_1^2 + (x_2 - \frac{1}{\kappa})^2 = \frac{1}{\kappa^2}$. In general, the *osculating circle* of $\operatorname{im}(\gamma)$ at $\gamma(s)$ is defined to be the circle in the osculating plane at $\gamma(s)$ that is tangent to $\operatorname{im}(\gamma)$ at $\gamma(s)$, has the same curvature as $\operatorname{im}(\gamma)$ at $\gamma(s)$, and lies toward the concave or inner side of $\operatorname{im}(\gamma)$. The osculating circle is centered at $\frac{1}{\kappa(s)} N(s)$ and has radius $\frac{1}{\kappa(s)}$ (recall that the osculating plane is a linear subspace, and not an affine plane containing $\gamma(s)$).

Exercise 5.56 (Geodesic). Let $I \subset \mathbf{R}$ denote an open interval, assume V to be a C^2 hypersurface in $\mathbf{R}^n$ for which a continuous choice $x \mapsto n(x)$ of a normal to V at $x \in V$ has been made, and let $\gamma : I \to V$ be a C^2 mapping. Then γ is said to be a *geodesic* in V if the acceleration $\gamma''(t) \in \mathbf{R}^n$ satisfies $\gamma''(t) \perp T_{\gamma(t)} V$, for all $t \in I$; that is, the curve γ has no acceleration in the direction of V, and therefore it goes "straight ahead" in V. In other words, $\gamma''(t)$ is a scalar multiple of the normal $(n \circ \gamma)(t)$ to V at $\gamma(t)$; and thus there is $\lambda(t) \in \mathbf{R}$ satisfying

$$\gamma''(t) = \lambda(t)\,(n \circ \gamma)(t) \qquad (t \in I).$$

(i) For every geodesic γ on the unit sphere S^{n-1} of dimension $n - 1$, show that there exist a number $a \in \mathbf{R}$ and an orthogonal pair of unit vectors v_1 and $v_2 \in \mathbf{R}^n$ such that $\gamma(t) = (\cos at)\, v_1 + (\sin at)\, v_2$. For $a \neq 0$ these are great circles on S^{n-1}.

(ii) Using $\gamma'(t) \in T_{\gamma(t)} V$, prove that we have, for a geodesic γ in V,

$$\lambda(t) = \langle\, \gamma'',\ (n \circ \gamma)\,\rangle(t) = -\langle\, \gamma',\ (n \circ \gamma)'\,\rangle(t) = -\langle\, \gamma',\ (Dn) \circ \gamma \cdot \gamma'\,\rangle(t);$$

and deduce the following identity in $C(I,\ \mathbf{R}^n)$:

$$\gamma'' + \langle\, \gamma',\ (Dn) \circ \gamma \cdot \gamma'\,\rangle (n \circ \gamma) = 0.$$

(iii) Verify that the equation from part (ii) is equivalent to the following system of second-order ordinary differential equations with variable coefficients:

$$\gamma_i'' + \sum_{1 \le j,\, k \le n} ((n_i\, D_j n_k) \circ \gamma)\, \gamma_j' \gamma_k' = 0 \qquad (1 \le i \le n).$$

Background. By the existence theorem for the solutions of such differential equations, for every point $x \in V$ and every vector $v \in T_x V$ there exists a geodesic $\gamma : I \to V$ with $0 \in I$ such that $\gamma(0) = x$ and $\gamma'(0) = v$; that is, through every point of V there is a geodesic passing in a prescribed direction.

(iv) Suppose $n = 3$. Show that any normal section of V is a geodesic.

Exercise 5.57 (Foucault's pendulum). The swing plane of an ideal pendulum with small oscillations suspended over a point of the Earth at latitude θ rotates uniformly about the direction of gravity by an angle of $-2\pi \sin\theta$ per day. This rotation is a consequence of the rotation of the Earth about its axis. This we shall prove in a number of steps.

We describe the surface of the Earth by the unit sphere S^2. The position of the swing plane is completely determined by the location x of the point of suspension and the unit tangent vector $X(x)$ (up to a sign) given by the swing direction; that is

$$(\star) \qquad x \in S^2 \qquad \text{and} \qquad X(x) \in T_x S^2 \subset \mathbf{R}^3, \qquad X(x) \in S^2.$$

Furthermore, after a suitable choice of signs we assume that $X : S^2 \to \mathbf{R}^3$ is a C^1 tangent vector field to S^2. Now suppose x is located on the parallel determined by a fixed $-\frac{\pi}{2} \le \theta \le \frac{\pi}{2}$. More precisely, we assume $x = \gamma(\alpha)$ at the time $\alpha \in \mathbf{R}$, where

$$\gamma : \mathbf{R} \to S^2 \qquad \text{and} \qquad \gamma(\alpha) = \phi(\alpha, \theta) = (\cos\alpha\cos\theta,\ \sin\alpha\cos\theta,\ \sin\theta).$$

In other words, the Earth is supposed to be stationary and the point of suspension to be moving with constant speed along the parallel. This angular motion being very slow, the force associated with the curvilinear motion and felt by the pendulum is negligible. Consequently, gravity is the only force felt by the pendulum and the sole cause of change in its swing direction. Therefore $(X \circ \gamma)'(\alpha) \in \mathbf{R}^3$ is perpendicular at $\gamma(\alpha)$ to the surface of the Earth, that is

$$(\star\star) \qquad (X \circ \gamma)'(\alpha) \perp T_{\gamma(\alpha)} S^2 \qquad (\alpha \in \mathbf{R}).$$

(i) Verify that the following two vectors in $\mathbf{R}^3$ form a basis for $T_{\gamma(\alpha)} S^2$, for $\alpha \in \mathbf{R}$:

$$\begin{aligned} v_1(\alpha) &= \frac{1}{\cos\theta}\frac{\partial\phi}{\partial\alpha}(\alpha,\theta) = (-\sin\alpha,\ \cos\alpha,\ 0) = \frac{1}{\cos\theta}\gamma'(\alpha), \\ v_2(\alpha) &= \frac{\partial\phi}{\partial\theta}(\alpha,\theta) = (-\cos\alpha\sin\theta,\ -\sin\alpha\sin\theta,\ \cos\theta). \end{aligned}$$

(ii) Prove that the functions $v_i : \mathbf{R} \to \mathbf{R}^3$ satisfy, for $1 \le i \le 2$,

$$\langle v_i, v_i \rangle = 1, \qquad \langle v_i', v_i \rangle = 0, \qquad \langle v_1, v_2 \rangle = 0,$$
$$\langle v_1', v_2 \rangle = -\langle v_1, v_2' \rangle = \sin\theta.$$

(iii) Deduce from ($\star$) and part (ii) that we can write

$$X \circ \gamma = (\cos\beta)\, v_1 + (\sin\beta)\, v_2 : \mathbf{R} \to \mathbf{R}^3,$$

where $\beta(\alpha)$ is the angle between the tangent vector $(X \circ \gamma)(\alpha)$ and the parallel.

(iv) Show

$$(X \circ \gamma)' = -(\beta' \sin\beta)\, v_1 + (\cos\beta)\, v_1' + (\beta' \cos\beta)\, v_2 + (\sin\beta)\, v_2'.$$

Now deduce from ($\star\star$) and part (ii) that $\beta : \mathbf{R} \to \mathbf{R}$ has to satisfy the differential equation

$$\beta'(\alpha) + \sin\theta = 0 \qquad (\alpha \in \mathbf{R}),$$

which has the solution

$$\beta(\alpha) - \beta(\alpha_0) = (\alpha_0 - \alpha)\sin\theta \qquad (\alpha \in \mathbf{R}).$$

In particular, $\beta(\alpha_0 + 2\pi) - \beta(\alpha_0) = -2\pi\, \sin\theta$.

Background. Angles are determined up to multiples of 2π. Therefore we may normalize the angle of rotation of the swing plane by the condition that the angle be small for small parallels; the angle then becomes $2\pi(1 - \sin\theta)$. According to Example 7.4.6 this is equal to the area of the cap of S^2 bounded by the parallel determined by θ. (See also Exercises 8.10 and 8.45.)

The directions of the swing planes moving along the parallel are said to be *parallel*, because the instantaneous change of such a direction has no component in the tangent plane to the sphere. More in general, let V be a manifold and $x \in V$. The element in the orthogonal group of the tangent space $T_x V$ corresponding to the change in direction of a vector in $T_x V$ caused by parallel transport from the point x along a closed curve in V is called the *holonomy* of V at x along the closed curve. This holonomy is determined by the *curvature form* of V (see Exercise 5.80).

Exercise 5.58 (Exponential of antisymmetric is special orthogonal in $\mathbf{R}^3$ – sequel to Exercises 4.22, 4.23 and 5.26 – needed for Exercises 5.59, 5.60, 5.67 and 5.68). The notation is that of Exercise 4.22. Let $a \in S^2 = \{ x \in \mathbf{R}^3 \mid \|a\| = 1 \}$, and define the *cross product operator* $r_a \in \mathrm{End}(\mathbf{R}^3)$ by

$$r_a\, x = a \times x.$$

(i) Verify that the matrix in $\mathrm{A}(3, \mathbf{R})$, the linear subspace in $\mathrm{Mat}(3, \mathbf{R})$ of the antisymmetric matrices, of r_a is given by (compare with Exercise 4.22.(iv))

$$\begin{pmatrix} 0 & -a_3 & a_2 \\ a_3 & 0 & -a_1 \\ -a_2 & a_1 & 0 \end{pmatrix}.$$

Prove by Grassmann's identity from Exercise 5.26.(ii), for $n \in \mathbf{N}$ and $x \in \mathbf{R}^3$,

$$r_a^{2n}\, x = (-1)^n (x - \langle a,\, x \rangle\, a), \qquad r_a^{2n+1}\, x = (-1)^n\, a \times x.$$

(ii) Let $\alpha \in \mathbf{R}$ with $0 \leq \alpha \leq \pi$. Show the following identity of mappings in $\mathrm{End}(\mathbf{R}^3)$ (see Example 2.4.10):

$$e^{\alpha\, r_a} = \sum_{n \in \mathbf{N}_0} \frac{\alpha^n}{n!} r_a^n : x \mapsto (1 - \cos\alpha)\langle a,\, x \rangle\, a + (\cos\alpha)\, x + (\sin\alpha)\, a \times x.$$

Prove $e^{\alpha\, r_a}\, a = a$. Select $x \in \mathbf{R}^3$ satisfying $\langle a, x \rangle = 0$ and $\|x\| = 1$ and show $\langle a, a \times x \rangle = 0$ and $\|a \times x\| = 1$. Deduce $e^{\alpha\, r_a} x = (\cos\alpha)\, x + (\sin\alpha)\, a \times x$. Use this to prove that

$$e^{\alpha\, r_a} = R_{\alpha, a},$$

the rotation in $\mathbf{R}^3$ by the angle α about the axis a. In other words, we have proved Euler's formula from Exercise 4.22.(iii). In particular, show that the exponential mapping $\mathrm{A}(3, \mathbf{R}) \to \mathbf{SO}(3, \mathbf{R})$ is surjective. Furthermore, verify the following identity of matrices:

$$\exp\alpha \begin{pmatrix} 0 & -a_3 & a_2 \\ a_3 & 0 & -a_1 \\ -a_2 & a_1 & 0 \end{pmatrix} = \cos\alpha\; I + (1 - \cos\alpha)\, aa^t + \sin\alpha\; r_a =$$

$$\begin{pmatrix} \cos\alpha + \;\; a_1^2\, c(\alpha) & -a_3 \sin\alpha + a_1 a_2\, c(\alpha) & a_2 \sin\alpha + a_1 a_3\, c(\alpha) \\ a_3 \sin\alpha + a_1 a_2\, c(\alpha) & \cos\alpha + \;\; a_2^2\, c(\alpha) & -a_1 \sin\alpha + a_2 a_3\, c(\alpha) \\ -a_2 \sin\alpha + a_1 a_3\, c(\alpha) & a_1 \sin\alpha + a_2 a_3\, c(\alpha) & \cos\alpha + \;\; a_3^2\, c(\alpha) \end{pmatrix},$$

where $c(\alpha) = 1 - \cos\alpha$.

(iii) Introduce $\boldsymbol{p} = (p_0, p) \in \mathbf{R} \times \mathbf{R}^3 \simeq \mathbf{R}^4$ by (compare with Exercise 5.65.(ii))

$$p_0 = \cos\frac{\alpha}{2}, \qquad p = \sin\frac{\alpha}{2}\, a.$$

Verify that $\boldsymbol{p} \in S^3 = \{\, \boldsymbol{p} \in \mathbf{R}^4 \mid p_0^2 + \|p\|^2 = 1 \,\}$. Prove

$$R_{\alpha,a} = R_{\boldsymbol{p}} : x \mapsto 2\langle p, x\rangle\, p + (p_0^2 - \|p\|^2)\, x + 2 p_0\, p \times x \qquad (x \in \mathbf{R}^3),$$

and also that we have the following *rational parametrization* of a matrix in $\mathbf{SO}(3, \mathbf{R})$ (compare with Exercise 5.65.(iv)):

$$R_{\alpha,a} = R_{\boldsymbol{p}} = (p_0^2 - \|p\|^2) I + 2(pp^t + p_0 r_p)$$

$$= \begin{pmatrix} p_0^2 + p_1^2 - p_2^2 - p_3^2 & 2(p_1 p_2 - p_0 p_3) & 2(p_1 p_3 + p_0 p_2) \\ 2(p_2 p_1 + p_0 p_3) & p_0^2 - p_1^2 + p_2^2 - p_3^2 & 2(p_2 p_3 - p_0 p_1) \\ 2(p_3 p_1 - p_0 p_2) & 2(p_3 p_2 + p_0 p_1) & p_0^2 - p_1^2 - p_2^2 + p_3^2 \end{pmatrix}.$$

Conversely, given $\boldsymbol{p} \in S^3$ we can reconstruct (α, a) with $R_{\alpha,a} = R_{\boldsymbol{p}}$ by means of

$$\alpha = 2 \arccos p_0 = \arccos(2p_0^2 - 1), \qquad a = \mathrm{sgn}(p_0)\frac{1}{\|p\|}p.$$

Note that $\boldsymbol{p}$ and $-\boldsymbol{p}$ give the same result. The calculation of $\boldsymbol{p} \in S^3$ from the matrix $R_{\boldsymbol{p}}$ above follows by means of

$$4p_0^2 = 1 + \mathrm{tr}\, R_{\boldsymbol{p}}, \qquad 4p_0 r_p = R_{\boldsymbol{p}} - R_{\boldsymbol{p}}^t.$$

If $p_0 = 0$, we use (write $R_{\boldsymbol{p}} = (R_{ij})$)

$$2p_i^2 = 1 + R_{ii}, \qquad 2p_i p_j = R_{ij} \qquad (1 \le i, j \le 3,\ i \ne j).$$

Because at least one $p_i \ne 0$, this means that $(p_0, p) \in \mathbf{R}^4$ is determined by $R_{\boldsymbol{p}}$ to within a factor ± 1.

(iv) Consider $A = (a_{ij}) \in \mathbf{SO}(3, \mathbf{R})$. Furthermore, let $A_{ij} \in \mathrm{Mat}(n-1, \mathbf{R})$ be the matrix obtained from A by deleting the i-th row and the j-th column, then $\det A_{ij}$ is the (i, j)-th minor of A. Prove that $a_{ij} = (-1)^{i+j} \det A_{ij}$ and verify this property for (some choice of) the matrix coefficients of $R_{\boldsymbol{p}}$ from part (iii).
Hint: Note that $A^t = A^\sharp$, in the notation of Cramer's rule (2.6).

Background. In the terminology of Section 5.9 we have now proved that the cross product operator $r_a \in \mathrm{A}(3, \mathbf{R})$ is the *infinitesimal generator* of the *one-parameter group of rotations* $\alpha \mapsto R_{\alpha,a}$ in $\mathbf{SO}(3, \mathbf{R})$. Moreover, we have given a solution of the differential equation in Formula (5.27) for a rotation, compare with Example 5.9.3.

Exercise 5.59 (Lie algebra of SO(3, R) – sequel to Exercises 5.26 and 5.58 – needed for Exercise 5.60). We use the notations from Exercise 5.58. In particular $\mathrm{A}(3, \mathbf{R}) \subset \mathrm{Mat}(3, \mathbf{R})$ is the linear subspace of the antisymmetric 3×3 matrices. We recall the Lie brackets $[\,\cdot, \cdot\,]$ on $\mathrm{Mat}(3, \mathbf{R})$ given by $[A_1, A_2] = A_1A_2 - A_2A_1 \in \mathrm{Mat}(3, \mathbf{R})$, for $A_1, A_2 \in \mathrm{Mat}(3, \mathbf{R})$ (compare with Exercise 2.41).

(i) Prove that $[\,\cdot, \cdot\,]$ induces a well-defined multiplication on $\mathrm{A}(3, \mathbf{R})$, that is $[A_1, A_2] \in \mathrm{A}(3, \mathbf{R})$ for $A_1, A_2 \in \mathrm{A}(3, \mathbf{R})$. Deduce that $\mathrm{A}(3, \mathbf{R})$ satisfies the definition of a Lie algebra.

(ii) Check that the mapping $a \mapsto r_a : \mathbf{R}^3 \to \mathrm{A}(3, \mathbf{R})$ is a linear isomorphism. Prove by Grassmann's identity from Exercise 5.26.(ii) that

$$r_{a \times b} = [r_a, r_b] \qquad (a, b \in \mathbf{R}^3).$$

Deduce that $a \mapsto r_a : \mathbf{R}^3 \to \mathrm{A}(3, \mathbf{R})$ is an isomorphism of Lie algebras between $(\mathbf{R}^3, \times)$ and $(\mathrm{A}(3, \mathbf{R}), [\,\cdot, \cdot\,])$, compare with Exercise 5.67.(ii).

From Example 4.6.2 we know that $\mathbf{SO}(3, \mathbf{R})$ is a C^∞ submanifold in $\mathbf{GL}(3, \mathbf{R})$ of dimension 3, while $\mathbf{SO}(3, \mathbf{R})$ is a group under matrix multiplication. Accordingly, $\mathbf{SO}(3, \mathbf{R})$ is a linear Lie group. Given $r_a \in \mathrm{A}(3, \mathbf{R})$, the mapping

$$\alpha \mapsto e^{\alpha\, r_a} : \mathbf{R} \to \mathbf{SO}(3, \mathbf{R})$$

is a C^∞ curve in the linear Lie group $\mathbf{SO}(3, \mathbf{R})$ through I (for $\alpha = 0$), and this curve has $r_a \in \mathrm{A}(3, \mathbf{R})$ as its tangent vector at the point I.

(iii) Show

$$\left.\frac{d}{ds}\right|_{s=0} (e^{sX})^t\, e^{sX} = X^t + X \qquad (X \in \mathrm{Mat}(n, \mathbf{R})).$$

Prove that $(\mathrm{A}(3, \mathbf{R}),\ [\,\cdot, \cdot\,]) \simeq (\mathbf{R}^3, \times)$ is the Lie algebra of $\mathbf{SO}(3, \mathbf{R})$.

Exercise 5.60 (Action of SO(3, R) on $C^\infty(\mathbf{R}^3)$ – sequel to Exercises 2.41, 3.9 and 5.59 – needed for Exercise 5.61). We use the notations from Exercise 5.59. We define an *action* of $\mathbf{SO}(3, \mathbf{R})$ on $C^\infty(\mathbf{R}^3)$ by assigning to $R \in \mathbf{SO}(3, \mathbf{R})$ and $f \in C^\infty(\mathbf{R}^3)$ the function

$$Rf \in C^\infty(\mathbf{R}^3) \qquad \text{given by} \qquad (Rf)(x) = f(R^{-1}x) \qquad (x \in \mathbf{R}^3).$$

(i) Check that this leads to a group action, that is $(RR')f = R(R'f)$, for R and $R' \in \mathbf{SO}(3, \mathbf{R})$ and $f \in C^\infty(\mathbf{R}^3)$.

For every $a \in \mathbf{R}^3$ we have the induced action ∂_{r_a} of the infinitesimal generator r_a in $\mathrm{A}(3, \mathbf{R})$ on $C^\infty(\mathbf{R}^3)$.

(ii) Prove the following, where L is the angular momentum operator from Exercise 2.41:

$$\partial_{r_a} f(x) = \langle a, Lf(x)\rangle \qquad (a \in \mathbf{R}^3,\ f \in C^\infty(\mathbf{R}^3)).$$

In particular one has $\partial_{r_{e_j}} = L_j$, with $e_j \in \mathbf{R}^3$ the standard basis vectors, for $1 \le j \le 3$.

(iii) Assume $f \in C^\infty(\mathbf{R}^3)$ is a function of the distance to the origin only, that is, f is invariant under the action of $\mathbf{SO}(3, \mathbf{R})$. Prove by the use of part (ii) that $Lf = 0$. (See Exercise 3.9.(ii) for the reverse of this property.)

(iv) Check that from part (ii) follows, by Taylor expansion with respect to the variable $\alpha \in \mathbf{R}$,

$$R_{\alpha,a} f = f + \langle \alpha a, Lf\rangle + \mathcal{O}(\alpha^2), \quad \alpha \to 0 \qquad (f \in C^\infty(\mathbf{R}^3)).$$

Background. For this reason the angular momentum operator L is known in *quantum physics* as the *infinitesimal generator* of the action of $\mathbf{SO}(3, \mathbf{R})$ on $C^\infty(\mathbf{R}^3)$.

(v) Prove by Exercise 5.59.(ii) the following *commutator relation*:

$$\partial_{r_{a\times b}} = [\,\partial_{r_a},\ \partial_{r_b}\,] = [\,\langle a, Lf\rangle,\ \langle b, Lf\rangle\,].$$

In other words, we have a homomorphism of Lie algebras

$$\mathbf{R}^3 \to \operatorname{End}(C^\infty(\mathbf{R}^3)) \qquad \text{with} \qquad a \mapsto \partial_{r_a} \mapsto \langle a, L\rangle = \sum_{1\le j\le 3} a_j L_j.$$

Hint: $\partial_{r_a}(\partial_{r_b} f) = \sum_{1\le j\le 3} a_j L_j(\partial_{r_b} f) = \sum_{1\le j,\,k\le 3} a_j b_k L_j L_k f.$

Exercise 5.61 (Action of SO(3, R) on harmonic homogeneous polynomials – sequel to Exercises 2.39, 2.41, 3.17 and 5.60). Let $l \in \mathbf{N}_0$ and let $\mathcal{H}_l$ be the linear space over $\mathbf{C}$ of the harmonic homogeneous polynomials on $\mathbf{R}^3$ of degree l, that is, for $p \in \mathcal{H}_l$ and $x \in \mathbf{R}^3$,

$$p(x) = \sum_{a+b+c=l} p_{abc} x_1^a x_2^b x_3^c, \qquad p_{abc} \in \mathbf{C}, \qquad \Delta p = 0.$$

(i) Show

$$\dim_{\mathbf{C}} \mathcal{H}_l = 2l + 1.$$

Hint: Note that $p \in \mathcal{H}_l$ can be written as

$$p(x) = \sum_{0\le j\le l} \frac{x_1^j}{j!} p_j(x_2, x_3) \qquad (x \in \mathbf{R}^3),$$

where p_j is a homogeneous polynomial of degree $l - j$ on $\mathbf{R}^2$. Verify that $\Delta p = 0$ if and only if

$$p_{j+2} = -\Big(\frac{\partial^2 p_j}{\partial x_2^2} + \frac{\partial^2 p_j}{\partial x_3^2}\Big) \qquad (0 \le j \le l - 2).$$

Therefore p is completely determined by the homogeneous polynomials p_0 of degree l and p_1 of degree $l - 1$ on $\mathbf{R}^2$.

(ii) Prove by Exercise 2.39.(v) that $\mathcal{H}_l$ is invariant under the restriction to $\mathcal{H}_l$ of the action of $\mathbf{SO}(3, \mathbf{R})$ on $C^\infty(\mathbf{R}^3)$ defined in Exercise 5.60.

We now want to prove that $\mathcal{H}_l$ is also invariant under the induced action of the angular momentum operators L_j, for $1 \le j \le 3$, and therefore also under the action on $\mathcal{H}_l$ of the differential operators H, X and Y from Exercise 2.41.(iii).

(iii) For arbitrary $R \in \mathbf{SO}(3, \mathbf{R})$ and $p \in \mathcal{H}_l$, verify that each coefficient of the polynomial $Rp \in \mathcal{H}_l$ is a polynomial function of the matrix coefficients of R. Now

$$\langle p, q \rangle = \sum_{a+b+c=l} p_{abc}\overline{q_{abc}}$$

defines an inner product on $\mathcal{H}_l$. Choose an orthonormal basis $(p_0, \dots, p_{2l})$ for $\mathcal{H}_l$, and write $Rp = \sum_j \lambda_j p_j$ with coefficients $\lambda_j \in \mathbf{C}$. Show that every λ_j is a polynomial function of the matrix coefficients of R.

(iv) Conclude by part (iii) that $\mathcal{H}_l$ is invariant under the action on $\mathcal{H}_l$ of the differential operators H, X and Y.

(v) Check that $p_l(x) := (x_1 + i x_2)^l \in \mathcal{H}_l$, while

$$H\, p_l = 2l\, p_l, \qquad X\, p_l = 0, \qquad Y\, p_l(x) = -2il\, x_3(x_1 + i x_2)^{l-1} \in \mathcal{H}_l.$$

Prove that, in the notation of Exercise 3.17,

$$p_l(\cos\alpha\cos\theta,\ \sin\alpha\cos\theta,\ \sin\theta) = e^{il\alpha}\cos^l\theta = \frac{2^l l!}{(2l)!}\, Y_l^l(\alpha, \theta).$$

Now let

$$\rho : p \mapsto p|_{S^2} : \mathcal{H}_l \to \mathcal{Y}$$

be the linear mapping of $\mathcal{H}_l$ into the space $\mathcal{Y}$ of continuous functions on the unit sphere $S^2 \subset \mathbf{R}^3$ defined by restriction of the function p on $\mathbf{R}^3$ to S^2. We want to prove that ρ followed by the identification ι^{-1} from Exercise 3.17 gives a linear isomorphism between $\mathcal{H}_l$ and the linear space $\mathcal{Y}_l$ of spherical functions from Exercise 3.17.

(vi) Prove that the restriction operator commutes with the actions of $\mathbf{SO}(3, \mathbf{R})$ on $\mathcal{H}_l$ and $\mathcal{Y}$, that is

$$\rho \circ R = R \circ \rho \qquad (R \in \mathbf{SO}(3, \mathbf{R})).$$

Conclude by differentiating that, for Y^* as in Exercise 3.9.(iii),

$$\iota^{-1} \circ \rho \circ Y = Y^* \circ \iota^{-1} \circ \rho.$$

(vii) Verify that part (v) now gives $\frac{(2l)!}{2^l l!}\,(\iota^{-1} \circ \rho)\, p_l = Y_l^l$. Deduce from this, using part (vi),

$$\frac{(2l)!}{2^l l!\, j!}\,(\iota^{-1} \circ \rho)(Y^j\, p_l) = \frac{1}{j!}(Y^*)^j\, Y_l^l = V_j \qquad (0 \le j \le 2l),$$

where $(V_0, \dots, V_{2l})$ forms the basis for $\mathcal{Y}_l$ from Exercise 3.17. In view of $\dim \mathcal{H}_l = \dim \mathcal{Y}_l$, conclude that $\iota^{-1} \circ \rho : \mathcal{H}_l \to \mathcal{Y}_l$ is a linear isomorphism. Consequently, the extension operator, described in Exercise 3.17.(iv), also is a linear isomorphism.

Background. According to Exercise 3.17.(iv), a function in $\mathcal{Y}_l$ is the restriction of a harmonic function on $\mathbf{R}^3$ which is homogeneous of degree l; however, this result does not yield that the function on $\mathbf{R}^3$ is a polynomial.

Exercise 5.62 (SL(2, R) – sequel to Exercises 2.44 and 4.24). We use the notation of the latter exercise, in particular, from its part (ii) we know that $\mathbf{SL}(n, \mathbf{R}) = \{ A \in \mathrm{Mat}(n, \mathbf{R}) \mid \det A = 1 \}$ is a C^∞ submanifold in $\mathrm{Mat}(n, \mathbf{R})$.

(i) Show that $\mathbf{SL}(n, \mathbf{R})$ is a linear Lie group, and use Exercise 2.44.(ii) to prove that its Lie algebra is equal to $\mathfrak{sl}(n, \mathbf{R}) := \{ X \in \mathrm{Mat}(n, \mathbf{R}) \mid \mathrm{tr}\, X = 0 \}$.

From now on we assume $n = 2$. Define $A_i(t) \in \mathbf{SL}(2, \mathbf{R})$, for $1 \le i \le 3$ and $t \in \mathbf{R}$, by

$$A_1(t) = \begin{pmatrix} 1 & t \\ 0 & 1 \end{pmatrix}, \qquad A_2(t) = \begin{pmatrix} e^t & 0 \\ 0 & e^{-t} \end{pmatrix}, \qquad A_3(t) = \begin{pmatrix} 1 & 0 \\ t & 1 \end{pmatrix}.$$

(ii) Verify that every $t \mapsto A_i(t)$ is a one-parameter subgroup and also a C^∞ curve in $\mathbf{SL}(2, \mathbf{R})$ for $1 \le i \le 3$. Define $X_i = A_i'(0) \in \mathfrak{sl}(2, \mathbf{R})$, for $1 \le i \le 3$. Show

$$X := X_1 = \begin{pmatrix} 0 & 1 \\ 0 & 0 \end{pmatrix}, \quad H := X_2 = \begin{pmatrix} 1 & 0 \\ 0 & -1 \end{pmatrix}, \quad Y := X_3 = \begin{pmatrix} 0 & 0 \\ 1 & 0 \end{pmatrix},$$

and that $A_i(t) = e^{tX_i}$, for $1 \le i \le 3$, $t \in \mathbf{R}$. Prove (compare with Exercise 2.41.(iii))

$$[\,H, X\,] = 2X, \qquad [\,H, Y\,] = -2Y, \qquad [\,X, Y\,] = H.$$

Denote by V_l the linear space of polynomials on $\mathbf{R}$ of degree $\le l \in \mathbf{N}$. Observe that V_l is linearly isomorphic with $\mathbf{R}^{l+1}$. Define, for $f \in V_l$ and $x \in \mathbf{R}$,

$$(\Phi(A)f)(x) = (cx+d)^l f\Big(\frac{ax+b}{cx+d}\Big) \qquad \Big(A = \begin{pmatrix} a & c \\ b & d \end{pmatrix} \in \mathbf{SL}(2, \mathbf{R})\Big).$$

(iii) Prove that $\Phi(A)f \in V_l$. Show that $\Phi : \mathbf{SL}(2, \mathbf{R}) \to \mathrm{End}(V_l)$ is a homomorphism of groups, that is, $\Phi(AB) = \Phi(A)\Phi(B)$ for A and $B \in \mathbf{SL}(2, \mathbf{R})$, and conclude that $\Phi(A) \in \mathrm{Aut}(V_l)$. Verify that $\Phi : \mathbf{SL}(2, \mathbf{R}) \to \mathrm{Aut}(V_l)$ is injective.

(iv) Deduce from (ii) and (iii) that we have one-parameter groups $(\Phi_i^t)_{t\in\mathbf{R}}$ of diffeomorphisms of V_l if $\Phi_i^t = \Phi \circ A_i(t)$. Verify, for t and $x \in \mathbf{R}$ and $f \in V_l$,

$$(\Phi_1^t f)(x) = (tx+1)^l f\Big(\frac{x}{tx+1}\Big), \qquad (\Phi_2^t f)(x) = e^{-lt} f(e^{2t}x),$$

$$(\Phi_3^t f)(x) = f(x+t).$$

Let $\phi_i \in \mathrm{End}(V_l)$ be the infinitesimal generator of $(\Phi_i^t)_{t\in\mathbf{R}}$, for $1 \le i \le 3$, and write $X_l = \phi_1$, $H_l = \phi_2$ and $Y_l = \phi_3$. Prove, for $f \in V_l$ and $x \in \mathbf{R}$,

$$(X_l f)(x) = \Big(lx - x^2\frac{d}{dx}\Big)f(x), \qquad (H_l f)(x) = \Big(2x\frac{d}{dx} - l\Big)f(x),$$

$$(Y_l f)(x) = \frac{d}{dx}f(x).$$

Verify that X_l, H_l and $Y_l \in \mathrm{End}(V_l)$ satisfy the same commutator relations as X, H and Y in (ii) (which is no surprise in view of (iii) and Theorem 5.10.6.(iii)).

Set

$$f_j(x) := \frac{1}{j!}(Y_l^j f_0)(x) = \binom{l}{j}x^{l-j} \qquad (0 \le j \le l).$$

(v) Determine the matrices in $\mathrm{Mat}(l+1, \mathbf{R})$ for H_l, X_l and Y_l with respect to the basis $\{ f_0, \dots, f_l \}$ for V_l. First, show, for $0 \le j \le l$,

$$H_l f_j = (l-2j)f_j, \qquad X_l f_j = (l-(j-1))f_{j-1} \quad (j \ne 0),$$

$$Y_l f_j = (j+1)f_{j+1} \quad (j \ne l),$$

while $X_l f_0 = Y_l f_l = 0$; and conclude that X_l and Y_l are given by, respectively,

$$\begin{pmatrix} 0 & l & 0 & \cdots & & 0 \\ 0 & 0 & l-1 & \ddots & & 0 \\ \vdots & & & & & \vdots \\ 0 & & & 0 & 2 & 0 \\ 0 & & & 0 & 0 & 1 \\ 0 & & \cdots & 0 & 0 & 0 \end{pmatrix}, \qquad \begin{pmatrix} 0 & 0 & 0 & \cdots & & 0 \\ 1 & 0 & 0 & & & 0 \\ 0 & 2 & 0 & & & 0 \\ \vdots & & & & & \vdots \\ 0 & & \ddots & l-1 & 0 & 0 \\ 0 & & \cdots & 0 & l & 0 \end{pmatrix}.$$

(vi) Prove that the *Casimir operator* (see Exercise 2.41.(vi)) acts as a scalar operator

$$C := \frac{1}{2}(X_lY_l + Y_lX_l + \frac{1}{2}H_l^2) = Y_lX_l + \frac{1}{2}H_l + \frac{1}{4}H_l^2 = \frac{l}{2}\Big(\frac{l}{2}+1\Big)I.$$

(vii) Verify that the formulae in part (v) also can be obtained using induction over $0 \le j \le l$ and the commutator relations satisfied by H_l, X_l and Y_l.
Hint: $jf_j = Y_l f_{j-1}$ and $H_lY_l = H_lY_l - Y_lH_l + Y_lH_l = Y_l(-2 + H_l)$, similarly $X_lY_l = H_l + Y_lX_l$, etc.

Background. The linear space V_l is said to be an *irreducible representation* of the linear Lie group $\mathbf{SL}(2, \mathbf{R})$ and of its Lie algebra $\mathfrak{sl}(2, \mathbf{R})$. This is because V_l is invariant under the action of the elements of $\mathbf{SL}(2, \mathbf{R})$ and under the action of the differential operators H_l, X_l and $Y_l \in \mathrm{End}(V_l)$, whereas every nontrivial subspace is not. Furthermore, V_l is a representation of *highest weight* l, because $H_l f_j = (l - 2j) f_j$ for $0 \le j \le l$, thus l is the largest among the eigenvalues of $H_l \in \mathrm{End}^+(V_l)$. Note that X_l is a *raising operator*, since it maps eigenvectors f_j of H_l to eigenvectors f_{j-1} of H_l corresponding to a larger eigenvalue, and it annihilates the basis vector f_0 with the highest eigenvalue l of H_l; that Y_l is a *lowering operator*; and that the set of eigenvalues of H_l is invariant under the symmetry $s \mapsto -s$ of $\mathbf{Z}$. Furthermore, we have $X_1 = X$, $H_1 = H$ and $Y_1 = Y$. Except for isomorphy, the irreducible representation V_l is uniquely determined by the highest weight $l \in \mathbf{N}_0$, and if l varies we obtain all irreducible representations of $\mathbf{SL}(2, \mathbf{R})$. See Exercises 3.17 and 5.61 for related results. The resemblance between the two cases, that of $\mathbf{SO}(3, \mathbf{R})$ and $\mathbf{SL}(2, \mathbf{R})$, is no coincidence; yet, there is the difference that for $\mathbf{SO}(3, \mathbf{R})$ the highest weight only assumes values in $2\mathbf{N}_0$.

Exercise 5.63 (Derivative of exponential mapping). Recall from Example 2.4.10 the formula

$$\frac{d}{dt} e^{tA} = e^{tA} A = A e^{tA} \qquad (t \in \mathbf{R},\ A \in \mathrm{End}(\mathbf{R}^n)).$$

Conclude, for $t \in \mathbf{R}$, A and $H \in \mathrm{End}(\mathbf{R}^n)$,

$$\frac{d}{dt}(e^{-tA} e^{t(A+H)}) = e^{-tA}(-A + (A+H)) e^{t(A+H)} = e^{-tA} H e^{t(A+H)}.$$

Apply the Fundamental Theorem of Integral Calculus 2.10.1 on $\mathbf{R}$ to obtain

$$e^{-A} e^{A+H} - I = \int_0^1 e^{-tA} H e^{t(A+H)}\, dt,$$

whence

$$e^{A+H} - e^A = \int_0^1 e^{(1-t)A} H e^{t(A+H)}\, dt.$$

Using the estimate $\|e^{A+H}\| \le e^{\|A+H\|}$ from Example 2.4.10 deduce from the latter equality the existence of a constant $c = c(A) > 0$ such that for $H \in \mathrm{End}(\mathbf{R}^n)$ with $\|H\|_{\mathrm{Eucl}} \le 1$ we have

$$\|e^{A+H} - e^A\| \le c(A) \|H\|.$$

Next, apply this estimate in order to replace the operator $e^{t(A+H)}$ in the integral above by e^{tA} plus an error term which is $\mathcal{O}(\|H\|)$, and conclude that exp is differentiable at $A \in \mathrm{End}(\mathbf{R}^n)$ with derivative

$$D\exp(A)H = \int_0^1 e^{(1-t)A} H e^{tA}\, dt \qquad (H \in \mathrm{End}(\mathbf{R}^n)).$$

In turn, this formula and the differentiability of exp can be used to show that the higher-order derivatives of exp exist at A. In other words, $\exp : \mathrm{End}(\mathbf{R}^n) \to \mathrm{Aut}(\mathbf{R}^n)$ is a C^∞ mapping. Furthermore, using the adjoint mappings from Section 5.10 rewrite the formula for the derivative as follows:

$$\begin{aligned} D\exp(A)H &= e^A \int_0^1 \mathrm{Ad}(e^{-tA})H\,dt = e^A \int_0^1 e^{-t\,\mathrm{ad}\,A}\,dt\,H \\ &= e^A \left[-\frac{e^{-t\,\mathrm{ad}\,A}}{\mathrm{ad}\,A}\right]_0^1 H = e^A \circ \frac{I - e^{-\,\mathrm{ad}\,A}}{\mathrm{ad}\,A} H \\ &= e^A \circ \sum_{k\in\mathbf{N}_0} \frac{(-1)^k}{(k+1)!}(\mathrm{ad}\,A)^k\,H. \end{aligned}$$

A proof of this formula without integration runs as follows. For $i \in \mathbf{N}_0$ and $A \in \mathrm{End}(\mathbf{R}^n)$ define

$$P_i,\ R_A : \mathrm{End}(\mathbf{R}^n) \to \mathrm{End}(\mathbf{R}^n) \qquad \text{by} \qquad P_i H = H^i, \qquad R_A H = HA.$$

Note that $\exp = \sum_{i\in\mathbf{N}_0} \frac{1}{i!} P_i$, that L_A and R_A commute, and that $\mathrm{ad}\,A = L_A - R_A$. Then

$$\begin{aligned} DP_i(A)H &= \left.\frac{d}{dt}\right|_{t=0} (A+tH)(A+tH)\cdots(A+tH) = \sum_{0\le j<i} A^{i-1-j} H A^j \\ &= \sum_{0\le j<i} L_A^{i-1-j} R_A^j H \qquad (H \in \mathrm{End}(\mathbf{R}^n)). \end{aligned}$$

But

$$R_A^j = (L_A - \mathrm{ad}\,A)^j = \sum_{0\le k\le j} (-1)^k \binom{j}{k} L_A^{j-k} (\mathrm{ad}\,A)^k.$$

Hence

$$\begin{aligned} DP_i(A) &= \sum_{0\le j<i}\ \sum_{0\le k\le j} (-1)^k \binom{j}{k} L_A^{i-1-k} (\mathrm{ad}\,A)^k \\ &= \sum_{0\le k<i} \Big(\sum_{k\le j<i} \binom{j}{k} \Big) (-1)^k L_A^{i-1-k} (\mathrm{ad}\,A)^k \\ &= \sum_{0\le k<i} \binom{i}{k+1} (-1)^k L_A^{i-1-k} (\mathrm{ad}\,A)^k. \end{aligned}$$

For the second equality interchange the order of summation and for the third use induction over i in order to prove the formula for the summation over j. Now this implies

$$\begin{aligned} D\exp(A) &= \sum_{i\in\mathbf{N}_0} \frac{1}{i!} DP_i(A) = \sum_{i\in\mathbf{N}_0} \frac{1}{i!} \sum_{0\le k<i} \binom{i}{k+1} (-1)^k L_A^{i-1-k} (\operatorname{ad} A)^k \\ &= \sum_{i\in\mathbf{N}_0} \sum_{0\le k<i} \frac{(-1)^k}{(k+1)!} (\operatorname{ad} A)^k \frac{1}{(i-1-k)!} L_A^{i-1-k} \\ &= \sum_{k\in\mathbf{N}_0} \frac{(-1)^k}{(k+1)!} (\operatorname{ad} A)^k \sum_{k+1\le i} \frac{1}{(i-1-k)!} L_A^{i-1-k} \\ &= \frac{1-e^{-\operatorname{ad} A}}{\operatorname{ad} A} \sum_{i\in\mathbf{N}_0} \frac{1}{i!} L_{A^i} = e^A \circ \frac{1-e^{-\operatorname{ad} A}}{\operatorname{ad} A}. \end{aligned}$$

Here the fourth equality arises from interchanging the order of summation.

Exercise 5.64 (Closed subgroup is a linear Lie group). In Theorem 5.10.2.(i) we saw that a linear Lie group is a closed subset of $\mathbf{GL}(n, \mathbf{R})$. We now prove the converse statement. Let $G \subset \mathbf{GL}(n, \mathbf{R})$ be a closed subgroup and define

$$\overline{\mathfrak{g}} = \{\, X \in \operatorname{Mat}(n, \mathbf{R}) \mid e^{tX} \in G, \text{ for all } t \in \mathbf{R} \,\}.$$

Then G is a submanifold of $\operatorname{Mat}(n, \mathbf{R})$ at I with $T_I G = \overline{\mathfrak{g}}$; more precisely, G is a linear Lie group with Lie algebra $\overline{\mathfrak{g}}$.

(i) Use Lie's product formula from Proposition 5.10.7.(iii) and the closedness of G to show that $\overline{\mathfrak{g}}$ is a linear subspace of $\operatorname{Mat}(n, \mathbf{R})$.

(ii) Prove that $Y \in \operatorname{Mat}(n, \mathbf{R})$ belongs to $\overline{\mathfrak{g}}$ if there exist sequences $(Y_k)_{k\in\mathbf{N}}$ in $\operatorname{Mat}(n, \mathbf{R})$ and $(t_k)_{k\in\mathbf{N}}$ in $\mathbf{R}$, such that

$$e^{Y_k} \in G, \qquad \lim_{k\to\infty} Y_k = 0, \qquad \lim_{k\to\infty} t_k Y_k = Y.$$

Hint: Let $t \in \mathbf{R}$. For every $k \in \mathbf{N}$ select $m_k \in \mathbf{Z}$ with $|t\, t_k - m_k| < 1$. Then, if $\|\cdot\|$ denotes the Euclidean norm on $\operatorname{Mat}(n, \mathbf{R})$,

$$\|m_k Y_k - tY\| \le \|(m_k - t\,t_k) Y_k\| + \|t\,t_k Y_k - tY\| \le \|Y_k\| + |t|\,\|t_k Y_k - Y\|.$$

Further, note $e^{m_k Y_k} \in G$ and use that G is closed.

Select a linear subspace $\mathfrak{h}$ in $\operatorname{Mat}(n, \mathbf{R})$ complementary to $\overline{\mathfrak{g}}$, that is, $\overline{\mathfrak{g}} \oplus \mathfrak{h} = \operatorname{Mat}(n, \mathbf{R})$, and define

$$\Phi : \overline{\mathfrak{g}} \times \mathfrak{h} \to \mathbf{GL}(n, \mathbf{R}) \qquad \text{by} \qquad \Phi(X, Y) = e^X e^Y.$$

(iii) Use the additivity of $D\Phi(0,0)$ to prove $D\Phi(0,0)(X,Y) = X + Y$, for $X \in \overline{\mathfrak{g}}$ and $Y \in \mathfrak{h}$, and deduce that Φ is a local diffeomorphism onto an open neighborhood of I in $\mathbf{GL}(n, \mathbf{R})$.

(iv) Show that $\exp \overline{\mathfrak{g}}$ is a neighborhood of I in G.
Hint: If not, it is possible to choose $X_k \in \overline{\mathfrak{g}}$ and $Y_k \in \mathfrak{h}$ satisfying

$$e^{X_k} e^{Y_k} \in G, \qquad Y_k \neq 0, \qquad \lim_{k\to\infty} (X_k, Y_k) = (0, 0).$$

Prove $e^{Y_k} \in G$. Next, consider $\overline{Y_k} = \frac{1}{\|Y_k\|} Y_k$. By compactness of the unit sphere in $\mathrm{Mat}(n, \mathbf{R})$ and by going over to a subsequence we may assume that there is $Y \in \mathrm{Mat}(n, \mathbf{R})$ with $\lim_{k\to\infty} \overline{Y_k} = Y$. Hence we have obtained sequences $(Y_k)_{k\in\mathbf{N}}$ in $\mathfrak{h}$ and $(\frac{1}{\|Y_k\|})_{k\in\mathbf{N}}$ in $\mathbf{R}$, such that

$$e^{Y_k} \in G, \qquad \lim_{k\to\infty} Y_k = 0, \qquad \lim_{k\to\infty} \frac{1}{\|Y_k\|} Y_k = Y \in \mathfrak{h}.$$

From part (ii) obtain $Y \in \overline{\mathfrak{g}}$ and conclude $Y = 0$, which is a contradiction.

(v) Deduce from part (iv) that G is a linear Lie group.

Exercise 5.65 (Reflections, rotations and Hamilton's Theorem – sequel to Exercises 2.5, 4.22, 5.26 and 5.27 – needed for Exercises 5.66, 5.67 and 5.72). For

$$\boldsymbol{p} = (p_0, p) \in S^3 = \{\, (p_0, p) \in \mathbf{R} \times \mathbf{R}^3 \simeq \mathbf{R}^4 \mid p_0^2 + \|p\|^2 = 1 \,\}$$

we define

$$R_{\boldsymbol{p}} \in \mathrm{End}(\mathbf{R}^3) \qquad \text{by} \qquad R_{\boldsymbol{p}}\, x = 2\langle p, x\rangle\, p + (p_0^2 - \|p\|^2)\, x + 2p_0\, p \times x.$$

Note that $R_{\boldsymbol{p}} = R_{-\boldsymbol{p}}$, so we may assume $p_0 \geq 0$.

(i) Suppose $p_0 = 0$. Show $-R_{\boldsymbol{p}}$ is the *reflection* of $\mathbf{R}^3$ in the two-dimensional linear subspace $N_p = \{\, x \in \mathbf{R}^3 \mid \langle x, p\rangle = 0 \,\}$.

Verify that $R_{\boldsymbol{p}} \in \mathbf{SO}(\mathbf{R}^3)$ in the following two ways. Conversely, prove that every element in $\mathbf{SO}(\mathbf{R}^3)$ is of the form $R_{\boldsymbol{p}}$ for some $\boldsymbol{p} \in S^3$.

(ii) Note that $(p_0^2 - \|p\|^2)^2 + (2p_0\|p\|)^2 = 1$. Therefore we can find $0 \leq \alpha \leq \pi$ such that $p_0^2 - \|p\|^2 = \cos\alpha$ and $2p_0\|p\| = \sin\alpha$. Hence there is $a \in S^2$ with $p = \sin\frac{\alpha}{2}\, a$ and thus $p_0 = \cos\frac{\alpha}{2}$, which implies that $R_{\boldsymbol{p}}$ takes the form as in Euler's formula in Exercise 4.22 or 5.58.

(iii) Use the results of Exercise 5.26 to show that $R_{\boldsymbol{p}}$ preserves the norm, and therefore the inner product on $\mathbf{R}^3$, and using that $\boldsymbol{p} \mapsto \det R_{\boldsymbol{p}}$ is a continuous mapping from the connected space S^3 to $\{\pm 1\}$ (see Lemma 1.9.3) prove $\det R_{\boldsymbol{p}} = 1$.

(iv) Verify that we get the following *rational parametrization* of a matrix in $\mathbf{SO}(3, \mathbf{R})$:

$$R_{\boldsymbol{p}} = (p_0^2 - \|p\|^2)I + 2(pp^t + p_0 r_p).$$

We now study relations between reflections and rotations in $\mathbf{R}^3$.

(v) Let S_i be the reflection of $\mathbf{R}^3$ in the two-dimensional linear subspace $N_i \subset \mathbf{R}^3$ for $1 \le i \le 2$. If α is the angle between N_1 and N_2 measured from N_1 to N_2, then $S_2 S_1$ (note the ordering) is the rotation of $\mathbf{R}^3$ about the axis $N_1 \cap N_2$ by the angle 2α. Prove this. Conversely, use Euler's Theorem in Exercise 2.5 to prove that every element in $\mathbf{SO}(\mathbf{R}^3)$ is the product of two reflections of $\mathbf{R}^3$. **Hint:** It is sufficient to verify the result for the reflections S_1 and S_2 of $\mathbf{R}^2$ in the one-dimensional linear subspace $\mathbf{R}e_1$ and $\mathbf{R}(\cos\alpha, \sin\alpha)$, respectively, satisfying, with $x \in \mathbf{R}^2$,

$$S_1 x = \begin{pmatrix} x_1 \\ -x_2 \end{pmatrix}, \qquad S_2 x = x + 2(x_1 \sin\alpha - x_2 \cos\alpha)\begin{pmatrix} -\sin\alpha \\ \cos\alpha \end{pmatrix}.$$

Let $\Delta(A)$ be a spherical triangle as in Exercise 5.27 determined by $a_i \in S^2$ and with angles α_i, and denote by S_i the reflection of $\mathbf{R}^3$ in the two-dimensional linear subspace containing a_{i+1} and a_{i+2} for $1 \le i \le 3$.

(vi) By means of (v) prove that $S_{i+1} S_{i+2} = R_{2\alpha_i, a_i}$, the rotation of $\mathbf{R}^3$ about the axis a_i and by the angle $2\alpha_i$ for $1 \le i \le 3$. Using $S_i^2 = I$ deduce *Hamilton's Theorem*

$$R_{2\alpha_1, a_1} R_{2\alpha_2, a_2} R_{2\alpha_3, a_3} = I.$$

Note that this is an analog of the fact that the sum of double the angles in a planar triangle equals 2π.

Exercise 5.66 (Reflections, rotations, Cayley–Klein parameters and Sylvester's Theorem – sequel to Exercises 5.26, 5.27 and 5.65 – needed for Exercises 5.67, 5.68, 5.69 and 5.72). Let $0 \le \alpha_1 \le \pi$ and $a_1 \in S^2$. According to Exercise 5.65.(v) the rotation R_{α_1, a_1} of $\mathbf{R}^3$ by the angle α_1 about the axis $\mathbf{R}a_1$ is the composition $S_2 S_1$ of reflections S_i in two-dimensional linear subspaces N_i, for $1 \le i \le 2$, of $\mathbf{R}^3$ that intersect in $\mathbf{R}a_1$ and make an angle of $\frac{\alpha_1}{2}$. We will prove that any such pair (S_1, S_2) or configuration (N_1, N_2) is determined by the parameter $\pm\boldsymbol{p}$ of R_{α_1, a_1} given by

$$\boldsymbol{p} = (p_0, p) = \Big(\cos\frac{\alpha_1}{2}, \sin\frac{\alpha_1}{2}\, a_1\Big) \in S^3 = \{\, (p_0, p) \in \mathbf{R} \times \mathbf{R}^3 \mid p_0^2 + \|p\|^2 = 1 \,\}.$$

Further, composition of rotations corresponds to the composition of parameters defined in $(\star)$ below.

(i) Set $N_p = \{\, x \in \mathbf{R}^3 \mid \langle x, p \rangle = 0 \,\}$. Define

$$\rho_{\boldsymbol{p}} \in \mathrm{End}(N_p) \qquad \text{by} \qquad \rho_{\boldsymbol{p}}\, x = p_0 x + p \times x.$$

Show that $\rho_{\boldsymbol{p}} = \rho_{-\boldsymbol{p}}$ is the counterclockwise (measured with respect to p) rotation in N_p by the angle $\frac{\alpha_1}{2}$, and furthermore, that R_{α_1,a_1} is the composition $S_2 S_1$ of reflections S_1 and S_2 in the two-dimensional linear subspaces N_1 and N_2 spanned by p and x, and p and $\rho_{\boldsymbol{p}}\, x$, respectively.

(ii) Now suppose R_{α_2,a_2} is a second rotation with corresponding parameter $\pm\boldsymbol{q}$ given by $\boldsymbol{q} = (q_0, q) = (\cos\frac{\alpha_2}{2}, \sin\frac{\alpha_2}{2}\, a_2) \in S^3$. Consider the triple of vectors in S^2

$$x_1 := \rho_{\boldsymbol{p}}^{-1}\, x_2 \in N_p, \qquad x_2 := \frac{1}{\|q \times p\|}\, q \times p \in N_p \cap N_q,$$
$$x_3 := \rho_{\boldsymbol{q}}\, x_2 \in N_q.$$

Then x_1 and x_2, and x_2 and x_3, determine a decomposition of R_{α_1,a_1} and R_{α_2,a_2} in reflections in two-dimensional linear subspaces N_1 and N_2, and N_3 and N_4, respectively, where $N_2 \cap N_3 = \mathbf{R}(q \times p)$. In particular,

$$x_2 = \rho_{\boldsymbol{p}}\, x_1 = p_0 x_1 + p \times x_1, \qquad x_3 = \rho_{\boldsymbol{q}}\, x_2 = q_0 x_2 + q \times x_2.$$

Eliminate x_2 from these equations to find

$$x_3 = (q_0 p_0 - \langle q, p\rangle) x_1 + (q_0 p + p_0 q + q \times p) \times x_1 =: r_0 x_1 + r \times x_1.$$

Using Exercise 5.26.(i) show that $\boldsymbol{r} = (r_0, r) \in S^3$. Prove that x_1 and $x_3 \in N_r$.

We have found the following rule of composition for the parameters of rotations:

$$(\star) \qquad \boldsymbol{q}\cdot\boldsymbol{p} = (q_0, q)\cdot(p_0, p) = (q_0 p_0 - \langle q, p\rangle,\ q_0 p + p_0 q + q \times p) =: (r_0, r) = \boldsymbol{r}.$$

Next we want to prove that the composite parameter $\pm\boldsymbol{r}$ is the parameter of the composition $R_{\alpha_3,a_3} = R_{\alpha_2,a_2} R_{\alpha_1,a_1}$.

(iii) Let $0 \le \alpha_3 \le \pi$ and $a_3 \in S^2$ be determined by $\boldsymbol{r}$, thus $(\cos\frac{\alpha_3}{2}, \sin\frac{\alpha_3}{2}\, a_3) = (r_0, r)$. Then $\Delta(x_1, x_2, x_3)$ is the spherical triangle with sides $\frac{\alpha_2}{2}$, $\frac{\alpha_3}{2}$ and $\frac{\alpha_1}{2}$, see Exercise 5.27. The polar triangle $\Delta'(x_1, x_2, x_3)$ has vertices a_2, $-a_3$ and a_1, and angles $\frac{2\pi-\alpha_2}{2}$, $\frac{2\pi-\alpha_3}{2}$ and $\frac{2\pi-\alpha_1}{2}$. Apply Hamilton's Theorem from Exercise 5.65.(vi) to this polar triangle to find

$$R_{2\pi-\alpha_2,a_2} R_{\alpha_3,a_3} R_{2\pi-\alpha_1,a_1} = I, \qquad \text{thus} \qquad R_{\alpha_3,a_3} = R_{\alpha_2,a_2} R_{\alpha_1,a_1}.$$

Note that the results in (ii) and (iii) give a geometric construction for the composition of two rotations, which is called *Sylvester's Theorem*; see Exercise 5.67.(xii) for a different construction.

The *Cayley–Klein parameter* of the rotation $R_{\alpha,a}$ is the pair of matrices $\pm\widehat{\boldsymbol{p}} \in \mathrm{Mat}(2, \mathbf{C})$ given by the following formula, where $i = \sqrt{-1}$,

$$\widehat{\boldsymbol{p}} = \begin{pmatrix} p_0 + ip_3 & -p_2 + ip_1 \\ p_2 + ip_1 & p_0 - ip_3 \end{pmatrix} = \begin{pmatrix} \cos\frac{\alpha}{2} + ia_3 \sin\frac{\alpha}{2} & (-a_2 + ia_1)\sin\frac{\alpha}{2} \\ (a_2 + ia_1)\sin\frac{\alpha}{2} & \cos\frac{\alpha}{2} - ia_3\sin\frac{\alpha}{2} \end{pmatrix}.$$

The Cayley–Klein parameter of a rotation describes how that rotation can be obtained as the composition of two reflections. In Exercise 5.67 we will examine in more detail the relation between a rotation and its Cayley–Klein parameter.

(iv) By computing the first column of the product matrix verify that the rule of composition in $(\star)$ corresponds to the usual multiplication of the matrices $\widehat{\boldsymbol{q}}$ and $\widehat{\boldsymbol{p}}$

$$\widehat{\boldsymbol{q}\cdot\boldsymbol{p}} = \widehat{\boldsymbol{q}}\,\widehat{\boldsymbol{p}} = (q_0p_0 - \langle q, p\rangle,\ q_0p + p_0q + q\times p)\widehat{\ } \qquad (\boldsymbol{q}, \boldsymbol{p} \in S^3).$$

Show that $\det\widehat{\boldsymbol{p}} = \|\boldsymbol{p}\|^2$, for all $\boldsymbol{p} \in S^3$. By taking determinants corroborate the fact $\|\boldsymbol{q}\cdot\boldsymbol{p}\| = \|\boldsymbol{q}\|\,\|\boldsymbol{p}\| = 1$ for all $\boldsymbol{q}$ and $\boldsymbol{q} \in S^3$, which we know already from part (ii). Replacing p_0 by $-p_0$, deduce the following *four-square identity*:

$$\begin{aligned}
&(q_0^2 + q_1^2 + q_2^2 + q_3^2)(p_0^2 + p_1^2 + p_2^2 + p_3^2)\\
&\quad = (q_0p_0 + q_1p_1 + q_2p_2 + q_3p_3)^2 + (q_0p_1 - q_1p_0 + q_2p_3 - q_3p_2)^2\\
&\quad +(q_0p_2 - q_2p_0 + q_3p_1 - q_1p_3)^2 + (q_0p_3 - q_3p_0 + q_1p_2 - q_2p_1)^2,
\end{aligned}$$

which is used in number theory, in the proof of *Lagrange's Theorem* that asserts that every natural number is the sum of four squares of integer numbers.

Note that $\boldsymbol{p} \mapsto \widehat{\boldsymbol{p}}$ gives an injection from S^3 into the subset of $\mathrm{Mat}(2, \mathbf{C})$ consisting of the Cayley–Klein parameters; we now determine its image. Define $\mathbf{SU}(2)$, the *special unitary group* acting in $\mathbf{C}^2$, as the subgroup of $\mathbf{GL}(2, \mathbf{C})$ given by

$$\mathbf{SU}(2) = \{\, U \in \mathrm{Mat}(2, \mathbf{C}) \mid U^*U = I,\ \det U = 1\,\}.$$

Here we write $U^* = \overline{U}^t = (\overline{u_{ji}}) \in \mathrm{Mat}(2, \mathbf{C})$ for $U = (u_{ij}) \in \mathrm{Mat}(2, \mathbf{C})$.

(v) Show that $\{\,\widehat{\boldsymbol{p}} \mid \boldsymbol{p} \in S^3\,\} = \mathbf{SU}(2)$.

(vi) For $\boldsymbol{p} = (p_0, p) \in S^3$ we set $a = p_0 + ip_3$ and $b = p_2 + ip_1 \in \mathbf{C}$. Verify that $R_{\boldsymbol{p}} \in \mathbf{SO}(3, \mathbf{R})$ in Exercise 5.65.(iv), and $\widehat{\boldsymbol{p}} \in \mathbf{SU}(2)$, respectively, take the form

$$R_{(a,b)} = \begin{pmatrix} \mathrm{Re}(a^2 - b^2) & -\mathrm{Im}(a^2 - b^2) & 2\,\mathrm{Re}(a\overline{b}) \\ \mathrm{Im}(a^2 + b^2) & \mathrm{Re}(a^2 + b^2) & 2\,\mathrm{Im}(a\overline{b}) \\ -2\,\mathrm{Re}(ab) & 2\,\mathrm{Im}(ab) & |a|^2 - |b|^2 \end{pmatrix},$$

$$U(a, b) := \begin{pmatrix} a & -\overline{b} \\ b & \overline{a} \end{pmatrix}.$$

Exercise 5.67 (Quaternions, SU(2), SO(3, R) and Rodrigues' Theorem – sequel to the Exercises 5.26, 5.27, 5.58, 5.65 and 5.66 – needed for Exercises 5.68, 5.70 and 5.71). The notation is that of Exercise 5.66. We now study $\mathbf{SU}(2)$ in more detail. We begin with $\mathbf{R} \cdot \mathbf{SU}(2)$, the linear space of matrices $\widehat{\boldsymbol{p}}$ for $\boldsymbol{p} \in \mathbf{R}^4$ with matrix multiplication. In particular, as in Exercise 5.66.(vi) we write $a = p_0 + ip_3$ and $b = p_2 + ip_1 \in \mathbf{C}$, for $\boldsymbol{p} = (p_0, p) \in \mathbf{R} \times \mathbf{R}^3 \simeq \mathbf{R}^4$, and we set

$$\widehat{\boldsymbol{p}} = \begin{pmatrix} p_0 + ip_3 & -p_2 + ip_1 \\ p_2 + ip_1 & p_0 - ip_3 \end{pmatrix} = U(a, b) = \begin{pmatrix} a & -\overline{b} \\ b & \overline{a} \end{pmatrix} \in \mathrm{Mat}(2, \mathbf{C}).$$

Note

$$\begin{aligned} \widehat{\boldsymbol{p}} &= p_0\begin{pmatrix} 1 & 0 \\ 0 & 1 \end{pmatrix} + p_1\begin{pmatrix} 0 & i \\ i & 0 \end{pmatrix} + p_2\begin{pmatrix} 0 & -1 \\ 1 & 0 \end{pmatrix} + p_3\begin{pmatrix} i & 0 \\ 0 & -i \end{pmatrix} \\ &=: p_0 e + p_1 i + p_2 j + p_3 k =: p_0 e + \widehat{p}. \end{aligned}$$

Here we abuse notation as the symbol i denotes the number $\sqrt{-1}$ as well as the matrix $\sqrt{-1}\begin{pmatrix} 0 & 1 \\ 1 & 0 \end{pmatrix}$ (for both objects, their square equals minus the identity); yet, in the following its meaning should be clear from the context.

(i) Prove that e, i, j and k all belong to $\mathbf{SU}(2)$ and satisfy

$$i^2 = j^2 = k^2 = -e,$$
$$ij = k = -ji, \qquad jk = i = -kj, \qquad ki = j = -ik.$$

Show

$$\widehat{\boldsymbol{p}}^{-1} = \frac{1}{p_0^2 + \|p\|^2}(p_0 e - \widehat{p}) \qquad (\boldsymbol{p} \in \mathbf{R}^4).$$

By computing the first column of the product matrix verify (compare with Exercise 5.66.(iv))

$$\widehat{\boldsymbol{p}}\widehat{\boldsymbol{q}} = (p_0 q_0 - \langle p, q \rangle,\ p_0 q + q_0 p + p \times q)^{\widehat{}} \qquad (\boldsymbol{p}, \boldsymbol{q} \in \mathbf{R}^4).$$

Deduce

$$\widehat{(0, p)}\widehat{(0, q)} + \widehat{(0, q)}\widehat{(0, p)} = -2(\langle p, q \rangle, 0)^{\widehat{}},$$
$$[\,\widehat{\boldsymbol{p}}, \widehat{\boldsymbol{q}}\,] := \widehat{\boldsymbol{p}}\widehat{\boldsymbol{q}} - \widehat{\boldsymbol{q}}\widehat{\boldsymbol{p}} = 2(0, p \times q)^{\widehat{}}.$$

Define the linear subspace $\mathfrak{su}(2)$ of $\mathrm{Mat}(2, \mathbf{C})$ by

$$\mathfrak{su}(2) = \{\, X \in \mathrm{Mat}(2, \mathbf{C}) \mid X^* + X = 0,\ \mathrm{tr}\, X = 0 \,\}.$$

Note that, in fact, $\mathfrak{su}(2)$ is the Lie algebra of the linear Lie group $\mathbf{SU}(2)$.

(ii) Prove i, j and $k \in \mathfrak{su}(2)$. Show that for every $X \in \mathfrak{su}(2)$ there exists a unique $x \in \mathbf{R}^3$ such that

$$X = \widehat{x} = \begin{pmatrix} ix_3 & -x_2 + ix_1 \\ x_2 + ix_1 & -ix_3 \end{pmatrix} \qquad \text{and} \qquad \det \widehat{x} = \|x\|^2.$$

Show that the mapping $x \mapsto \widehat{x}$ is a linear isomorphism of vector spaces $\mathbf{R}^3 \to \mathfrak{su}(2)$. Deduce from (i), for $x, y \in \mathbf{R}^3$,

$$\widehat{x}\widehat{y} + \widehat{y}\widehat{x} = -2\langle x, y\rangle e, \qquad [\frac{1}{2}\widehat{x}, \frac{1}{2}\widehat{y}] = \frac{1}{2}\widehat{x \times y},$$

$$\widehat{x}\widehat{y} = -\langle x, y\rangle e + \widehat{x \times y}.$$

In particular, $\widehat{x}\widehat{y} = -\widehat{y}\widehat{x} = \widehat{x \times y}$ if $\langle x, y\rangle = 0$. Verify that the mapping $(\mathbf{R}^3, \times) \to (\mathfrak{su}(2), [\cdot, \cdot])$ given by $x \mapsto \frac{1}{2}\widehat{x}$, is an isomorphism of Lie algebras, compare with Exercise 5.59.(ii).

In the following we will identify $\mathbf{R}^3$ and $\mathfrak{su}(2)$ via the mapping $x \leftrightarrow \widehat{x}$. Note that this way we introduce a product for two vectors in $\mathbf{R}^3$ which is called *Clifford multiplication*; however, the product does not necessarily belong to $\mathbf{R}^3$ since $\widehat{x}^2 = -\|x\|^2 e \notin \mathfrak{su}(2)$ for all $x \in \mathbf{R}^3 \setminus \{0\}$.

Let $x_0 \in S^2$ and set $N_{x_0} = \{ x \in \mathbf{R}^3 \mid \langle x, x_0\rangle = 0 \}$. The reflection of $\mathbf{R}^3$ in the two-dimensional linear subspace N_{x_0} is given by $x \mapsto x - 2\langle x, x_0\rangle x_0$.

(iii) Prove that in terms of the elements of $\mathfrak{su}(2)$ this reflection corresponds to the linear mapping

$$\widehat{x} \mapsto \widehat{x_0}\,\widehat{x}\,\widehat{x_0} = -\widehat{x_0}\,\widehat{x}\,\widehat{x_0}^{-1} : \mathfrak{su}(2) \to \mathfrak{su}(2).$$

Let $\boldsymbol{p} \in S^3$ and $x_1 \in N_p \cap S^2$. According to Exercise 5.66.(i) we can write $R_{\boldsymbol{p}} = S_2 S_1$ with S_i the reflection in N_{x_i} where $x_2 = \rho_{\boldsymbol{p}}\, x_1 = p_0 x_1 + p \times x_1$. Use (ii) to show

$$\widehat{x_2} = \widehat{\boldsymbol{p}}\,\widehat{x_1}, \qquad \text{thus} \qquad \widehat{x_2}\widehat{x_1} = -\widehat{\boldsymbol{p}}, \qquad \widehat{x_1}\widehat{x_2} = (\widehat{x_2}\widehat{x_1})^{-1} = -\widehat{\boldsymbol{p}}^{-1}.$$

Note that the expression for $\widehat{x_2}\widehat{x_1}$ is independent of the choice of x_1 and entirely in terms of $\widehat{\boldsymbol{p}}$. Deduce that in terms of elements in $\mathfrak{su}(2)$ the rotation $R_{\boldsymbol{p}}$ takes the form

$$\widehat{x} \mapsto \widehat{x_2}(\widehat{x_1}\,\widehat{x}\,\widehat{x_1})\widehat{x_2} = \widehat{\boldsymbol{p}}\,\widehat{x}\,\widehat{\boldsymbol{p}}^{-1} : \mathfrak{su}(2) \to \mathfrak{su}(2).$$

Using that $\widehat{p}\widehat{x} + \widehat{x}\widehat{p} = -2\langle p, x\rangle e$ implies $\widehat{p}\,\widehat{x}\,\widehat{p} = \|p\|^2\widehat{x} - 2\langle p, x\rangle\widehat{p}$, verify

$$\widehat{R_{\boldsymbol{p}}\, x} = 2\langle p, x\rangle\,\widehat{p} + (p_0^2 - \|p\|^2)\,\widehat{x} + 2p_0\,\widehat{p \times x} = \widehat{\boldsymbol{p}}\,\widehat{x}\,\widehat{\boldsymbol{p}}^{-1}.$$

Once the idea has come up of using the mapping $\widehat{x} \mapsto \widehat{\boldsymbol{p}}\,\widehat{x}\,\widehat{\boldsymbol{p}}^{-1}$, the computation above can be performed in a less explicit fashion as follows.

(iv) Verify $UXU^{-1} \in \mathfrak{su}(2)$ for $U \in \mathbf{SU}(2)$ and $X \in \mathfrak{su}(2)$. Given $U \in \mathbf{SU}(2)$, show that

$$\mathrm{Ad}\, U : \mathfrak{su}(2) \to \mathfrak{su}(2) \qquad \text{defined by} \qquad (\mathrm{Ad}\, U)X = UXU^{-1}$$

belongs to Aut ($\mathfrak{su}(2)$). Note that for every $x \in \mathbf{R}^3$ there exists a uniquely determined $R_U(x) \in \mathbf{R}^3$ with

$$U \widehat{x} U^{-1} = (\operatorname{Ad} U)\widehat{x} = \widehat{R_U x}.$$

Prove that $R_U : \mathbf{R}^3 \to \mathbf{R}^3$ is a linear mapping satisfying $\|R_U x\|^2 = \det \widehat{R_U x} = \det \widehat{x} = \|x\|^2$ for all $x \in \mathbf{R}^3$, and conclude $R_U \in \mathbf{O}(3, \mathbf{R})$ for $U \in \mathbf{SU}(2)$ using the polarization identity from Lemma 1.1.5.(iii). Thus $\det R_U = \pm 1$. In fact, $R_U \in \mathbf{SO}(3, \mathbf{R})$ since $U \mapsto \det R_U$ is a continuous mapping from the connected space $\mathbf{SU}(2)$ to $\{\pm 1\}$, see Lemma 1.9.3.

(v) Given $U \in \mathbf{SU}(2)$ determine the matrix of R_U with respect to the standard basis (e_1, e_2, e_3) in $\mathbf{R}^3$.
Hint: According to (ii) we have $\widehat{e_l}^2 = -e$ for $1 \leq l \leq 3$, therefore $\widehat{x} = \sum_{1 \leq l \leq 3} x_l \widehat{e_l}$ implies $\widehat{e_l}\widehat{x} = -x_l e + \cdots$. Hence $x_l = -\frac{1}{2}\operatorname{tr}(\widehat{e_l}\widehat{x})$, and this gives, for $1 \leq m \leq 3$,

$$\widehat{R_U e_m} = U \widehat{e_m} U^{-1} = \sum_{1 \leq l \leq 3} -\frac{1}{2}\operatorname{tr}(\widehat{e_l} U \widehat{e_m} U^*) \widehat{e_l}.$$

Therefore the matrix is $-\frac{1}{2}(\operatorname{tr}(\widehat{e_l} U \widehat{e_m} U^*))_{1 \leq l, m \leq 3}$.

In view of the identification of $\mathfrak{su}(2)$ and $\mathbf{R}^3$ we may consider the *adjoint mapping*

$$\operatorname{Ad} : \mathbf{SU}(2) \to \mathbf{SO}(3, \mathbf{R}) \qquad \text{given by} \qquad U \mapsto \operatorname{Ad} U = R_U.$$

We now study the properties of this mapping.

(vi) Show that the adjoint mapping is a homomorphism of groups, in other words, $\operatorname{Ad}(U_1 U_2) = \operatorname{Ad} U_1 \operatorname{Ad} U_2$ for U_1 and $U_2 \in \mathbf{SU}(2)$.

(vii) Deduce from part (iii) that $(\operatorname{Ad} \widehat{\boldsymbol{p}})\, x = R_{\boldsymbol{p}} x$, for $\boldsymbol{p} \in S^3$ and $x \in \mathbf{R}^3$. Now apply Exercise 5.58.(iii) or 5.65 to find that $\operatorname{Ad} : \mathbf{SU}(2) \to \mathbf{SO}(3, \mathbf{R})$ in fact is a surjection satisfying

$$\pm \operatorname{Ad} \widehat{\boldsymbol{p}} = R_{\boldsymbol{p}}.$$

This formula is another formulation of the relation between a rotation in $\mathbf{SO}(3, \mathbf{R})$ and its *Cayley–Klein parameter* in $\mathbf{SU}(2)$, see Exercise 5.66. A geometric interpretation of the Cayley–Klein parameter is given in Exercise 5.68.(vi) below.

(viii) Suppose $\widehat{\boldsymbol{p}} \in \ker \operatorname{Ad} = \{ U \in \mathbf{SU}(2) \mid \operatorname{Ad} U = I \}$, the kernel of the adjoint mapping. Then $(\operatorname{Ad} \widehat{\boldsymbol{p}})\, x = x$ for all $x \in \mathbf{R}^3$. In particular, for $0 \neq x \in \mathbf{R}^3$ with $\langle x, p \rangle = 0$ this gives $(p_0^2 - \|p\|^2)\, x + 2 p_0\, p \times x = x$. By taking the inner product of this equality with x deduce $\ker \operatorname{Ad} = \{\pm e\}$. Next apply the *Isomorphism Theorem for groups* in order to obtain the following isomorphism of groups:

$$\mathbf{SU}(2)/\{\pm e\} \to \mathbf{SO}(3, \mathbf{R}).$$

(ix) It follows from (ii) that $\widehat{a}^2 = -e$ if $a \in S^2$. Deduce the following formula for the Cayley–Klein parameter $\pm\widehat{p} \in \mathbf{SU}(2)$ of $R_{\alpha,a} \in \mathbf{SO}(3, \mathbf{R})$ for $0 \le \alpha \le \pi$ (compare with Exercise 5.66)

$$\widehat{p} = e^{\frac{\alpha}{2}\widehat{a}} := \sum_{n \in \mathbf{N}_0} \frac{1}{n!}\Big(\frac{\alpha}{2}\widehat{a}\Big)^n = \cos\frac{\alpha}{2}\, e + \sin\frac{\alpha}{2}\, \widehat{a}.$$

Thus, in view of (vii)

$$\pm \operatorname{Ad}(e^{\frac{\alpha}{2}\widehat{a}}) = R_{\alpha,a}.$$

Furthermore, conclude that $\exp : \mathfrak{su}(2) \to \mathbf{SU}(2)$ is surjective.

(x) Because of (vi) we have the one-parameter group $t \mapsto \operatorname{Ad}(e^{tX}) \in \mathbf{SO}(3, \mathbf{R})$ for $X \in \mathfrak{su}(2)$, consider its infinitesimal generator

$$\operatorname{ad} X = \frac{d}{dt}\Big|_{t=0} \operatorname{Ad}(e^{tX}) \in \operatorname{End}(\,\mathfrak{su}(2)) \simeq \operatorname{Mat}(3, \mathbf{R}).$$

Prove that $\operatorname{ad} X \in \operatorname{End}^-(\,\mathfrak{su}(2))$, for $X \in \mathfrak{su}(2)$, see Lemma 2.1.4. Deduce from (ix) and Exercise 5.58.(ii), with $r_a \in \mathrm{A}(3, \mathbf{R})$ as in that exercise,

$$\operatorname{ad}\widehat{a} = \frac{d}{d\alpha}\Big|_{\alpha=0} R_{2\alpha,a} = 2r_a \in \mathrm{A}(3, \mathbf{R}) \qquad (a \in \mathbf{R}^3),$$

and verify this also by computing the matrix of $\operatorname{ad}\widehat{a} \in \operatorname{End}^-(\,\mathfrak{su}(2))$ directly. Conclude once more $[\,\frac{1}{2}\widehat{a}, \frac{1}{2}\widehat{b}\,] = \frac{1}{2}\widehat{a \times b}$ for $a, b \in \mathbf{R}^3$ (see part (ii)). Further, prove that the mapping

$$\operatorname{ad} : (\,\mathfrak{su}(2), [\cdot, \cdot]) \to (\mathrm{A}(3, \mathbf{R}), [\,\cdot, \cdot\,]), \qquad \frac{1}{2}\widehat{a} \mapsto \operatorname{ad}\frac{1}{2}\widehat{a} = r_a$$

is a homomorphism of Lie algebras, see the following diagrams.

$$\begin{array}{ccccc} & & (\,\mathfrak{su}(2), [\cdot, \cdot]) & & \\ & \overset{\frac{1}{2}\widehat{\ }}{\nearrow} & & \overset{\operatorname{ad}}{\searrow} & \\ (\mathbf{R}^3, \times) & & \xrightarrow{\quad \times \quad} & & (\mathrm{A}(3, \mathbf{R}), [\cdot, \cdot]) \end{array} \qquad \begin{array}{ccc} \frac{1}{2}\widehat{a} & \longmapsto & \operatorname{ad}\frac{1}{2}\widehat{a} \\ \uparrow & & \| \\ a & \longmapsto & r_a \end{array}$$

Show using (ix) (compare with Theorem 5.10.6.(iii))

$$\pm \operatorname{Ad}(e^{\frac{\alpha}{2}\widehat{a}}) = R_{\alpha,a} = e^{\alpha r_a} = e^{\frac{\alpha}{2}\operatorname{ad}\widehat{a}}.$$

Hence

$$\operatorname{Ad} \circ \exp = \exp \circ \operatorname{ad} : \mathfrak{su}(2) \to \mathbf{SO}(3, \mathbf{R}).$$

(xi) Conclude from (ix) that $R_{\alpha_3,a_3} = R_{\alpha_2,a_2} \circ R_{\alpha_1,a_1}$ if $e^{\frac{\alpha_3}{2}\widehat{a_3}} = e^{\frac{\alpha_2}{2}\widehat{a_2}}\, e^{\frac{\alpha_1}{2}\widehat{a_1}}$. Using (ii) show

$$\cos\frac{\alpha_3}{2} = \cos\frac{\alpha_2}{2}\cos\frac{\alpha_1}{2} - \langle a_2, a_1\rangle \sin\frac{\alpha_2}{2}\sin\frac{\alpha_1}{2};$$

$$\widetilde{a_3} = \frac{1}{1-\langle\widetilde{a_2},\widetilde{a_1}\rangle}(\widetilde{a_2}+\widetilde{a_1}+\widetilde{a_2}\times\widetilde{a_1}) \qquad \text{with} \qquad \widetilde{a_i} = \tan\frac{\alpha_i}{2}\, a_i.$$

Note that the term with the cross product is responsible for the noncommutativity of $\mathbf{SO}(3, \mathbf{R})$.

Example. In particular, because

$$(\frac{1}{2}\sqrt{2}\,e+\frac{1}{2}\sqrt{2}\,i)(\frac{1}{2}\sqrt{2}\,e+\frac{1}{2}\sqrt{2}\,j) = \frac{1}{2}(e+i+j+k) = \frac{1}{2}\,e+\frac{1}{2}\sqrt{3}\frac{1}{\sqrt{3}}(i+j+k),$$

the result of rotating first by $\frac{\pi}{2}$ about the axis $\mathbf{R}(0, 1, 0)$, and then by $\frac{\pi}{2}$ about the axis $\mathbf{R}(1, 0, 0)$ is tantamount to rotation by $\frac{2\pi}{3}$ about the axis $\mathbf{R}(1, 1, 1)$.

(xii) Deduce from (xi) and Exercise 5.27.(viii) that, in the terminology of that exercise, the spherical triangle determined by a_1, a_2 and $a_3 \in S^2$ has angles $\frac{\alpha_1}{2}$, $\frac{\alpha_2}{2}$ and $\pi - \frac{\alpha_3}{2}$. Conversely, given (α_1, a_1) and (α_2, a_2), we can obtain (α_3, a_3) satisfying $R_{\alpha_3,a_3} = R_{\alpha_2,a_2} \circ R_{\alpha_1,a_1}$ in the following geometrical fashion, which is known as *Rodrigues' Theorem.* Consider the spherical triangle $\Delta(a_2, a_1, a_3)$ corresponding to a_2, a_1 and a_3 (note the ordering, in particular $\det(a_2 a_1 a_3) > 0$) determined by the segment $\overline{a_2 a_1}$ and the angles at a_i which are positive and equal to $\frac{\alpha_i}{2}$ for $1 \le i \le 2$, if the angles are measured from the successive to the preceding segment. According to the formulae above the segments $\overline{a_2 a_3}$ and $\overline{a_1 a_3}$ meet in a_3 (as notation suggests) and α_3 is read off from the angle $\pi - \frac{\alpha_3}{2}$ at a_3. The latter fact also follows from Hamilton's Theorem in Exercise 5.65.(vi) applied to $\Delta(a_2, a_1, a_3)$ with angles $\frac{\alpha_2}{2}$, $\frac{\alpha_1}{2}$, and $\frac{\gamma}{2}$ say, since $R_{\alpha_2,a_2} R_{\alpha_1,a_1} R_{\gamma,a_3} = I$ gives $R_{\alpha_3,a_3} = R_{\gamma,a_3}^{-1} = R_{2\pi-\gamma,a_3}$, which implies $\frac{\gamma}{2} = \pi - \frac{\alpha_3}{2}$. See Exercise 5.66.(ii) and (iii) for another geometric construction for the composition of two rotations.

Background. The space $\mathbf{R}\cdot\mathbf{SU}(2)$ forms the noncommutative field $\mathbf{H}$ of the *quaternions.* The idea of introducing **anticommuting** quantities i, j and k that satisfy the rules of multiplication $i^2 = j^2 = k^2 = ijk = -1$ occurred to W. R. Hamilton on October 16, 1843 near Brougham Bridge at Dublin. The geometric construction in Exercise 5.66.(ii) and (iii) gives the rule of composition for the Cayley–Klein parameters, and therefore the group structure of $\mathbf{SU}(2)$ and of the quaternions too. The Cayley–Klein parameter $\widehat{\boldsymbol{p}} \in \mathbf{SU}(2)$ describes the decomposition of $R_{\boldsymbol{p}} \in \mathbf{SO}(3, \mathbf{R})$ into reflections. This characterization is consistent with Hamilton's description of the quaternion $\widehat{\boldsymbol{p}} = r(\cos\alpha\, e + \sin\alpha\, \widehat{a}) \in \mathbf{H}$, where $r \ge 0$, $0 \le \alpha \le \pi$ and $a \in S^2$, as parametrizing pairs of vectors x_1 and $x_2 \in \mathbf{R}^3$ such that $\frac{\|x_1\|}{\|x_2\|} = r$, $\angle(x_1, x_2) = \alpha$, x_1 and x_2 are perpendicular to a, and $\det(a\, x_1\, x_2) > 0$. Furthermore, by considering reflections one is naturally led to the adjoint mapping, which in fact is the adjoint representation of the linear Lie group $\mathbf{SU}(2)$ in the linear space of automorphisms of its Lie algebra $\mathfrak{su}(2)$ (see Formula (5.36) and Theorem 5.10.6.(iii)). Another way of formulating the result in (viii) is that $\mathbf{SU}(2)$ is the two-fold *covering group* of $\mathbf{SO}(3, \mathbf{R})$. A straightforward argument from group theory proves the impossibility of "sensibly" choosing in any way representatives in $\mathbf{SU}(2)$ for the elements of $\mathbf{SO}(3, \mathbf{R})$. The mapping $\mathrm{Ad}^{-1} : \mathbf{SO}(3, \mathbf{R}) \to \mathbf{SU}(2)$ is called the (double-valued) *spinor representation* of $\mathbf{SO}(3, \mathbf{R})$. The equality $\widehat{x}\widehat{y} + \widehat{y}\widehat{x} = -2\langle x, y\rangle e$ is fundamental in the theory of *Clifford algebras*: by means of the $\widehat{e_i} \in \mathfrak{su}(2)$ the quadratic form $\sum_{1\le i\le 3} x_i^2$ is turned into minus the square of the linear form $\sum_{1\le i\le 3} x_i\widehat{e_i}$. In the context of the Clifford algebras the construction above of the group $\mathbf{SU}(2)$ and the adjoint mapping $\mathrm{Ad} : \mathbf{SU}(2) \to \mathbf{SO}(3, \mathbf{R})$ are generalized to the construction of the *spinor groups* $\mathbf{Spin}(n)$ and the short exact sequence

$$e \longrightarrow \{\pm e\} \longrightarrow \mathbf{Spin}(n) \xrightarrow{\mathrm{Ad}} \mathbf{SO}(n, \mathbf{R}) \longrightarrow I \qquad (n \ge 3).$$

The case of $n = 3$ is somewhat special as the $\widehat{x}$ can be realized as matrices, in the general case their existence requires a more abstract construction.

Exercise 5.68 (SU(2), Hopf fibration and slerp – sequel to Exercises 4.26, 5.58, 5.66 and 5.67 – needed for Exercise 5.70). We take another look at the *Hopf fibration* from Exercise 4.26. We have the subgroup $T = \{\, e^{\frac{\alpha}{2}k} = U(e^{i\frac{\alpha}{2}}, 0) \mid \alpha \in \mathbf{R}\,\} \simeq S^1$ of $\mathbf{SU}(2)$, and furthermore $\mathbf{SU}(2) \simeq S^3$ according to Exercise 5.66.(v).

(i) Use Exercise 5.67.(vii) to prove that $\{\, (\mathrm{Ad}\, U)\, n \mid U \in \mathbf{SU}(2)\,\} = S^2$ if $n = (0, 0, 1) \in S^2$.

Now consider the following diagram:

$$\begin{array}{ccccc} T \simeq S^1 \longrightarrow & \mathbf{SU}(2) \simeq S^3 & \xrightarrow{\quad h \quad} & \mathrm{Ad}\,(\mathbf{SU}(2))\, n \simeq S^2 \\ & {\scriptstyle f_+}\searrow & & \swarrow{\scriptstyle \Phi_+} \\ & & \mathbf{C} & \end{array}$$

Here we have, for U, $U(\alpha, \beta) \in \mathbf{SU}(2)$ with $\beta \ne 0$ and $c \in S^2 \setminus \{n\}$,

$$h : U \mapsto (\mathrm{Ad}\, U)\, n : \mathbf{SU}(2) \to S^2, \qquad f_+(U(\alpha, \beta)) = \frac{\alpha}{\beta},$$

$$\Phi_+(c) = \frac{c_1 + i c_2}{1 - c_3}.$$

(ii) Verify by means of Exercise 5.66.(vi) that the mapping h above coincides with the one in Exercise 4.26. In particular, deduce that $\Phi_+ : S^2 \setminus \{n\} \to \mathbf{C}$ is the stereographic projection satisfying

$$\Phi_+(\,\mathrm{Ad}\, U(\alpha, \beta)\, n) = f_+(U(\alpha, \beta)) = \frac{\alpha}{\beta} \qquad (U(\alpha, \beta) \in \mathbf{SU}(2),\ \beta \ne 0).$$

(iii) Use Exercise 5.67.(ix) to show $h^{-1}(\{n\}) = T$, and part (vi) of that exercise to show that $h^{-1}(\{(\mathrm{Ad}\, U)\, n\}) = U\,T$ given $U \in \mathbf{SU}(2)$. Thus the coset space $G/T \simeq S^2$.

(iv) Prove that the subgroup T is a great circle on $\mathbf{SU}(2)$. Given $U = U(a, b) \in \mathbf{SU}(2)$ verify that $U(\alpha, \beta) \mapsto U\,U(\alpha, \beta)$ for every $U(\alpha, \beta) \in \mathbf{R} \cdot \mathbf{SU}(2)$ gives an isometry of $\mathbf{R} \cdot \mathbf{SU}(2) \simeq \mathbf{R}^4$ because of

$$|\alpha|^2+|\beta|^2 = \det U(\alpha, \beta) = \det\,(U(a, b)U(\alpha, \beta)) = |a\alpha-\overline{b}\beta|^2+|b\alpha+\overline{a}\beta|^2.$$

For any $U \in \mathbf{SU}(2)$ deduce that the coset $U\,T$ is a great circle on $\mathbf{SU}(2)$, and compare this result with Exercise 4.26.(v).

Note that $U \in \mathbf{SU}(2)$ acts on $\mathbf{SU}(2)$ by left multiplication; on S^2 through $\mathrm{Ad}\, U$, that is,

$$\mathrm{Ad}\, U(\alpha, \beta)\, n \mapsto \mathrm{Ad}\, U\, \mathrm{Ad}\, U(\alpha, \beta)\, n = \mathrm{Ad}\, U U(\alpha, \beta)\, n \qquad (U(\alpha, \beta) \in \mathbf{SU}(2));$$

and on $\mathbf{C}$ by the *fractional linear* or *homographic* transformation

$$z \mapsto U \cdot z = U(a, b) \cdot z = \frac{az - \overline{b}}{bz + \overline{a}} \qquad (z \in \mathbf{C}).$$

(v) Verify that the fractional linear transformations of this kind form a group G under composition of mappings, and that the mapping $\mathbf{SU}(2) \to G$ with $U \mapsto U\cdot$ is a homomorphism of groups with kernel $\{\pm e\}$; thus $G \simeq \mathbf{SU}(2)/\{\pm e\} \simeq \mathbf{SO}(3, \mathbf{R})$ in view of Exercise 5.67.(viii).

(vi) Using Exercise 5.67.(vi) and part (ii) above show that, for any $U(a, b) \in \mathbf{SU}(2)$ and $c = \mathrm{Ad}\, U(\alpha, \beta)\, n \in S^2$ with $\beta \neq 0$,

$$\begin{aligned}
\Phi_+(\,\mathrm{Ad}\, U(a, b)\, c) \;&= \Phi_+(\,\mathrm{Ad}\,(U(a, b)U(\alpha, \beta))\, n) \\
&= f_+(U(a\alpha - \overline{b}\beta, b\alpha + \overline{a}\beta)) = \frac{a\alpha - \overline{b}\beta}{b\alpha + \overline{a}\beta} \\
&= \frac{a\frac{\alpha}{\beta} - \overline{b}}{b\frac{\alpha}{\beta} + \overline{a}} = U(a, b) \cdot \frac{\alpha}{\beta} = U(a, b) \cdot \Phi_+(c).
\end{aligned}$$

Deduce $\Phi_+ \circ \mathrm{Ad}\, U = U \cdot \Phi_+ : S^2 \setminus \{n\} \to \mathbf{C}$ and $f_+ \circ U = U \cdot f_+ : \mathbf{SU}(2) \to \mathbf{C}$ for $U \in \mathbf{SU}(2)$. We say that the mappings Φ_+ and f_+ are *equivariant* with respect to the actions of $\mathbf{SU}(2)$. Conclude that moving a point on S^2 by a rotation with a given Cayley–Klein parameter corresponds to applying the Cayley–Klein parameter, acting as a fractional linear transformation, to the stereographic projection of that point.

See Exercise 5.70.(xvi) for corresponding properties of the general fractional linear transformation $\mathbf{C} \to \mathbf{C}$ given by $z \mapsto \frac{az+c}{bz+d}$ for a, b, c and $d \in \mathbf{C}$ with $ad - bc = 1$.

(vii) Let U_1 and $U_2 \in \mathbf{SU}(2) \simeq S^3 \subset \mathbf{R}^4$ and suppose $\langle U_1, U_2\rangle = \frac{1}{2}\operatorname{tr}(U_1 U_2^*) = \cos\theta$. Consider $U \in \mathbf{SU}(2)$ belonging to the great circle in $\mathbf{SU}(2)$ connecting U_1 and U_2. Because U lies in the plane in $\mathbf{R}^4$ spanned by U_1 and U_2, there exist λ and $\mu \in \mathbf{R}$ satisfying $U = \lambda U_1 + \mu U_2$ and $\|\lambda U_1 + \mu U_2\| = 1$. Writing $\langle U, U_1\rangle = \cos t\theta$, for suitable $t \in [\,0, 1\,]$, deduce

$$U = U(t) = \frac{\sin(1-t)\theta}{\sin\theta}\,U_1 + \frac{\sin t\theta}{\sin\theta}\,U_2.$$

In *computer graphics* the curve $[\,0,1\,] \to \mathbf{SU}(2)$ with $t \mapsto U(t)$ is known as the *slerp* (= spherical linear interpolation) between U_1 and U_2; the corresponding rotations interpolate smoothly between the rotations with Cayley-Klein parameter U_1 and U_2, respectively.

Exercise 5.69 (Cartan decomposition of SL(2, C) – sequel to Exercises 2.44 and 5.66 – needed for Exercises 5.70 and 5.71). Let $\mathbf{SU}(2)$ and $\mathfrak{su}(2)$ be as in Exercise 5.67. Set $\mathbf{SL}(2, \mathbf{C}) = \{\, A \in \mathbf{GL}(2, \mathbf{C}) \mid \det A = 1 \,\}$ and

$$\mathfrak{sl}(2, \mathbf{C}) = \{\, X \in \operatorname{Mat}(2, \mathbf{C}) \mid \operatorname{tr} X = 0 \,\}, \qquad \mathfrak{p} = i\,\mathfrak{su}(2) = \mathfrak{sl}(2, \mathbf{C}) \cap \mathrm{H}(2, \mathbf{C}).$$

Here $\mathrm{H}(2, \mathbf{C}) = \{\, A \in \operatorname{Mat}(2, \mathbf{C}) \mid A^* = A \,\}$ where $A^* = \overline{A}^t$ as usual denotes the linear subspace of *Hermitian* matrices. Note that $\mathfrak{sl}(2, \mathbf{C}) = \mathfrak{su}(2) \oplus \mathfrak{p}$ as linear spaces over $\mathbf{R}$.

(i) For every $A \in \mathbf{SL}(2, \mathbf{C})$ there exist $U \in \mathbf{SU}(2)$ and $X \in \mathfrak{p}$ such that the following *Cartan decomposition* holds:

$$A = U\,e^X.$$

Indeed, write $D(a, b)$ for the diagonal matrix in $\operatorname{Mat}(2, \mathbf{C})$ with coefficients a and $b \in \mathbf{C}$. Since $A^*A \in \mathrm{H}(2, \mathbf{C})$ is positive–definite, by the version over $\mathbf{C}$ of the Spectral Theorem 2.9.3 there exist $V \in \mathbf{SU}(2)$ and $\lambda > 0$ with $A^*A = V D(\lambda, \lambda^{-1})V^{-1}$. Now $D(\lambda, \lambda^{-1}) = \exp 2D(\mu, -\mu)$ with $\mu = \frac{1}{2}\log\lambda \in \mathbf{R}$. Set

$$X = V D(\mu, -\mu)V^{-1} \in \mathfrak{p}, \qquad U = A\,e^{-X},$$

and verify $U \in \mathbf{SU}(2)$.

(ii) Verify that $e^X \in \mathbf{SL}(2, \mathbf{C}) \cap \mathrm{H}(2, \mathbf{C})$ is positive definite for $X \in \mathfrak{p}$, and conversely, that every positive definite Hermitian element in $\mathbf{SL}(2, \mathbf{C})$ is of this form, for a suitable $X \in \mathfrak{p}$ (see Exercise 2.44.(ii)).

(iii) Define $\pi : \mathbf{SU}(2) \times \mathfrak{p} \to \mathbf{SL}(2, \mathbf{C})$ by $\pi(U, X) = U\,e^X$. Show that π is continuous and deduce from Exercise 5.66.(v) and Theorem 1.9.4 that $\mathbf{SL}(2, \mathbf{C})$ is a connected set.

(iv) Verify that $\mathbf{SL}(2, \mathbf{C}) \cap \mathrm{H}(2, \mathbf{C})$ is not a subgroup of $\mathbf{SL}(2, \mathbf{C})$ by considering the product of $\begin{pmatrix} 2 & -i \\ i & 1 \end{pmatrix}$ and $\begin{pmatrix} 1 & 1 \\ 1 & 2 \end{pmatrix}$.

Background. The Cartan decomposition is a version over $\mathbf{C}$ of the polar decomposition of an automorphism from Exercise 2.68. Moreover, it is the global version of the decomposition (at the infinitesimal level) of an arbitrary matrix into Hermitian and anti-Hermitian matrices, the complex analog of the Stokes decomposition from Definition 8.1.4.

Exercise 5.70 (Hopf fibration, $\mathbf{SL}(2, \mathbf{C})$ and Lorentz group – sequel to Exercises 4.26, 5.67, 5.68, 5.69 and 8.32 – needed for Exercise 5.71). Important properties of the Lorentz group $\mathbf{Lo}(4, \mathbf{R})$ from Exercise 8.32 can be obtained using an extension of the Hopf mapping from Exercise 4.26. In this fashion $\mathbf{SL}(2, \mathbf{C}) = \{ A \in \mathbf{GL}(2, \mathbf{C}) \mid \det A = 1 \}$ arises naturally as the two-fold *covering group* of the proper Lorentz group $\mathbf{Lo}^\circ(4, \mathbf{R})$ which is defined as follows. We denote by C the *(light) cone* in $\mathbf{R}^4$, and by C^+ the *forward (light) cone*, respectively, given by

$$C = \{ (x_0, x) \in \mathbf{R}^4 \mid x_0^2 = \|x\|^2 \}, \qquad C^+ = \{ (x_0, x) \in C \mid x_0 \geq 0 \}.$$

Note that any $L \in \mathbf{Lo}(4, \mathbf{R})$ preserves C, i.e., satisfies $L(C) \subset C$ (and therefore $L(C) = C$). We define $\mathbf{Lo}^\circ(4, \mathbf{R})$, the *proper Lorentz group*, to be the subgroup of $\mathbf{Lo}(4, \mathbf{R})$ consisting of elements preserving C^+ and having determinant equal to 1. Elements of $\mathbf{Lo}^\circ(4, \mathbf{R})$ are said to be *proper Lorentz transformations*. In this exercise we identify linear transformations and matrices using the standard basis in $\mathbf{R}^4$.

We begin by introducing the Hopf mapping in a natural way. To this end, set $\mathrm{H}(2, \mathbf{C}) = \{ A \in \mathrm{Mat}(2, \mathbf{C}) \mid A^* = A \}$ where $A^* = \overline{A}^t$ as usual, and define $\kappa : \mathbf{C}^2 \to \mathrm{H}(2, \mathbf{C})$ by

$$\kappa(z) = 2\, z\, z^* \qquad (z \in \mathbf{C}^2),$$

in other words

$$\kappa\begin{pmatrix} a \\ b \end{pmatrix} = 2\begin{pmatrix} a\bar{a} & a\bar{b} \\ b\bar{a} & b\bar{b} \end{pmatrix} \qquad (a, b \in \mathbf{C}).$$

Note that κ is neither injective nor surjective.

(i) Verify $\det \kappa(z) = 0$ and $\kappa \circ A(z) = A\, \kappa(z)\, A^*$ for $z \in \mathbf{C}^2$ and $A \in \mathrm{Mat}(2, \mathbf{C})$.

Define for all $\boldsymbol{x} = (x_0, x) \in \mathbf{R} \times \mathbf{R}^3 \simeq \mathbf{R}^4$ (see Exercise 5.67.(ii) for the definition of $\widehat{x} \in \mathfrak{su}(2)$)

$$\widetilde{\boldsymbol{x}} = x_0 e - i\,\widehat{x} = \begin{pmatrix} x_0 + x_3 & x_1 + i x_2 \\ x_1 - i x_2 & x_0 - x_3 \end{pmatrix} \in \mathrm{H}(2, \mathbf{C}).$$

(ii) Show that the mapping $x \mapsto \widetilde{x}$ is a linear isomorphism of vector spaces $\mathbf{R}^4 \to \mathrm{H}(2, \mathbf{C})$; and denote its inverse by $\iota : \mathrm{H}(2, \mathbf{C}) \to \mathbf{R}^4$. Prove

$$\det \widetilde{x} = x_0^2 - \|x\|^2 =: \lceil x \rceil^2, \qquad \mathrm{tr}\, \widetilde{x} = 2x_0 \qquad (x \in \mathbf{R}^4).$$

(iii) Prove that $h := \iota \circ \kappa : \mathbf{C}^2 \to \mathbf{R}^4$ is given by (compare with Exercise 4.26)

$$h\begin{pmatrix} a \\ b \end{pmatrix} = \begin{pmatrix} |a|^2 + |b|^2 \\ 2\,\mathrm{Re}(a\overline{b}) \\ 2\,\mathrm{Im}(a\overline{b}) \\ |a|^2 - |b|^2 \end{pmatrix} \qquad (a, b \in \mathbf{C}).$$

Deduce from (i) that $(h \circ A(z))\widetilde{\ } = A\,\kappa(z)\,A^*$ for $A \in \mathrm{Mat}(2, \mathbf{C})$ and $z \in \mathbf{C}^2$. Further show that $h(\lambda z) = |\lambda|^2 h(z)$, for all $\lambda \in \mathbf{C}$ and $z \in \mathbf{C}^2$.

(iv) Using (ii) show $\det \kappa(z) = \lceil h(z) \rceil^2$ and using (i) deduce $h(z) \in C^+$ for all $z \in \mathbf{C}^2$. More precisely, prove that $h : \mathbf{C}^2 \to C^+$ is surjective, and that

$$h^{-1}(\{h(z)\}) = \{\, e^{i\alpha} z \mid \alpha \in \mathbf{R} \,\} \qquad (z \in \mathbf{C}^2).$$

Compare this result with the Exercises 4.26 and 5.68.(iii).

Next we show that the natural action of $A \in \mathbf{SL}(2, \mathbf{C})$ on $\mathbf{C}^2$ induces a Lorentz transformation $L_A \in \mathbf{Lo}^\circ(4, \mathbf{R})$ of $\mathbf{R}^4$.

(v) Since $h \circ A : \mathbf{C}^2 \to C^+$ for all $A \in \mathrm{Mat}(2, \mathbf{C})$, and $h \circ A \circ e^{i\alpha} = h \circ A$ for all $\alpha \in \mathbf{R}$, we have the well-defined mapping

$$L_A : C^+ \to C^+ \qquad \text{given by} \qquad L_A \circ h = h \circ A : \mathbf{C}^2 \to C^+.$$

Deduce from (iii) that $(L_A \circ h(z))\widetilde{\ } = A\,\kappa(z)\,A^*$. Given $x \in C^+$, part (iv) now implies that we can find $z \in \mathbf{C}^2$ with $h(z) = x$, and thus $\kappa(z) = \widetilde{x}$. This gives

$$\widetilde{L_A(x)} = A\,\widetilde{x}\,A^* \qquad (x \in C^+).$$

C^+ spans all of $\mathbf{R}^4$ as the linearly independent vectors $e_0 + e_1$, $e_0 + e_2$ and $e_0 \pm e_3$ in $\mathbf{R}^4$ all belong to C^+. Therefore the mapping L_A in fact is the restriction to C^+ of a unique linear mapping, also denoted by L_A,

$$L_A \in \mathrm{End}(\mathbf{R}^4) \qquad \text{satisfying} \qquad \widetilde{L_A x} = A\,\widetilde{x}\,A^* \qquad (x \in \mathbf{R}^4).$$

Using (ii) prove

$$\lceil L_A x \rceil^2 = \det(A\,\widetilde{x}\,A^*) = \det \widetilde{x} = \lceil x \rceil^2 \qquad (A \in \mathbf{SL}(2, \mathbf{C})).$$

By means of the analog of the polarization identity from Lemma 1.1.5.(iii) deduce $L_A \in \mathbf{Lo}(4, \mathbf{R})$ for $A \in \mathbf{SL}(2, \mathbf{C})$; thus $\det L_A = \pm 1$. In fact, $L_A \in$

$\mathbf{Lo}^\circ(4, \mathbf{R})$ since $A \mapsto \det L_A$ is a continuous mapping from the connected space $\mathbf{SL}(2, \mathbf{C})$ to $\{\pm 1\}$, see Exercise 5.69.(iii) and Lemma 1.9.3. Note we have obtained the mapping

$$L : \mathbf{SL}(2, \mathbf{C}) \to \mathbf{Lo}^\circ(4, \mathbf{R}) \qquad \text{given by} \qquad A \mapsto L_A.$$

(vi) Show that the mapping L is a homomorphism of groups, that is, $L_{A_1 A_2} = L_{A_1} L_{A_2}$ for $A_1, A_2 \in \mathbf{SL}(2, \mathbf{C})$.

(vii) Any $L \in \mathbf{Lo}^\circ(4, \mathbf{R})$ is determined by its restriction to C^+ (see part (v)). Hence the surjectivity of the mapping $\mathbf{SL}(2, \mathbf{C}) \to \mathbf{Lo}^\circ(4, \mathbf{R})$ follows by showing that there is $A \in \mathbf{SL}(2, \mathbf{C})$ such that $L_A|_{C^+} = L|_{C^+}$, which in view of (v) comes down to $h \circ A = L_A \circ h = L \circ h$. Evaluation at e_1 and $e_2 \in \mathbf{C}^2$, respectively, gives the following equations for $A = (a_1\, a_2)$, where a_1 and $a_2 \in \mathbf{C}^2$ are the column vectors of A:

$$h(a_1) = L(e_0 + e_3) \in C^+, \qquad h(a_2) = L(e_0 - e_3) \in C^+.$$

Because $h : \mathbf{C}^2 \to C^+$ is surjective we can find a solution $A \in \mathrm{Mat}(2, \mathbf{C})$. On the other hand, as $\widetilde{e_0} = e$, we have in view of (v)

$$1 = \lceil e_0 \rceil^2 = \lceil L_A\, e_0 \rceil^2 = \det(AA^*) = |\det A|^2.$$

Hence $|\det A| = 1$. Since the first column a_1 of A is determined up to a factor $e^{i\alpha}$ with $\alpha \in \mathbf{R}$, we can find a solution $A \in \mathrm{Mat}(2, \mathbf{C})$ with $\det A = 1$; that is, $L = L_A \in \mathbf{Lo}^\circ(4, \mathbf{R})$ with $A \in \mathbf{SL}(2, \mathbf{C})$.

(viii) Show $\ker L = \{\pm I\}$. Indeed, if $L_A = I$, then $AXA^* = X$ for every $X \in \mathrm{H}(2, \mathbf{C})$. By applying this relation successively with X equal to I, $\begin{pmatrix} 0 & 1 \\ 1 & 0 \end{pmatrix}$ and $\begin{pmatrix} 1 & 0 \\ 0 & -1 \end{pmatrix}$, deduce $A = a\, I$ with $a \in \mathbf{C}$. Then $\det A = 1$ implies $A = \pm I$.

Apply the Isomorphism Theorem for groups in order to obtain the following isomorphism of groups:

$$\mathbf{SL}(2, \mathbf{C})/\{\pm I\} \to \mathbf{Lo}^\circ(4, \mathbf{R}).$$

(ix) For $U \in \mathbf{SU}(2) = \{\, U \in \mathbf{SL}(2, \mathbf{C}) \mid U^*U = I \,\}$ show, in the notation from Exercise 5.67.(ii),

$$\widetilde{L_U\, \mathbf{x}} = U(x_0 e - i\, \widehat{x})U^{-1} = (x_0 e - i\, U\widehat{x}U^{-1}) = (x_0 e - i\, \widehat{R_U(x)}) = (\overline{R_U}(\mathbf{x}))\widetilde{\ }.$$

Here $R \in \mathbf{SO}(3, \mathbf{R})$ induces $\overline{R} \in \mathbf{Lo}^\circ(4, \mathbf{R})$ by $\overline{R}(x_0, x) = (x_0, Rx)$. Conversely, every $L \in \mathbf{Lo}^\circ(4, \mathbf{R})$ satisfying $L\, e_0 = e_0$ necessarily is of the form $\overline{R}$ for some $R \in \mathbf{SO}(3, \mathbf{R})$, since L leaves $\mathbf{R}^3$ invariant. Deduce that $L|_{\mathbf{SU}(2)} : \mathbf{SU}(2) \to \mathbf{Lo}^\circ(4, \mathbf{R})$ coincides with the adjoint mapping from Exercise 5.67.(vi), and that

$$\{ \overline{R} \in \mathbf{Lo}^\circ(4, \mathbf{R}) \mid R \in \mathbf{SO}(3, \mathbf{R}) \} \subset \mathrm{im}(L).$$

If $\overline{L_A} \in \mathbf{SO}(3, \mathbf{R})$, then $L_A\, e_0 = e_0$ and thus $\widetilde{L_A\, e_0} = \widetilde{e_0}$, which implies $AA^* = I$, thus $A \in \mathbf{SU}(2)$. Note this gives a proof of the surjectivity of $\mathrm{Ad} : \mathbf{SU}(2) \to \mathbf{SO}(3, \mathbf{R})$ different from the one in Exercise 5.67.(vii).

(x) Let $A = \begin{pmatrix} a & c \\ b & d \end{pmatrix} \in \mathbf{SL}(2, \mathbf{C})$. Use (iii) and (v), respectively, to prove

$$h(e_1) = e_0 + e_3, \qquad h(e_1 + e_2) = 2(e_0 + e_1),$$

$$h(e_2) = e_0 - e_3, \qquad h(e_1 - i e_2) = 2(e_0 + e_2);$$

$$L_A(e_0 + e_3) = h\begin{pmatrix} a \\ b \end{pmatrix}, \qquad L_A(e_0 + e_1) = \tfrac{1}{2} h\begin{pmatrix} a + c \\ b + d \end{pmatrix},$$

$$L_A(e_0 - e_3) = h\begin{pmatrix} c \\ d \end{pmatrix}, \qquad L_A(e_0 + e_2) = \tfrac{1}{2} h\begin{pmatrix} a - ic \\ b - id \end{pmatrix},$$

in order to show that the matrix of L_A with respect to the standard basis (e_0, e_1, e_2, e_3) in $\mathbf{R}^4$ is given by

$$\begin{pmatrix} \frac{1}{2}(a\bar{a} + b\bar{b} + c\bar{c} + d\bar{d}) & \mathrm{Re}(a\bar{c} + b\bar{d}) & \mathrm{Im}(\bar{a}c + \bar{b}d) & \frac{1}{2}(a\bar{a} + b\bar{b} - c\bar{c} - d\bar{d}) \\ \mathrm{Re}(a\bar{b} + c\bar{d}) & \mathrm{Re}(a\bar{d} + c\bar{b}) & \mathrm{Im}(\bar{a}d + \bar{b}c) & \mathrm{Re}(a\bar{b} - c\bar{d}) \\ \mathrm{Im}(a\bar{b} + c\bar{d}) & \mathrm{Im}(a\bar{d} + c\bar{b}) & \mathrm{Re}(\bar{a}d - \bar{b}c) & \mathrm{Im}(a\bar{b} - c\bar{d}) \\ \frac{1}{2}(a\bar{a} - b\bar{b} + c\bar{c} - d\bar{d}) & \mathrm{Re}(a\bar{c} - b\bar{d}) & \mathrm{Im}(\bar{a}c - \bar{b}d) & \frac{1}{2}(a\bar{a} - b\bar{b} - c\bar{c} + d\bar{d}) \end{pmatrix}$$

Prove that this matrix also is given by $\frac{1}{2}(\,\mathrm{tr}(\widetilde{e_l} A \widetilde{e_m} A^*))_{0 \le l, m \le 3}$.
Hint: As $\widetilde{e_l} = -i\widehat{e_l}$ for $1 \le l \le 3$, it follows from Exercise 5.67.(ii) that $\widetilde{e_l}^2 = e$ for $0 \le l \le 3$. Now imitate the argument of Exercise 5.67.(v).

$\mathbf{SL}(2, \mathbf{C})$ is the two-fold *covering group* of $\mathbf{Lo}^\circ(4, \mathbf{R})$. Furthermore, the mapping $L^{-1} : \mathbf{Lo}^\circ(4, \mathbf{R}) \to \mathbf{SL}(2, \mathbf{C})$ is called the (double-valued) *spinor representation* of $\mathbf{Lo}^\circ(4, \mathbf{R})$; in this context the elements of $\mathbf{C}^2$ are referred to as *spinors*.

Now we analyze the structure of $\mathbf{Lo}^\circ(4, \mathbf{R})$ using properties of the mapping $A \mapsto L_A$.

(xi) From (x) we know that $L_A(e_0 + e_3) = h(A\, e_1)$ for $A \in \mathbf{SL}(2, \mathbf{C})$. Since $\mathbf{SL}(2, \mathbf{C})$ acts transitively on $\mathbf{C}^2 \setminus \{0\}$ and $h : \mathbf{C}^2 \to C^+$ is surjective, there exists for every $0 \neq \boldsymbol{x} \in C^+$ an $A \in \mathbf{SL}(2, \mathbf{C})$ such that $L_A(e_0 + e_3) = \boldsymbol{x}$. Use (vi) to show that $\mathbf{Lo}^\circ(4, \mathbf{R})$ acts transitively on $C^+ \setminus \{0\}$.

Set

$$\mathfrak{sl}(2, \mathbf{C}) = \{ X \in \mathrm{Mat}(2, \mathbf{C}) \mid \mathrm{tr}\, X = 0 \}, \qquad \mathfrak{p} = i\, \mathfrak{su}(2) = \mathfrak{sl}(2, \mathbf{C}) \cap \mathrm{H}(2, \mathbf{C}).$$

(xii) Given $\widetilde{p} \in \mathbf{SL}(2, \mathbf{C}) \cap \mathrm{H}(2, \mathbf{C})$ prove using Exercise 5.69.(i) and (ii) that there exists $v \in \mathbf{R}^3$ satisfying

$$-\frac{i}{2}\widehat{v} \in i\ \mathfrak{su}(2) = \mathfrak{p} \qquad \text{and} \qquad \widetilde{p} = e^{-i\frac{1}{2}\widehat{v}}.$$

In part (xiii) below we shall prove that $L_{\boldsymbol{p}} := L_{\widetilde{\boldsymbol{p}}} \in \mathbf{Lo}^\circ(4, \mathbf{R})$ is a *hyperbolic screw* or *boost*. Next use Exercise 5.69.(i) to write an arbitrary $A \in \mathbf{SL}(2, \mathbf{C})$ as $A = U\widetilde{p}$ with $U \in \mathbf{SU}(2)$ and $\widetilde{p} \in \mathbf{SL}(2, \mathbf{C}) \cap \mathrm{H}(2, \mathbf{C})$. Now apply L to A, and use parts (vi) and (ix) to obtain the Cartan decomposition for elements in $\mathbf{Lo}^\circ(4, \mathbf{R})$

$$L_A = \overline{R_U} L_{\boldsymbol{p}}.$$

That is, every proper Lorentz transformation is a composition of a boost followed by a space rotation.

(xiii) Let $\boldsymbol{p} \in \mathbf{R}^4$ and use Exercise 5.67.(ii), and its part (iii) for $\widehat{p}\widehat{x}\widehat{p}$, to show

$$\widetilde{p}\widetilde{x}\widetilde{p} = ((p_0^2+\|p\|^2)x_0+2p_0\langle p, x\rangle)e-i\,((p_0^2-\|p\|^2)x+2(p_0x_0+\langle p, x\rangle)p)\widehat{\ }.$$

From $\widehat{p} \in \mathfrak{su}(2)$ (see Exercise 5.67.(ii)) deduce $i\,\widehat{p} \in \mathrm{H}(2, \mathbf{C})$. Now define the *unit hyperboloid* H^3 in $\mathbf{R}^4$ by $H^3 = \{\, \boldsymbol{p} \in \mathbf{R}^4 \mid \lceil \boldsymbol{p} \rceil = 1 \,\}$. Verify that the matrix of $L_{\boldsymbol{p}}$ for $\boldsymbol{p} \in H^3$ with respect to the standard basis in $\mathbf{R}^4$ has the following *rational parametrization* (compare with the rational parametrization of an orthogonal matrix in Exercise 5.65.(iv)):

$$L_{\boldsymbol{p}} = \begin{pmatrix} p_0^2 + \|p\|^2 & 2p_0 p^t \\ 2p_0 p & I_3 + 2pp^t \end{pmatrix}.$$

Note that $\widehat{a}^2 = -e$ for $a \in S^2$ according to Exercise 5.67.(ii); use this to prove for $\alpha \in \mathbf{R}$ (compare with Exercise 5.67.(ix))

$$\widetilde{p} \ := e^{-i\,\frac{\alpha}{2}\widehat{a}} := \sum_{n\in\mathbf{N}_0} \frac{1}{n!}\Big(-i\,\frac{\alpha}{2}\widehat{a}\Big)^n = \cosh\frac{\alpha}{2}\ e - i\ \sinh\frac{\alpha}{2}\ \widehat{a}$$

$$= \begin{pmatrix} \cosh\dfrac{\alpha}{2} + a_3 \sinh\dfrac{\alpha}{2} & (a_1 + ia_2)\sinh\dfrac{\alpha}{2} \\ (a_1 - ia_2)\sinh\dfrac{\alpha}{2} & \cosh\dfrac{\alpha}{2} - a_3\sinh\dfrac{\alpha}{2} \end{pmatrix}.$$

Note that the substitution $\alpha \mapsto i\alpha$ produces the matrix $\widehat{p}$ from Exercise 5.66 above part (iv). Using (ii) verify

$$\det\widetilde{p} = 1, \qquad p_0^2 + \|p\|^2 = \cosh\alpha,$$
$$2p_0 p = \sinh\alpha\ a, \qquad 2pp^t = (-1 + \cosh\alpha)\,aa^t.$$

By comparison with Exercise 8.32.(vi) find

$$L_{e^{-i\frac{\alpha}{2}\widehat{a}}} = L_{\boldsymbol{p}} = B_{\alpha,a} \qquad (\alpha \in \mathbf{R},\ a \in S^2),$$

the hyperbolic screw or boost in the direction $a \in S^2$ with rapidity α. Conclude that

$$\frac{d}{d\alpha}\Big|_{\alpha=0} L_{e^{-i\frac{\alpha}{2}\widehat{a}}} = DL(I)\Big(-\frac{i}{2}\widehat{a}\Big) = \begin{pmatrix} 0 & a^t \\ a & 0 \end{pmatrix} \qquad (a \in \mathbf{R}^3).$$

(xiv) For $\boldsymbol{p} \in \mathbf{R}^4 \setminus C$ define the *hyperbolic reflection* $S_{\boldsymbol{p}} \in \mathrm{End}(\mathbf{R}^4)$ by

$$S_{\boldsymbol{p}}\,\boldsymbol{x} = \boldsymbol{x} - 2\frac{\lceil \boldsymbol{x}, \boldsymbol{p} \rceil}{\lceil \boldsymbol{p}, \boldsymbol{p} \rceil}\boldsymbol{p}.$$

Show that $S_{\boldsymbol{p}} \in \mathbf{Lo}(4, \mathbf{R})$. Now let $\boldsymbol{p} \in H^3$, the unit hyperboloid, and prove by means of (xiii)

$$L_{\boldsymbol{p}} = S_{\boldsymbol{p}} S_{e_0}.$$

The *Cayley–Klein parameter* $\pm\widetilde{p} \in \mathbf{SL}(2, \mathbf{C})$ of the hyperbolic screw or boost $L_{\boldsymbol{p}}$ describes how $L_{\boldsymbol{p}}$ can be obtained as the composition of two hyperbolic reflections. Because of Exercise 5.69.(iv) there are no direct counterparts for the composition of two boosts of the formulae in Exercise 5.67.(xi).

In the last part of this exercise we use the terminology of Exercise 5.68, in particular from its parts (v) and (vi). There the fractional linear transformations of $\mathbf{C}$ determined by elements of $\mathbf{SU}(2)$ were associated with rotations of the unit sphere S^2 in $\mathbf{R}^3$. Now we shall give a description of all the fractional linear transformations of $\mathbf{C}$.

(xv) Consider the following diagram:

$$\begin{array}{ccccc} \mathbf{SL}(2,\mathbf{C}) & \xrightarrow{\ \widetilde{h}\ } & C^+ \setminus \{0\} & \xrightarrow{\ p\ } & S^2 \\ & \searrow^{f_+} & & \swarrow_{\Phi_+} & \\ & & \mathbf{C} \cup \{\infty\} & & \end{array}$$

Here $\widetilde{h} : \mathbf{SL}(2, \mathbf{C}) \to C^+ \setminus \{0\}$, $f_+ : \mathbf{SL}(2, \mathbf{C}) \to \mathbf{C} \cup \{\infty\}$, $p : C^+ \setminus \{0\} \to S^2$, and the stereographic projection $\Phi_+ : S^2 \to \mathbf{C} \cup \{\infty\}$, respectively, are given by

$$\widetilde{h}(A) = L_A(e_0 + e_3) = (h \circ A)(e_1), \qquad f_+\begin{pmatrix} a & c \\ b & d \end{pmatrix} = \frac{a}{b},$$

$$p(x) = \frac{1}{x_0}x, \qquad \Phi_+(c) = \frac{c_1 + ic_2}{1 - c_3}.$$

Note that $h(e_1) = e_0 + e_3$ and $p(e_0 + e_3) = n \in S^2$. Deduce from (xi) and (iii) that, for $A = \begin{pmatrix} a & c \\ b & d \end{pmatrix} \in \mathbf{SL}(2, \mathbf{C})$,

$$p \circ L_A(e_0 + e_3) = p \circ h\begin{pmatrix} a \\ b \end{pmatrix} = \frac{1}{|a|^2 + |b|^2}\begin{pmatrix} 2\,\mathrm{Re}(a\overline{b}) \\ 2\,\mathrm{Im}(a\overline{b}) \\ |a|^2 - |b|^2 \end{pmatrix}.$$

Prove that the diagram is commutative by establishing

$$\Phi_+ \circ p \circ L_A(e_0 + e_3) = \frac{a}{b} = f_+(A) \qquad (A \in \mathbf{SL}(2, \mathbf{C})),$$

where we admit the value ∞ in the case of $b = 0$.

Let $A \in \mathbf{SL}(2, \mathbf{C})$ act on $\mathbf{SL}(2, \mathbf{C})$ by left multiplication, on $C^+ \setminus \{0\}$ by L_A, and on $\mathbf{C}$ by the fractional linear transformation $z \mapsto A \cdot z = \frac{az+c}{bz+d}$ for $z \in \mathbf{C}$. Let $0 \neq x \in C^+$ be arbitrary. According to (xi) we can find $\mathcal{A} = \begin{pmatrix} \alpha & \gamma \\ \beta & \delta \end{pmatrix} \in \mathbf{SL}(2, \mathbf{C})$ satisfying $x = L_{\mathcal{A}}(e_0 + e_3)$.

(xvi) Using (vi) verify for all $A \in \mathbf{SL}(2, \mathbf{C})$

$$\Phi_+ \circ p(L_A x) = \Phi_+ \circ p \circ L_{A\mathcal{A}}(e_0 + e_3) = \frac{a\alpha + c\beta}{b\alpha + d\beta} = A \cdot (\Phi_+ \circ p)(x).$$

Deduce

$$f_+ \circ A = A \cdot f_+ : \mathbf{SL}(2, \mathbf{C}) \to \mathbf{C},$$

$$(\Phi_+ \circ p) \circ L_A = A \cdot (\Phi_+ \circ p) : C^+ \setminus \{0\} \to \mathbf{C}.$$

Thus f_+ and $\Phi_+ \circ p$ are equivariant with respect to the actions of $\mathbf{SL}(2, \mathbf{C})$. Now consider $L \in \mathbf{Lo}^\circ(4, \mathbf{R})$, and recall that L preserves C^+. Being linear, L induces a transformation of the set of half-lines in C^+, which can be represented by the points of $\{ (1, x) \mid x \in \mathbf{R}^3 \} \cap C^+$. This intersection of an affine hyperplane by the forward light cone is the two-dimensional sphere $S = \{ (1, x) \mid \|x\| = 1 \} \subset \mathbf{R}^4$, and hence we obtain a mapping from S into itself induced by L. Next the sphere $S \simeq S^2$ can be mapped to $\mathbf{C}$ by stereographic projection, and in this way $L \in \mathbf{Lo}^\circ(4, \mathbf{R})$ defines a transformation of $\mathbf{C}$. Conclude that the resulting set of transformations of $\mathbf{C}$ consists exactly of the group of all fractional linear transformations of $\mathbf{C}$, which is isomorphic with $\mathbf{SL}(2, \mathbf{C})/\{\pm I\}$. This establishes a geometrical interpretation for the two-fold covering group $\mathbf{SL}(2, \mathbf{C})$ of the proper Lorentz group $\mathbf{Lo}^\circ(4, \mathbf{R})$.

Exercise 5.71 (Lie algebra of Lorentz group – sequel to Exercises 5.67, 5.69, 5.70 and 8.32). Let the notation be as in these exercises.

(i) Using Exercise 8.32.(i) or 5.70.(v) show that the Lie algebra $\mathfrak{lo}(4)$ of the Lorentz group $\mathbf{Lo}(4, \mathbf{R})$ satisfies

$$\mathfrak{lo}(4) = \{ X \in \mathrm{Mat}(4, \mathbf{R}) \mid X^t J_4 + J_4 X = 0 \}.$$

Prove $\dim \mathfrak{lo}(4) = 6$ and that $X \in \mathfrak{lo}(4)$ if and only if $X = \begin{pmatrix} 0 & v^t \\ v & r_a \end{pmatrix}$, with v and $a \in \mathbf{R}^3$, and $r_a \in \mathrm{A}(3, \mathbf{R})$ as in Exercise 5.67.(x). Show that we obtain linear subspaces of $\mathfrak{lo}(4)$ if we define

$$\begin{aligned} \mathfrak{so}(3, \mathbf{R}) &= \{ r_a := \begin{pmatrix} 0 & 0^t \\ 0 & r_a \end{pmatrix} \mid a \in \mathbf{R}^3 \}, \\ \mathfrak{b} &= \{ b_v := \begin{pmatrix} 0 & v^t \\ v & 0 \end{pmatrix} \mid v \in \mathbf{R}^3 \}, \end{aligned}$$

and furthermore, that $\mathfrak{lo}(4) = \mathfrak{so}(3, \mathbf{R}) \oplus \mathfrak{b}$ is a direct sum of vector spaces.

From the isomorphism of groups $\overline{L} : \mathbf{SL}(2, \mathbf{C})/\{\pm I\} \to \mathbf{Lo}^\circ(4, \mathbf{R})$ from Exercise 5.70.(viii) we obtain the isomorphism of Lie algebras $\lambda := D\overline{L}(I) : \mathfrak{sl}(2, \mathbf{R}) = \mathfrak{su}(2) \oplus \mathfrak{p} \to \mathfrak{lo}(4)$ (see Exercise 5.69) with

$$\begin{aligned} \mathfrak{su}(2) \oplus \mathfrak{p} &= \mathfrak{su}(2) \oplus \{ X \in \mathfrak{sl}(2, \mathbf{C}) \mid X^* = X \} \\ &= \{ \frac{1}{2}\widehat{a} \mid a \in \mathbf{R}^3 \} \oplus \{ -\frac{i}{2}\widehat{v} \mid v \in \mathbf{R}^3 \}. \end{aligned}$$

(ii) From Exercise 5.67.(ii) obtain the following commutator relations in $\mathfrak{sl}(2, \mathbf{C})$, for a_1, a_2, a, v_1, v_2 and $v \in \mathbf{R}^3$,

$$\Big[\frac{1}{2}\widehat{a_1}, \frac{1}{2}\widehat{a_2}\Big] = \frac{1}{2}\widehat{a_1 \times a_2}, \qquad \Big[\frac{1}{2}\widehat{a}, -\frac{i}{2}\widehat{v}\Big] = -\frac{i}{2}\widehat{a \times v},$$
$$\Big[-\frac{i}{2}\widehat{v_1}, -\frac{i}{2}\widehat{v_2}\Big] = -\frac{1}{2}\widehat{v_1 \times v_2}.$$

(iii) Obtain from Exercise 5.67.(x) and Exercise 5.70.(xiii), respectively,

$$\lambda\Big(\frac{1}{2}\widehat{a}\Big) = r_a \quad (a \in \mathbf{R}^3), \qquad \lambda\Big(-\frac{i}{2}\widehat{v}\Big) = b_v \quad (v \in \mathbf{R}^3).$$

By application of λ to the identities in part (ii) conclude that the Lie algebra structure of $\mathfrak{lo}(4)$ is given by

$$[\, r_{a_1}, r_{a_2} \,] = r_{a_1 \times a_2}, \qquad [\, b_{v_1}, b_{v_2} \,] = -r_{v_1 \times v_2}, \qquad [\, r_a, b_v \,] = b_{a \times v}.$$

Show that $\mathfrak{so}(3, \mathbf{R})$ is a Lie subalgebra of $\mathfrak{lo}(4)$, whereas $\mathfrak{b}$ is not.

(iv) Define the linear isomorphism of vector spaces $\mathfrak{lo}(4, \mathbf{R}) \simeq \mathbf{R}^3 \times \mathbf{R}^3$ by $r_a + b_v \leftrightarrow (a, v)$. Using part (iii) prove that the corresponding Lie algebra structure on $\mathbf{R}^3 \times \mathbf{R}^3$ is given by

$$[\, (a_1, v_1), (a_2, v_2) \,] = (a_1 \times a_2 - v_1 \times v_2, \; a_1 \times v_2 - a_2 \times v_1).$$

Exercise 5.72 (Quaternions and spherical trigonometry – sequel to Exercises 4.22, 5.27, 5.65, and 5.66). The formulae of spherical trigonometry, in the form associated with the polar triangle, follow from the quaternionic formulation of Hamilton's Theorem by a single computation.

Let $\Delta(A)$ be a spherical triangle as in Exercise 5.27 determined by $a_i \in S^2$, with angles α_i and sides A_i, for $1 \le i \le 3$. Hamilton's Theorem from Exercise 5.65.(vi) applied to $\Delta(A)$ gives

$$(\star) \qquad R_{2\alpha_{i+1},a_{i+1}} R_{2\alpha_{i+2},a_{i+2}} = R_{2\alpha_i,a_i}^{-1}.$$

According to Exercise 4.22.(iv) the pair $(2\alpha, a) \in V := \,]\,0, \pi\,[\, \times S^2$ is uniquely determined by $R_{2\alpha,a}$ (but $R_{\pi,a} = R_{\pi,-a}$ for all $a \in S^2$). Therefore, if $(2\alpha, a) \in V$, we can choose a representative $-U(2\alpha, a) \in \mathbf{SU}(2)$ of the Cayley–Klein parameter for $R_{2\alpha,a}$ as in Exercise 5.66 that depends continuously on $(2\alpha, a) \in V$, where

$$\begin{aligned} U(2\alpha, a) &= \begin{pmatrix} \cos\alpha + ia_3 \sin\alpha & (-a_2 + ia_1)\sin\alpha \\ (a_2 + ia_1)\sin\alpha & \cos\alpha - ia_3\sin\alpha \end{pmatrix} \\ &= \cos\alpha\, e + \sin\alpha(a_1 i + a_2 j + a_3 k). \end{aligned}$$

Now $(\alpha_1, a_1, \dots, \alpha_3, a_3) \mapsto \prod_{1\le i\le 3} -U(2\alpha_i, a_i)$ is a continuous mapping $V^3 \to \mathbf{SU}(2)$ which takes values in the two-point set $\{\pm I\}$ according to $(\star)$. Thus by Lemma 1.9.3 it has a constant value since V^3 is connected. In fact, this value is I, as can be seen from the special case of $\Delta(A)$ with $a_i = e_i$, the i-th standard basis vector in $\mathbf{R}^3$, for which we have $\alpha_i = \frac{\pi}{2}$. Thus we find for all nontrivial spherical triangles

$$-U(2\alpha_{i+1}, a_{i+1}) \cdot -U(2\alpha_{i+2}, a_{i+2}) = -U(2\alpha_i, a_i)^{-1} = -U(-2\alpha_i, a_i).$$

Verify that explicit computation of this equality yields the following identities of elements in $\mathbf{R}$ and $\mathbf{R}^3$, respectively,

$$\begin{aligned} &\cos\alpha_{i+1}\cos\alpha_{i+2} - \langle a_{i+1}, a_{i+2}\rangle \sin\alpha_{i+1}\sin\alpha_{i+2} = -\cos\alpha_i, \\ (\star\star) \qquad &\sin\alpha_{i+1}\cos\alpha_{i+2}\, a_{i+1} + \cos\alpha_{i+1}\sin\alpha_{i+2}\, a_{i+2} \\ &\qquad + \sin\alpha_{i+1}\sin\alpha_{i+2}\, a_{i+1} \times a_{i+2} = \sin\alpha_i\, a_i. \end{aligned}$$

Note that combination of the two identities gives, if $\alpha_i \ne \frac{\pi}{2}$ (compare with Exercise 5.67.(xi)),

$$\overline{a_i} = \frac{1}{\langle \overline{a_{i+1}}, \overline{a_{i+2}}\rangle - 1}(\overline{a_{i+1}} + \overline{a_{i+2}} + \overline{a_{i+1}} \times \overline{a_{i+2}}), \qquad \text{with} \qquad \overline{a_i} = \tan\alpha_i\, a_i.$$

Now $\langle a_{i+1}, a_{i+2}\rangle = \cos A_i$ according to Exercise 5.27, and so we obtain the spherical rule of cosines applied to the polar triangle $\Delta'(A)$, compare with Exercise 5.27.(viii),

$$\cos\alpha_i = \sin\alpha_{i+1}\sin\alpha_{i+2}\cos A_i - \cos\alpha_{i+1}\cos\alpha_{i+2},$$

By taking the inner product of $(\star\star)$ with $a_{i+1} \times a_{i+2}$ and with the a_i, respectively, we derive additional equalities. In fact

$$\|a_{i+1} \times a_{i+2}\|^2 \sin \alpha_{i+1} \sin \alpha_{i+2} = \langle a_i, a_{i+1} \times a_{i+2} \rangle \sin \alpha_i = \det A \, \sin \alpha_i.$$

From Exercise 5.27.(i) we get $\|a_{i+1} \times a_{i+2}\| = \sin A_i$, and hence we have for $1 \leq i \leq 3$

$$\left(\frac{\sin A_i}{\sin \alpha_i} \right)^2 = \frac{\det A}{\prod_{1 \leq i \leq 3} \sin \alpha_i}.$$

This gives the spherical rule of sines since $\frac{\sin A_i}{\sin \alpha_i} \geq 0$. Taking the inner product of $(\star\star)$ with a_{i+2} we obtain

$$\sin \alpha_i \cos A_{i+1} = \cos A_i \sin \alpha_{i+1} \cos \alpha_{i+2} + \cos \alpha_{i+1} \sin \alpha_{i+2}.$$

This is the *analog formula* applied to the polar triangle $\Delta'(A)$, relating three angles and two sides. Further, taking the inner product of $(\star\star)$ with a_i we see

$$\begin{aligned} \sin \alpha_i &= \sin \alpha_{i+1} \cos \alpha_{i+2} \cos A_{i+2} + \cos \alpha_{i+1} \sin \alpha_{i+2} \cos A_{i+1} \\ &\quad + \det A \sin \alpha_{i+1} \sin \alpha_{i+2}. \end{aligned}$$

In view of Exercise 5.27.(iv) we can rewrite this in the form

$$\begin{aligned} &(1 - \sin \alpha_{i+1} \sin \alpha_{i+2} \sin A_{i+1} \sin A_{i+2}) \sin \alpha_i \\ &\quad = \sin \alpha_{i+1} \cos \alpha_{i+2} \cos A_{i+2} + \cos \alpha_{i+1} \sin \alpha_{i+2} \cos A_{i+1}. \end{aligned}$$

Furthermore, for $\Delta(A)$ we have the following equality of elements in $\mathbf{SO}(3, \mathbf{R})$:

$$(\star\star\star) \qquad R_{-A_{i+1}, e_1} R_{\alpha_i, e_3} R_{A_{i+2}, e_1} = R_{\pi - \alpha_{i+2}, e_3} R_{A_i, e_1} R_{-\alpha_{i+1}, e_3}.$$

By going over to the polar triangle $\Delta'(A)$ this identity is equivalent to

$$R_{\alpha_{i+1} - \pi, e_1} R_{\pi - A_i, e_3} R_{\pi - \alpha_{i+2}, e_1} - R_{A_{i+2}, e_3} R_{\pi - \alpha_i, e_1} R_{A_{i+1} - \pi, e_3} = 0.$$

Prove this by using the spherical rule of cosines and of sines for $\Delta(A)$ and $\Delta'(A)$, the analogue formula for $\Delta(A)$, $\Delta'(A)$, $\Delta(a_1, a_3, a_2)$ and $\Delta'(a_1, a_3, a_2)$, as well as *Cagnoli's formula*

$$\begin{aligned} &\sin \alpha_{i+1} \sin \alpha_{i+2} - \sin A_{i+1} \sin A_{i+2} \\ &\quad = \cos \alpha_i \cos A_{i+1} \cos A_{i+2} + \cos A_i \cos \alpha_{i+1} \cos \alpha_{i+2}. \end{aligned}$$

By lifting the equality $(\star\star\star)$ in $\mathbf{SO}(3, \mathbf{R})$ to $\mathbf{SU}(2)$, show (the symbol i occurs as an index as well as a quaternion)

$$\begin{aligned} &\left(\cos \frac{A_{i+1}}{2} - i \sin \frac{A_{i+1}}{2} \right) \left(\cos \frac{\alpha_i}{2} + k \sin \frac{\alpha_i}{2} \right) \left(\cos \frac{A_{i+2}}{2} + i \sin \frac{A_{i+2}}{2} \right) \\ &\quad = \left(\sin \frac{\alpha_{i+2}}{2} + k \cos \frac{\alpha_{i+2}}{2} \right) \left(\cos \frac{A_i}{2} + i \sin \frac{A_i}{2} \right) \left(\cos \frac{\alpha_{i+1}}{2} - k \sin \frac{\alpha_{i+1}}{2} \right). \end{aligned}$$

Equate the coefficients of e, i, j and $k \in \mathbf{SU}(2)$ in this equality and deduce the following *analogs of Delambre-Gauss*, which link all angles and sides of $\Delta(A)$:

$$\cos\frac{\alpha_i}{2}\cos\frac{A_{i+1}-A_{i+2}}{2} = \sin\frac{\alpha_{i+1}+\alpha_{i+2}}{2}\cos\frac{A_i}{2},$$

$$\cos\frac{\alpha_i}{2}\sin\frac{A_{i+1}-A_{i+2}}{2} = \sin\frac{\alpha_{i+1}-\alpha_{i+2}}{2}\sin\frac{A_i}{2},$$

$$\sin\frac{\alpha_i}{2}\sin\frac{A_{i+1}+A_{i+2}}{2} = \cos\frac{\alpha_{i+1}-\alpha_{i+2}}{2}\sin\frac{A_i}{2},$$

$$\sin\frac{\alpha_i}{2}\cos\frac{A_{i+1}+A_{i+2}}{2} = \cos\frac{\alpha_{i+1}+\alpha_{i+2}}{2}\cos\frac{A_i}{2}.$$

Exercise 5.73 (Transversality). Let $V \subset \mathbf{R}^n$ be given.

(i) Prove that V is a submanifold in $\mathbf{R}^n$ of dimension 0 if and only if V is *discrete*, that is, every $x \in V$ possesses an open neighborhood U in $\mathbf{R}^n$ such that $U \cap V = \{x\}$.

(ii) Prove that V is a submanifold in $\mathbf{R}^n$ of codimension 0 if and only if V is open in $\mathbf{R}^n$.

Assume that V_1 and V_2 are C^k submanifolds in $\mathbf{R}^n$ of codimension c_1 and c_2, respectively. One says that V_1 and V_2 have *transverse intersection* if $T_x V_1 + T_x V_2 = \mathbf{R}^n$, for all $x \in V_1 \cap V_2$.

(iii) Prove that $V_1 \cap V_2$ is a C^k submanifold in $\mathbf{R}^n$ of codimension $c_1 + c_2$ if V_1 and V_2 have transverse intersection, and, in that case $T_x(V_1 \cap V_2) = T_x V_1 \cap T_x V_2$, for $x \in V_1 \cap V_2$.

(iv) If V_1 and V_2 have transverse intersection and $c_1 + c_2 = n$, then $V_1 \cap V_2$ is discrete and one has $T_x V_1 \oplus T_x V_2 = \mathbf{R}^n$, for all $x \in V_1 \cap V_2$. Prove this, and also discuss what happens if $c_1 = 0$.

(v) Finally, give an example where $n = 3$, $c_1 = c_2 = 1$ and $V_1 \cap V_2$ consists of a single point x. In this case one necessarily has $T_x V_1 = T_x V_2$; not, therefore, $T_x(V_1 \cap V_2) = T_x V_1 \cap T_x V_2$.

Exercise 5.74 (Tangent mapping – sequel to Exercise 4.31). Let the notation be as in Section 5.2; in particular, $V \subset \mathbf{R}^n$ and $W \subset \mathbf{R}^p$ are both C^∞ submanifolds, of dimension d and f, respectively, and Φ is a C^∞ mapping $\mathbf{R}^n \to \mathbf{R}^p$ such that $\Phi(V) \subset W$. We want to prove that the tangent mapping of $\Phi : V \to W$ at a point $x \in V$ is independent of the behavior of Φ outside V, and is therefore completely determined by the restriction of Φ to V. To do so, consider $\widetilde{\Phi} : \mathbf{R}^n \to \mathbf{R}^p$ such that $\widetilde{\Phi}|_V = \Phi|_V$, and define $F = \widetilde{\Phi} - \Phi$.

(i) Verify that the problem is a local one; in such a situation we may assume that $V = N(g, 0)$, with $g : \mathbf{R}^n \to \mathbf{R}^{n-d}$ a C^∞ submersion. Now apply Exercise 4.31 to the component functions $F_i : \mathbf{R}^n \to \mathbf{R}$, for $1 \le i \le p$, of F. Conclude that, for every $x^0 \in V$, there exist a neighborhood U of x^0 in $\mathbf{R}^n$ and C^∞ mappings $F^{(i)} : U \to \mathbf{R}^p$, with $1 \le i \le n-d$, such that on U

$$F = \sum_{1 \le i \le n-d} g_i F^{(i)}.$$

(ii) Use Theorem 5.1.2 to prove, for every $x \in V$, the following equality of mappings:

$$D\Phi(x) = D\widetilde{\Phi}(x) \in \mathrm{Lin}(T_x V, T_{\Phi(x)} W).$$

Exercise 5.75 ("Intrinsic" description of tangent space). Let V be a C^k submanifold in $\mathbf{R}^n$ of dimension d and let $x \in V$. The definition of $T_x V$, the tangent space to V at x, is independent of the description of V as a graph, parametrized set, or zero-set. Once any such characterization of V has been chosen, however, Theorem 5.1.2 gives a corresponding description of $T_x V$. We now want to give an "intrinsic" description of $T_x V$. To find this, begin by assuming $\phi : \mathbf{R}^d \supset\!\to \mathbf{R}^n$ and $\psi : \mathbf{R}^d \supset\!\to \mathbf{R}^n$ are both C^k embeddings, with image $V \cap U$, where U is an open neighborhood of x in $\mathbf{R}^n$. Note that

$$\phi(\phi^{-1}(x)) = \psi \circ (\psi^{-1} \circ \phi)(\phi^{-1}(x)).$$

(i) Prove by immersivity of ψ in $\psi^{-1}(x)$ that, for two vectors $v, w \in \mathbf{R}^d$, the equality of vectors in $\mathbf{R}^n$

$$D\phi(\phi^{-1}(x))v = D\psi(\psi^{-1}(x))w$$

holds if and only if

$$w = D(\psi^{-1} \circ \phi)(\phi^{-1}(x))v.$$

Next, define the coordinatizations $\kappa := \phi^{-1} : V \cap U \to \mathbf{R}^d$ and $\lambda := \psi^{-1} : V \cap U \to \mathbf{R}^d$.

(ii) Prove

$$(\star) \qquad \sum_{1 \le j \le d} v_j \, D_j\phi(\kappa(x)) = \sum_{1 \le k \le d} w_k \, D_k\psi(\lambda(x))$$

holds if and only if

$$(\star\star) \qquad w = D(\lambda \circ \kappa^{-1})(\kappa(x))v.$$

Note that $v = (v_1, \dots, v_d)$ are the coordinates of the vector above in $T_x V$, with respect to the basis vectors $D_j\phi(\kappa(x))$ $(1 \le j \le d)$ for $T_x V$ determined by κ. Condition $(\star)$ means that the pairs (v, κ) and (w, λ) represent the same vector in $T_x V$; and this is evidently the case if and only if $(\star\star)$ holds. Accordingly, we now define the relation $\sim_x$ on the set $\mathbf{R}^d \times \{$ coordinatizations of a neighborhood of x $\}$ by

$$(v, \kappa) \sim_x (w, \lambda) \qquad \Longleftrightarrow \qquad \text{relation } (\star\star) \text{ holds.}$$

(iii) Prove that $\sim_x$ is an equivalence relation.

Denote the equivalence class of (v, κ) by $[v, \kappa]_x$, and define $\mathcal{T}_x V$ as the collection of these equivalence classes.

(iv) Verify that addition and scalar multiplication in $\mathcal{T}_x V$

$$[v, \kappa]_x + [v', \kappa]_x = [v + v', \kappa]_x, \qquad r[v, \kappa]_x = [rv, \kappa]_x$$

are defined independent of the choice of the representatives; and that these operations make $\mathcal{T}_x V$ into a vector space.

(v) Prove that $\alpha_x : \mathcal{T}_x V \to T_x V$ is a linear isomorphism, if

$$\alpha_x([v, \kappa]_x) = \sum_{1 \le j \le d} v_j \, D_j\phi(\kappa(x)).$$

Exercise 5.76 (Dual vector space, cotangent bundle, and differentials – needed for Exercises 5.77 and 8.46). Let A be a vector space. We define the *dual vector space* A^* as $\mathrm{Lin}(A, \mathbf{R})$.

(i) Prove that A^{**} is linearly isomorphic with A.

(ii) Assume B is another vector space, and $L \in \mathrm{Lin}(A, B)$. Prove that the *adjoint linear operator* $L^* : B^* \to A^*$ is a well-defined linear operator if

$$(L^*\mu)(a) = \mu(La) \qquad (\mu \in B^*, \ a \in A).$$

Contrary to Section 2.1, in this case we do not use inner products to define the adjoint of a linear operator, which explains the different notation.

(iii) Let $\dim A = d$. Prove that $\dim A^* = d$ and conclude that A is linearly isomorphic with A^* (in a noncanonical way).
Hint: Choose a basis $(a_1, \dots, a_d)$ for A, and define $\lambda_1, \dots, \lambda_d \in A^*$ by $\lambda_i(a_j) = \delta_{ij}$, for $1 \le i, \ j \le d$, where δ_{ij} stands for the Kronecker delta. Check that $(\lambda_1, \dots, \lambda_d)$ forms a basis for A^*.

Assume that V is a C^∞ submanifold in $\mathbf{R}^n$ of dimension d. As in Definition 4.7.3 a function $f : V \to \mathbf{R}$ is said to be a C^∞ *function on* V if for every $x \in V$ there exist a neighborhood U of x in $\mathbf{R}^n$, an open set D in $\mathbf{R}^d$, and a C^∞ embedding $\phi : D \to V \cap U$ such that $f \circ \phi : D \to \mathbf{R}$ is a C^∞ function. Let $\mathcal{O}_V = \mathcal{O}$ be the collection of all C^∞ functions on V.

(iv) Prove by Lemma 4.3.3.(iii) that this definition of $\mathcal{O}$ is independent of the choice of the embeddings ϕ. Also prove that $\mathcal{O}$ is an algebra, for the operations of pointwise addition and multiplication of functions, and pointwise multiplication by scalars.

Let us assume for the moment that $\dim V = n$; that is, $V \subset \mathbf{R}^n$ is an open set in $\mathbf{R}^n$. Then $\dim T_x V = n$, and therefore $T_x V$ in this case equals $\mathbf{R}^n$, for all $x \in V$. If $f \in \mathcal{O}, x \in V$, this gives for the derivative $Df(x)$ of f at x

$$Df(x) = (D_1 f(x), \dots, D_n f(x)) \in \mathrm{Lin}(T_x V, \mathbf{R}).$$

We now write $T_x^* V$ for the dual vector space of $T_x V$; the vector space $T_x^* V$ is also known as the *cotangent space* of V at x. Consequently $Df(x) \in T_x^* V$. Furthermore, we write $T^* V = \coprod_{x \in V} T_x^* V$ for the disjoint union over all $x \in V$ of the cotangent spaces of V at x. Here we ignore the fact that all these cotangent spaces are actually copies of the same space $\mathbf{R}^n$, taking for every point $x \in V$ a different copy of $\mathbf{R}^n$. Then $T^* V$ is said to be the *cotangent bundle* of V. We can now assert that the total derivative Df is a mapping from the manifold V to the cotangent bundle $T^* V$,

$$Df : V \to T^* V \qquad \text{with} \qquad Df : x \mapsto Df(x) \in T_x^* V.$$

We note that, in contrast to the present treatment, in Definition 2.2.2 of the derivative one does not take the disjoint union of all cotangent spaces ($\simeq \mathbf{R}^n$), but one copy only of $\mathbf{R}^n$.

For the general case of $\dim V = d \le n$ we take inspiration from the result above. Let $x = \phi(y)$, with $y \in D \subset \mathbf{R}^d$. We then have $T_x V = \mathrm{im}\,(D\phi(y))$. Given $f \in \mathcal{O}, x \in V$, we define

$$d_x f \in T_x^* V \qquad \text{by} \qquad d_x f \circ D\phi(y) = D(f \circ \phi)(y) : \mathbf{R}^d \to \mathbf{R}.$$

The linear functional $d_x f \in T_x^* V$ is said to be the *differential at* x of the C^∞ function f on V.

(v) Check that $d_x f = Df(x)$, for all $x \in V$, if $\dim V = n$.

Again we write

$$T^* V = \coprod_{x \in V} T_x^* V$$

for the disjoint union of the cotangent spaces. The reason for taking the disjoint union will now be apparent: the cotangent spaces might differ from each other, and we do not wish to water this fact down by including into the union just one point from among the points lying in each of these different cotangent spaces. Finally, define the *differential* df of the C^∞ function f on V as the mapping from the manifold V to the cotangent bundle T^*V, given by

$$df : V \to T^*V \qquad \text{with} \qquad df : x \mapsto d_x f \in T_x^*V.$$

A mapping $V \to T^*V$ that assigns to $x \in V$ an element of T_x^*V is said to be a *section* of the cotangent bundle T^*V, or, alternatively, a *differential form* on V. Accordingly, df is such a section of T^*V, or a differential form on V. We finally introduce, for application in Exercise 5.77, the linear mapping

$$(\star) \qquad d_x : \mathcal{O} \to T_x^*V \qquad \text{by} \qquad d_x : f \mapsto d_x f \qquad (x \in V).$$

(vi) Once more consider the case $\dim V = n$. Verify that $\mathcal{O}$ contains the coordinate functions $x_i : x \mapsto x_i$, and that $d_x x_i$ equals the $1 \times n$ matrix having a 1 in the i-th column and 0 elsewhere, for $x \in V$ and $1 \leq i \leq n$. Conclude that for $f \in \mathcal{O}$ one has

$$df = \sum_{1 \leq i \leq n} D_i f \, dx_i.$$

Background. In Sections 8.6 – 8.8 we develop a more complete theory of differential forms.

Exercise 5.77 (Algebraic description of tangent space – sequel to Exercise 5.76 – needed for Exercise 5.78). The notation is as in that exercise. Let $x \in V$ be fixed. Then define $\mathcal{M}_x \subset \mathcal{O}$ as the linear subspace of the $f \in \mathcal{O}$ with $f(x) = 0$; and define $\mathcal{M}_x^2$ as the linear subspace in $\mathcal{M}_x$ generated by the products $fg \in \mathcal{M}_x$, where $f, g \in \mathcal{M}_x$.

(i) Check that $\mathcal{M}_x/\mathcal{M}_x^2$ is a vector space.

(ii) Verify that d_x from $(\star)$ in Exercise 5.76 induces $\widetilde{d_x} \in \mathrm{Lin}(\mathcal{M}_x/\mathcal{M}_x^2,\ T_x^*V)$ (that is, $\mathcal{M}_x^2 \subset \ker d_x$) such that the following diagram is commutative:

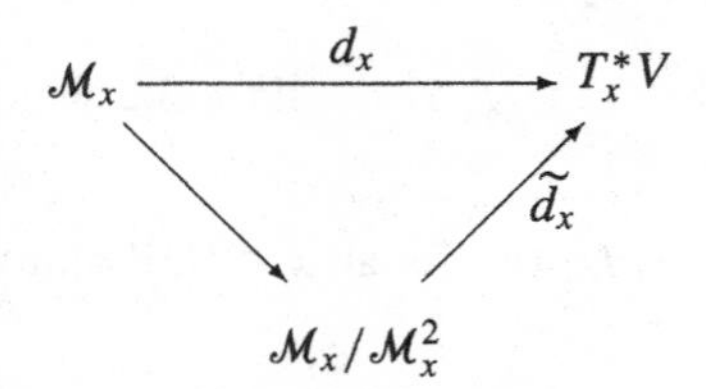

(iii) Prove that $\widetilde{d_x} \in \mathrm{Lin}(\mathcal{M}_x/\mathcal{M}_x^2,\ T_x^*V)$ is surjective.
Hint: For $\lambda \in T_x^*V$, consider the (locally defined) affine function $f : V \to \mathbf{R}$, given by

$$f \circ \phi(y') = \lambda \circ D\phi(y)(y' - y) \qquad (y' \in D).$$

Finally, we want to prove that $\widetilde{d_x} \in \mathrm{Lin}(\mathcal{M}_x/\mathcal{M}_x^2,\ T_x^*V)$ is a linear isomorphism of vector spaces, that is, one has (see Exercise 5.76.(i))

$$T_x V \simeq (\mathcal{M}_x/\mathcal{M}_x^2)^*.$$

Therefore, what we have to prove is $\ker d_x \subset \mathcal{M}_x^2$. Note that $f \circ \phi(y) = 0$ if $f \in \mathcal{M}_x$.

(iv) Prove that for every $f \in \mathcal{M}_x$ there exist functions $g_i \in \mathcal{O}$, with $1 \le i \le d$, such that for $y' \in D$ and $g = (g_1, \dots, g_d)$,

$$f \circ \phi(y') = \langle g \circ \phi(y'),\ (y' - y) \rangle.$$

Hint: Compare with Exercise 4.31.(ii) and with the proof of the Fourier Inversion Theorem 6.11.6.

(v) Verify that for $f \in \ker d_x$ one has $D_j(f \circ \phi)(y) = 0$, for $1 \le j \le d$, and conclude by (iv) that this implies $\ker d_x \subset \mathcal{M}_x^2$.

Background. The importance of this result is that it proves

$$(\star) \qquad d = \dim_x V = \dim T_x V = \dim T_x^* V = \dim(\mathcal{M}_x/\mathcal{M}_x^2);$$

and what is more, the algebra $\mathcal{O}$ contains complete information on the tangent spaces $T_x V$. The following serves to illustrate this property.

Let V and V' be C^∞ submanifolds in $\mathbf{R}^n$ and $\mathbf{R}^{n'}$ of dimension d and d', respectively; let $\mathcal{O}$ and $\mathcal{O}'$, respectively, be the associated algebras of C^∞ functions. As in Definition 4.7.3 we say that a mapping of manifolds $\Phi : V \to V'$ is a C^∞ mapping if for every $x \in V$ there exist neighborhoods U of x in $\mathbf{R}^n$ and U' of $\Phi(x)$ in $\mathbf{R}^{n'}$, open sets $D \subset \mathbf{R}^d$ and $D' \subset \mathbf{R}^{d'}$, and C^∞ coordinatizations $\kappa : V \cap U \to D$ and $\kappa' : V' \cap U' \to D'$, respectively, such that $\kappa' \circ \Phi \circ \kappa^{-1} : D \supset\!\to \mathbf{R}^{d'}$ is a C^∞ mapping. Given such a mapping Φ, we have for every $x \in V$ an *induced homomorphism of rings*

$$\Phi_x^* : \mathcal{M}'_{\Phi(x)} \to \mathcal{M}_x \qquad \text{given by} \qquad \Phi_x^*(f) = f \circ \Phi.$$

(vi) Assume that for $x \in V$ the mapping Φ_x^* is a homomorphism of rings. Then prove that Φ near x is a C^∞ diffeomorphism of manifolds.
Hint: By means of $(\star)$ one readily checks $d = d'$. Without loss of generality we may assume that $\kappa : V \cap U \to D \subset \mathbf{R}^d$ is a C^∞ coordinatization with $\kappa(x) = 0$, that is, $\kappa_i \in \mathcal{M}_x$, for $1 \le i \le d$. Consequently, then, there exist functions $\lambda_i \in \mathcal{M}'_{\Phi(x)}$ such that $\lambda_i \circ \Phi = \Phi_x^*(\lambda_i) = \kappa_i$, for $1 \le i \le d$. Hence $\lambda \circ \Phi = \kappa$ in a neighborhood of x, if $\lambda = (\lambda_1, \dots, \lambda_d)$. Now, let $\kappa' : V' \cap U' \to \mathbf{R}^d$ be a C^∞ coordinatization of a neighborhood of $\Phi(x)$. Then

$$(\lambda \circ \kappa'^{-1}) \circ (\kappa' \circ \Phi \circ \kappa^{-1})(\kappa(x)) = \kappa(x);$$

and so, by the chain rule,

$$D(\lambda \circ \kappa'^{-1})(\kappa'(\Phi(x))) \circ D(\kappa' \circ \Phi \circ \kappa^{-1})(\kappa(x)) = I.$$

In particular, $D(\lambda \circ \kappa'^{-1})(\kappa'(\Phi(x))) : \mathbf{R}^d \to \mathbf{R}^d$ is surjective, and therefore invertible. According to the Local Inverse Function Theorem 3.2.4, $\lambda \circ \kappa'^{-1}$ locally is a C^∞ diffeomorphism; and therefore $\kappa' \circ \Phi \circ \kappa^{-1}$ locally is a C^∞ diffeomorphism.

Exercise 5.78 (Lie derivative, vector field, derivation and tangent bundle – sequel to Exercise 5.77 – needed for Exercise 5.79). From Exercise 5.77 we know that, for every $x \in V$, there exists a linear isomorphism

$$T_x V \simeq (\mathcal{M}_x/\mathcal{M}_x^2)^*.$$

We will now give such an isomorphism in explicit form, in a way independent of the arguments from Exercise 5.77. Note that in Exercise 5.77 functions act on tangent vectors; we now show how tangent vectors act on functions.

We assume $T_x V = \operatorname{im}(D\phi(y))$. For $h = D\phi(y)v \in T_x V$ we define

$$L_{x,h} : \mathcal{O} \to \mathbf{R} \qquad \text{by} \qquad L_{x,h}(f) = d_x f(h) = D(f \circ \phi)(y)v.$$

Then $L_{x,h}$ is called the *Lie derivative at* x in the direction of the tangent vector $h \in T_x V$.

(i) Verify that $L_{x,h} : \mathcal{O} \to \mathbf{R}$ is a linear mapping satisfying

$$L_{x,h}(f\,g) = f(x)L_{x,h}(g) + g(x)L_{x,h}(f) \qquad (f,\ g \in \mathcal{O}).$$

We now abstract the properties of this Lie derivative in the following definition. Let $\delta_x : \mathcal{O} \to \mathbf{R}$ be a linear mapping with the property

$$\delta_x(f\,g) = f(x)\delta_x(g) + g(x)\delta_x(f) \qquad (f,\ g \in \mathcal{O});$$

then δ_x is said to be a *derivation* at the point x. In particular, therefore, $L_{x,h}$ is a derivation at x, for all $h \in T_x V$.

(ii) Prove that $\delta_x(c) = 0$, for every constant function $c \in \mathcal{O}$, and conclude that, for every $f \in \mathcal{O}$,

$$\delta_x(f) = \delta_x(f - f(x)).$$

This result means that we may assume, without loss of generality, that a derivation at x acts on functions in $\mathcal{M}_x$, instead of those in $\mathcal{O}$.

(iii) Demonstrate

$$\delta_x(f\,g) = 0 \qquad (f,\ g \in \mathcal{M}_x); \qquad \text{hence} \qquad \delta_x|_{\mathcal{M}_x^2} = 0.$$

We have, therefore, obtained

$$\widetilde{\delta_x} \in \mathrm{Lin}(\mathcal{M}_x/\mathcal{M}_x^2, \mathbf{R}); \qquad \text{that is,} \qquad \widetilde{\delta_x} \in (\mathcal{M}_x/\mathcal{M}_x^2)^*.$$

Conversely, we may now conclude that, for a $\widetilde{\delta_x} \in \mathrm{Lin}(\mathcal{M}_x/\mathcal{M}_x^2, \mathbf{R})$, there is exactly one derivation $\delta_x : \mathcal{O} \to \mathbf{R}$ at x such that (with δ_x acting on representatives)

$$(\star) \qquad \delta_x\big|_{\mathcal{M}_x} = \widetilde{\delta_x}.$$

The existence of δ_x is found as follows. Given $f \in \mathcal{O}$, one has $f - f(x) \in \mathcal{M}_x$, and hence we can define

$$\delta_x(f) := \widetilde{\delta_x}([f - f(x)]).$$

In particular, therefore, $\delta_x(c) = 0$, for every constant function $c \in \mathcal{O}$. Now, for f, $g \in \mathcal{O}$,

$$f\,g = (f - f(x))(g - g(x)) + f(x)g + g(x)f - f(x)g(x);$$

and because $(f - f(x))(g - g(x)) \in \mathcal{M}_x^2$, it follows that

$$\delta_x(f\,g) = f(x)\delta_x(g) + g(x)\delta_x(f).$$

In other words, δ_x is a derivation at x. Because a derivation at x always vanishes on the constant functions, we have that the derivation δ_x at x is uniquely defined by $(\star)$.

Finally, we prove that $\delta_x = L_{x,\,X(x)}$, for a unique element $X(x) \in T_x V$. This then yields a linear isomorphism

$$X(x) \leftrightarrow L_{x,\,X(x)} = \delta_x \leftrightarrow \widetilde{\delta_x} \qquad \text{so that} \qquad T_x V \simeq (\mathcal{M}_x/\mathcal{M}_x^2)^*.$$

To prove this, note that, as in Exercise 5.77.(iv), one has $f - f(x) = \langle g, \kappa - \kappa(x)\rangle$, with $\kappa := \phi^{-1}$.

(iv) Now verify that

$$\begin{aligned}\delta_x(f) &= \sum_{1\le i\le n} g_i(x)\,\delta_x(\kappa_i - \kappa_i(x)) = \sum_{1\le i\le n} \delta_x(\kappa_i - \kappa_i(x))D_i(f\circ\phi)(y)\\ &= \sum_{1\le i\le n} \delta_x(\kappa_i - \kappa_i(x))L_{x,\,e_i}(f) = L_{x,\,X(x)}(f),\end{aligned}$$

where $X(x) = \sum_{1\le i\le n} \delta_x(\kappa_i - \kappa_i(x))e_i \in T_x V$.

Next, we want to give a global formulation of the preceding results, that is, for all points $x \in V$ at the same time. To do so, we define the *tangent bundle* TV of V by

$$TV = \coprod_{x\in V} T_x V,$$

where, as in Exercise 5.76, we take the disjoint union, but now of the tangent spaces. Further define a *vector field* on V as a *section* of the tangent bundle, that is, as a mapping $X : V \to TV$ which assigns to $x \in V$ an element $X(x) \in T_x V$. Additionally, introduce the *Lie derivative* L_X in the direction of the vector field X by

$$L_X : \mathcal{O} \to \mathcal{O} \qquad \text{with} \qquad L_X(f)(x) = L_{x,\, X(x)}(f) \qquad (f \in \mathcal{O},\ x \in V).$$

And finally, define a *derivation* in $\mathcal{O}$ to be

$$\delta \in \mathrm{End}(\mathcal{O}) \qquad \text{satisfying} \qquad \delta(f\,g) = f\delta(g) + g\delta(f).$$

(v) Now proceed to demonstrate that, for every derivation δ in $\mathcal{O}$, there exists a unique vector field X on V with

$$\delta = L_X, \qquad \text{that is,} \qquad \delta(f)(x) = L_X(f)(x) \qquad (f \in \mathcal{O},\ x \in V).$$

We summarize the preceding as follows. Differential forms (sections of the cotangent bundle) and vector fields (sections of the tangent bundle) are dual to each other, and can be paired to a function on V. In particular, if df is the differential of a function $f \in \mathcal{O}$ and X is a vector field on V, then

$$df(X) = L_X(f) : V \to \mathbf{R} \qquad \text{with} \qquad df(X)(x) = L_X(f)(x) = d_x f(X(x)).$$

Exercise 5.79 (Pullback and pushforward under diffeomorphism – sequel to Exercises 3.8 and 5.78). The notation is as in Exercise 5.78. Assume that V and U are both n-dimensional C^∞ submanifolds (that is, open subsets) in $\mathbf{R}^n$ and that $\Psi : V \to U$ is a C^∞ diffeomorphism with $\Psi(y) = x$. We introduce the linear operator Ψ^* of *pullback under the diffeomorphism* Ψ between the linear spaces of C^∞ functions on U and V by (compare with Definition 3.1.2)

$$\Psi^* : \mathcal{O}_U \to \mathcal{O}_V \qquad \text{with} \qquad \Psi^*(f) = f \circ \Psi \qquad (f \in \mathcal{O}_U).$$

Further, let W be open in $\mathbf{R}^n$ and let $\Xi : W \to V$ be a C^∞ diffeomorphism.

(i) Prove that $(\Psi \circ \Xi)^* = \Xi^* \circ \Psi^* : \mathcal{O}_U \to \mathcal{O}_W$.

We now define the induced linear operator Ψ_* of *pushforward under the diffeomorphism* Ψ between the linear spaces of derivations in $\mathcal{O}_V$ and $\mathcal{O}_U$ by means of *duality*, that is

$$\Psi_* : \mathrm{Der}(V) \to \mathrm{Der}(U), \qquad \Psi_*(\delta)(f) = \delta(\Psi^*(f)) \qquad (\delta \in \mathrm{Der}(V),\ f \in \mathcal{O}_U).$$

(ii) Prove that $(\Psi \circ \Xi)_* = \Psi_* \circ \Xi_* : \mathrm{Der}(W) \to \mathrm{Der}(U)$.

Via the linear isomorphism $L_Y \leftrightarrow Y$ between the linear space $\mathrm{Der}(V)$ and the linear space $\Gamma(TV)$ of the vector fields on V, and analogously for U, the operator Ψ_* induces a linear operator

$$\Psi_* : \Gamma(TV) \to \Gamma(TU) \qquad \text{via} \qquad \Psi_* L_Y = L_{\Psi_* Y} \qquad (Y \in \Gamma(TV)).$$

The definition of Ψ_* then takes the following form:

$$(\star) \qquad ((\Psi_* Y) f)(\Psi(y)) = (Y(f \circ \Psi))(y) \qquad (f \in \mathcal{O}_U,\ y \in V).$$

(iii) Assume $Y \in \Gamma(TV)$ is given by $y \mapsto Y(y) = \sum_{1 \le j \le n} Y_j(y) e_j \in T_y V$. Prove

$$\begin{aligned} ((\Psi_* Y) f)(x) &= \sum_{1 \le j \le n} Y_j(y)\, D_j (f \circ \Psi)(y) \\ &= \sum_{1 \le i \le n} \Big(\sum_{1 \le j \le n} D_j \Psi_i(y)\, Y_j(y) \Big) D_i f(x) = \sum_{1 \le i \le n} (D\Psi(y) Y(y))_i\, D_i f(x). \end{aligned}$$

Thus we find that

$$\Psi_* Y = (\Psi^{-1})^* (D\Psi \circ Y) \in \Gamma(TU)$$

is the vector field on U with $x = \Psi(y) \mapsto D\Psi(\Psi^{-1}(x)) Y(\Psi^{-1}(x)) \in T_x U$ (compare with Exercise 3.15).

(iv) Now write $X = \Psi_* Y \in \Gamma(TU)$. Prove that $Y = (\Psi_*)^{-1} X = (\Psi^{-1})_* X$, by part (ii). Conclude that the identity $(\star)$ implies the following identity of linear operators acting on the linear space of functions $\mathcal{O}_U$:

$$(\star\star) \qquad \Psi^* \circ X = \Psi_*^{-1} X \circ \Psi^* \qquad (X \in \Gamma(TU)).$$

Here one has $\Psi_*^{-1} X : y = \Psi^{-1}(x) \mapsto D\Psi(y)^{-1} X(x) \in T_y V$.

(v) In particular, let $X \in \Gamma(TU)$ with $x = \Psi(y) \mapsto e_j$, with $1 \le j \le n$. Verify that

$$\begin{aligned} D\Psi(y)^{-1} e_j &= \sum_{1 \le k \le n} (D\Psi(y)^{-1})_{kj} e_k = \sum_{1 \le k \le n} (D\Psi(y)^{-1})^t_{jk} e_k \\ &=: \sum_{1 \le k \le n} \psi_{jk}(y) e_k. \end{aligned}$$

Show that $(\star\star)$ leads to the formula from Exercise 3.8.(ii)

$$(D_j f) \circ \Psi = \sum_{1 \le k \le n} \psi_{jk}\, D_k (f \circ \Psi) \qquad (1 \le j \le n).$$

Finally, we define the induced linear operator Ψ^* of *pullback under the diffeomorphism* Ψ between the linear spaces of differential forms on U and V

$$\Psi^* : \Omega^1(U) \to \Omega^1(V)$$

by *duality*, that is, we require the following equality of functions on V:

$$\Psi^*(df)(Y) = df(\Psi_* Y) \circ \Psi \qquad (f \in \mathcal{O}_U,\ Y \in \Gamma(TV)).$$

It follows that $\Psi^*(df)(Y)(y) = df(\Psi(y))(D\Psi(y)Y(y))$.

(vi) Show that $(\Psi \circ \Xi)^* = \Xi^* \circ \Psi^* : \Omega^1(U) \to \Omega^1(W)$.

(vii) Prove by the chain rule, for $f \in \mathcal{O}_U$, $Y \in \Gamma(TV)$ and $y \in V$,

$$\Psi^*(df)(Y)(y) = d(f \circ \Psi)(Y)(y) = d(\Psi^* f)(Y)(y).$$

In other words, we have the following identity of linear operators acting on the linear space of functions $\mathcal{O}_U$:

$$\Psi^* \circ d = d \circ \Psi^*;$$

that is, the linear operators of pullback under Ψ and of exterior differentiation commute.

(viii) Verify

$$\Psi^*(dx_i) = \sum_{1 \le j \le n} D_j \Psi_i \, dy_j \qquad (1 \le i \le n).$$

Conclude

$$\Psi^*(df) = \sum_{1 \le i \le n} \Psi^*(D_i f)\, \Psi^*(dx_i) \qquad (f \in \mathcal{O}_U).$$

Exercise 5.80 (Covariant differentiation, connection and Theorema Egregium – needed for Exercise 8.25). Let $U \subset \mathbf{R}^n$ be an open set and $Y : U \to \mathbf{R}^n$ a C^∞ vector field. The directional derivative D_Y in the direction of Y acts on $f \in C^\infty(U)$ by means of

$$(D_Y f)(x) = Df(x)Y(x) \qquad (x \in U).$$

If $X : U \to \mathbf{R}^n$ is a C^∞ vector field too, we define the C^∞ vector field $D_X Y : U \to \mathbf{R}^n$ by

$$(D_X Y)(x) = DY(x)X(x) \qquad (x \in U),$$

the directional derivative of Y at the point x in the direction $X(x)$. Verify

$$D_X(D_Y f)(x) = D^2 f(x)(X(x), Y(x)) + Df(x)\,(D_X Y)(x),$$

and, using the symmetry of $D^2 f(x) \in \mathrm{Lin}^2(\mathbf{R}^n, \mathbf{R})$, prove

$$[D_X, D_Y]f(x) := (D_X D_Y - D_Y D_X)f(x) = Df(x)(D_X Y - D_Y X)(x).$$

We introduce the C^∞ vector field $[X, Y]$, the *commutator* of X and Y, on U by

$$[X, Y] = D_X Y - D_Y X : U \to \mathbf{R}^n,$$

and we have the identity $[D_X, D_Y] = D_{[X,Y]}$ of directional derivatives on U. In standard coordinates on $\mathbf{R}^n$ we find

$$X = \sum_{1 \le j \le n} X_j\, e_j, \qquad Y = \sum_{1 \le j \le n} Y_j\, e_j,$$

$$[X, Y] = \sum_{1 \le j \le n} (\langle X, \operatorname{grad} Y_j \rangle - \langle Y, \operatorname{grad} X_j \rangle)\, e_j.$$

Furthermore,

$$\begin{aligned} D_X(fY)(x) &= D(fY)(x)X(x) = f(x)(D_X Y)(x) + Df(x)X(x)\,Y(x) \\ &= (f\,D_X Y + (D_X f)\,Y)(x); \end{aligned}$$

and therefore

$$D_X(fY) = f\,D_X Y + (D_X f)\,Y.$$

Now let V be a C^∞ submanifold in U of dimension d and assume $X|_V$ and $Y|_V$ are vector fields tangent to V. Then we define $\nabla_X Y : V \to \mathbf{R}^n$, the *covariant derivative* relative to V of Y in the direction of X, as the vector field tangent to V given by

$$(\nabla_X Y)(x) = (D_X Y)(x)^{\parallel} \qquad (x \in V),$$

where the right–hand side denotes the orthogonal projection of $DY(x)X(x) \in \mathbf{R}^n$ onto the tangent space $T_x V$. We obtain

$$(\nabla_X Y - \nabla_Y X)(x) = (D_X Y - D_Y X)(x)^{\parallel} = [X, Y](x)^{\parallel} = [X, Y](x) \qquad (x \in V),$$

since the commutator of two vector fields tangent to V again is a vector field tangent to V, as one can see using a local description $Y \times \{0_{\mathbf{R}^{n-d}}\}$ of the image of V under a diffeomorphism, where $Y \subset \mathbf{R}^d$ is an open subset (see Theorem 4.7.1.(iv)). Therefore

$$(\star) \qquad \nabla_X Y - \nabla_Y X = [X, Y].$$

Because an orthogonal projection is a linear mapping, we have the following properties for the covariant differentiation relative to V:

$$(\star\star) \qquad \begin{aligned} &\nabla_{fX+X'} Y = f\nabla_X Y + \nabla_{X'} Y \\ &\nabla_X(Y + Y') = \nabla_X Y + \nabla_X Y', \qquad \nabla_X(fY) = f\nabla_X Y + (D_X f)\,Y. \end{aligned}$$

Moreover, we find, if $Z|_V$ also is a vector field tangent to V,

$$\begin{aligned} D_X\langle\, Y, Z\,\rangle &= \langle\, \nabla_X Y, Z\,\rangle + \langle\, Y, \nabla_X Z\,\rangle,\\ D_Y\langle\, Z, X\,\rangle &= \langle\, \nabla_Y Z, X\,\rangle + \langle\, Z, \nabla_Y X\,\rangle,\\ D_Z\langle\, X, Y\,\rangle &= \langle\, \nabla_Z X, Y\,\rangle + \langle\, X, \nabla_Z Y\,\rangle. \end{aligned}$$

Adding the first two identities, subtracting the third, and using ($\star$) gives

$$\begin{aligned} 2\langle\, \nabla_X Y, Z\,\rangle \ &= D_X\langle\, Y, Z\,\rangle + D_Y\langle\, Z, X\,\rangle - D_Z\langle\, X, Y\,\rangle\\ &\quad +\langle\, [X, Y], Z\,\rangle + \langle\, [Z, X], Y\,\rangle - \langle\, [Y, Z], X\,\rangle. \end{aligned}$$

This result of Levi–Civita shows that covariant differentiation relative to V can be defined in terms of the manifold V in an *intrinsic* fashion, that is, independently of the ambient space $\mathbf{R}^n$.

Background. Let $\Gamma(TV)$ be the linear space of C^∞ sections of the tangent bundle TV of V (see Exercise 5.78). A mapping ∇ which assigns to every $X \in \Gamma(TV)$ a linear mapping ∇_X, the covariant differentiation in the direction of X, of $\Gamma(TV)$ into itself such that ($\star\star$) is satisfied, is called a *connection* on the tangent bundle of V.

Classically, this theory was formulated relative to coordinate systems. Therefore, consider a C^∞ embedding $\phi : D \to V$ with $D \subset \mathbf{R}^d$ an open set. Then the $E_i := D_i\phi(\phi^{-1}(x)) \in \mathbf{R}^n$, for $1 \le i \le d$, form a basis for $T_x V$, with $x \in V$. Thus there exist X_j and $Y_k \in C^\infty(V)$ with

$$X = \sum_{1\le j\le d} X_j E_j, \qquad Y = \sum_{1\le k\le d} Y_k E_k.$$

The properties ($\star\star$) then give

$$\begin{aligned} \nabla_X Y \ &= \sum_{1\le j\le d} X_j \sum_{1\le k\le d} \nabla_{E_j}(Y_k E_k)\\ &= \sum_{1\le j\le d} X_j \sum_{1\le k\le d} ((D_{E_j} Y_k)\, E_k + Y_k\, \nabla_{E_j} E_k)\\ &= \sum_{1\le i,\, j\le d} \Big(X_j D_j(Y_i \circ \phi) \circ \phi^{-1} + \sum_{1\le k\le d} \Gamma^i_{jk} X_j Y_k\Big) E_i, \end{aligned}$$

where we have used

$$D_{E_j} f(x) = Df(x) D_j\phi(\phi^{-1}(x)) = Df(x) D\phi(\phi^{-1}(x))\, e_j = D_j(f \circ \phi)(\phi^{-1}(x)).$$

Furthermore, the *Christoffel symbols* $\Gamma^i_{jk} : V \to \mathbf{R}$ associated with V, for $1 \le i, j, k \le d$, are defined via

$$\nabla_{E_j} E_k = \sum_{1\le i\le d} \Gamma^i_{jk} E_i.$$

Apparently, when differentiating Y covariantly relative to V in the direction of the tangent vector $D_j\phi$, one has to correct the Euclidean differentiations $D_j(Y_i \circ \phi)$ by contributions $\sum_{1\le k\le d} \Gamma^i_{jk} Y_k$ in terms of the Christoffel symbols associated with V.

From $D_{E_i} E_j = (D_i D_j \phi)\circ\phi^{-1}$ we get $[E_i, E_j] = 0$, and so, for $1 \le i, j, k \le d$, with $g_{ij} = \langle E_i, E_j \rangle$ and $(g^{ij}) = (g_{ij})^{-1}$,

$$\Gamma^i_{jk} = \frac{1}{2} \sum_{1\le l\le d} g^{il}(D_j(g_{lk}\circ\phi) + D_k(g_{jl}\circ\phi) - D_l(g_{jk}\circ\phi))\circ\phi^{-1}.$$

As an example consider the unit sphere S^2, which occurs as an image under the embedding $\phi(\alpha, \theta) = (\cos\alpha\cos\theta,\ \sin\alpha\cos\theta,\ \sin\theta)$. Then $D_1 D_2\phi(\alpha,\theta) = -(\tan\theta) D_1\phi(\alpha,\theta)$, from which

$$\Gamma^1_{12}(\phi(\alpha,\theta)) = -\tan\theta, \qquad \Gamma^2_{12}(\phi(\alpha,\theta)) = 0.$$

Or, equivalently

$$\Gamma^1_{12}(x) = \Gamma^1_{21}(x) = -\frac{x_3}{\sqrt{1-x_3^2}}, \qquad \Gamma^2_{12}(x) = \Gamma^2_{21}(x) = 0 \qquad (x\in S^2).$$

Finally, let $d = n-1$ and assume that a continuous choice $x \mapsto n(x)$ of a normal to V at $x \in V$ has been made. Let $1 \le i, j \le n-1$ and let X be a vector field tangent to V, then

$$\begin{aligned} D_{E_i} D_{E_j} X &= D_{E_i}(\nabla_{E_j} X + \langle D_{E_j} X, n\rangle n) \\ &= \nabla_{E_i}\nabla_{E_j} X + \langle D_{E_j} X, n\rangle D_{E_i} n + \text{scalar multiple of } n \\ &= \nabla_{E_i}\nabla_{E_j} X - \langle X, D_{E_j} n\rangle D_{E_i} n + \text{scalar multiple of } n. \end{aligned}$$

Now $[D_{E_i}, D_{E_j}] = D_{[E_i, E_j]} = 0$. Interchanging the roles of i and j, subtracting, and taking the tangential component, we obtain

$$(\star\star\star) \qquad (\nabla_{E_i}\nabla_{E_j} - \nabla_{E_j}\nabla_{E_i})X = \langle X, D_{E_j} n\rangle D_{E_i} n - \langle X, D_{E_i} n\rangle D_{E_j} n.$$

Since the left–hand side is defined intrinsically in terms of V, the same is true of the right–hand side in $(\star\star\star)$. The latter being trilinear, we obtain for every triple of vector fields X, Y and Z tangent to V the following, intrinsically defined, vector field tangent to V:

$$\langle X, D_Y n\rangle D_Z n - \langle X, D_Z n\rangle D_Y n.$$

Now apply the preceding to $V \subset \mathbf{R}^3$. Select Y and Z such that $Y(x)$ and $Z(x)$ form an orthonormal basis for $T_x V$, for every $x \in V$, and let $X = Y$. This gives for the Gaussian curvature K of the surface V

$$K = \det Dn = \langle D_Y n, Y\rangle\langle D_Z n, Z\rangle - \langle D_Y n, Z\rangle\langle D_Z n, Y\rangle.$$

Thus we have obtained Gauss' *Theorema Egregium* (egregius = excellent) which asserts that the Gaussian curvature of V is intrinsic.

Background. In the special case where $\dim V = n - 1$ the identity $(\star\star\star)$ suggests that the following mapping, where X, Y and Z are vector fields on V:

$$(X, Y, Z) \mapsto R(X, Y)Z = (\nabla_X \nabla_Y - \nabla_Y \nabla_X - \nabla_{[X,Y]})Z$$

is trilinear over the functions on V, as is also true in general. Hence we can consider R as a mapping which assigns to tangent vectors $X(x)$ and $Y(x) \in T_x V$ the mapping $Z(x) \mapsto R(X(x), Y(x))Z(x)$ belonging to $\text{Lin}(T_x V,\ T_x V)$; and the mapping acting on $X(x)$ and $Y(x)$ is bilinear and antisymmetric. Therefore R may be considered as a differential 2-form on V (see the Remark following Proposition 8.1.12) with values in $\text{Lin}\,(\Gamma(TV),\ \Gamma(TV))$; and this justifies calling R the *curvature form* of the connection ∇.

Notation

$d(\cdot, \cdot)$	Euclidean distance	5
div	divergence	166, 268
dom	domain space of mapping	108
$\mathrm{End}(\mathbf{R}^n)$	linear space of linear mappings $\mathbf{R}^n \to \mathbf{R}^n$	38
$\mathrm{End}^+(\mathbf{R}^n)$	linear subspace in $\mathrm{End}(\mathbf{R}^n)$ of self-adjoint operators	41
$\mathrm{End}^-(\mathbf{R}^n)$	linear subspace in $\mathrm{End}(\mathbf{R}^n)$ of anti-adjoint operators	41
$f^{-1}(\{\cdot\})$	inverse image under f	12
$\mathbf{GL}(n, \mathbf{R})$	general linear group, of invertible matrices in $\mathrm{Mat}(n, \mathbf{R})$	38
grad	gradient	59
graph	graph of mapping	107
$\mathrm{H}(n, \mathbf{C})$	linear subspace of Hermitian matrices in $\mathrm{Mat}(n, \mathbf{C})$	385
I	identity mapping	44
im	image under mapping	108
inf	infimum	21
int	interior	7
$L^\perp$	orthocomplement of linear subspace L	201
lim	limit	6, 12
$\mathrm{Lin}(\mathbf{R}^n, \mathbf{R}^p)$	linear space of linear mappings $\mathbf{R}^n \to \mathbf{R}^p$	38
$\mathrm{Lin}^k(\mathbf{R}^n, \mathbf{R}^p)$	linear space of k-linear mappings $\mathbf{R}^n \to \mathbf{R}^p$	63
$\mathbf{Lo}^\circ(4, \mathbf{R})$	proper Lorentz group	386
$\mathrm{Mat}(n, \mathbf{R})$	linear space of $n \times n$ matrices with coefficients in $\mathbf{R}$	38
$\mathrm{Mat}(p \times n, \mathbf{R})$	linear space of $p \times n$ matrices with coefficients in $\mathbf{R}$	38
$N(c)$	level set of c	15
$\mathcal{O} = \{ O_i \mid i \in I \}$	open covering of set	30
$\mathbf{O}(\mathbf{R}^n)$	orthogonal group, of orthogonal operators in $\mathrm{End}(\mathbf{R}^n)$	73
$\mathbf{O}(n, \mathbf{R})$	orthogonal group, of orthogonal matrices in $\mathrm{Mat}(n, \mathbf{R})$	124
$\mathbf{R}^n$	Euclidean space of dimension n	2
S^{n-1}	unit sphere in $\mathbf{R}^n$	124
$\mathfrak{sl}(n, \mathbf{C})$	Lie algebra of $\mathbf{SL}(n, \mathbf{C})$	385
$\mathbf{SL}(n, \mathbf{R})$	special linear group, of matrices in $\mathrm{Mat}(n, \mathbf{R})$ with determinant 1	303
$\mathbf{SO}(3, \mathbf{R})$	special orthogonal group, of orthogonal matrices in $\mathrm{Mat}(3, \mathbf{R})$ with determinant 1	219
$\mathbf{SO}(n, \mathbf{R})$	special orthogonal group, of orthogonal matrices in $\mathrm{Mat}(n, \mathbf{R})$ with determinant 1	302
$\mathfrak{su}(2)$	Lie algebra of $\mathbf{SU}(2)$	385
$\mathbf{SU}(2)$	special unitary group, of unitary matrices in $\mathrm{Mat}(2, \mathbf{C})$ with determinant 1	377
sup	supremum	21
$T_x V$	tangent space of submanifold V at point x	134
$T^* V$	cotangent bundle of submanifold V	399
$T_x^* V$	cotangent space of submanifold V at point x	399
tr	trace	39
$V(a; \delta)$	closed ball of center a and radius δ	8

Index

Printed in the United States

Printed in the United States
By Bookmasters